Nonlinear Filters for Image Processing

SPIE/IEEE SERIES ON IMAGING SCIENCE & ENGINEERING

The SPIE/IEEE Series on Imaging Science & Engineering publishes original books on theoretical and applied electronic and optical imaging.

Series Editor
Edward R. Dougherty
Texas A&M University

Advisory Board

David P. Casasent
Carnegie Mellon University

Robert Haralick
University of Washington

Murat Kunt
Swiss Federal Institute of Technology

Azriel Rosenfeld
University of Maryland

Associate Editors

Digital Image Processing and Analysis
Jaakko Astola
Tampere University of Technology

Hardware Architecture
Charles C. Weems
University of Massachusetts

Image Acquisition and Display
Henri Maître
Telecom-Paris

Machine Vision
Mohan M. Trivedi
University of California/San Diego

Optical Image Formation
Jack D. Gaskill
Optical Sciences Center,
University of Arizona

Optical Image Processing
Bahram Javidi
University of Connecticut

Visual Communications
Mark J. Smith
Georgia Institute of Technology

Titles in the series:

Fundamentals of Electronic Image Processing,
Arthur R. Weeks, Jr.

Random Processes for Image and Signal Processing,
Edward R. Dougherty

Nonlinear Filters for Image Processing,
Edward R. Dougherty and Jaakko T. Astola, Editors

Nonlinear Filters for Image Processing

Edward R. Dougherty
Texas A&M University

Jaakko T. Astola
Tampere University of Technology

SPIE Optical Engineering Press
A Publication of SPIE—The International Society for Optical Engineering
Bellingham, Washington USA

IEEE PRESS

The Institute of Electrical and Electronics Engineers, Inc., New York

Library of Congress Cataloging-in-Publication Data

Nonlinear filters for image processing / Edward R. Dougherty, editor
[and] Jaakko T. Astola, editor.
p. cm. – (SPIE/IEEE series on imaging science & engineering)
Includes bibliographical references and index.
ISBN 0-8194-3033-1
1. Image processing–Digital techniques. 2. Digital filters (Mathematics).
3. Nonlinear theories. I. Dougherty, Edward R. II. Astola, Jaakko. III. Series.
TA1637.N66 1999
621.36'7—dc21 99-17646
CIP

Copublished by

SPIE—The International Society for Optical Engineering
P.O. Box 10
Bellingham, Washington 98227-0010
Phone: 360/676-3290
Fax: 360/647-1445
Email: spie@spie.org
WWW: http://www.spie.org
SPIE Press Volume PM59
ISBN 0-8194-3033-1

IEEE Press
445 Hoes Lane
P.O. Box 1331
Piscataway, NJ 08855-1331
Phone: 1-800/678-IEEE
Fax: 732/981-9334
Email: ieeepress@ieee.org
WWW: http://www.ieee.org
IEEE Order No. PC5825
ISBN 0-7803-5385-4

Printed in the United States of America.

CONTENTS

PREFACE

Classical signal processing is dominated by linear filtering. There are many reasons for this, but perhaps these can best be summarized by noting that many tasks can be accomplished with an acceptable level of performance by linear filters, and the mathematical analysis of linear operators is much easier than the analysis of nonlinear operators. The problem is that performance loss is often beyond what is acceptable. A filter is a transformation between random processes, the goal being to estimate a desired output from the input. If one considers the entire class of operators between two random processes, then restriction to linearity can result in being forced to choose a filter from a class of filters that is inappropriate to the task at hand. While an optimal linear filter may perform well, rarely will the optimal filter be linear, and very often the optimal linear filter will perform poorly compared to some nonlinear filters. In the latter case, one needs to enlarge, or change, the class from which filters are to be selected.

Two fundamental issues arise at once: from what class should we select the filter, and, given a class of nonlinear filters, how do we choose a good one? There are three factors implicit in the problem of choosing a class: understanding the mathematical properties of the filter class, having suitable filter representations, and quantifying filter performance over random processes of interest. For linear filters, these problems have been addressed for decades. Regarding selection of a good filter from a given class, one needs to have design tools that lead to optimal or acceptable suboptimal performance relative to relevant error measures. For linear filters under the criterion of mean-square error, filter optimization can be carried out in the framework of Hilbert spaces. We have analytic characterization of optimality via an integral equation, analytic formulation of the mean-square error, and, given a canonical expansion of the input random process, analytic expression of the optimal filter. Moreover, all of this probabilistic analysis involves only second moments of the random processes. Not only does this greatly simplify the analysis, it also reduces statistical estimation to second moments. The price for these conveniences is heavy: not only can linearity represent a stringent constraint on optimality, it also only involves second-order information. These restrictions are often unacceptable for image processing.

The two main thrusts of nonlinear image filtering have been addressed to the two aforementioned fundamental issues. First, operator classes have been defined, mathematical properties studied, representations developed, and effects on images investigated; second, methods for optimal filter design have been developed. As would naturally be expected, properties and representations have come before optimization, and prior to the recent decade the bulk of the research was deterministic. But the balance has shifted during the present decade and in recent years both deterministic and random analyses have made significant progress.

As it applies to image processing, nonlinear filtering has two historical roots. First, beginning in the 1960's, there is mathematical morphology, in its early application in the context of random sets. The basic erosion and opening representations of nonlinear filtering go back to G. Matheron's classic 1975 work, *Random Sets and Integral Geometry*. One can hardly overemphasize the impact of this book, not only on image processing, but on the theory of random sets. The basic operator representations formulated by Matheron have provided the seed for the algebraic representations of nonlinear filtering, which in turn have provided the framework for filter optimization. But probability, not algebraic representation, is Matheron's central issue. Concerning his endeavors, he writes, "The formulation and the very choice of problems for solution are directly inspired by experimental techniques of texture analysis that currently permit the thorough study of such [porous] media...." Matheron is not talking about the historical problems of signal estimation; rather, he is laying the foundations for the random analysis of geometric forms and texture. Today, his study remains the fundament for both random geometry and texture analysis.

The second historical root for nonlinear filtering is classical digital signal processing. Here, signal values in a window are treated as a random sample and the value at the window center is estimated by some nonlinear estimator. The effort was to move away from linear operators, especially in the case of noise suppression, where linear filters have serious detrimental effects on edge information, which is crucial for image processing. Early studies involved the median, beginning with the median filtering of time series by J. Tukey in the early 1970's, and then stack filters, which are increasing filters that commute with thresholding, of which the median is a particular case. Medians are attractive from the perspective of robust statistics, and are readily viewed in terms of maximum-likelihood estimation.

In fact, the two roots are not distinct. Morphological representation provides the link. Matheron's representation of increasing filters applied only to binary operators. This was extended to increasing gray-scale operators, to nonincreasing binary images, and finally to all translation-invariant lattice operators, and in all cases special effort was attached to minimal representations (see Section 2.12). But the whole matter goes quite far back, with the erosion representation of the binary median and commutation of the gray-scale median with thresholding being recognized by 1980. In any event, the latter half of the last decade and the first part of the present saw a large effort at operator representation.

To use the representations, there need to be ways of synthesizing operators in accord with desired outcomes and the probabilistic characteristics of the random processes or sets upon which they are acting. Perhaps we can paraphrase R. Haralick, who, at a talk during the *1986 IEEE CVPR Conference*, stated that the near future would see the publication of a large number of papers on the design of optimal morphological filters to parallel the developments of the Wiener-Kolmogorov theory of linear filtering. Haralick's prediction proved prophetic because shortly thereafter came early attempts to design optimal increasing filters, including the special cases of stack filters, and optimal openings. Henceforth, the subject would find itself more in line with classical signal processing, where the algebraic structures (in that case, linear spaces) are applied in conjunction with the theory of random processes.

Currently there appears to be a split that is reflective of the dual roots of nonlinear filtering, and perhaps more substantive than the original differences. The movement in the direction of statistics has become so pronounced that nonlinear filtering may be becoming more associated with nonlinear statistical estimation and inverse problems, whereas mathematical morphology may be becoming associated with the theory of lattice operators, nonlinear equations of evolution, and other mathematical issues. Random set theory, from which the basics sprung, is a subject in its own right with applications to various physical sciences, as well as to image processing. Such fragmentation in conjunction with maturation is natural, if sometimes disheartening.

The present text remains, for the most part, in the unified tradition, no doubt because the editors prefer it that way. We continue to believe that there remains an integral whole involving nonlinear estimation, mathematical morphology, and random processes (sets), and that this whole has both discrete and continuous components. Indeed, *fundamental separation is impossible because nonlinear filtering involves nonlinear estimation of random processes using operators that possess morphological representations*. The goal of the book is to provide a collection of chapters on topics of contemporary interest, in addition to sufficient basic material to keep the book self-contained and connected to its roots in translation-invariant set operators, stack filters, and random sets. The mathematical treatment is intended to be rigorous, but not so abstract as to make it inhospitable to engineers possessing good mathematical training. We hope you enjoy the book and find it useful for your studies.

We would like to thank all the contributors to this book, SPIE Press and IEEE Press for publishing this work, and especially our wives Terry and Ulla for their ongoing support.

Edward R. Dougherty
College Station, Texas

Jaakko Astola
Tampere, Finland

Nonlinear Filters for Image Processing

CHAPTER 1

LOGICAL IMAGE OPERATORS

Edward R. Dougherty
Texas A&M University
College Station, Texas

Junior Barrera
University of Sao Paulo
Sao Paulo, Brazil

All digital image processing algorithms involve logical functions operating on vectors of logical variables, where the logical vectors represent binary encodings of image data. *Ipso facto*, characterization of image operators is naturally set in a logical framework, and this is especially straightforward in the case of binary image operators. The central tasks of any applied operator theory are analysis and synthesis of operators. Analysis concerns operator effects and synthesis the design of operators to perform desired tasks. Since digital images are modeled as discrete random sets, operator design involves synthesis governed by probabilistic criteria relating input and output image processes. The methodology is to find suitable operator representations, examine the manner in which these representations interact with the image probability structure to characterize optimal operators, and develop tools that translate them into efficient image processing procedures. The approach taken in the present chapter is to use examples (collections of input-output pairs of images) as the knowledge source for the representation formalism.

1.1 BOOLEAN FUNCTIONS

A binary-valued function $\psi(x_1, x_2, \ldots, x_n)$ of binary variables $x_1, x_2, \ldots, x_n$ is called a *Boolean function*. We will denote binary variables and Boolean functions by lower-case italic and Greek letters, respectively. As a logical function, ψ possesses a logical sum-of-products *disjunctive-normal-form* representation in terms of the n variables $x_1, x_2, \ldots, x_n$:

$$\psi(x_1, x_2, \ldots, x_n) = \sum_i x_1^{p(i,1)} x_2^{p(i,2)} \cdots x_n^{p(i,n)} \tag{1–1}$$

where the "sum" denotes OR, the "product" denotes AND, and $p(i, k)$ is either $'$ (prime) or null, depending on whether the variable is complemented or not complemented. There are at most 2^n products in the expansion, and each product is called a *minterm*. The representation can be (nonuniquely) reduced to a sum of products containing a minimal number of logic gates; that is,

$$\psi(x_1, x_2, \ldots, x_n) = \sum_i x_{i,1}^{p(i,1)} x_{i,2}^{p(i,2)} \cdots x_{i,n(i)}^{p(i,n(i))} \tag{1–2}$$

Disjunctive normal form will be used for operator design, with reduction being employed to cut the logic cost of implementation.

The truth table formulation of ψ corresponds directly to the disjunctive normal form of Eq. 1–1. ψ is defined by a 2^n-row truth table of n variables in which each string $t_1t_2\cdots t_n$ of 0s and 1s is assigned a binary value $\psi(t_1t_2\cdots t_n)$. The correspondence between Eq. 1–1 and the truth table is given by the following rule: the minterm $x_1^{p(i,1)} x_2^{p(i,2)} \cdots x_n^{p(i,n)}$ appears in the expansion of Eq. 1–1 if and only if there is a 1-valued string $t_1t_2\cdots t_n$ in the truth table with $p(i, j)$ null if $t_j = 1$ and $p(i, j) ='$ if $t_j = 0$.

The product set $\{0, 1\}^n$ is composed of binary n-vectors and is a finite lattice under the partial-order relation $(x_1, x_2, \ldots, x_n) \leqslant (y_1, y_2, \ldots, y_n)$ if and only if $x_j \leqslant y_j$ for $j = 1, 2, \ldots, n$. We denote vectors of binary variables by bold-face lower-case letters, such as $\mathbf{x} = (x_1, x_2, \ldots, x_n)$ and $\mathbf{y} = (y_1, y_2, \ldots, y_n)$. For any $\mathcal{A} \subset \{0, 1\}^n$, the *upper set* of $\mathcal{A}$ is defined by

$$\mathcal{U}[\mathcal{A}] = \{\mathbf{y}\text{: there exists } \mathbf{x} \in \mathcal{A} \text{ with } \mathbf{x} \leqslant \mathbf{y}\} \tag{1–3}$$

$\mathcal{A}^-$ denotes the set of minimal elements in $\mathcal{A}$, and $\mathcal{U}[\mathcal{A}] = \mathcal{U}[\mathcal{A}^-]$. In terms of $\{0, 1\}^n$, a Boolean function is a mapping $\psi\colon \{0, 1\}^n \to \{0, 1\}$. ψ is defined by specifying either of the subsets

$$\begin{aligned} \mathcal{S}_0[\psi] &= \left\{\mathbf{x} \in \{0, 1\}^n\colon \psi(\mathbf{x}) = 0\right\} \\ \mathcal{S}_1[\psi] &= \left\{\mathbf{x} \in \{0, 1\}^n\colon \psi(\mathbf{x}) = 1\right\} \end{aligned} \tag{1–4}$$

$\mathcal{S}_0[\psi]$ and $\mathcal{S}_1[\psi]$ are called the 0-*set* and 1-*set* (0-*slice* and 1-*slice*) of ψ, respectively, and $\mathcal{S}_0[\psi]$ is the complement of $\mathcal{S}_1[\psi]$ in $\{0, 1\}^n$.

A Boolean function ψ is *increasing* if $\mathbf{x} \leqslant \mathbf{y}$ implies $\psi(\mathbf{x}) \leqslant \psi(\mathbf{y})$. If ψ is increasing, then it is a called *positive* Boolean function. ψ is positive if and only if it can be represented as a logical sum of products having no complemented variables,

$$\psi(x_1, x_2, \ldots, x_n) = \sum_i x_{i,1}x_{i,2}\cdots x_{i,n(i)} \tag{1–5}$$

A complementation-free expansion is called a *positive* expansion. If the variable set in any product of the expansion contains as a subset the set of variables in a distinct

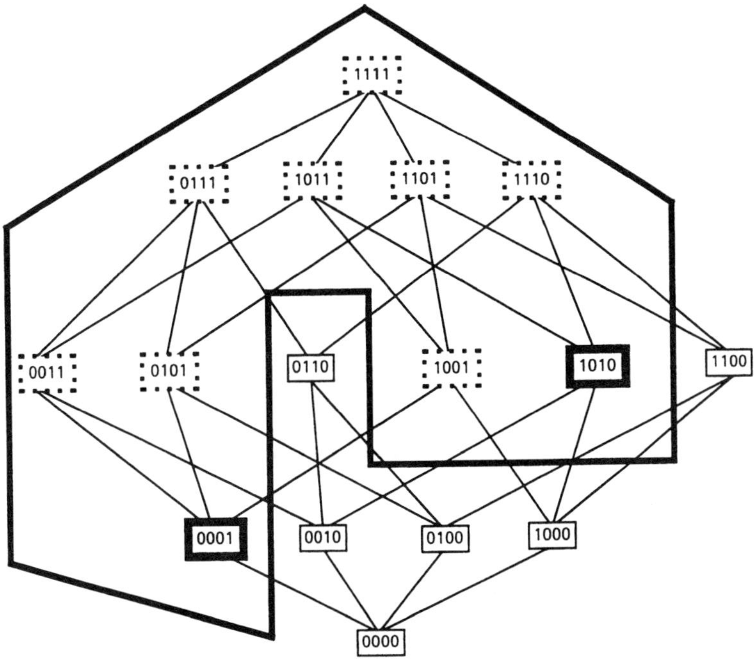

Figure 1–1. Increasing Boolean function.

product, then, whenever the former product has value 1, so too does the latter. Thus, inclusion of the former product in the expansion is redundant and it can be deleted from the expansion without changing ψ. No product whose variable set does not contain the variable set of a distinct product can be deleted without changing ψ. Performing the permitted deletions produces a unique *minimal representation* of ψ. Unless otherwise stated, it is convention to assume that positive Boolean functions are represented by positive expansions.

A Boolean function ψ is increasing if and only if there do not exist vectors $\mathbf{x} \in \mathcal{S}_1[\psi]$ and $\mathbf{y} \in \mathcal{S}_0[\psi]$ such that $\mathbf{x} \leqslant \mathbf{y}$. In terms of upper sets, ψ is increasing if and only if $\mathcal{U}[\mathcal{S}_1[\psi]^-] = \mathcal{U}[\mathcal{S}_1[\psi]] = \mathcal{S}_1[\psi]$. $\mathcal{S}_1[\psi]$ is called the *kernel* of ψ and, if ψ is increasing, then $\mathcal{S}_1[\psi]^-$ is called the *increasing basis* (or, commonly, just the *basis*) of ψ. We denote the kernel and basis by $\mathcal{K}[\psi]$ and $\mathcal{B}[\psi]$, respectively, or just $\mathcal{K}$ and $\mathcal{B}$ when not specifying the function. A four-variable increasing Boolean function is shown in Fig. 1–1. The enclosed vectors compose the kernel, and minimal elements are shown in solid black boxes. Figure 1–2 corresponds to a nonincreasing Boolean function; again the enclosed vectors compose the kernel. Since the upper set of the minimal elements does not equal the kernel, the Boolean function is nonincreasing. It becomes increasing if 0111 is switched into the kernel.

Suppose ψ is an increasing Boolean function, its positive representation according to Eq. 1–5 is assumed to be minimal, and there are m products in the expansion

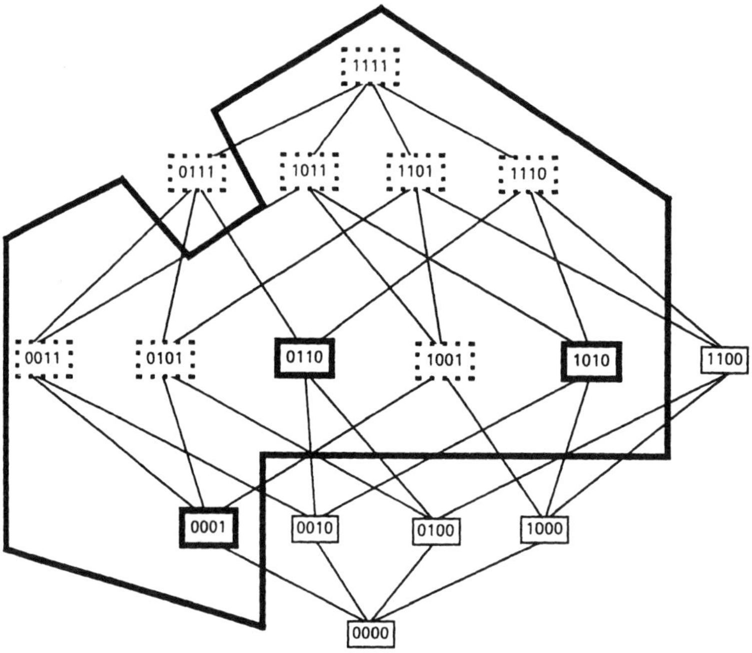

Figure 1–2. Nonincreasing Boolean function.

made up of the m variable sets

$$\begin{aligned} V_1 &= \{x_{1,1}, x_{1,2}, \ldots, x_{1,n(1)}\} \\ V_2 &= \{x_{2,1}, x_{2,2}, \ldots, x_{2,n(2)}\} \\ &\vdots \\ V_m &= \{x_{m,1}, x_{m,2}, \ldots, x_{m,n(m)}\} \end{aligned} \tag{1–6}$$

Then $\mathcal{B}[\psi] = \{\mathbf{y}_1, \mathbf{y}_2, \ldots, \mathbf{y}_m\}$, where the components of $\mathbf{y}_i = (y_{i,1}, y_{i,2}, \ldots, y_{i,n})$ are

$$y_{i,k} = \begin{cases} 1, & \text{if } x_k \in V_i \\ 0, & \text{if } x_k \notin V_i \end{cases} \tag{1–7}$$

for $k = 1, 2, \ldots, n$. Hence, the product terms of the minimal expansion are referred to as the *basis elements* of ψ.

If ψ and ξ are two n-variable Boolean functions, their switching set is defined by

$$\mathcal{Z}[\psi, \xi] = \{\mathbf{x}\colon \psi(\mathbf{x}) \neq \xi(\mathbf{x})\} = (\mathcal{S}_1[\psi] \cap \mathcal{S}_0[\xi]) \cup (\mathcal{S}_0[\psi] \cap \mathcal{S}_1[\xi]) \tag{1–8}$$

If we were to switch (change the value of) $\xi(\mathbf{x})$ for every $\mathbf{x} \in \mathcal{Z}[\psi, \xi]$ or switch $\psi(\mathbf{x})$ for every $\mathbf{x} \in \mathcal{Z}[\psi, \xi]$, then we would have $\psi = \xi$. Many filtering problems are concerned with switching one Boolean function into another where there is a cost associated with switching. To formulate a switching cost, we postulate a real-valued cost function Co defined on $\{0, 1\}^n$ and define the *cost* of switching ψ into ξ or ξ into ψ by

$$Co[\psi, \xi] = \sum_{\mathbf{x} \in \mathcal{Z}[\psi, \xi]} Co(\mathbf{x}) \tag{1–9}$$

As a convention, the cost function is defined so that if the switching cost is negative, the switch is advantageous; if it is positive, the switch is disadvantageous (relative to Co). The cost of switching a function ψ into a class $\Im$ of functions is defined to be

$$Co_{\Im}[\psi] = \min\{Co[\psi, \xi]: \xi \in \Im\} \tag{1–10}$$

As will be discussed subsequently, for filter design an important switching cost involves constraining a filter to a particular filter class. In this case, the cost function is an error probability and the switching cost gives the increased error owing to the constraint.

1.2 MORPHOLOGICAL REPRESENTATION

Boolean functions are used to define translation-invariant windowed operators on binary digital images. These images are modeled as subsets of the Cartesian grid Z^2. Each point is called a *pixel* and is an ordered pair of integers. An image S consists of a set of pixels $z \in S$. As sets, binary images are operated upon by the usual set operations: union, intersection, complement, set subtraction, etc. Logically, an image is represented as 0s and 1s. The image value is 1 if a pixel is in the image and 0 if it is not: if $S(z)$ denotes the value of S at z, then $z \in S$ if and only if $S(z) = 1$ and $z \notin S$ if and only if $S(z) = 0$. The translation of S by pixel z is defined by $S_z = \{u + z: u \in S\}$.

To define a windowed operator, let $W = \{w_1, w_2, \ldots, w_n\}$ be an n-pixel window and ψ be an n-variable Boolean function. The corresponding set operator Ψ is defined by

$$\Psi(S)(z) = \psi(S \cap W_z) = \psi\big(S(w_1 + z), S(w_2 + z), \ldots, S(w_n + z)\big) \tag{1–11}$$

(Fig. 1–3). Note that we are simultaneously treating S and $\Psi(S)$ as subsets of the digital plane and as binary-valued functions: in the first instance, $S \cap W_z$ is the intersection between the sets S and W_z; in the second, $S \cap W_z$ is the $\{0, 1\}$-valued function S restricted to W_z. Ψ is translation-invariant, meaning $\Psi(S_z) = \Psi(S)_z$, because the same Boolean function is applied at every pixel. We call Ψ a *W-operator* and ψ its *window function*. The representation of Ψ corresponds directly to the logical representation of ψ.

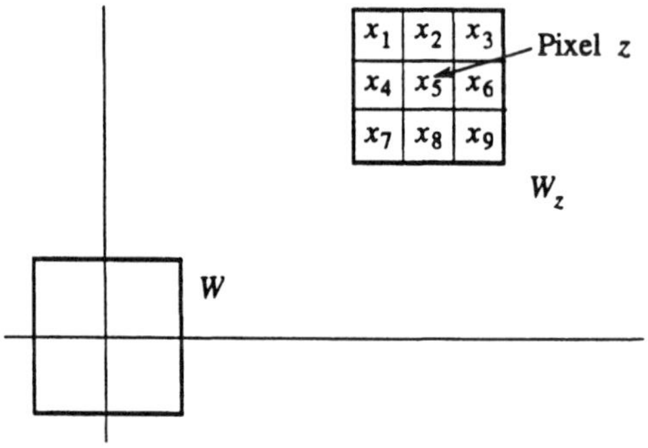

Figure 1–3. Window W applied at pixel z.

A W-operator Ψ_i defined by a single-product Boolean function

$$\psi_i(x_1, x_2, \ldots, x_n) = x_{i,1}x_{i,2}\cdots x_{i,n(i)} \tag{1–12}$$

is called an *erosion*. According to the correspondence between a binary image being defined as a subset of the grid and as a binary-valued function, $z \in \Psi_i(S)$ if and only if $\Psi_i(S)(z) = 1$. In terms of window logic, this means that the pixels in the translated window W_z corresponding to the pixels $w_{i,1}, w_{i,2}, \ldots, w_{i,n(i)}$ in W must all have value 1 so that the product of Eq. 1–12 is 1; that is, so that $x_{i,1} = x_{i,2} = \cdots = x_{i,n(i)} = 1$. The pixels in W_z corresponding to $w_{i,1}, w_{i,2}, \ldots, w_{i,n(i)}$ are $w_{i,1} + z, w_{i,2} + z, \ldots, w_{i,n(i)} + z$, and the product of Eq. 1–12 is 1 if and only if all of these pixel translates lie in S. Letting $B^i = \{w_{i,1}, w_{i,2}, \ldots, w_{i,n(i)}\}$, $z \in \Psi_i(S)$ if and only if $B^i_z \subset S$. Ψ_i is called an *erosion* operator.

Erosion of set S by set B, called a *structuring element*, is denoted by $\mathrm{E}_B(S)$ and the window function corresponding to E_B is denoted by ε_B. Figure 1–4 shows a set A and a structuring element B (with origin marked). Part (a) shows a translate of B to a pixel z for which $B_z \subset A$, so that $z \in \mathrm{E}_B(A)$; part (b) shows a translate of B to a pixel w for which $B_w \not\subset A$, so that $w \notin \mathrm{E}_B(A)$; part (c) shows $\mathrm{E}_B(A)$. Morphological image processing is based on the representation of image operators in terms of primary image operators, the most fundamental being erosion.

A set operator is said to be *increasing* if and only if $S_1 \subset S_2$ implies $\Psi(S_1) \subset \Psi(S_2)$. A W-operator Ψ with window function ψ is increasing if and only if ψ is a positive Boolean function. It follows at once from the logical representation of Eq. 1–5 that a W-operator is increasing if and only if it possesses an erosion representation [1] of the form

$$\Psi(S) = \bigcup_i \mathrm{E}_{B^i}(S) \tag{1–13}$$

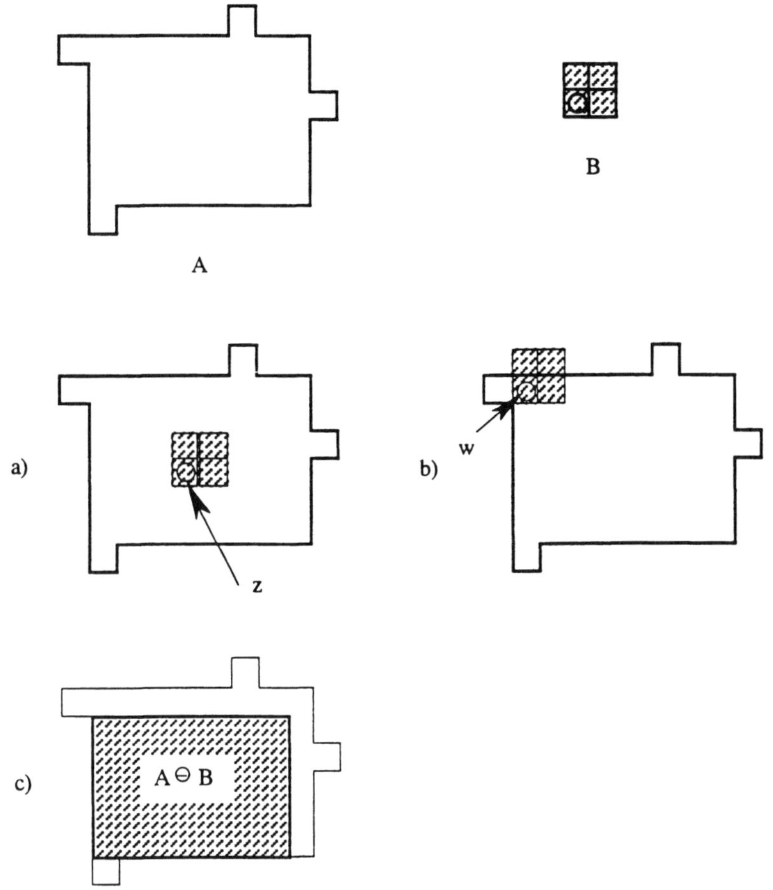

Figure 1–4. Erosion: (a) pixel in eroded set; (b) pixel not in eroded set; (c) eroded set [with $A \ominus B$ denoting erosion].

where the correspondence between Eqs. 1–5 and 1–13 is given by

$$x_{i,1}x_{i,2}\cdots x_{i,n(i)} \leftrightarrow B^i \quad (1\text{–}14)$$

The erosion representation is minimal if no structuring element is a subset of another. This corresponds directly to a minimal logical representation [2–4] and the corresponding structuring elements comprise the *basis*, $\mathcal{B}[\Psi]$, of Ψ.

Consider the binary moving median M defined over a window W containing an odd number of pixels. The window function μ is defined by $\mu(x_1, x_2, \ldots, x_n) = 1$ if and only if more than $n/2$ pixels in W are 1-valued. Hence a row of the truth table defining μ is 1-valued if and only if at least $(n+1)/2$ of the variables are 1-valued, which means that an n-variable product is a minterm in the disjunctive normal form for μ if and only if at least $(n+1)/2$ variables are uncomplemented. The minimal positive expansion for μ consists of all products of exactly $(n+1)/2$ variables. For instance, suppose W consists of the origin together with the pixels immediately below, above, to the right, and to the left of it. Let x_1, x_2, x_3, x_4, and x_5 denote the five variables corresponding to the five pixels. The minimal representation of μ is given by

$$\begin{aligned}\mu(x_1, x_2, x_3, x_4, x_5) &= x_1x_2x_3 + x_1x_2x_4 + x_1x_3x_4 + x_1x_2x_5 + x_1x_3x_5 \\ &+ x_1x_4x_5 + x_2x_3x_4 + x_2x_3x_5 + x_2x_4x_5 + x_3x_4x_5 \qquad (1\text{–}15)\end{aligned}$$

The basis of M consists of ten structuring elements:

$$\begin{pmatrix}0&1&0\\0&\mathbf{0}&1\\0&1&0\end{pmatrix}\begin{pmatrix}0&0&0\\1&\mathbf{1}&1\\0&0&0\end{pmatrix}\begin{pmatrix}0&0&0\\1&\mathbf{1}&0\\0&1&0\end{pmatrix}\begin{pmatrix}0&0&0\\1&\mathbf{0}&1\\0&1&0\end{pmatrix}\begin{pmatrix}0&0&0\\0&\mathbf{1}&1\\0&1&0\end{pmatrix}$$
$$\begin{pmatrix}0&1&0\\1&\mathbf{1}&0\\0&0&0\end{pmatrix}\begin{pmatrix}0&1&0\\1&\mathbf{0}&1\\0&0&0\end{pmatrix}\begin{pmatrix}0&1&0\\0&\mathbf{1}&1\\0&0&0\end{pmatrix}\begin{pmatrix}0&1&0\\1&\mathbf{0}&0\\0&1&0\end{pmatrix}\begin{pmatrix}0&1&0\\0&\mathbf{1}&0\\0&1&0\end{pmatrix}$$

Note that a bold character indicates the origin and including extra 0s in the matrix representation of a structuring element does not change the structuring element.

Morphological representation of arbitrary (nonincreasing) W-operators is achieved via the *hit-or-miss transform* [5], which is defined, for any pair of disjoint subsets $E, F \subset W$, by

$$\Lambda_{E,F}(S) = \mathrm{E}_E(S) \cap \mathrm{E}_F(S^c) \qquad (1\text{–}16)$$

Pixelwise, as a binary-valued function, $\Lambda_{E,F}$ is given by

$$\begin{aligned}\Lambda_{E,F}(S)(z) &= \mathrm{E}_E(S)(z) \wedge \mathrm{E}_F(S^c)(z) \\ &= \begin{cases}1, & \text{if } E_z \subset S \cap W_z \text{ and } F_z \subset S^c \cap W_z \\ 0, & \text{otherwise}\end{cases} \\ &= \begin{cases}1, & \text{if } E_z \subset S \cap W_z \text{ and } F_z \cap (S \cap W_z) = \emptyset \\ 0, & \text{otherwise}\end{cases} \qquad (1\text{–}17)\end{aligned}$$

Corresponding to the logical representation of Eq. 1–2 is the *standard morphological representation* for a W-operator [6],

$$\Psi(S) = \bigcup_i \Lambda_{E^i,F^i}(S) \qquad (1\text{–}18)$$

The correspondence between Eqs. 1–2 and 1–18 is given by

$$(E^i, F^i) \leftrightarrow x_1^{p(i,1)} x_2^{p(i,2)} \cdots x_{n(i)}^{p(i,n(i))} \qquad (1\text{–}19)$$

where x_k corresponds to the pixel $w_k \in W$, x_k appears uncomplemented if $w_k \in E^i$, x_k appears complemented if $w_k \in F^i$, and x_k does not appear if $x_k \notin E^i \cup F^i$. A structuring pair is said to be *canonical* if $E^i \cup F^i = W$ and is canonical if and only if the corresponding logical product contains all n variables in the window. The representation of Eq. 1–18 is said to be canonical if all structuring pairs are canonical, in which case it corresponds to the disjunctive normal form of Eq. 1–1. If Ψ happens to be increasing, then the hit-or-miss expansion of Eq. 1–18 reduces to the erosion expansion of Eq. 1–13.

There is a useful alternative form of the hit-or-miss tranform. If L and U are binary functions defined on W such that $L \leqslant U$, then the *interval* defined by the pair (L, U) is the set of all binary functions V defined on W such that $L \leqslant V \leqslant U$. This interval is denoted by $[L, U]$ and L and U are called the *lower and upper endpoints*, respectively. For any binary image S, the *hit-or-miss transform* corresponding to (L, U) is defined at pixel z by

$$\Lambda_{L,U}(S)(z) = \begin{cases} 1, & \text{if } L_z \leqslant S \cap W_z \leqslant U_z \\ 0, & \text{otherwise} \end{cases} \tag{1–20}$$

where L_z and U_z denote L and U operating on the translate W_z. If $L = U$, then the pair is said to be canonical and $\Lambda_{L,U}(S)(z) = \Lambda_{U,U}(S)(z) = 1$ if and only if $S \cap W_z = U_z$, which means there is an exact pattern match. To see that the two definitions agree, let E and F be the sets of all pixels at which L is 1-valued and U is 0-valued, respectively; that is, $E = L$ and $F = U^c$. Then $L_z \leqslant S \cap W_z \leqslant U_z$ if and only if $E_z \subset S \cap W_z$ and $F_z \cap (S \cap W_z) = \emptyset$.

For an illustration, let W be the 3×3 square centered at the origin and

$$E = L = \begin{pmatrix} 1 & 1 & 1 \\ 0 & \mathbf{1} & 0 \\ 0 & 0 & 0 \end{pmatrix} \quad U = \begin{pmatrix} 1 & 1 & 1 \\ 1 & \mathbf{1} & 1 \\ 0 & 0 & 0 \end{pmatrix} \quad F = U^c = \begin{pmatrix} 0 & 0 & 0 \\ 0 & \mathbf{0} & 0 \\ 1 & 1 & 1 \end{pmatrix} \tag{1–21}$$

Then

$$[L, U] = \left\{ \begin{pmatrix} 1 & 1 & 1 \\ 0 & \mathbf{1} & 0 \\ 0 & 0 & 0 \end{pmatrix}, \begin{pmatrix} 1 & 1 & 1 \\ 1 & \mathbf{1} & 0 \\ 0 & 0 & 0 \end{pmatrix}, \begin{pmatrix} 1 & 1 & 1 \\ 0 & \mathbf{1} & 1 \\ 0 & 0 & 0 \end{pmatrix}, \begin{pmatrix} 1 & 1 & 1 \\ 1 & \mathbf{1} & 1 \\ 0 & 0 & 0 \end{pmatrix} \right\} \tag{1–22}$$

E and F are often expressed in a single matrix, with 1 and 0 denoting the pixels of E and F, respectively, and "×" denoting a (*don't care*) pixel in neither E nor F. In this notation, the hit-or-miss-transform structuring pair corresponding to (L, U) is expressed by

$$[E, F] = \begin{pmatrix} 1 & 1 & 1 \\ \times & \mathbf{1} & \times \\ 0 & 0 & 0 \end{pmatrix} \tag{1–23}$$

1.3 SYSTEM MODEL

A key task of binary operator theory is automatic filter design. We desire an operator to optimally estimate an image when it is observed after going through some system. To frame the problem, binary digital images are modeled as discrete random sets (i.e., a collection of subsets associated with a probability distribution) [7]. If Ψ is a set operator, then for each input random set $\mathbf{S}$ there is an output random set $\Psi(\mathbf{S})$. If S is any observed realization of $\mathbf{S}$, then $\Psi(S)$ is a realization of $\Psi(\mathbf{S})$. The task is to design an operator Ψ so that, given $\mathbf{S}$, $\Psi(\mathbf{S})$ is probabilistically close to some desired process $\mathbf{I}$. We call $\mathbf{S}$, $\mathbf{I}$, and $\Psi(\mathbf{S})$ the *observation*, *ideal*, and *estimator* processes, respectively.

The distance between the ideal and estimator processes is measured by some probabilistic error measure $Er[\mathbf{I}, \Psi(\mathbf{S})]$. Assuming the operator belongs to some operator family $\Im$, an *optimal operator* relative to $\Im$ is an operator $\Psi_{opt} \in \Im$ for which

$$Er[\mathbf{I}, \Psi_{opt}(\mathbf{S})] \leqslant Er[\mathbf{I}, \Psi(\mathbf{S})] \tag{1–24}$$

for all $\Psi \in \Im$. If $\Im$ is characterized by some operator representation, then every operator $\Psi \in \Im$ has a representation and optimization can be viewed as finding the representation defining an operator possessing minimum error $Er[\mathbf{I}, \Psi(\mathbf{S})]$.

The optimization model includes a *system transformation* Ξ such that $\mathbf{S} = \Xi(\mathbf{I})$; that is, the observation process is assumed to be the output of some system operating on the ideal process [8]. Optimization involves minimizing the error $Er[\mathbf{I}, \Psi(\Xi(\mathbf{I}))]$. In general, Ξ is a multivalued image operator, meaning that, given a realization I of the ideal image, there are many possible output realizations $\Xi(I)$. The task of finding an optimal operator Ψ constitutes an inverse problem: find Ψ to best invert Ξ.

A much-studied application of the method is when Ξ is a degradation (noise) transformation — the ideal image is obscured by noise [9, 10]. The form of Ξ depends on the physical system causing the degradation. If $\mathbf{N}$ is some other binary process, then the *signal-union-noise model* is defined by the degradation transformation $\mathbf{S} = \Xi(\mathbf{I}) = \mathbf{I} \cup \mathbf{N}$. Each observed realization is of the form $S = I \cup N$, where I and N are realizations of $\mathbf{I}$ and $\mathbf{N}$, respectively. Unless there are no restrictions placed on the ideal and noise processes, the union $\mathbf{I} \cup \mathbf{N}$ does not fully define the system transformation. For instance, it might be required that $\mathbf{I}$ and $\mathbf{N}$ are statistically independent or that the union is restricted in such a way that there is no intersection between signal and noise. In document processing, the noise is often restricted to character edges, so that the noise process is strongly dependent on the signal process. Rather than union noise with the signal, one might subtract the noise, thereby having the degradation transformation $\Xi(\mathbf{I}) = \mathbf{I} - \mathbf{N}$. There could be two noise processes, $\mathbf{N}_1$ and $\mathbf{N}_2$, with the degradation both adjoining and deleting pixels and the system transformation being $\Xi(\mathbf{I}) = (\mathbf{I} \cup \mathbf{N}_1) - \mathbf{N}_2$.

Examples of morphological system models are the dilation and erosion models (see Section 3.3 for discussion of basic morphological operators). If B is a fixed structuring element and we define the system transformation by dilation or erosion, $\Xi(\mathbf{I}) = \Delta_B(\mathbf{I})$ or $\Xi(\mathbf{I}) = \mathrm{E}_B(\mathbf{I})$, then Ξ is single-valued: for each realization of the ideal process, a single determined observation results from the system. If the designed operator needs to be applied across various possible dilations or erosions, then it is better to treat the structuring element as a random set $\mathbf{B}$, in which case the dilation and erosion system transformations take the forms $\Xi(\mathbf{I}) = \Delta_{\mathbf{B}}(\mathbf{I})$ and $\Xi(\mathbf{I}) = \mathrm{E}_{\mathbf{B}}(\mathbf{I})$, respectively, both of which are multivalued. System models can be defined by other morphological operators. If $\mathbf{A}$ and $\mathbf{B}$ are random structuring elements, then the system transformation might be erosion followed by dilation, $\Xi(\mathbf{I}) = \Delta_{\mathbf{A}}(\mathrm{E}_{\mathbf{B}}(\mathbf{I}))$, or dilation followed by erosion, $\Xi(\mathbf{I}) = \mathrm{E}_{\mathbf{A}}(\Delta_{\mathbf{B}}(\mathbf{I}))$. If $\mathbf{A} = \mathbf{B}$, then these system transformations are, respectively, *opening* and *closing* by $\mathbf{B}$.

To characterize binary edge detection in terms of a system model, assume there exists a canonical edge detector: given a binary deterministic image L, there is an edge operator Θ such that $\Theta(L)$ is by definition the edge of L. For the system model, the ideal image process is a process of edges, the class of edges consisting of all possible edges that might be observed. To avoid undue mathematical complexity, assume that, for each ideal edge realization I, there exists a unique binary image R_I having edge $I[\Theta(L_I) = I]$ and we know the algorithm H that produces L_I from I. For instance, Θ might be 3×3 dilation of the image minus the image and H a contour filler. If $\Xi = \mathrm{H}$, then perfect edge construction would result from applying Θ and there would be no operator-design problem. A more practical model results by assuming the system model is H followed by noise degradation. For union noise, the system transformation takes the form $\Xi(\mathbf{I}) = \mathrm{H}(\mathbf{I}) \cup \mathbf{N}$. An optimal edge detector minimizes the error $Er[\mathbf{I}, \Psi(\mathrm{H}(\mathbf{I}) \cup \mathbf{N})]$.

For matched filtering, the ideal image is a set of points at which the object of interest is located. We assume there exists an operator that places an object to be recognized at each point of the ideal image process. An object to be recognized is a random set (shape) $\mathbf{A}$. Given a realization $I = \{z_1, z_2, \ldots, z_n\}$ of the ideal image (point set at which instances of the object are located), a realization of the observed image is given by a union of realizations $A_1, A_2, \ldots, A_n$ of $\mathbf{A}$ translated to the points $z_1, z_2, \ldots, z_n$, respectively. As a random process, the observation image is defined by the system transformation

$$\mathbf{S} = \Xi_{\text{match}}(\mathbf{I}) = \bigcup_{i=1}^{N} \mathbf{A}_i + \mathbf{z}_i \tag{1–25}$$

where N is a random variable giving the number of points in the ideal image; $\mathbf{A}_1, \mathbf{A}_2, \ldots, \mathbf{A}_N$ are random sets identically distributed to the primary shape $\mathbf{A}$;

and $\mathbf{z}_1, \mathbf{z}_2, \ldots, \mathbf{z}_N$ are the random points composing the ideal image. The system is generally not completely characterized by these minimal conditions. For instance, there may be a constraint that objects cannot intersect or that intersection is constrained. There are also independence assumptions. Are $\mathbf{A}_1, \mathbf{A}_2, \ldots, \mathbf{A}_N$ mutually independent? Are they independent of the points at which they are located? The randomness of $\mathbf{A}$ also needs to be modeled. It might be that $\mathbf{A}$ is a fixed shape distorted by union noise, $\mathbf{A} = A_0 \cup \mathbf{N}$; it might be that $\mathbf{A}$ is analytically described with random parameters, such as an ellipse with random axes and angle of rotation.

The system transformation of Eq. 1–25 can itself be more general. In character recognition, there exist random primary shapes (characters) different from $\mathbf{A}$, say $\mathbf{A}^1, \mathbf{A}^2, \ldots, \mathbf{A}^m$ and random point images $\mathbf{F}^1, \mathbf{F}^2, \ldots, \mathbf{F}^m$ such that

$$\Xi_{\text{match},m}(\mathbf{I}) = \left(\bigcup_{i=1}^{N} \mathbf{A}_i + \mathbf{z}_i\right) \cup \left(\bigcup_{k=1}^{m}\bigcup_{j=1}^{N(k)} \mathbf{A}_j^k + \mathbf{z}_{k,j}\right) \tag{1–26}$$

where $N(1), N(2), \ldots, N(m)$ are the numbers of points in $\mathbf{F}^1, \mathbf{F}^2, \ldots, \mathbf{F}^m$, for $k = 1, 2, \ldots, m$, $\mathbf{F}^k = \{\mathbf{z}_{k,1}, \mathbf{z}_{k,2}, \ldots, \mathbf{z}_{k,N(k)}\}$, and $\mathbf{A}_j^k$ is identically distributed to $\mathbf{A}^k$ for $j = 1, 2, \ldots, N(k)$.

Finally, suppose we desire a W-operator to emulate a given algorithm A. If A operates on a random image process $\mathbf{R}$, then the ideal image process is $\mathbf{I} = \mathrm{A}(\mathbf{R})$ and the system transformation is $\Xi_{\mathrm{A}}(\mathbf{I}) = \mathrm{A}^{-1}(\mathbf{I})$, where Ξ_{A} is multivalued because many realizations of $\mathbf{R}$ might yield the same ideal realization. The optimal emulator of A minimizes the emulation error $Er[\mathbf{I}, \Psi(\mathrm{A}^{-1}(\mathbf{I}))]$. If A is a segmentation algorithm, then $\mathbf{I}$ is composed of segmented images from $\mathbf{R}$ and $\mathrm{A}^{-1}(\mathbf{I})$ consists of unsegmented images. More generally, consider adapting a given algorithm to a noisy environment. Suppose $\mathrm{A}^{-1}(\mathbf{I})$ is the observation process resulting from algorithm inversion. As designed originally, A is meant to be applied to this process; however, suppose the observations have been degraded. Then the system transformation is given by $\Xi_{\mathrm{N}}(\mathbf{I}) = \mathrm{N}(\mathrm{A}^{-1}(\mathbf{I}))$, where N is a noise-degradation operator. For segmentation, if the presegmented images are degraded by erosion by a random structuring element $\mathbf{B}$, then $\Xi_{\mathrm{N}}(\mathbf{I}) = \mathrm{E}_{\mathbf{B}}(\mathrm{A}^{-1}(\mathbf{I}))$ and the optimal emulator minimizes the error $Er[\mathbf{I}, \Psi(\mathrm{E}_{\mathbf{B}}(\mathrm{A}^{-1}(\mathbf{I})))]$. With direct emulation using system function Ξ_{A}, the only advantage is translation of the algorithm into a W-operator. When emulation involves a degraded version of $\mathrm{A}^{-1}(\mathbf{I})$, there is the added advantage that the original algorithm may not work well in the degraded environment, whereas the emulation will estimate the algorithm as if it were applied in a nondegraded environment.

1.4 Optimal W-Operators

Estimation of $\mathbf{I}$ from $\Xi(\mathbf{I})$ by a W-operator Ψ requires finding a Boolean function ψ to minimize error. Since Ψ is translation-invariant, we make the modeling assumption that **I** and $\Xi(\mathbf{I})$ are jointly strict-sense stationary. This means that, if **X** is the random vector of binary values in W_z and $Y = \mathbf{I}(z)$, then the joint probability distribution for **X** and Y is independent of z, so that estimating Y from **X** yields a translation-invariant filter. We will denote random variables and random vectors by upper-case italic and bold-face letters, respectively. Realizations of the random variable Y and the random vector **X** will be denoted by y and **x**, respectively.

For operator optimization we require a loss function l: $\{0, 1\}^2 \to [0, \infty)$, where $l(a, b)$ measures the cost of the difference between a and b, with $l(0, 0) = l(1, 1) = 0$. Relative to the loss function (and owing to stationarity), filter error, $Er\langle\Psi\rangle$, is given by the expected loss from estimating $\mathbf{I}(z)$ by $\Psi(\Xi(\mathbf{I}))(z)$,

$$Er\langle\Psi\rangle = Er\big[\mathbf{I}, \Psi(\Xi(\mathbf{I}))\big] = E\big[l\big(\mathbf{I}(z), \Psi(\Xi(\mathbf{I}))(z)\big)\big] \tag{1–27}$$

where z is an arbitrary pixel. In terms of the Boolean function ψ for Ψ,

$$\begin{aligned}
Er\langle\Psi\rangle &= E\big[l\big(Y, \psi(\mathbf{X})\big)\big] \\
&= \sum_{\mathbf{x}} E\big[l\big(Y, \psi(\mathbf{x})\big) \mid \mathbf{x}\big] P(\mathbf{x}) \\
&= \sum_{\{\mathbf{x}:\ \psi(\mathbf{x})=0\}} E\big[l(Y, 0) \mid \mathbf{x}\big] P(\mathbf{x}) + \sum_{\{\mathbf{x}:\ \psi(\mathbf{x})=1\}} E\big[l(Y, 1) \mid \mathbf{x}\big] P(\mathbf{x}) \\
&= \sum_{\{\mathbf{x}:\ \psi(\mathbf{x})=0\}} l(1, 0) P(Y = 1 \mid \mathbf{x}) P(\mathbf{x}) \\
&\quad + \sum_{\{\mathbf{x}:\ \psi(\mathbf{x})=1\}} l(0, 1) P(Y = 0 \mid \mathbf{x}) P(\mathbf{x})
\end{aligned} \tag{1–28}$$

where $P(\mathbf{x})$ denotes $P(\mathbf{X} = \mathbf{x})$. An optimal image filter is one whose Boolean function ψ minimizes Eq. 1–28. Although there can be more than one filter achieving minimal error, we shall denote "the" optimal filter and its window function by Ψ_{opt} and ψ_{opt}, respectively, the convention being that, from the standpoint of filter optimization, all filters possessing minimal error are equivalent.

The *mean-absolute-error* (*MAE*) loss function is defined by

$$l\big(y, \psi(\mathbf{x})\big) = \big|y - \psi(\mathbf{x})\big| \tag{1–29}$$

Since y and $\psi(\mathbf{x})$ are binary-valued, the loss function is given by $l(1, 0) = l(0, 1) = 1$ and $l(0, 0) = l(1, 1) = 0$. The associated error is the mean-absolute error and is denoted by $MAE\langle\Psi\rangle$. Because $l(1, 0) = l(0, 1)$, it follows from Eq. 1–28 that the optimal Boolean function and the error of the corresponding optimal set filter are given by

$$\psi_{MAE}(\mathbf{x}) = \begin{cases} 1, & \text{if } P(Y = 1 \mid \mathbf{x}) > 0.5 \\ 0, & \text{if } P(Y = 1 \mid \mathbf{x}) \leqslant 0.5 \end{cases} \tag{1–30}$$

$$\begin{aligned} MAE\langle\Psi_{MAE}\rangle &= E\big[|Y-\psi(X)|\big] \\ &= \sum_{\{\mathbf{x}:P(Y=1|\mathbf{x})>0.5\}} P(\mathbf{x})P(Y=0\mid\mathbf{x}) \\ &\quad + \sum_{\{\mathbf{x}:P(Y=1|\mathbf{x})\leqslant 0.5\}} P(\mathbf{x})P(Y=1\mid\mathbf{x}) \end{aligned} \tag{1–31}$$

If we list the possible realizations $\mathbf{x}$ of $\mathbf{X}$ in a table along with the prior probabilities $P(\mathbf{x})$ and conditional probabilities $P(Y=0\mid\mathbf{x})$ and $P(Y=1\mid\mathbf{x})$, then $MAE\langle\Psi_{MAE}\rangle$ is obtained by summing the joint probabilities corresponding to the values (0 or 1) not chosen for ψ_{MAE}. Because images are binary,

$$E[Y\mid\mathbf{X}]=P(Y=1\mid\mathbf{X}) \tag{1–32}$$

so that

$$\psi_{MAE}(\mathbf{x})=\begin{cases}1, & \text{if } E[Y\mid\mathbf{x}]>0.5\\ 0, & \text{if } E[Y\mid\mathbf{x}]\leqslant 0.5\end{cases} \tag{1–33}$$

the *binary conditional expectation*. Relative to the input image $\mathbf{S}=\Xi(\mathbf{I})$, the optimal MAE filter is defined pixelwise by

$$\Psi_{MAE}(\mathbf{S})(z)=\begin{cases}1, & \text{if } P(\mathbf{I}(z)=1\mid\mathbf{S}\cap W_z)>0.5\\ 0, & \text{if } P(\mathbf{I}(z)=1\mid\mathbf{S}\cap W_z)\leqslant 0.5\end{cases} \tag{1–34}$$

The MAE loss function is used in many applications. Consider object recognition in a model in which $\mathbf{I}$ gives the locations of a shape A. The conditional probability in Eq. 1–34 gives the probability of A being at pixel z given the observation $\mathbf{S}\cap W_z$. Equivalently, the shape occurrence probability is given by $P(Y=1\mid\mathbf{X})$. Now, suppose there exists a single observation $\mathbf{x}_A$ for which $P(Y=1\mid\mathbf{x}_A)>0.5$ and $P(Y=1\mid\mathbf{x})<0.5$ for $\mathbf{x}\neq\mathbf{x}_A$. Then, according to Eq. 1–30, the optimal MAE filter is defined by the Boolean function

$$\psi_{MAE}(\mathbf{x})=\begin{cases}1, & \text{if } \mathbf{x}=\mathbf{x}_A\\ 0, & \text{if } \mathbf{x}\neq\mathbf{x}_A\end{cases} \tag{1–35}$$

In particular, suppose $P(Y=1\mid\mathbf{x}_A)=1$ and $P(Y=1\mid\mathbf{x})<0.5$ for $\mathbf{x}\neq\mathbf{x}_A$. Then $\psi_{MAE}(\mathbf{x})=1$ if, based on the observation $\mathbf{x}$, we are almost sure that shape A occurs. If, however, there were to exist $\mathbf{x}_0\neq\mathbf{x}_A$ such that $P(Y=1\mid\mathbf{x}_0)>0.5$, then we would also have $\psi_{MAE}(\mathbf{x}_0)=1$.

A different loss function can achieve a more general shape recognition result. Let $\mathbf{x}_1,\mathbf{x}_2,\ldots,\mathbf{x}_m$ denote the $m=2^n$ possible realizations of $\mathbf{X}$. Let $\mathbf{x}_1,\mathbf{x}_2,\ldots,\mathbf{x}_r$, $r<m$, correspond to the patterns which when observed guarantee (almost surely)

that the shape A occurs at z and $\mathbf{x}_{r+1}, \mathbf{x}_{r+2}, \ldots, \mathbf{x}_m$ correspond to patterns which when observed do not guarantee (almost surely) that A occurs at z. In terms of conditional probabilities, $P(Y = 1 \mid \mathbf{x}_j) = 1$ for $j = 1, 2, \ldots, r$ and $P(Y = 1 \mid \mathbf{x}_j) < 1$ for $j = r + 1, \ldots, m$. Let

$$\tau = \max_{j>r} P(Y = 1 \mid \mathbf{x}_j)P(\mathbf{x}_j) < 1 \tag{1–36}$$

and define a loss function by $l(0, 0) = l(1, 1) = 0$, $l(1, 0) = 1$, and $l(0, 1) = (1 - \tau)^{-1}$. We claim that the optimal shape-recognition filter relative to l is given by

$$\psi_{SR}(\mathbf{x}) = \begin{cases} 1, & \text{if } \mathbf{x} = \mathbf{x}_j,\ j = 1,\ 2, \ldots, r \\ 0, & \text{if } \mathbf{x} = \mathbf{x}_j,\ j = r + 1, \ldots, m \end{cases} \tag{1–37}$$

which is precisely the filter we desire. From Eq. 1–28, the error for ψ_{SR} is given by

$$\varepsilon_{SR} = \sum_{j=r+1}^{m} P(Y = 1 \mid \mathbf{x}_j)P(\mathbf{x}_j) \tag{1–38}$$

Now, suppose ψ_{SR} is changed so that some of the first r vectors, say $\mathbf{u}_1, \mathbf{u}_2, \ldots, \mathbf{u}_b$, are redefined to be 0-valued and some of the originally 0-valued vectors, say $\mathbf{v}_1, \mathbf{v}_2, \ldots, \mathbf{v}_c$, are redefined to be 1-valued. The error for the new function, call it ξ, is given by

$$\begin{aligned} \varepsilon_\xi &= \varepsilon_{SR} - \sum_{j=1}^{c} P(Y = 1 \mid \mathbf{v}_j)P(\mathbf{v}_j) \\ &\quad + \frac{1}{1-\tau}\sum_{j=1}^{c} P(Y = 0 \mid \mathbf{v}_j)P(\mathbf{v}_j) + \sum_{j=1}^{b} P(\mathbf{u}_j) \end{aligned} \tag{1–39}$$

The first sum is bounded above by $c\tau$, the second sum is bounded below by $c(1 - \tau)$, and the third sum is nonnegative. Hence,

$$\varepsilon_\xi \geqslant \varepsilon_{SR} + c(1 - \tau) + \sum_{j=1}^{b} P(\mathbf{u}_j) > \varepsilon_{SR} \tag{1–40}$$

thereby proving that ψ_{SR} is optimal relative to the new loss function.

Every W-operator Ψ can be canonically represented as

$$\Psi(S) = \bigcup_{(E,F)\in\mathcal{C}_\Psi} \Lambda_{E,F}(S) \tag{1–41}$$

where $\mathcal{C}_\Psi$ is the set of canonical structuring pairs defining Ψ. Owing to canonicalness, $\Psi(S)(z) = 1$ if and only if there is a structuring pair $(E, F) \in \mathcal{C}_\Psi$ such that $E = S \cap W_z$ and $F = S^c \cap W_z$. We denote the structuring-pair set defining an optimal filter Ψ_{opt} by $\mathcal{C}_{opt}$.

For an arbitrary loss function [8] $(E, F) \in \mathcal{C}_{opt}$ if and only if

$$E\big[l(\mathbf{I}(z), 1) \mid \mathbf{S} \cap W_z = E\big] < E\big[l(\mathbf{I}(z), 0) \mid \mathbf{S} \cap W_z = E\big] \tag{1–42}$$

where $\mathbf{S} \cap W_z = E$ means that the observed windowed process equals E and, because the structuring pair is canonical, $\mathbf{S}^c \cap W_z = F$. Each canonical pair corresponds to the binary n-vector $\mathbf{x}$ of values in (E, F); for each set operator Ψ with window function ψ, $\mathcal{C}_\Psi$ corresponds to the kernel $\mathcal{K}[\psi]$; and Eq. 1–42 is equivalent to $\mathbf{x} \in \mathcal{K}[\psi_{opt}]$ if and only if

$$E\big[l(Y, 1) \mid \mathbf{x}\big] < E\big[l(Y, 0) \mid \mathbf{x}\big] \tag{1–43}$$

Once the disjunctive normal form for ψ_{opt} (canonical expansion for Ψ_{opt}) has been determined, logic reduction can be employed to reduce the expansion to a non-canonical sum of products (hit-or-miss expansion) involving fewer logic gates. For estimation purposes, we continue to assume canonical representation.

For the MAE loss function [11] $\mathcal{C}_{opt}$ is denoted by $\mathcal{C}_{MAE}$ and

$$E\big[l(\mathbf{I}(z), 1) \mid \mathbf{S} \cap W_z = E\big] = 1 - P\big(\mathbf{I}(z) = 1 \mid \mathbf{S} \cap W_z = E\big) \tag{1–44}$$

According to Eq. 1–42, $(E, F) \in \mathcal{C}_{MAE}$ if and only if

$$P\big(\mathbf{I}(z) = 1 \mid \mathbf{S} \cap W_z = E\big) > 0.5 \tag{1–45}$$

Equivalently, $\mathbf{x} \in \mathcal{K}[\psi_{MAE}]$ if and only if

$$P(Y = 1 \mid \mathbf{x}) > 0.5 \tag{1–46}$$

We often use a suboptimal filter Ψ instead of the optimal filter Ψ_{opt}, with a concomitant increase in error. The expansion of Eq. 1–41 is taken over $\mathcal{C}_\Psi$ instead of $\mathcal{C}_{opt}$ and, equivalently, $\mathcal{K}[\psi]$ replaces $\mathcal{K}[\psi_{opt}]$. To quantify the error increase, if $P(\mathbf{x}) > 0$, we define the *advantage* of $\mathbf{x}$ by

$$Ad_l(\mathbf{x}) = \big(E\big[l(Y, 0) \mid \mathbf{x}\big] - E\big[l(Y, 1) \mid \mathbf{x}\big]\big) P(\mathbf{x}) \tag{1–47}$$

According to Eq. 1–43, $Ad_l(\mathbf{x}) > 0$ if and only if $\mathbf{x} \in \mathcal{K}[\psi_{opt}]$. An increase in error can arise in two ways from using Ψ instead of Ψ_{opt}: $\mathbf{x} \in \mathcal{K}[\psi_{opt}]$ but $\mathbf{x} \notin \mathcal{K}[\psi]$, or $\mathbf{x} \notin \mathcal{K}[\psi_{opt}]$ but $\mathbf{x} \in \mathcal{K}[\psi]$. This error increase is the switching cost with cost

Table 1.1. MAE cost of using the median instead of the optimal filter.

$\mathbf{x}$	$P(\mathbf{x})$	$P(Y=1 \mid \mathbf{x})$	$P(Y=0 \mid \mathbf{x})$	$Ad_{MAE}(\mathbf{x})$
000	0.30	0.10	0.90	−0.24
001	0.05	0.10	0.90	−0.04
010	0.20	0.60	0.40	0.04
011	0.05	0.70	0.30	0.02
100	0.05	0.10	0.90	−0.04
101	0.05	0.30	0.70	−0.02
110	0.10	0.70	0.30	0.04
111	0.20	0.90	0.10	0.16

function $Co(\mathbf{x}) = |Ad_l(\mathbf{x})|$. According to Eq. 1–9, the total error increase from using Ψ instead of Ψ_{opt} is

$$\begin{aligned} Er\langle\Psi\rangle - Er\langle\Psi_{opt}\rangle &= Co[\psi, \psi_{opt}] \\ &= \sum_{\mathbf{x}\in\mathcal{Z}[\psi,\psi_{opt}]} \left|Ad_l(\mathbf{x})\right| = \sum_{\mathbf{x}\in\mathcal{K}[\psi_{opt}]\Delta\mathcal{K}[\psi]} \left|Ad_l(\mathbf{x})\right| \end{aligned} \quad (1\text{–}48)$$

where the last sum is over the symmetric difference between the kernels. $Ad_l(\mathbf{x})$ can also be written as $Ad_l(E, F)$, where the canonical pair (E, F) corresponds to $\mathbf{x}$.

Using a particular filter without concern for optimization almost always leads to suboptimality and often the error increase is excessive. To illustrate the cost of *ad hoc* filter selection, we consider a three-point horizontal window centered at the origin and the MAE loss function. There are eight observation vectors. For the probabilities and advantages in Table 1.1, $\mathcal{K}[\psi_{opt}] = \{010, 011, 110, 111\}$. Suppose, instead of using Ψ_{opt}, one were to apply the three-point median M, whose window function has kernel $\mathcal{K}[\mu] = \{011, 101, 110, 111\}$. Since $\mathcal{K}[\psi_{opt}]\Delta\mathcal{K}[\mu] = \{010, 101\}$, Eq. 1–48 gives the MAE cost of using the median instead of the optimal filter as 0.06.

Suboptimality can be introduced when there is too much logic for implementation, even after reduction. All pairs in C_{opt} contribute to error reduction but some contribute very little and can be deleted with a small loss in filter performance. If $C_\Psi \subset C_{opt}$, then

$$Er\langle\Psi\rangle - Er\langle\Psi_{opt}\rangle = \sum_{\mathbf{x}\in\mathcal{K}[\psi_{opt}]-\mathcal{K}[\psi]} \left|Ad_l(\mathbf{x})\right| \quad (1\text{–}49)$$

In practice, vectors with positive advantage are listed from largest to smallest advantage. Filter logic is reduced by deleting those at the bottom of the list prior to logic reduction.

For the MAE loss function, advantage and absolute advantage are given by

$$Ad_{MAE}(\mathbf{x}) = \big[P(Y=1 \mid \mathbf{x}) - P(Y=0 \mid \mathbf{x})\big]P(\mathbf{x}) \tag{1–50}$$

$$\big|Ad_{MAE}(\mathbf{x})\big| = \big|1 - 2P(Y=1 \mid \mathbf{x})P(\mathbf{x})\big| \tag{1–51}$$

In image restoration, $Ad_{MAE}(\mathbf{x})$ $[Ad_{MAE}(E, F)]$ has been called [9, 10] the *restoration effect* of $\mathbf{x}$ $[(E, F)]$. The advantage for the SR loss function is

$$Ad_{SR}(\mathbf{x}) = \big[P(Y=1 \mid \mathbf{x}) - (1-\tau)^{-1}P(Y=0 \mid \mathbf{x})\big]P(\mathbf{x}) \tag{1–52}$$

If $P(Y=1 \mid \mathbf{x}) = 1$, then $Ad_{SR}(\mathbf{x}) = 1$; if $P(Y=1 \mid \mathbf{x}) < 1$, then $Ad_{SR}(\mathbf{x}) \leqslant \tau - 1 < 0$.

1.5 Estimation of Optimal W-Operators

In practice, the optimal filter is statistically estimated from image realizations by estimating the conditional expectations composing the decision criterion of Eq. 1–43. This is accomplished by taking image-pair realizations (I_1, S_1), $(I_2, S_2), \ldots, (I_m, S_m)$ of $\mathbf{I}$ and $\mathbf{S} = \Xi(\mathbf{I})$ and forming estimators

$$\begin{aligned} \hat{e}_{l,\mathbf{x}}(0) &= \widehat{E}\big[l(Y,0) \mid \mathbf{x}\big] \\ \hat{e}_{l,\mathbf{x}}(1) &= \widehat{E}\big[l(Y,1) \mid \mathbf{x}\big] \end{aligned} \tag{1–53}$$

The designed estimate of the optimal filter, $\widehat{\Psi}_{opt}$ (with window function $\widehat{\psi}_{opt}$), is determined by the set $\mathcal{K}[\widehat{\psi}_{opt}]$ of observation vectors $\mathbf{x}$ for which

$$\hat{e}_{l,\mathbf{x}}(1) < \hat{e}_{l,\mathbf{x}}(0) \tag{1–54}$$

For a given $\mathbf{x}$ there is an increase in error owing to estimation of the optimal filter if and only if the inequality of Eq. 1–43 holds but the inequality of Eq. 1–54 does not, or if Eq. 1–43 does not hold but Eq. 1–54 does. These cases correspond to two types of estimation error: $\mathbf{x} \in \mathcal{K}[\psi_{opt}]$ but $\mathbf{x} \notin \mathcal{K}[\widehat{\psi}_{opt}]$ and $\mathbf{x} \notin \mathcal{K}[\psi_{opt}]$ but $\mathbf{x} \in \mathcal{K}[\widehat{\psi}_{opt}]$. This suboptimality situation is covered by Eq. 1–48 with $\widehat{\Psi}_{opt}$ being the suboptimal filter used in place of Ψ. Hence, $Er\langle\widehat{\Psi}_{opt}\rangle - Er\langle\Psi_{opt}\rangle$, the error increase owing to estimation of the optimal filter is given by Eq. 1–48 with $\mathcal{K}[\widehat{\psi}_{opt}]$ in place of $\mathcal{K}[\psi]$.

For the MAE loss function, we can employ Eq. 1–42 or the respective equivalent conditions of Eqs. 1–44 and 1–45. For the latter, we can use the probability estimator

$$\widehat{P}(Y=k \mid \mathbf{x}) = \frac{Card[Y=k \mid \mathbf{x}]}{Card[\mathbf{x}]} \tag{1–55}$$

for $k = 0, 1$, where *Card* denotes set cardinality and the numerator and denominator give the number of times the sample ideal images are k-valued given $\mathbf{x}$ and the number of times $\mathbf{x}$ is observed across the sample, respectively. $\mathbf{x} \in \mathcal{K}[\widehat{\psi}_{MAE}]$ if and only if Eq. 1–46 holds with the estimate of Eq. 1–55 used in place of the true conditional probability.

As thus far stated, the error increase corresponds to a given designed kernel $\mathcal{K}[\widehat{\psi}_{opt}]$; in fact, $\mathcal{K}[\widehat{\psi}_{opt}]$ is a random collection depending on the realizations selected. Thus, $Er\langle\widehat{\Psi}_{opt}\rangle - Er\langle\Psi_{opt}\rangle$ is a random variable and we measure the precision with which $\widehat{\Psi}_{opt}$ estimates Ψ_{opt} by the expected error increase,

$$\rho_{opt} = E\big[Er\langle\widehat{\Psi}_{opt}\rangle - Er\langle\Psi_{opt}\rangle\big] \tag{1–56}$$

Note that $\mathbf{x} \in \mathcal{K}[\psi_{opt}] - \mathcal{K}[\widehat{\psi}_{opt}]$ if and only if $Ad_l(\mathbf{x}) > 0$ and $\hat{e}_{l,\mathbf{x}}(1) \geqslant \hat{e}_{l,\mathbf{x}}(0)$. Moreover, $\mathbf{x} \in \mathcal{K}[\widehat{\psi}_{opt}] - \mathcal{K}[\psi_{opt}]$ if and only if $Ad_l(\mathbf{x}) \leqslant 0$ and $\hat{e}_{l,\mathbf{x}}(1) < \hat{e}_{l,\mathbf{x}}(0)$. Hence,

$$\begin{aligned}\rho_{opt} = & \sum_{\mathbf{x}\in\mathcal{K}[\psi_{opt}]} Ad_l(\mathbf{x})P\big(\hat{e}_{l,\mathbf{x}}(1) \geqslant \hat{e}_{l,\mathbf{x}}(0)\big) \\ & - \sum_{\mathbf{x}\notin\mathcal{K}[\psi_{opt}]} Ad_l(\mathbf{x})P\big(\hat{e}_{l,\mathbf{x}}(1) < \hat{e}_{l,\mathbf{x}}(0)\big)\end{aligned} \tag{1–57}$$

This error has been studied in the context of binary-image restoration and the analysis applies to the general MAE theory [9, 10, 12]. For the MAE loss function,

$$\begin{aligned}\rho_{MAE} = & \sum_{\{\mathbf{x}\in P(Y=1|\mathbf{x})>0.5\}} |2P(Y=1 \mid \mathbf{x}) - 1|P\big(\widehat{P}(Y=1 \mid \mathbf{x}) \leqslant 0.5\big)P(\mathbf{x}) \\ + & \sum_{\{\mathbf{x}\in P(Y=1|\mathbf{x})\leqslant 0.5\}} |2P(Y=1 \mid \mathbf{x}) - 1|P\big(\widehat{P}(Y=1 \mid \mathbf{x}) > 0.5\big)P(\mathbf{x})\end{aligned} \tag{1–58}$$

As the number of observations increases to infinity, $\rho_{MAE} \to 0$.

The preceding estimation methodology is somewhat idealized. If $P(\mathbf{x}) > 0$, then we obtain good estimates $\hat{e}_{l,\mathbf{x}}(0)$ and $\hat{e}_{l,\mathbf{x}}(1)$ for sufficiently large observation samples and the error analysis applies. In practice, however, a particular structuring pair $\mathbf{x}$ may not be observed during the estimation procedure, in which case $\hat{e}_{l,\mathbf{x}}(0)$ and $\hat{e}_{l,\mathbf{x}}(1)$ are not defined. Then, based solely on our statistical knowledge, it is irrelevant whether $\mathbf{x}$ is placed into $\mathcal{K}[\widehat{\psi}_{opt}]$. If $\hat{e}_{l,\mathbf{x}}(0)$ and $\hat{e}_{l,\mathbf{x}}(1)$ are not defined, then the design procedure discussed in the next section decides whether to place $\mathbf{x}$ into $\mathcal{K}[\widehat{\psi}_{opt}]$ on the basis of algorithm efficiency. In fact, if $P(\mathbf{x}) > 0$, then whether or not $\mathbf{x} \in \mathcal{K}[\widehat{\psi}_{opt}]$ does affect the error of estimation, the arbitrariness of our decision coming from lack of statistical knowledge (too small a sample). Extra criteria can be employed to decide whether an unobserved structuring pair is placed into $\mathcal{K}[\widehat{\psi}_{opt}]$. Such criteria produce a different design procedure and therefore different errors of estimation. A key design method for document processing is the use of differencing representation, which produces the estimate $\widehat{\Psi}_{opt}$ from a different representation than the standard hit-or-miss expansion [9]. For some system models, error of estimation is less for differencing design; for others, it is less for standard hit-or-miss design. Even if a structuring pair is observed, it may be observed so few times (a single observation not being unusual) that we lack confidence in the estimates $\hat{e}_{l,\mathbf{x}}(0)$ and $\hat{e}_{l,\mathbf{x}}(1)$. In such a case we might employ a conditonal decision criterion of the following form: if $\mathbf{x}$ is observed at least r times, then decide whether to place $\mathbf{x}$ into $\mathcal{K}[\widehat{\psi}_{opt}]$ on the basis of Eq. 1–54; otherwise, apply some other stated criterion. Such conditional decision criteria have their own errors of estimation [9].

1.6 Design Procedure

The design procedure of set operators for binary image analysis is composed of two main steps: (1) estimation of conditional probabilities and (2) computation of an operator of minimal cost in accordance with a given loss function and the estimated conditional probabilities. The detailed design procedure may be outlined as follows:

1. Shift the window to all pixel locations within the observation image.
2. At each location, record the observed canonical structuring pair.
3. At each location, record the value of the pixel in the ideal image that is colocated with the window origin in the observation image.
4. For each structuring pair, tally the number of times a 1 is observed in the ideal image and the number of times a 0 is observed.
5. For the operator representation select, for each canonical structuring pair observed, the value (0 or 1) of minimal cost.
6. To reduce the representation cost, perform logic minimization assuming that the unobserved templates are "don't cares".

Except for the fact that a large sample may be required, the computational cost of the first five steps is low; however, the computational cost of the sixth step may be high.

Step 5 creates a truth table defining a family of statistically equivalent Boolean functions. Each function in the family possesses a large number of representations. An ideal procedure for logical reduction would give the best representation (i.e., the one using a minimal number of logic gates) among all possible representations of all equivalent functions. In practical applications, such a procedure is usually not available. Typically, the canonical representation of one (or, at best, several) of the equivalent Boolean functions is chosen and the best representation for this function is found.

In this more modest context, simplification of a Boolean function consists of transforming a given expression into another expression with fewer terms (products) or literals (variables). A Boolean expression is considered a *minimal expression* if: (1) there does not exist an equivalent Boolean expression with a smaller number of terms or (2) there does not exist an equivalent expression with an equal number of terms but with a smaller number of literals. Boolean expressions can be simplified or minimized via the laws of Boolean algebra. We consider two reduction algorithms. Intervals in $\{0, 1\}^n$ play a key role. If $\mathbf{a}, \mathbf{b} \in \{0, 1\}^n$ and $\mathbf{a} \leqslant \mathbf{b}$, then the *interval (cube)* $[\mathbf{a}, \mathbf{b}] = \{\mathbf{x} \in \{0, 1\}^n\colon \mathbf{a} \leqslant \mathbf{x} \leqslant \mathbf{b}\}$. The dimension of $[\mathbf{a}, \mathbf{b}]$, $dim([\mathbf{a}, \mathbf{b}])$, is the difference between the number of 1-valued components in $\mathbf{b}$ and $\mathbf{a}$. A cube of dimension n is called an n-cube.

A classical simplification algorithm for Boolean expressions is the *Quine-McCluskey* (*QM*) tabular algorithm [13, 14]. This algorithm provides a minimal expression and is well suited for computer implementation since it systematizes the simplification process. The QM algorithm is composed of three basic steps:

1. Complete with 1s the table describing the Boolean function.
2. Take the set of 1-valued patterns as vertices (points) in an n-dimensional space and perform a systematic minimization: from the set of vertices (0-cubes), try to find all adjacent vertices (1-cubes); from the set of 1-cubes, try to find all adjacent 1-cubes (2-cubes); and so on. The set of cubes resulting at the final step of the process is called the set of *prime implicants*.
3. Select the essential prime implicants from the set of prime implicants generated in step 2. A term is not essential in the set of prime implicants if it can be represented by two other terms of the set. The central point in this step is to compute the smallest subset of the set of prime implicants that is sufficient to represent the function.

Figure 1–5 illustrates the generation of the prime-implicant set in a simple example. The drawback of the QM algorithm is that step 1 generates an enormous amount of data, even when the number of known points in the function is small relative to the size of the domain. The QM algorithm cannot be practically used for functions of more than twelve variables.

Incremental splitting of intervals (*ISI algorithm*) has been proposed for functions having large numbers of variables and relatively small numbers of fixed values, a common occurrence when learning set operators [8, 15]. The main idea of the

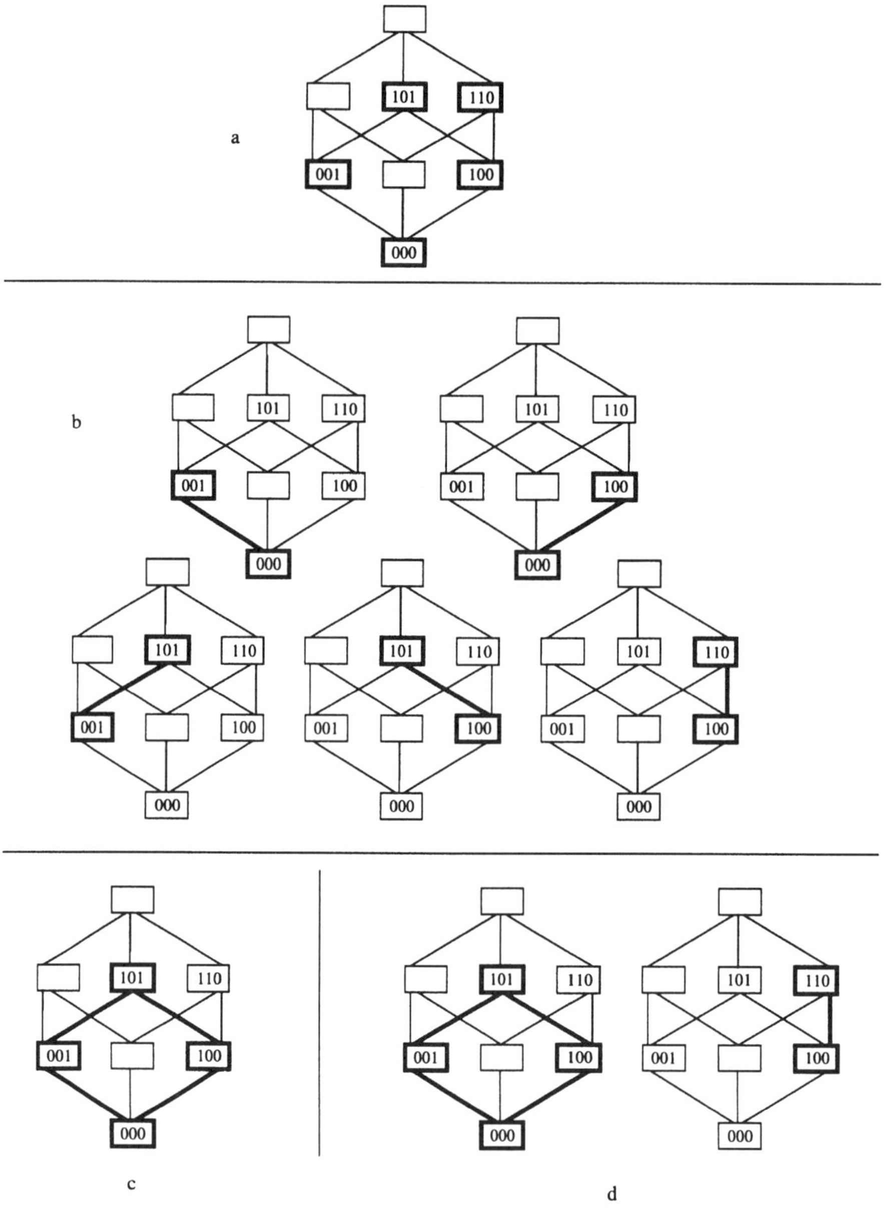

Figure 1–5. Generation of prime-implicant set by QM algorithm.

QM algorithm is repetition of a process that joins cubes (from the 0-cubes) to obtain cubes as large as possible. The main idea of the ISI algorithm is dual to this: it is a repeated process that splits cubes into lower order cubes while there are negative examples that must be satisfied. Successive application of this procedure, with some basic caveats, results in a minimal expression.

The ISI algorithm is composed of five basic steps:

1. Let $\mathcal{T}$ be an initial set of prime implicants. Usually, in a space of dimension n, $\mathcal{T}$ is the set whose single element is the n-cube (the set $\{0, 1\}^n$). If there are no negative examples, then the algorithm stops and $\mathcal{T}$ itself is the output.
2. Separate the cubes in $\mathcal{T}$ according to the following criterion: put the cubes of $\mathcal{T}$ that cover the next negative example (shape associated to 0) to be extracted in $\mathcal{N}$ and the others in $\mathcal{P}$. Set $\mathcal{M} = \emptyset$.
3. Split each cube in $\mathcal{N}$ and represent it by its low-order cubes. The low-order cubes that were not covered by some cube in $\mathcal{P}$ are put in $\mathcal{M}$.
4. Put $\mathcal{T} = \mathcal{P} \cup \mathcal{M}$ and repeat steps 2 and 3 for the next negative example, while negative examples exist.
5. Select the essential prime implicants by the method used in the QM algorithm.

Symbolically, the first four steps of the algorithm can be expressed in the following manner, where all vectors have n binary components, $\mathbf{0}$ is the zero vector, $\mathbf{1}$ is the vector of all 1s, and $[\mathbf{a}, \mathbf{b}]$ is the cube in $\{0, 1\}^n$ determined by $\mathbf{a}$ and $\mathbf{b}$:

1. Set $\mathcal{T} = \{[\mathbf{0}, \mathbf{1}]\}$;
2. Set $\mathcal{X} = \{\mathbf{x} \in \{0, 1\}^n\colon (\mathbf{x}, 0)$ is a line of the truth table$\}$;
3. If $\mathcal{X} \neq \emptyset$, select $\mathbf{x} \in \mathcal{X}$, else go to 11;
4. Set $\mathcal{N} = \{[\mathbf{a}, \mathbf{b}] \in \mathcal{T}\colon \mathbf{x} \in [\mathbf{a}, \mathbf{b}]\}$;
5. Set $\mathcal{P} = \mathcal{T} - \mathcal{N}$;
6. Set $\mathcal{M} = \emptyset$;
7. For each $[\mathbf{a}, \mathbf{b}] \in \mathcal{N}$:
   - Set $\mathcal{S}$ = split of $[\mathbf{a}, \mathbf{b}]$ by $\mathbf{x}$;
   - Set $\mathcal{M} = \mathcal{M} \cup \{[\mathbf{c}, \mathbf{d}] \in \mathcal{S}\colon \exists$ no $[\mathbf{u}, \mathbf{v}] \in \mathcal{P}$ such that $[\mathbf{c}, \mathbf{d}] \subset [\mathbf{u}, \mathbf{v}]\}$;
8. Set $\mathcal{T} = \mathcal{P} \cup \mathcal{M}$;
9. Set $\mathcal{X} = \mathcal{X} - \{\mathbf{x}\}$;
10. Go to 3;
11. Return $\mathcal{T}$;
12. END.

A fundamental aspect of the ISI algorithm is splitting a cube by a negative example. A cube $[\mathbf{a}, \mathbf{b}]$ is *split* by a negative example $\mathbf{x} \in [\mathbf{a}, \mathbf{b}]$ into a collection of intervals $\mathcal{S}$ in the following manner:

$$\mathcal{S} = \left\{\left[\mathbf{a}, \mathbf{b} \wedge \mathbf{z}_1'\right]\colon \mathbf{z}_1 \leqslant \mathbf{x}\right\} \cup \left\{[\mathbf{a} \vee \mathbf{z}_1, \mathbf{b}]\colon \mathbf{z}_1 \leqslant \mathbf{x}'\right\} \tag{1–59}$$

where $\mathbf{z}_1$ denotes a generic vector having exactly one 1-valued component. It can be proved that

$$Card(\mathcal{S}) = dim\big([\mathbf{a}, \mathbf{b}]\big) \tag{1–60}$$

where the number on the left is a count of intervals and the number on the right is a count of elements. Moreover, for any interval $[\mathbf{u}, \mathbf{v}] \in \mathcal{S}$,

$$dim([\mathbf{u}, \mathbf{v}]) = dim([\mathbf{a}, \mathbf{b}]) - 1 \tag{1–61}$$

Thus, the split of an n-cube by a negative example produces n cubes of dimension $n-1$. It can be also proved that the intervals of S are maximal.

The ISI algorithm is recursive. Each execution of step 8 gives a set of prime implicants. Iterations of the algorithm can be visualized via a tree, where nodes at a given level are cubes resulting after the execution of the iteration. If we consider the root at level zero, then the nodes in the first level are the cubes resulting after the extraction of the first negative example and so on. The dynamics for minimization of a function of three variables are shown in Fig. 1–6. The same 3-space process can be viewed in Fig. 1–7.

For small numbers of variables the ISI algorithm uses much less storage space than the QM algorithm; however, relative execution time depends on the quantity and distribution of the examples given. The ISI algorithm can be applied for minimizing expressions with large numbers of variables that cannot be treated by the QM algorithm so long as the number of don't cares is also large [8].

We now present some example applications [8] of optimal design of set operators in binary image analysis using the MAE loss function and the ISI learning algorithm. We begin with an example that illustrates the complete design process.

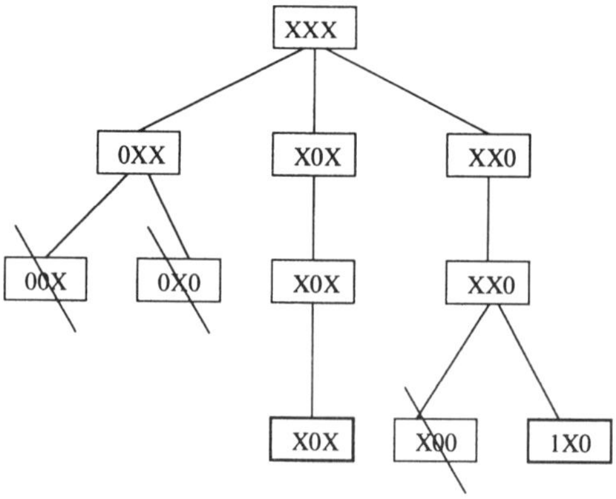

Figure 1–6. Minimization of a function of three variables.

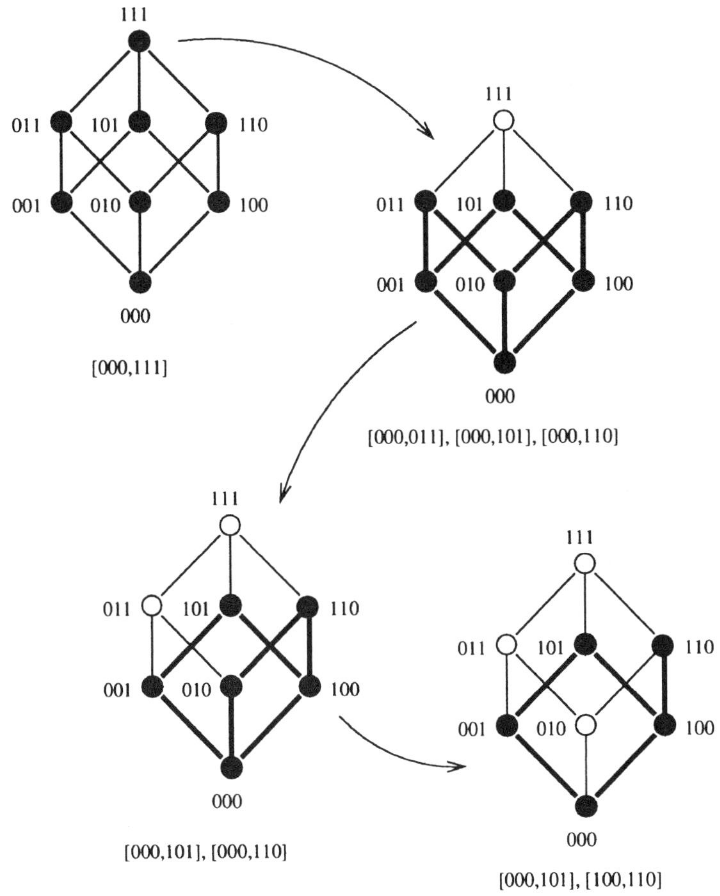

Figure 1–7. Visualization of 3-space process of the ISI algorithm.

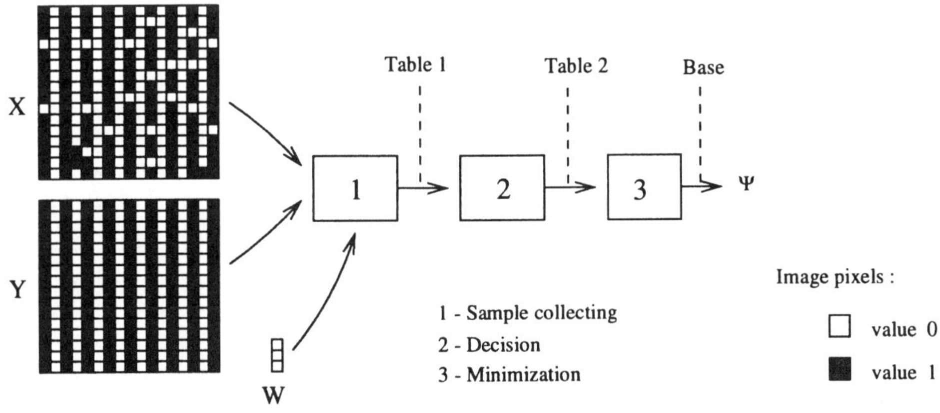

Figure 1–8. Design process.

Table 1.2. Observed statistics.

$x_1x_2x_3$	0	1
000	108	0
001	2	0
011	1	18
101	0	19
110	1	18
111	0	71

Table 1.3. Designed filter.

$x_1x_2x_3$	$f(x_1x_2x_3)$
000	0
001	0
011	1
101	1
110	1
111	1
100	$\times$
010	$\times$

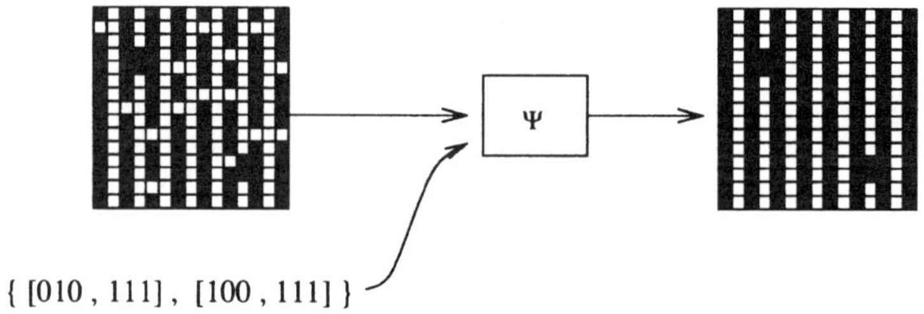

Figure 1–9. Application of designed operator.

EXAMPLE 1–1. We design a 3-variable operator to minimize additive and subtractive point noise in an image composed of vertical stripes. Figure 1–8 shows realizations of the observed and ideal images, along with a diagram with all the stages composing the design process. Table 1.2 shows the statistics obtained from the data of the figure and Table 1.3 shows the result of the decision under the MAE loss function from the data of Table 1.2. Observe that Table 1.3 defines a class of four operators that are statistically equivalent. These are obtained by specifying the don't cares (denoted by $\times$) as 0 or 1. Between these four operators we have chosen one of minimal computational cost. Figure 1–9 shows the basis of the designed operator and application of the operator to a new realization.

EXAMPLE 1–2 (*edge detection*). The window is the 3×3 square and the training sample is taken from the images of Fig. 1–10(a). The training sample size was 3,844. The number of distinct observed examples was 46: 24 positives and 22 negatives. Learning time was 1s. The error measure was zero. The resulting basis is composed of the following maximal intervals:

$$\begin{bmatrix} 0 & \times & \times \\ \times & 1 & \times \\ \times & \times & \times \end{bmatrix}, \begin{bmatrix} \times & \times & 0 \\ \times & 1 & \times \\ \times & \times & \times \end{bmatrix}, \begin{bmatrix} \times & \times & \times \\ \times & 1 & \times \\ \times & \times & 0 \end{bmatrix}, \begin{bmatrix} \times & \times & \times \\ \times & 1 & \times \\ 0 & \times & \times \end{bmatrix}.$$

Figure 1–10(b) shows application of the learned operator.

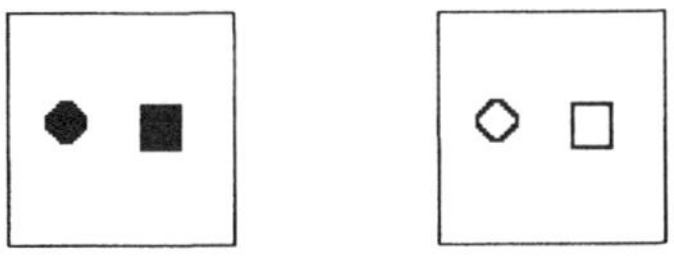

(a)

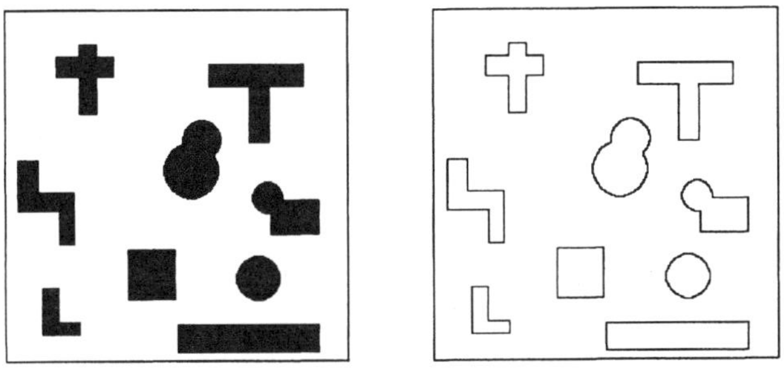

(b)

Figure 1–10. Edge detection: (a) training images; (b) application of learned operator.

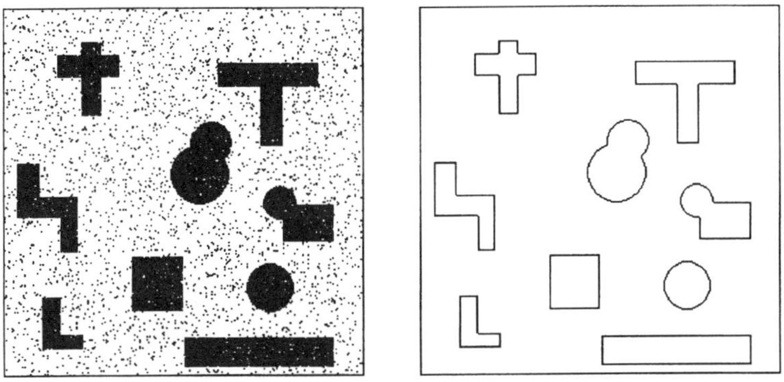

(a)

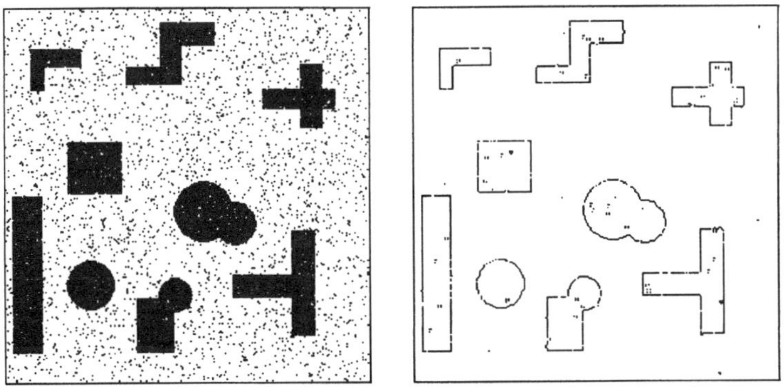

(b)

Figure 1–11. Edge detection in noise: (a) training images; (b) application of learned operator.

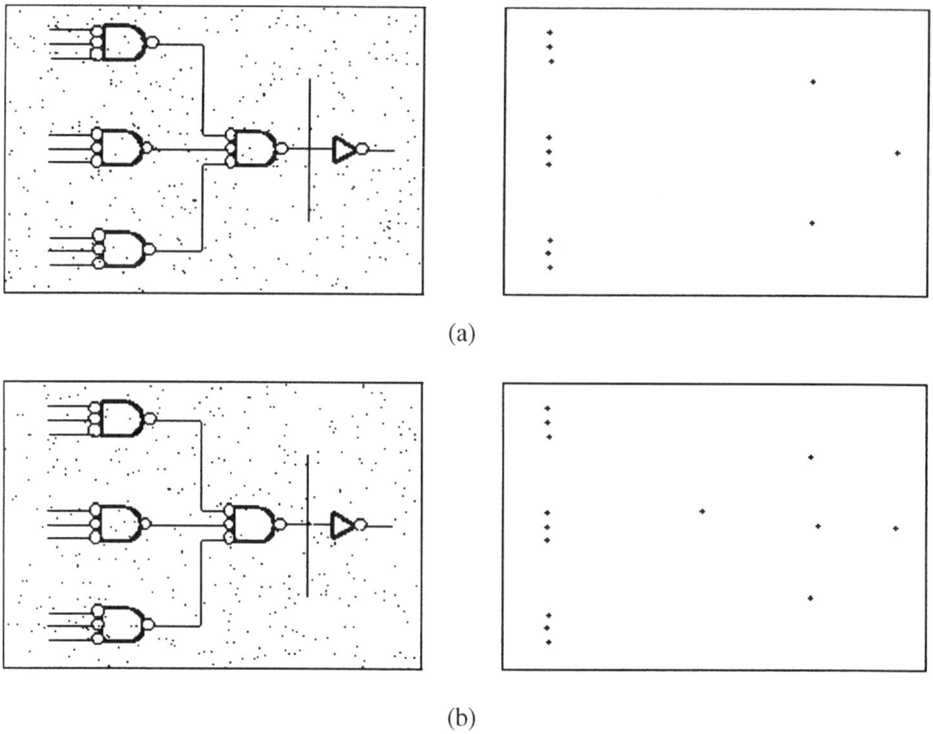

Figure 1–12. End-point detection in noise: (a) training images; (b) application of learned operator.

EXAMPLE 1–3 (*edge detection in noise*). The window is the 3×3 square. The training sample of size 80,392 was taken from the images of Fig. 1–11(a). There were 363 distinct examples observed: 278 negative and 85 positive. Learning time was 1s. The resulting basis is composed of 44 intervals. The noise is punctual, additive and subtractive, and uniformly distributed, with density 5%. The error measure was 0.48%. Figure 1–11(b) shows application of the learned operator.

EXAMPLE 1–4 (*end-point detection in noise*). The window is

$$W = \begin{pmatrix} 0 & 1 & 1 & 1 & 0 \\ 1 & 1 & 1 & 1 & 1 \\ 1 & 1 & 1 & 1 & 1 \\ 1 & 1 & 1 & 1 & 1 \\ 0 & 1 & 1 & 1 & 0 \end{pmatrix},$$

where the 0s in the corners mean that these pixels are not part of the window. The training sample of size 36,290 was taken from two image pairs, one being shown in Fig. 1–12(a). There were 1,545 distinct observed examples: 1,538 negative and

7 positive. Learning time was 1s. The resulting basis is composed of 4 intervals:

$$\begin{pmatrix} \times & 0 & 1 & 0 & \times \\ 0 & \times & 1 & 0 & 0 \\ 0 & \times & 1 & 0 & \times \\ \times & 0 & 0 & \times & \times \\ \times & \times & \times & \times & \times \end{pmatrix}, \quad \begin{pmatrix} \times & 0 & 0 & 0 & \times \\ 0 & \times & \times & 0 & \times \\ 1 & 1 & 1 & 0 & \times \\ \times & \times & \times & 0 & \times \\ \times & \times & \times & \times & \times \end{pmatrix},$$

$$\begin{pmatrix} \times & \times & 0 & \times & \times \\ \times & 0 & 0 & \times & \times \\ \times & 0 & 1 & 1 & 1 \\ \times & 0 & 0 & \times & \times \\ \times & \times & \times & \times & \times \end{pmatrix}, \quad \begin{pmatrix} \times & \times & \times & \times & \times \\ 0 & 0 & 0 & 0 & 0 \\ 0 & \times & 1 & 0 & \times \\ 0 & \times & 1 & \times & \times \\ \times & \times & 1 & \times & \times \end{pmatrix}.$$

Note that, since the corners are not part of the window, each corner pixel in the basis must be a don't-care pixel. The noise is punctual, additive and subtractive, and uniformly distributed, with densities of 1% and 0.1%. The error measure was 0.005%. Figure 1–12(b) shows application of the learned operator.

EXAMPLE 1–5 (*defect lines*). Detection of defect lines in an image of a transversal section of an eutectic alloy is a classical problem in mathematical morphology [5, 16]. To apply automatic design, first we have performed a homotopic transformation on the image and a shrink so we can observe defect lines in a low-resolution image. The transformed image has been divided into two halves, the first to be used as training data and the second for testing the result. The window is the 5×5 square. Figure 1–13(a) shows the images used for training and Fig. 1–13(b) shows application of the operator. The size of the training sample was 7,260. There were 3,812 distinct observed examples: 3,519 negatives and 293 positives. Learning time was 30s. The resulting basis is composed of 127 intervals.

1.7 CONSTRAINED OPTIMIZATION

Suboptimality often results from requiring that the chosen filter come from some subclass of all possible filters: rather than optimize over all logical expansions of the kind given in Eq. 1–1, optimize over some subclass of logical expansions. We implicitly assume that constraints are deterministic, meaning that optimization is over a filter class defined without reference to the conditional probabilities $P(Y = 1 \mid \mathbf{x})$. For unconstrained optimization, the kernel $\mathcal{K}[\psi]$ can be any subset of $\{0, 1\}^n$; for constrained optimization, there exists a family $\mathcal{Q}$ of subsets of $\{0, 1\}^n$ that is a proper subfamily of the family of all subsets of $\{0, 1\}^n$ such that $\mathcal{K}[\psi] \in \mathcal{Q}$. If ψ_{con} is the optimal logical function over the constrained filter class, then the increase in error owing to constraint is given by Eq. 1–48 with ψ_{con} in place of ψ.

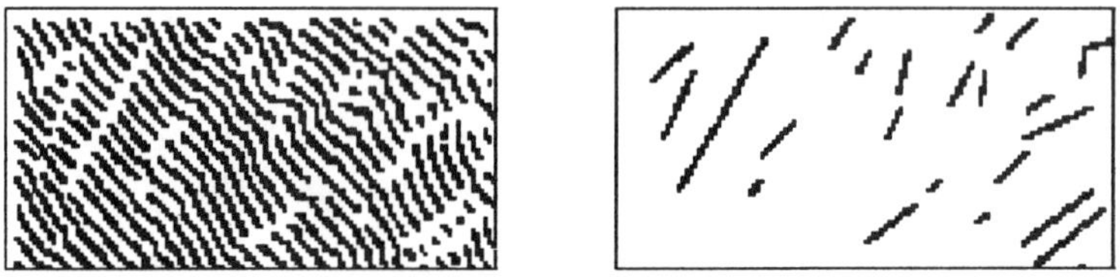

(a)

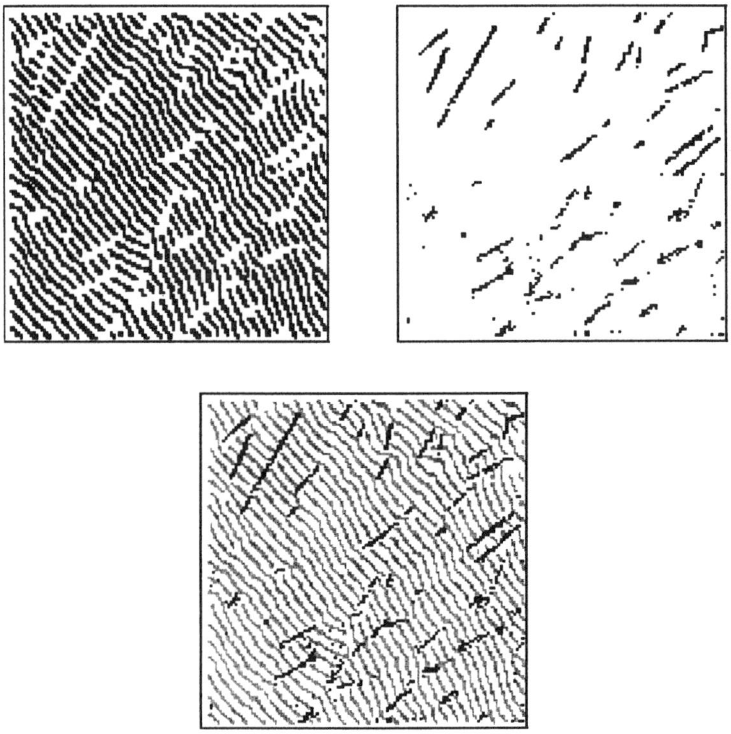

(b)

Figure 1–13. Defect lines: (a) training images; (b) application of learned operator.

The error increase is zero if and only if there exists a filter in the constrained class possessing minimum MAE among all filters.

An *independent* constraint is one for which the decision whether to place a vector **x** in the kernel is constrained by a condition involving only **x** itself, and no other vectors. A number of filter conditions can result in independent constraints. Independent constraints can result from geometric conditions. Letting W be the 3×3 window and reading vectors in the usual raster method, the condition that a pixel be 1-valued if it is interior to a vertical or horizontal line of pixels produces the independent constraints

$$\begin{aligned} C_1&: \mathbf{x} \in \mathcal{K}[\psi] \quad \text{if } \mathbf{x} \geqslant 010010010 \\ C_2&: \mathbf{x} \in \mathcal{K}[\psi] \quad \text{if } \mathbf{x} \geqslant 000111000 \end{aligned} \tag{1–62}$$

For these independent constraints, a set $\mathcal{A}$ of vectors lies in $\mathcal{Q}$ if and only if

$$\mathcal{U}[010010010] \cup \mathcal{U}[000111000] \subset \mathcal{A} \tag{1–63}$$

A more refined constraint set occurs if the condition is changed to state that a pixel must be 1-valued if it is interior to a vertical or horizontal line of pixels and there are no other 1-valued pixels in the window. This yields the independent constraints

$$\begin{aligned} C_1&: \mathbf{x} \in \mathcal{K}[\psi] \quad \text{if } \mathbf{x} = 010010010 \\ C_2&: \mathbf{x} \in \mathcal{K}[\psi] \quad \text{if } \mathbf{x} = 000111000 \end{aligned} \tag{1–64}$$

For these independent constraints, a set $\mathcal{A}$ of vectors lies in $\mathcal{Q}$ if and only if

$$\{010010010, 000111000\} \subset \mathcal{A} \tag{1–65}$$

Independent constraints can arise from algebraic conditions. If we demand that Ψ be antiextensive, then there exists a single constraint:

$$C_1: \mathbf{x} \notin \mathcal{K}[\psi] \text{ if } \mathbf{x} \not\geqslant 000010000 \tag{1–66}$$

Independent constraints occur from placing a filter bound on the designed filter Ψ. For instance, $\Psi(S) \supset \Phi(S)$ $[\Psi(S) \subset \Phi(S)]$ for all S, meaning $\Psi \geqslant \Phi$ $[\Psi \leqslant \Phi]$. If Φ is increasing with $\mathcal{B}[\phi] = \{\mathbf{x}_1, \mathbf{x}_2, \ldots, \mathbf{x}_m\}$, then $\Phi \leqslant \Psi$ yields m independent constraints:

$$\begin{aligned} C_1&: \mathbf{x} \in \mathcal{K}[\psi] \quad \text{if } \mathbf{x} \geqslant \mathbf{x}_1 \\ C_2&: \mathbf{x} \in \mathcal{K}[\psi] \quad \text{if } \mathbf{x} \geqslant \mathbf{x}_2 \\ &\vdots \\ C_m&: \mathbf{x} \in \mathcal{K}[\psi] \quad \text{if } \mathbf{x} \geqslant \mathbf{x}_m \end{aligned} \tag{1–67}$$

If Ω is increasing with $\mathcal{B}[\omega] = \{\mathbf{y}_1, \mathbf{y}_2, \ldots, \mathbf{y}_k\}$ and we desire the bounding constraint $\Phi \leqslant \Psi \leqslant \Omega$, then, in addition to the constraints $C_1, C_2, \ldots, C_m$, we have the independent constraint

$$C_{m+1}: \mathbf{x} \notin \mathcal{K}[\psi] \quad \text{if there does not exist } \mathbf{y}_j \text{ such that } \mathbf{x} \geqslant \mathbf{y}_j \tag{1–68}$$

For the bounding constraint $\Phi \leqslant \Psi \leqslant \Omega$, one can design an unconstrained estimate Ψ' of the optimal filter and then the estimate of the constrained filter is

$$\Psi = \left(\Phi \vee \Psi'\right) \wedge \Omega \tag{1–69}$$

This transformation can take place on the truth table defining Ψ', in which case it effects the following changes: if $\psi'(\mathbf{x}) = 0$ and $\phi(\mathbf{x}) = 1$, then $\psi(\mathbf{x}) = 1$; if $\psi'(\mathbf{x}) = 1$ and $\omega(\mathbf{x}) = 0$, then $\psi(\mathbf{x}) = 0$; if $\psi'(\mathbf{x})$ is undetermined (because $\mathbf{x}$ has not been seen in training), then $\psi(\mathbf{x})$ can be arbitrarily chosen so long as $\phi(\mathbf{x}) \leqslant \psi(\mathbf{x}) \leqslant \omega(\mathbf{x})$. The ISI algorithm can then be employed. Owing to the constraints, we do not concern ourselves with any conditional probability $P(Y = 1 \mid \mathbf{x})$ for which $\phi(\mathbf{x}) = 1$ or $\omega(\mathbf{x}) = 0$.

To illustrate filter-bound constraint we use opening, an increasing morphological filter to be discussed in Chapter 3. Given set A, the *opening* of set S by structuring element A, $\Gamma_A(S)$, is defined as the union of all translates of A that are subsets of S. If W is a 3×3 window and A is a 4-pixel square, then the basis of γ_A (the Boolean function for Γ_A) is

$$\mathcal{B}[\gamma_A] = \{110110000, 011011000, 000110110, 000011011\} \tag{1–70}$$

For the bounding constraint $\Gamma_A \leqslant \Psi$, we need not concern ourselves with any conditional probability $P(Y = 1 \mid \mathbf{x})$ for $\mathbf{x} \in \mathcal{U}[\mathcal{B}[\gamma_A]]$.

The error increase resulting from an independent constraint can be computed from Eq. 1–48 by letting $\mathcal{A}_0^{con}$ and $\mathcal{A}_1^{con}$ be the sets of vectors deterministically constrained to the 0-set and 1-set (kernel), respectively:

$$Er\langle\Psi_{con}\rangle - Er\langle\Psi_{opt}\rangle = \sum_{\mathbf{x} \in (\mathcal{A}_0^{con} - \mathcal{S}_0[\psi_{opt}]) \cup (\mathcal{A}_1^{con} - \mathcal{S}_1[\psi_{opt}])} \left|Ad_l(\mathbf{x})\right| \tag{1–71}$$

For the order constraints of Eq. 1–67, $\mathcal{A}_0^{con} = \emptyset$ and $\mathcal{A}_1^{con} = \mathcal{U}[\text{Bas}[\phi]]$.

Independent constraints can reduce design complexity. If prior probabilities for a class $\mathcal{C}$ of vectors are negligible, then vectors in $\mathcal{C}$ can be arbitrarily independently constrained with $Er\langle\Psi_{con}\rangle - Er\langle\Psi_{opt}\rangle \leqslant \beta$, where β is the sum of the prior probabilities in $\mathcal{C}$. In fact, $Er\langle\Psi_{con}\rangle - Er\langle\Psi_{opt}\rangle$ is likely to be much less than β since many vectors may be correctly constrained and the error factor $|1 - 2P(Y = 1 \mid \mathbf{x})|$ may often be substantially less than 1. Should $Er\langle\Psi_{con}\rangle - Er\langle\Psi_{opt}\rangle$ be small, the designed constrained filter can outperform the designed unconstrained filter because the difference in estimation errors between the constrained and unconstrained filters can exceed $Er\langle\Psi_{con}\rangle - Er\langle\Psi_{opt}\rangle$.

So far we have considered independent constraints; a *dependent* constraint is one that cannot be applied to each vector independently. This means there are required

Table 1.4. Weighted medians.

$\mathbf{x}$	$\mathbf{a}_1$	$\mathbf{a}_2$	$\mathbf{a}_3$	$\mathbf{a}_4$
0 0 0 0	0	0	0	0
0 0 0 1	0	0	0	0
0 0 1 0	0	0	0	0
0 0 1 1	0	0	1	1
0 1 0 0	0	0	0	0
0 1 0 1	0	1	0	1
0 1 1 0	0	1	1	0
0 1 1 1	1	1	1	1
1 0 0 0	0	0	0	0
1 0 0 1	1	0	0	1
1 0 1 0	1	0	1	0
1 0 1 1	1	1	1	1
1 1 0 0	1	1	0	0
1 1 0 1	1	1	1	1
1 1 1 0	1	1	1	1
1 1 1 1	1	1	1	1

relations among vectors that do not reduce to independent constraints. A simple constraint is that there are two vectors $\mathbf{x}$ and $\mathbf{y}$ such that $\psi(\mathbf{x}) \vee \psi(\mathbf{y}) = 1$. Equivalently, $\mathbf{x} \in \mathcal{K}[\psi]$ or $\mathbf{y} \in \mathcal{K}[\psi]$. Typically, dependent constraints result from requiring that the designed filter belong to a given filter class (and it may not be easy to deduce the dependency relations).

As an illustration, consider the weighted median, which, for the binary values $x_1, x_2, \ldots, x_n$ with positive-integer weights $a_1, a_2, \ldots, a_n$, is defined by

$$\mu(x_1, x_2, \ldots, x_n) = \begin{cases} 1, & \text{if } \sum_{i=1}^{n} a_i x_i \geqslant \left(\sum_{i=1}^{n} a_i\right)/2 \\ 0, & \text{otherwise} \end{cases} \tag{1–72}$$

Constrained optimality occurs when we desire the optimal weighted median with sum of weights a (which we assume to be odd). Consider the four-point weighted median with the sum of the weights being 5. There are four possible weight vectors: $\mathbf{a}_1 = (2, 1, 1, 1)$, $\mathbf{a}_2 = (1, 2, 1, 1)$, $\mathbf{a}_3 = (1, 1, 2, 1)$, and $\mathbf{a}_4 = (1, 1, 1, 2)$. These lead to four possible filters defined by the following minimal Boolean functions (basis expansions):

$$\begin{aligned} \mu_1(x_1, x_2, x_3, x_4) &= x_1x_2 + x_1x_3 + x_1x_4 + x_2x_3x_4 \\ \mu_2(x_1, x_2, x_3, x_4) &= x_1x_2 + x_2x_3 + x_2x_4 + x_1x_3x_4 \\ \mu_3(x_1, x_2, x_3, x_4) &= x_1x_3 + x_2x_3 + x_3x_4 + x_1x_2x_4 \\ \mu_4(x_1, x_2, x_3, x_4) &= x_1x_4 + x_2x_4 + x_3x_4 + x_1x_2x_3 \end{aligned} \tag{1–73}$$

The four filters are defined by Table 1.4. Having found all MAEs, filter errors can be found from the table and the optimal filter in the class is the one possessing minimum MAE. The increase in error owing to suboptimality is given by Eq. 1–48. Here it is easy to find the optimal constrained filter because only four filters must be checked. More generally, the number of filters in a given subclass can be enormous.

1.8 Optimal Increasing Filters

The most studied dependent constraint is that the filter be increasing [17–20]. The relationship among the vectors is given by: if $\mathbf{x} \leqslant \mathbf{y}$ and $\mathbf{x} \in \mathcal{K}[\psi]$, then $\mathbf{y} \in \mathcal{K}[\psi]$. If $\Im_{inc}$ is the class of all increasing operators, then the optimal increasing filter, Ψ_{inc}, is the operator in $\Im_{inc}$ possessing minimal MAE. If, for a particular model, Ψ_{opt} is increasing, then $\Psi_{inc} = \Psi_{opt}$ and reduction of the logical expansion for ψ_{opt} yields a positive expansion for ψ_{inc}. From a purely probabilistic standpoint, derivation of an increasing optimal filter via the conditional expectation and logic reduction is a valid approach. Nonetheless, given prior knowledge that Ψ_{opt} is increasing, it can be beneficial to directly design an increasing filter; that is, design Ψ_{opt} by means of a procedure that finds the optimal increasing filter, in this case that being Ψ_{opt}, itself. Furthermore, there are reasons why it might be beneficial to design the best increasing filter even when it is not fully optimal (when $\Psi_{inc} \neq \Psi_{opt}$): (1) minimal positive representations can significiantly reduce the logic cost in comparison to nonpositive representations; (2) most important statistically, typically many fewer realizations need be observed during training to obtain good estimates of optimal increasing filters compared to the number of realizations required for equivalently good estimates of nonincreasing filters. The present section briefly describes direct design of optimal MAE increasing filters. One way to proceed is to apply a switching algorithm that derives Ψ_{inc} from Ψ_{opt} by switching vectors between $\mathcal{S}_0[\psi_{opt}]$ and $\mathcal{S}_1[\psi_{opt}]$ to obtain an increasing filter for which the switching error is minimal [21, 22]. This approach has a drawback: if Ψ_{inc} cannot be obtained with a small number of switches, then switching algorithms can be prohibitively computational.

If $B \subset W$, then the single-erosion filter E_B is defined by the Boolean function

$$\varepsilon_B(\mathbf{x}) = \min\{x_i\colon\, b_i = 1\} \tag{1–74}$$

where $\mathbf{b} = (b_1, b_2, \ldots, b_n)$, and $b_i = 1$ if the ith pixel of W is in B and $b_i = 0$ otherwise. Filter MAE, denoted by $MAE\langle B\rangle$, is given by

$$MAE\langle B\rangle = E\big[|Y - \varepsilon_{\mathbf{b}}(\mathbf{X})|\big] = \sum_{\{(\mathbf{x},y)\colon\, y \neq \varepsilon_{\mathbf{b}}(\mathbf{x})\}} P(\mathbf{x}, y) \tag{1–75}$$

For practical design, $MAE\langle B\rangle$ is estimated from realizations of the ideal and observed images. The observed realization is eroded pixelwise and compared to the ideal-image realization. An estimate of $MAE\langle B\rangle$ is obtained by dividing the number of pixels at which the eroded observation and ideal disagree by the total number of pixels considered.

For an m-erosion filter $\Psi_{\mathcal{B}}$ with basis $\mathcal{B} = \{B_1, B_2, \ldots, B_m\}$, filter error is given by

$$\begin{aligned} MAE\langle\Psi_{\mathcal{B}}\rangle &= E\big[|Y - \psi_{\mathcal{B}}(\mathbf{X})|\big] \\ &= E\big[Y - (\varepsilon_{\mathbf{b}_1}(\mathbf{X}) \vee \varepsilon_{\mathbf{b}_2}(\mathbf{X}) \vee \cdots \vee \varepsilon_{\mathbf{b}_m}(\mathbf{X})\big] \\ &= \sum_{\{(\mathbf{x},y):\ y \neq \max_i \varepsilon_{\mathbf{b}_i}(\mathbf{x})\}} P(\mathbf{x}, y) \end{aligned} \tag{1–76}$$

An estimate of Ψ_{inc} can be found by estimating $MAE\langle\Psi_{\mathcal{B}}\rangle$ over all possible bases and choosing Ψ_{inc} as the filter corresponding to the basis generating minimal MAE. If the window has n pixels, then it has 2^n subsets, but many are eliminated from consideration owing to the minimality condition for a basis. Nonetheless, except for relatively small windows, constraints must be imposed, thereby (it is hoped slightly) increasing MAE. Constraints include limiting the basis size and constraining the search to a library (subclass) of all possible structuring elements [18]. Library constraint, requires some method of choosing the library. In practice, two techniques have dominated. *Expert libraries* are collections of structuring elements whose effects are well known (to experts) or which are basis members of popular filters known to work reasonably well for similar image models. *First-order libraries* are found by placing into the library some number of structuring elements possessing the smallest MAEs as single-erosion filters.

As expressed in Eq. 1–76, it would appear that filter design must include obtaining realization-based statistics for every basis, a prohibitive task. In fact, one need only obtain MAE estimates for single-erosion filters and then recursively obtain MAE estimates for multiple-erosion filters. According to the *morphological MAE theorem* [19], the MAE of an m-erosion filter Ψ_m can be expressed in terms of a single-erosion filter with structuring element B_m and two $(m-1)$-erosion filters Ψ_{m-1} and Φ_{m-1}:

$$MAE\langle\Psi_m\rangle = MAE\langle\Psi_{m-1}\rangle - MAE\langle\Phi_{m-1}\rangle + MAE\langle B_m\rangle \tag{1–77}$$

where the bases are given by

$$\begin{aligned} &\mathcal{B}[\Psi_{m-1}] = \{B_1, B_2, \ldots, B_{m-1}\} \\ &\mathcal{B}[\Psi_m] = \mathcal{B}[\Psi_{m-1}] \cup \{B_m\} = \{B_1, B_2, \ldots, B_m\} \\ &\mathcal{B}[\Phi_{m-1}] = \{B_1 \cup B_m, B_2 \cup B_m, \ldots, B_{m-1} \cup B_m\} \end{aligned} \tag{1–78}$$

To see how Eq. 1–77 can be used in filter design, suppose we wish to optimize by selecting bases from some structuring-element collection $\mathcal{C} = \{B_1, B_2, \ldots, B_q\}$.

For $p = 1, 2, \ldots, q$, let $\mathcal{C}_p$ be the closure of $\mathcal{C}$ under unions of p or less elements within $\mathcal{C}$. Then $\mathcal{C}_1 = \mathcal{C}$ and $\mathcal{C}_1 \subset \mathcal{C}_2 \subset \cdots \subset \mathcal{C}_q$. If we know the MAE for each structuring element in $\mathcal{C}_q$, then we can proceed recursively. For any 2-element filter Ψ_2 with basis $\mathcal{B}[\Psi_2] = \{B_{i1}, B_{i2}\}$,

$$\begin{aligned} MAE\langle\Psi_2\rangle &= MAE\langle B_{i1}, B_{i2}\rangle \\ &= MAE\langle B_{i1}\rangle + MAE\langle B_{i2}\rangle - MAE\langle B_{i1} \cup B_{i2}\rangle \end{aligned} \tag{1–79}$$

If Ψ_3 is a 3-erosion filter with basis $\mathcal{B}[\Psi_3] = \{B_{i1}, B_{i2}, B_{i3}\}$, then

$$\begin{aligned} MAE\langle\Psi_3\rangle &= MAE\langle B_{i1}, B_{i2}, B_{i3}\rangle \\ &= MAE\langle B_{i1}, B_{i2}\rangle + MAE\langle B_{i3}\rangle \\ &\quad - MAE\langle B_{i1} \cup B_{i3}, B_{i2} \cup B_{i3}\rangle \end{aligned} \tag{1–80}$$

All terms on the right-hand side of the equation can be obtained from the previous stage of the iteration. The MAE theorem allows for evaluation of redundant structuring-element combinations but reduces to nonredundant forms.

Basis-size constraint, library constraint, and estimation of filter MAEs affect the relationship between the optimal increasing filter and the designed filter meant to estimate it. If we simply impose basis-size constraint, say k structuring elements, then optimization is over the class of increasing operators having at most k terms in their minimal representations, thereby yielding a suboptimal increasing filter $\Psi_{inc}^{(k)}$. If there is also first-order-library constraint using the r structuring elements possessing minimum single-erosion MAE, then there is further suboptimality yielding a filter $\Psi_{inc}^{(k,r)}$ and

$$MAE\langle\Psi_{inc}\rangle \leqslant MAE\big\langle\Psi_{inc}^{(k)}\big\rangle \leqslant MAE\big\langle\Psi_{inc}^{(k,r)}\big\rangle \tag{1–81}$$

Since the designed filter is based on estimated MAEs, even without basis-size or library constraints, it is an estimate $\widehat{\Psi}_{inc}$ of Ψ_{inc}, and $MAE\langle\widehat{\Psi}_{inc}\rangle \geqslant MAE\langle\Psi_{inc}\rangle$. According to theory [12], if the number of observations is large, then $MAE\langle\widehat{\Psi}_{inc}\rangle \approx MAE\langle\Psi_{inc}\rangle$; from experience, a single 1024×1024 realization is usually sufficient for good estimation. If both basis-size and library constraint are employed, then we actually estimate $\Psi_{inc}^{(k,r)}$ from data, so that the designed filter is a statistical estimate $\widehat{\Psi}_{inc}^{(k,r)}$ of $\Psi_{inc}^{(k,r)}$, and $MAE\langle\widehat{\Psi}_{inc}^{(k,r)}\rangle \geqslant MAE\langle\Psi_{inc}^{(k,r)}\rangle$. Experience has shown that basis-size- and library-constrained estimates are close to optimal in many situations. Therefore we consider $\widehat{\Psi}_{inc}^{(k,r)}$ to be a reasonably good estimate of Ψ_{inc} and when we speak of the designed filter we are referring to $\widehat{\Psi}_{inc}^{(k,r)}$. Finally, estimation of Ψ_{inc} typically requires far less data than equivalently good estimation of Ψ_{opt}. Owing to error of estimation, $\widehat{\Psi}_{inc}^{(k,r)}$ can outperform $\widehat{\Psi}_{opt}$ even when the optimal filter is not increasing.

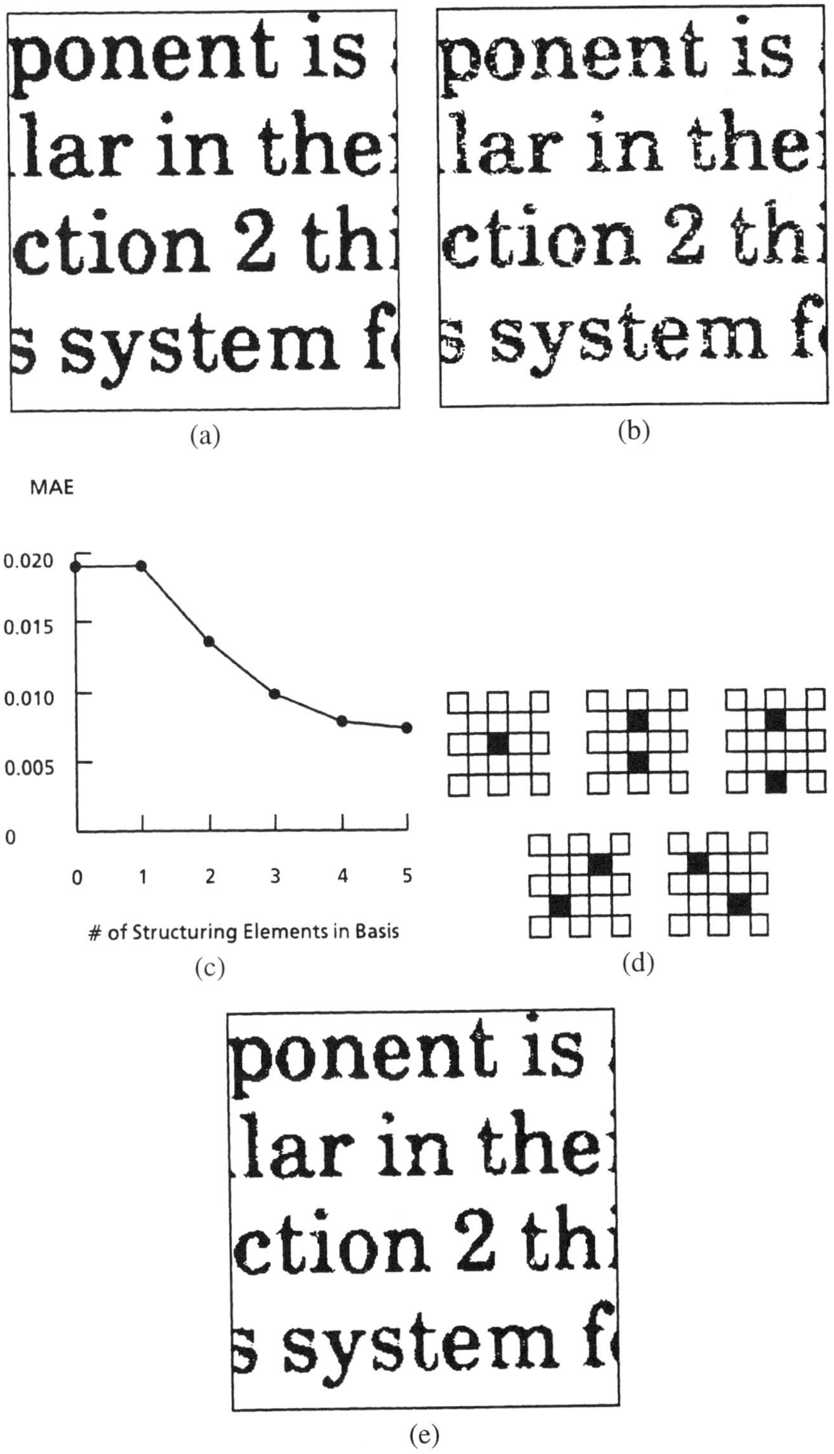

Figure 1–14. Text restoration: (a) realization of ideal process; (b) realization of degraded process; (c) MAE-vs-basis-size curve; (d) optimal 5-erosion basis; (e) restored image.

EXAMPLE 1–6 (*text restoration* [10]). Figure 1–14(a) shows a realization of an ideal text process and Fig. 1–14(b) shows a realization of the degraded process that results from pixel dropouts and for which $MAE = 0.0190$. A 300-element first-order library over a 17-pixel window has been employed in designing an increasing filter. The MAE vs. basis-size curve, optimal 5-erosion basis, and restored image are shown in Figs. 1–14(c), 1–14(d), and 1–14(e), respectively. MAE has been reduced to 0.0074 and the characters have been reconnected without joining characters.

1.9 ITERATIVE FILTERS

Another important dependent constraint is that the filter possess an iterative decomposition. Iterative filter design involves finding a statistically optimal filter for a small observation window that minimizes the error between the desired ideal image and the filtered observed image, finding a second filter that minimizes the error between the desired ideal image and the filtered output from the first filter, and then cascading the two filters to obtain a filter whose effective window is the dilation of the original small window with itself [18, 23]. The class of composite filters that can be constructed by such cascading is much smaller than the class of all filters over the large window. Hence, relative to direct optimization over the large window, an iteratively designed filter is suboptimal. But computation is greatly reduced by iterative design, often to the point where large-window optimization is computationally impossible whereas iterative design has no computational impediments. Moreover, iteratively designed filters often provide only marginally reduced performance than filters optimally designed over a large window. Finally, by employing several iterations, it is possible to achieve better filters that take less design time than could be achieved by a single-iteration method taking greater design time. We focus on increasing operators.

The key to iterative design is composition of logical sums of products. Let Ψ_1 and Ψ_2 be increasing W-operators with respective window functions

$$\begin{aligned} \psi_1(x_1, x_2, \ldots, x_n) &= \sum_k x_{k,1}x_{k,2}\cdots x_{k,m(k)} \\ \psi_2(x_1, x_2, \ldots, x_n) &= \sum_i x_{i,1}x_{i,2}\cdots x_{i,n(i)} \end{aligned} \tag{1–82}$$

The *iterative (composite) filter* $\Psi^2 = \Psi_2\Psi_1$ is defined by $\Psi_2\Psi_1(S) = \Psi_2(\Psi_1(S))$. Owing to translation invariance, we need only examine $\Psi_2\Psi_1(S)$ at 0 to arrive at the form of the window function, ψ^2, for Ψ^2. Specifically, we consider

$$\Psi_2\Psi_1(S)(0) = \psi_2\big(\psi_1(S \cap W_{x_1}), \psi_1(S \cap W_{x_2}), \ldots, \psi_1(S \cap W_{x_m})\big) \tag{1–83}$$

The functional expression on the right defines ψ^2 and depends on values of S in

$$\bigcup_{l=1}^{m} W_{x_i} = W \oplus W \tag{1–84}$$

which is the Minkowski sum of W with itself. Denoting the varibles in $W \oplus W$ by $z_1, z_2, \ldots, z_N$ and applying the representations of Eq. 1–82 to ψ^2 yields

$$\begin{aligned}\psi^2(z_1, z_2, \ldots, z_N) &= \sum_i \psi_1(S \cap W_{x_{i,1}})\psi_1(S \cap W_{x_{i,2}}) \cdots \psi_1(S \cap W_{x_{i,n(i)}}) \\ &= \sum_i \left(\sum_k x_{i,1,k,1} x_{i,1,k,2} \cdots x_{i,1,k,m(k)} \right) \\ &\quad \times \left(\sum_k x_{i,2,k,1} x_{i,2,k,2} \cdots x_{i,2,k,m(k)} \right) \\ &\quad \cdots \times \left(\sum_k x_{i,n(i),k,1} x_{i,n(i),k,2} \cdots x_{i,n(i),k,m(k)} \right)\end{aligned} \tag{1–85}$$

where the variables of $S \cap W_{x_{i,j}}$ appearing in the kth product forming $\psi_1(S \cap W_{x_{i,j}})$ are $x_{i,j,k,1}, x_{i,j,k,2}, \ldots, x_{i,j,k,m(k)}$ for $j = 1, 2, \ldots, n(i)$. The resulting sum of products over $W \oplus W$ defines the window function for the iterative filter $\Psi^2 = \Psi_2\Psi_1$. Logic reduction yields product terms corresponding to $\mathcal{B}[\psi^2]$. Since the product terms for ψ_1 and ψ_2 correspond to $\mathcal{B}[\psi_1]$ and $\mathcal{B}[\Psi_2]$, respectively, derivation of $\mathcal{B}[\psi^2]$ from $\mathcal{B}[\psi_1]$ and $\mathcal{B}[\psi_2]$ is automatically accomplished via logic software.

If we desire an optimal MAE increasing filter over window $W \oplus W$ but window size makes the problem too computationally intensive, then one way to proceed is to find an optimal iterative filter. Suppose $\mathbf{S}$ is the observed image process and $\mathbf{I}$ is the ideal. An optimal increasing W-filter Ψ_1 is found to minimize MAE for $\Psi_1(\mathbf{S})$ as an estimator of $\mathbf{I}$. Next, an optimal increasing W-filter Ψ_2 is found to minimize MAE for $\Psi_2(\Psi_1(\mathbf{S}))$ as an estimator of $\mathbf{I}$. Relative to Ψ_2, $\Psi_1(\mathbf{S})$ is the observed image. This *optimal iterative filter* $\Psi_2\Psi_1$ is an increasing $(W \oplus W)$-filter. $\Psi_2\Psi_1$ is very likely not optimal over all $(W \oplus W)$-filters, since this would require that $\Psi_2\Psi_1$ be decomposable into W-filters; nonetheless, since filter design is computationally limited to relatively small windows, if iteration provides good suboptimal results, then it permits automatic design for larger windows than could be accomplished by direct noniterative design.

More generally, we can consider an iteration of n filters,

$$\Psi^n = \Psi_n \Psi_{n-1} \Psi_{n-2} \cdots \Psi_2 \Psi_1 \tag{1–86}$$

Recursively, $\Psi^n = \Psi_n \Psi^{n-1}$. If each filter composing Ψ^n is an increasing W-filter, then Ψ^n is an increasing nW-filter, where

$$nW = W^1 \oplus W^2 \oplus \cdots \oplus W^n \tag{1–87}$$

and $W^i = W$ for $i = 1, 2, \ldots, n$. If we optimize recursively to obtain Ψ^n, then Ψ^n is an nW-filter that is suboptimal relative to the optimal filter, $\Psi_{opt,n}$, over nW. Since Ψ_k can be the identity filter, it must be that $MAE\langle\Psi^k\rangle \leqslant MAE\langle\Psi^{k-1}\rangle$. Hence,

$$MAE\langle\Psi_{opt,n}\rangle \leqslant MAE\langle\Psi^n\rangle \leqslant MAE\langle\Psi^{n-1}\rangle \leqslant \cdots \leqslant MAE\langle\Psi^1\rangle \tag{1–88}$$

Because the MAEs for the iterative filters form a decreasing sequence, they must possess a limit; however, the inequality of Eq. 1–88 does not imply that this limiting MAE is equal to $MAE\langle\Psi_{opt,n}\rangle$ for some n. Moreover, unless an optimal W^n-filter is decomposable, the leftmost inequality is strict for each n. Indeed, a basic problem of iteration is to determine the number of iterations necessary for further iterations to produce neglible improvement in MAE. Both the degree of optimality and the number of iterations necessary to be close to minimal iteratively achievable MAE are dependent on the window and the signal-noise model.

A measure of difference is needed to compare Ψ^n and $\Psi_{opt,n}$. From a purely logical perspective, the extent by which two operators disagree can be measured by the size of their switching set relative to the total number of observation vectors. The *logical difference* between operators Ψ and Φ is defined by $Card(\mathcal{Z}[\Psi, \Phi])/2^n$.

From the standpoint of filtering random sets, logical difference is not the key issue. If $\psi(\mathbf{x}) \neq \phi(\mathbf{x})$, but the probability of observing $\mathbf{x}$ is very small relative to other observation probabilities, then it matters little that $\psi(\mathbf{x}) \neq \phi(\mathbf{x})$. Their *probabilistic difference* is defined by $P(\mathcal{Z}[\Psi, \Phi])$. Applying Eq. 1–28 with the MAE loss function yields

$$\begin{aligned} MAE\langle\Psi\rangle - MAE\langle\Phi\rangle &= \sum_{\mathbf{x}\in\mathcal{S}_1[\psi]} P(\mathbf{x})P(Y=0\mid\mathbf{x}) + \sum_{\mathbf{x}\in\mathcal{S}_0[\Psi]} P(\mathbf{x})P(Y=1\mid\mathbf{x}) \\ &\quad - \sum_{\mathbf{x}\in\mathcal{S}_1[\phi]} P(\mathbf{x})P(Y=0\mid\mathbf{x}) - \sum_{\mathbf{x}\in\mathcal{S}_0[\phi]} P(\mathbf{x})P(Y=1\mid\mathbf{x}) \\ &= \sum_{\mathbf{x}\in\mathcal{S}_1[\psi]-\mathcal{S}_1[\phi]} P(\mathbf{x})\big[P(Y=0\mid\mathbf{x}) - P(Y=1\mid\mathbf{x})\big] \\ &\quad + \sum_{\mathbf{x}\in\mathcal{S}_1[\phi]-\mathcal{S}_1[\psi]} P(\mathbf{x})\big[P(Y=1\mid\mathbf{x}) - P(Y=0\mid\mathbf{x})\big] \end{aligned} \tag{1–89}$$

Each term in both of the latter sums is bounded by $P(\mathbf{x})$, so that

$$\big|MAE\langle\Psi\rangle - MAE\langle\Phi\rangle\big| \leqslant P\big(\mathcal{Z}[\Psi, \Phi]\big) \tag{1–90}$$

and the probabilistic difference between two filters serves as an upper bound on the difference between their MAEs.

In designing iterative filters, it is not unusual for the designed estimates of Ψ^n and $\Psi_{opt,n}$ to be substantially different while their probabilistic difference is negligible. Hence, they differ significantly as logical operators but insignficantly as filters on the observed random set.

EXAMPLE 1–7 (*text restoration* [24]). To illustrate increasing-filter iterative design, we consider document restoration. Because we wish to compare optimal filtering over $W \oplus W$ with iterative filtering over W, we employ the small window

$$W = \begin{pmatrix} 0 & 1 & 0 \\ 1 & 1 & 1 \\ 0 & 1 & 0 \end{pmatrix}$$

for which

$$W \oplus W = \begin{pmatrix} 0 & 0 & 1 & 0 & 0 \\ 0 & 1 & 1 & 1 & 0 \\ 1 & 1 & 1 & 1 & 1 \\ 0 & 1 & 1 & 1 & 0 \\ 0 & 0 & 1 & 0 & 0 \end{pmatrix}$$

An ideal document image is subjected to dilation and random pepper noise. Iterative filters using both W and $W \oplus W$ are designed to restore the original. Using W, two iterations achieve essentially the same degree of restoration as a single application of the filter designed over $W \oplus W$. Not only is design much faster for the iterative filter, were we to employ a 5×5 iterative filter, two stages would approximate a 9×9 optimal filter, which could not be designed by the increasing-filter methodology. Figure 1–15 shows a realization of the ideal image, a degraded version of the realization, restoration after a single-stage W-filter, restoration after a two-stage W-filter, and restoration after a single-stage $(W \oplus W)$-filter.

EXAMPLE 1–8 (*connected thinning*). We desire a thinning (skeletonization) algorithm based on iterative filtering over a 3×3 window to thin connected components. Figure 1–16 shows a realization of the input process and the results of four iterations of the designed filter. Note how most of the transformation is accomplished in the first iteration and how small "corrections" are made subsequently. This is not unusual for iterative filtering, where later-stage filters can "correct over filtering" of early stages [24]. Figure 1–17 shows the realization of Fig. 1–16(a) with 5% random pepper noise and the result of the designed iterative filter after four interations. Table 1.5 gives error percentages for each iteration.

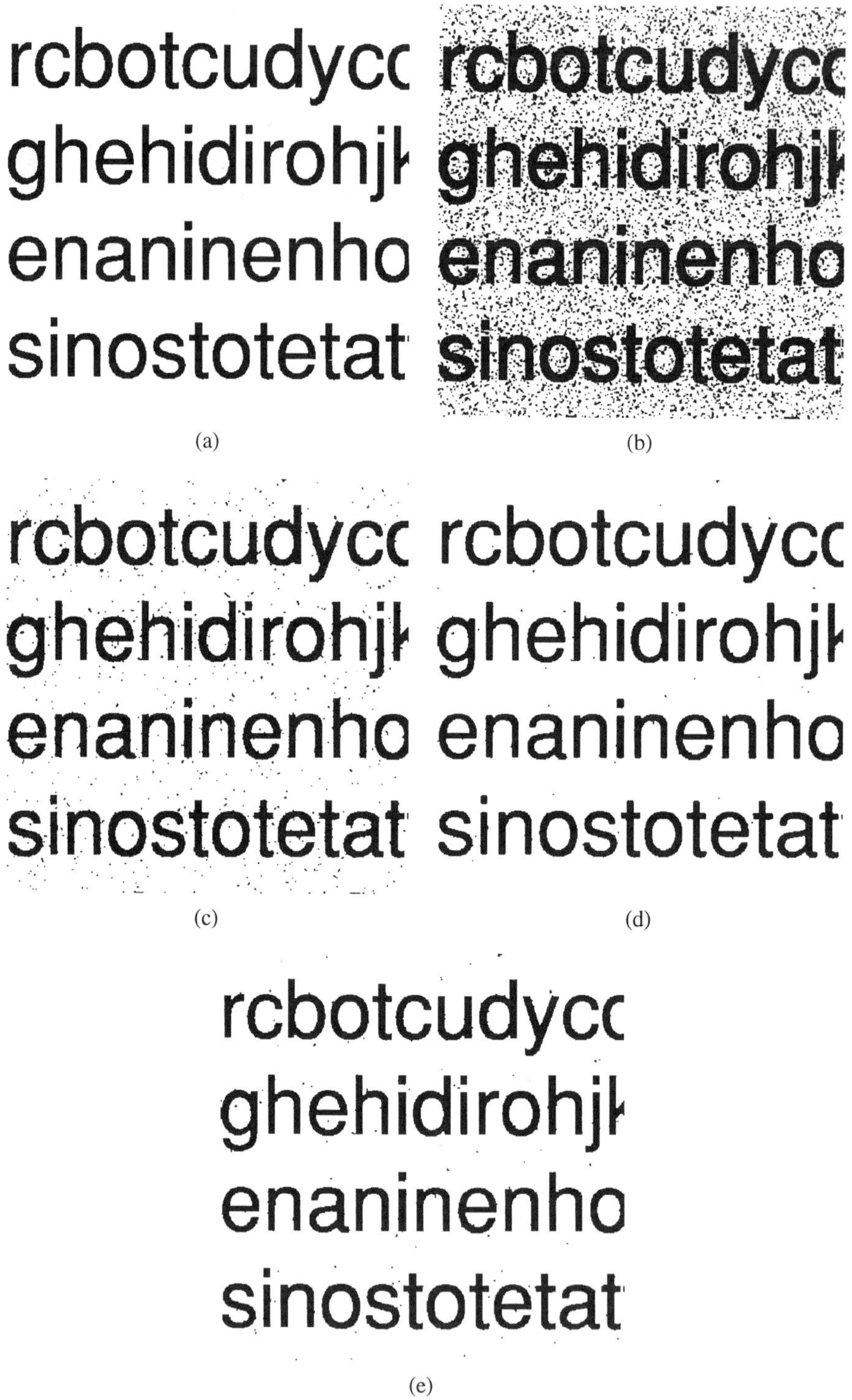

Figure 1–15. Text restoration: (a) ideal realization; (b) degraded realization; (c) restoration with single-stage W-filter; (d) restoration with two-stage W-filter; (e) restoration with single stage $(W \oplus W)$-filter.

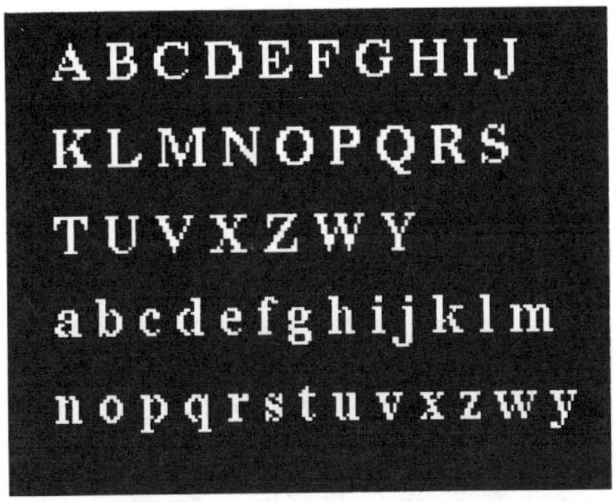

(a)

(b)

(c)

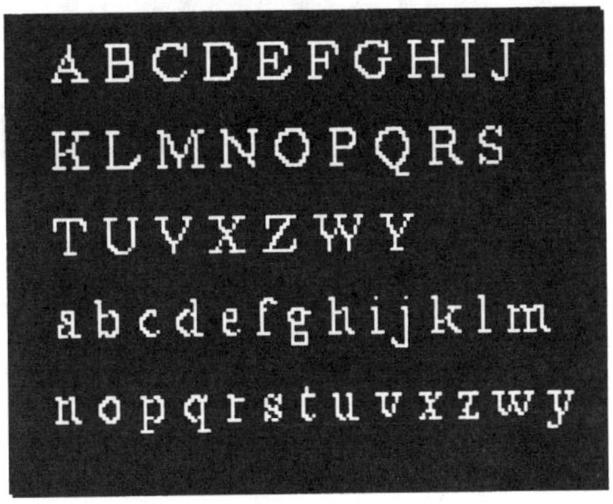

(d)

(e)

Figure 1–16. Connected thinning: (a) realization of input process; (b) result of first iteration; (c) result of second iteration; (d) result of third iteration; (e) result of fourth iteration.

Table 1.5. Error percentages for connected thinning.

Iteration	Nonnoisy	Noisy
1	0.40%	1.20%
2	0.20%	0.96%
3	0.18%	0.88%
4	0.17%	0.83%

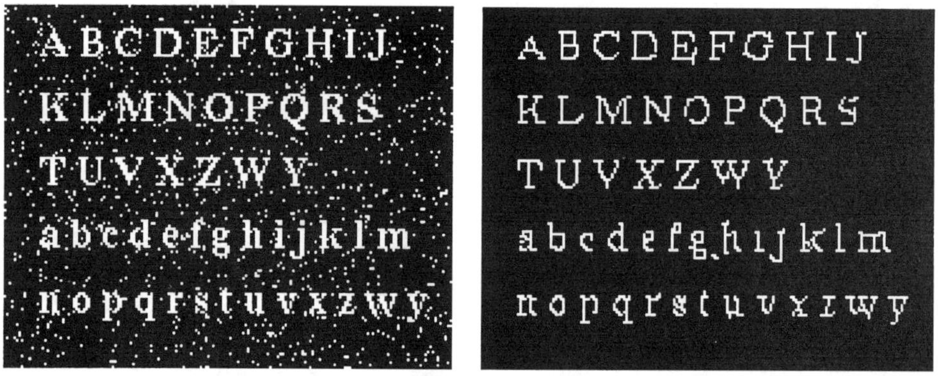

Figure 1–17. Realization of Fig. 1–16(a) with 5% random pepper noise and the result of the designed iterative filter after four interations.

EXAMPLE 1–9 (*character recognition* [15]). The present example combines algebraic constraint and iterative filtering for character recognition. It has been performed on an Intel-586 processor and processing time is measured in hours (h), minutes (m), and seconds (s). The input image is text and the output a set of markers for a desired character. Operators are constrained to be antiextensive [meaning $S \supset \Psi(S)$], the first training stage uses the SR loss function, subsequent stages use the MAE loss function, the first filter is over either a 7×7 or 9×9 window, and each subsequent stage uses a window reduced by 2 pixels per side compared to its preceding stage. Figure 1–18 shows an input image (gray) with markers for the character 'a' superimposed. Table 1.6 provides some sample results for recognition of 'a'.

1.10 MACHINE LEARNING THEORY AND OPTIMAL OPERATOR DESIGN

Computational learning theory is one of the first attempts to construct a mathematical model for the cognitive concept-learning process. It provides a framework for studying a variety of algorithmic processes. We briefly review basic elements of

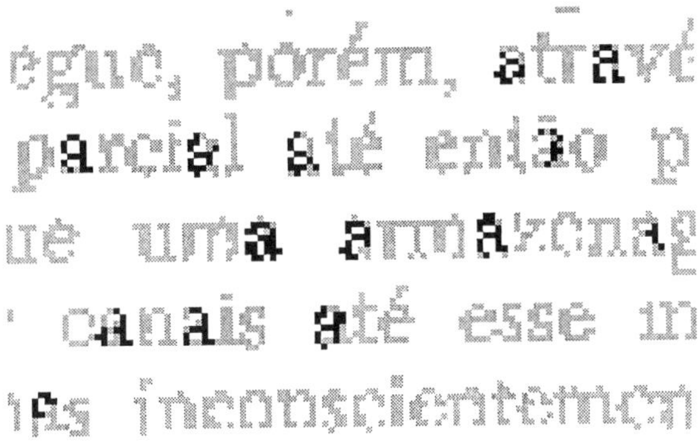

Figure 1–18. Character recognition: input gray image with markers for the character 'a' superimposed.

Table 1.6. Training data.

First training stage	# Examples (thousands)	# of Stages	Basis size	Training time	Relative error (%)
7×7	79	1	644	2 hour/47 min	10.4
7×7	88	2	726	2 hour/48 min	1.4
7×7	96	3	762	2 hour/48 min	0.5
9×9	79	1	551	3 hour/45 min	12.8
9×9	88	2	700	3 hour/48 min	0.8
9×9	96	3	760	3 hour/48 min	0.5
9×9	104	4	781	3 hour/48 min	0.4

the model and then discuss the relationship between automatic design of optimal W-operators and Haussler's paradigm for learning Boolean concepts in the context of machine learning theory [8, 25, 26].

Consider a finite set (*domain*) of objects D structured by an unknown distribution μ. A *concept* c is a subset of objects in a predefined domain D or, equivalently, a Boolean function from $D \rightarrow \{0, 1\}$. An *example* of a concept c is a pair (x, b), where x is an object in D and b is a binary *label* (0 or 1) indicating whether or not $x \in c$. If $b = 1$, then $x \in c$ and the example is called *positive*; otherwise, $x \notin c$ and the example is called *negative*. An object is taken randomly from the domain and a *teacher*, who knows the concept, classifies the object as a positive or negative example. In more sophisticated models, the teacher is assumed to be imperfect, that is, the teacher may erroneously classify some objects. In the imperfect case, the teacher is modeled by a conditional distribution $P(\mathbf{B} \mid \mathbf{X})$, where $\mathbf{X}$ is the random process, with distribution μ, representing the domain, and $\mathbf{B}$ is a binary random process labeling a domain object when it is observed.

Concept learning is the process by which a learner constructs a good approximation to an unknown concept from a number of examples and possibly some prior information about the concept to be learned. After observing labels of a sequence of objects taken randomly from the domain, the learner obtains a Boolean function that decides whether an object taken from the domain is an element of the concept with a small probability of error. The set of all possible concepts to be learned is called the *hypothesis space* and denoted by H. The concept $t \in H$ to be determined is called the *target concept*. The task is to find a concept $h \in H$, called a *hypothesis*, that is a good approximation of t.

A *training sample* $\mathbf{s}$ of length m is a sequence of m examples,

$$\mathbf{s} = \big((x_1, b_1), (x_2, b_2), \ldots, (x_m, b_m)\big) \tag{1–91}$$

where, for $i = 1, 2, \ldots, m$, x_i is an object and b_i a label. Assuming that object selection from the domain is independent, it may happen that an object occurs more than once in the examples that constitute $\mathbf{s}$. When the teacher does not make mistakes, the training sample is said to be *consistent*; that is, if $x_i = x_j$, then $b_i = b_j$. When the teacher may err, the training sample is said to be *nonconsistent*; that is, it may happen that $b_i \neq b_j$ when $x_i = x_j$. For fixed m, the class $\mathbf{S}$ of training samples is a random process with

$$P(\mathbf{S} = \mathbf{s}) = \prod_{i=1}^{m} P(\mathbf{X} = x_i) P(\mathbf{B} = b_i \mid x_i) \tag{1–92}$$

The probability mass for $\mathbf{S}$ is induced by the probability mass on the domain in conjunction with the conditional labeling probabilities. A *learning algorithm* is a function L that assigns to any training sample $\mathbf{s}$, for a target concept $t \in H$, a *hypothesis* $h \in H$. We write $h = L(\mathbf{s})$.

Consider a fixed target concept $t \in H$ and a given loss function l. Let $\mathbf{B}$ be the Boolean random process describing the label attributed to a domain object when it is observed in the learning procedure. For any hypothesis $h \in H$, the *risk* $r(h, t)$ of choosing hypothesis h for the concept t is defined by

$$r(h, t) = E\big[l\big(\mathbf{B}, h(\mathbf{X})\big)\big] \tag{1–93}$$

where $\mathbf{X}$ is the random process modeling the domain. Because $\mathbf{B}$ depends on $\mathbf{X}$, the distributions of both $\mathbf{B}$ and $h(\mathbf{X})$ depend on the distribution of $\mathbf{X}$. Let $h^* \in H$ be the hypothesis of minimum risk for t, meaning $r(h, t) \geqslant r(h^*, t)$ for all $h \in H$.

Algorithm L is called a *probably approximately correct* (*PAC*) learning algorithm for the hypothesis space H if, given real numbers ε $(0 < \varepsilon < 1)$ and δ $(0 < \delta < 1)$, there exists a positive integer $m(\varepsilon, \delta)$ such that $m \geqslant m(\varepsilon, \delta)$ implies

$$P\left(\left|r(L(\mathbf{S}), t) - r\left(h^*, t\right)\right| < \varepsilon\right) > 1 - \delta \tag{1–94}$$

for any target concept $t \in H$, distribution μ on D, and conditional distribution $P(\mathbf{B} \mid \mathbf{X})$. The value $m(\varepsilon, \delta)$ and the pair (ε, δ) are called, respectively, the *sample complexity* and *precision* of the learning algorithm. If H is finite, then the sample complexity of any PAC learning algorithm L is bounded [26] by

$$m(\varepsilon, \delta) = \frac{1}{\varepsilon^2}\left(\log Card(H) + \log \frac{1}{\delta}\right) \tag{1–95}$$

The formulation of PAC learning we have presented is a simplification of Haussler's formulation to concepts understood in the sense defined here. If the loss function is

$$l\big(\mathbf{B}, h(\mathbf{X})\big) = \left|\mathbf{B} - h(\mathbf{X})\right| \tag{1–96}$$

and the training sample for a given target concept t is consistent, then $r(h^*, t) = r(t, t) = 0$,

$$r(h, t) = \mu\big(\{x \in D\colon\ t(x) \neq h(x)\}\big) \tag{1–97}$$

the formulation reduces to the original formulation of PAC learning algorithms [27] and the bound for the sample complexity reduces to

$$m(\varepsilon, \delta) = \frac{1}{\varepsilon}\left(\log Card(H) + \log \frac{1}{\delta}\right) \tag{1–98}$$

The complete procedure of estimation of set operators we have proposed (estimation of the conditional probabilities and estimation of the best operator by optimization) is equivalent to the learning-concept formulation just presented. W-operators are equivalent to Boolean functions, and concepts (in the sense we have defined) are Boolean functions defined on a given domain. When interpreting W-operators as concepts, the domain is the set of patterns observed in W; the distribution μ gives the relative proportions of observed patterns and is determined by the probabilities $P(\mathbf{X} = \mathbf{x})$; and the conditional probability $P(\mathbf{B} \mid \mathbf{X})$ results from nondeterminacy in the ideal image given an observed pattern and noise affecting the images.

In the hit-or-miss representation, for a hypothesis (Boolean function) h (determining a W-operator Ψ), a loss occurs if and only if $(E, F) \in C_\Psi$, where (E, F) is the canonical structuring pair equivalent to $\mathbf{x}$, and $\mathbf{B} = 0$ when $\mathbf{x}$ is observed, or

$(E, F) \notin \mathcal{C}_\Psi$ and $\mathbf{B} = 1$ when $\mathbf{x}$ is observed. Since $(E, F) \in \mathcal{C}_\Psi$ can be equivalently expressed as $\mathbf{x} \in \mathcal{C}_\Psi$ and $h(\mathbf{x}) = 1$ if and only if $\mathbf{x} \in \mathcal{C}_\Psi$,

$$\begin{aligned} r(h,t) &= E\big[l\big(\mathbf{B}, h(\mathbf{X})\big)\big] \\ &= \sum_{\mathbf{x}} E\big[l\big(\mathbf{B}, h(\mathbf{x})\big) \mid \mathbf{x}\big] P(\mathbf{X}=\mathbf{x}) \\ &= \sum_{\mathbf{x}\in\mathcal{C}_\Psi} E\big[l(\mathbf{B},1) \mid \mathbf{x}\big] P(\mathbf{X}=\mathbf{x}) + \sum_{\mathbf{x}\notin\mathcal{C}_\Psi} E\big[l(\mathbf{B},0) \mid \mathbf{x}\big] P(\mathbf{X}=\mathbf{x}) \end{aligned} \tag{1–99}$$

Assuming the target function t is a possible hypothesis,

$$\begin{aligned} r(h,t) - r(t,t) = &\sum_{\mathbf{x}\in\mathcal{C}_\Psi - \mathcal{C}_l} \big(E\big[l(\mathbf{B},1) \mid \mathbf{x}\big] - E\big[l(\mathbf{B},0) \mid \mathbf{x}\big]\big) P(\mathbf{X}=\mathbf{x}) \\ &+ \sum_{\mathbf{x}\notin\mathcal{C}_l - \mathcal{C}_\Psi} \big(E\big[l(\mathbf{B},0) \mid \mathbf{x}\big] - E\big[l(\mathbf{B},1) \mid \mathbf{x}\big]\big) P(\mathbf{X}=\mathbf{x}) \end{aligned} \tag{1–100}$$

where $\mathcal{C}_l$ is the class of structuring pairs corresponding to the W-operator defined by the target Boolean function t. Since the $r(h,t)$ is minimized for $h = t$, according to Eq. 1–99 applied to t, $E[l(\mathbf{B},1) \mid \mathbf{x}] \leqslant E[l(\mathbf{B},0) \mid \mathbf{x}]$ if $\mathbf{x} \in \mathcal{C}_l$ and $E[l(\mathbf{B},0) \mid \mathbf{x}] \leqslant E[l(\mathbf{B},1) \mid \mathbf{x}]$ if $\mathbf{x} \notin \mathcal{C}_l$. Hence, Eq. 1–100 reduces to

$$r(h,t) - r(t,t) = \sum_{\mathbf{x}\in\mathcal{C}_\Psi \Delta \mathcal{C}_l} \big|E\big[l(\mathbf{B},1) \mid \mathbf{x}\big] - E\big[l(\mathbf{B},0) \mid \mathbf{x}\big]\big| P(\mathbf{X}=\mathbf{x}) \tag{1–101}$$

which is equivalent to the suboptimality error-increase expression given in terms of absolute advantages in Eq. 1–48. Interpreting Eq. 1–48 relative to sampling, the concept-learning precision inequality of Eq. 1–94 takes the form

$$1 - \delta < P\big(\big|Er\langle\widehat{\Psi}_l\rangle - Er\langle\Psi_l\rangle\big| < \varepsilon\big) = P\Bigg(\sum_{(E,F)\in\widehat{\mathcal{C}}_l \Delta \mathcal{C}_l} \big|Ad_l(E,F)\big| < \varepsilon\Bigg) \tag{1–102}$$

for $m \geqslant m(\varepsilon, \delta)$.

General machine-learning sample complexity bounds are unrealistically loose when compared with practical results found in the literature [9, 12, 15]. A salient reason is use of prior information, which is information that a learner has about the domain or concept to be learned. If this information is used properly, then the training-sample size needed to obtain a given precision (ε, δ) can become smaller or, equivalently, training samples of a fixed size can give sharper estimates.

A key issue in learning W-operators is the choice of a window W [18]. Window size should be as small as possible since the size of the domain D is affected exponentially by the size of W, $Card(D) = 2^{Card(W)}$. In this task, prior information plays an important role. Often knowledge of some geometrical properties expressed by the concepts is sufficient to properly choose W. For instance, if the target concept is an operator to detect edges, an edge in digital topology depends on just a small neighborhood of 4 or 8 pixels, depending on the connectivity required [28]; if the target concept is a filter of connected components or holes, then shape recognition can be described by canonical hit-or-miss operators that depend on the smallest window containing all components to be filtered [29–31].

Window size is related to image resolution: higher resolutions require larger windows. Low resolutions require less training data to achieve a given estimation precision. For instance, consider a training-data set consisiting of N pairs of square images with n lines and a square window of r lines, with $r \ll n$. There are $m = 2^{r^2}$ pattern vectors $\mathbf{x}_1, \mathbf{x}_2, \ldots, \mathbf{x}_m$ defining the full observation domain. The training data yield Nn^2 examples (observations) and, owing to repetitions, many fewer than Nn^2 of the potential m patterns are observed to cover the full observation domain of size m. If images with half the resolution are sufficient for the task, then we could employ N square images with $n/2$ lines and a square window with $r/2$ lines. In this case, we observe $Nn^2/4$ examples and the domain size is $2^{r^2/4}$. At half resolution, the number of examples is divided by 4, as is the exponent of the domain size. Choosing a minimum resolution sufficient to solve an imaging problem is a basic preliminary step.

Practical image analysis is usually restricted to specific contexts, that is, to particular classes of images for which the operators should perform well. Restriction to a given context implies that the domain D becomes a subset of the power set of W. It is typically not easy to estimate the number of patterns in a given context, but the number is often significantly smaller than $2^{Card(W)}$. The worst probabilistic structure for the domain is the uniform distribution, which models the most disorganized space. This distribution is poor for concept learning because to obtain small risks for a hypothesis h that estimates a target concept t, a large portion of the domain must be covered by examples of the training sample. Thus, very large training samples are needed. The best probability structure for the domain is the deterministic one, where there is just one pattern with probability 1. Practical applications involve neither of these extremes and are usually far from both.

Information concerning properties of the target operator is also useful. If the target operator is a marker for shape recognition, then it is antiextensive; if it is a size classifier, then it is increasing, antiextensive and idempotent; etc. These kinds of properties can be interpreted as constraints that characterize families of operators in the hypothesis space. Under knowledge of target-operator properties, the hypothesis space will be the intersection of the families of operators defined by the constraints.

Another kind of prior knowledge is a good initial hypothesis for the target operator [32]. Instead of learning the target operator directly, we can learn the symmetric difference between the target operator and the initial hypothesis. The nearer the initial hypothesis to the target operator, the easier will be the task of learning the symmetric difference between them. Use of differencing representation for document restoration is an example of this kind of prior knowledge and, for it, the identity is used as the initial hypothesis. When there is a very small amount of noise affecting just the edges of text characters, the identity operator is a reasonable initial hypothesis.

In general, learning a target concept corresponds to generating a function (the hypothesis) by the specifications of values (labels) for the most probable subsets of W. For generating the hypothesis, patterns are chosen randomly from the domain. If the number of patterns with unknown classification is large, then it will be necessary to have large training samples to get good precision; otherwise the training samples may be smaller to get the same precision. Prior information can reduce the number of subsets of W with an unknown label, thereby yielding smaller sample complexities. A small window reduces domain size. Context can determine subsets of W that should be labeled; others are taken as don't-cares. Distributional knowledge allows patterns with small probabilities to be treated as don't-cares. Operator properties facilitate deduction of some pattern labels from knowledge of other pattern labels. An initial hypothesis may label a large number of patterns.

1.11 Robustness

A fundamental aspect of any filter is the degree to which its performance degrades when it is applied to random processes different than the one for which it has been designed. Qualitatively, a filter is said to be *robust* when its performance degradation is acceptable for processes statistically close to the design process. Robustness is crucial for application because a filter will surely be applied in nondesign settings. This may occur because it is applied to different stationary random processes or to nonstationary random processes. For instance, an optimal document filter may be applied to documents using different fonts than the design fonts. Robustness depends on both the ideal and observed images.

To define robustness, consider a parameterized ideal discrete random set $\mathbf{I_r}$, where $\mathbf{r}$ is a parameter vector determining the probability law governing $\mathbf{I_r}$, and a parameterized system transformation $\Xi_\mathbf{t}$, whose probability law is determined by the parameter $\mathbf{t}$. The observed random set is $\mathbf{S_a} = \Xi_\mathbf{t}(\mathbf{I_r})$, where $\mathbf{a} = (\mathbf{r}, \mathbf{t})$, and the filter problem is to design an optimal filter $\Psi_\mathbf{a}$ to restore $\mathbf{I_r}$ by means of the estimator $\Psi_\mathbf{a}(\mathbf{S_a})$. Taken together, $\mathbf{I_r}$, $\Xi_\mathbf{t}$, and $\mathbf{S_a}$ form a parameterized system model $\mathcal{M}_\mathbf{a}$, and $\Psi_\mathbf{a}$ is optimal relative to $\mathcal{M}_\mathbf{a}$. To unify notation, for $\mathbf{a} = (\mathbf{r}, \mathbf{t})$ we notate all aspects of the model by $\mathbf{a}$, namely, $\mathbf{I_a}$, $\Xi_\mathbf{a}$, and $\mathbf{S_a}$, where it is understood that $\mathbf{I_a}$ and $\Xi_\mathbf{a}$ are each parameterized by separate components of $\mathbf{a}$.

For a filter Ψ, the error of Eq. 1–27 must indexed by the model parameter. Let $Er_{\mathbf{a}}\langle\Psi\rangle$ denote the error for Ψ relative to the system model $\mathcal{M}_{\mathbf{a}}$. If $\Psi_{\mathbf{b}}$ is optimal relative to $\mathcal{M}_{\mathbf{b}}$, then $Er_{\mathbf{a}}\langle\Psi_{\mathbf{b}}\rangle \geqslant Er_{\mathbf{a}}\langle\Psi_{\mathbf{a}}\rangle$: $\Psi_{\mathbf{a}}$ is a better estimator of $\mathbf{I_a}$ than is $\Psi_{\mathbf{b}}$. Intuitively, for $\Psi_{\mathbf{b}}$ to be robust, the inequality should not be too great when $\mathbf{b}$ is close to $\mathbf{a}$. Hence, robustness of the optimal filter $\Psi_{\mathbf{b}}$ for model $\mathcal{M}_{\mathbf{b}}$ relative to $\mathcal{M}_{\mathbf{a}}$ is defined by

$$\kappa(\mathbf{b};\,\mathbf{a}) = Er_{\mathbf{a}}\langle\Psi_{\mathbf{b}}\rangle - Er_{\mathbf{a}}\langle\Psi_{\mathbf{a}}\rangle \tag{1–103}$$

For the MAE loss function,

$$\kappa(\mathbf{b};\,\mathbf{a}) = E\big[|Y_{\mathbf{a}} - \psi_{\mathbf{b}}(\mathbf{X_a})|\big] - E\big[|Y_{\mathbf{a}} - \psi_{\mathbf{a}}(\mathbf{X_a})|\big] \tag{1–104}$$

where $Y_{\mathbf{a}}$ and $\mathbf{X_a}$ are the ideal value and observation random vector for model $\mathcal{M}_{\mathbf{a}}$, and $\psi_{\mathbf{b}}$ and $\psi_{\mathbf{a}}$ are the optimal Boolean functions for $\mathcal{M}_{\mathbf{b}}$ and $\mathcal{M}_{\mathbf{a}}$, respectively [33]. $\kappa(\mathbf{b};\,\mathbf{a}) \geqslant 0$ and $\kappa(\mathbf{a};\,\mathbf{a}) = 0$. $\kappa(\mathbf{b};\,\mathbf{a})$ is not symmetric with respect to $\mathbf{a}$ and $\mathbf{b}$, since except in special circumstances $\kappa(\mathbf{b};\,\mathbf{a}) \neq \kappa(\mathbf{a};\,\mathbf{b})$.

Practically, we desire some degree of robustness. For robustness of a particular designed filter $\Psi_{\mathbf{b}}$, we say $\Psi_{\mathbf{b}}$ is *robust to the degree* (ε, δ) if $\kappa(\mathbf{b};\,\mathbf{a}) \leqslant \varepsilon$ when $|\mathbf{b} - \mathbf{a}| \leqslant \delta$. If the vector $(\mathbf{a}, \mathbf{b})$ can vary over some region R, then we say that the optimal filter is *uniformly robust to the degree* (ε, δ) over R if $\kappa(\mathbf{b};\,\mathbf{a}) \leqslant \varepsilon$ when $|\mathbf{b} - \mathbf{a}| \leqslant \delta$ and $(\mathbf{a}, \mathbf{b}) \in R$.

Focusing on MAE, some probabilistic calculations show that for a given vector $\mathbf{x}$, the increase in MAE owing to using the filter $\Psi_{\mathbf{b}}$ instead of $\Psi_{\mathbf{a}}$ for the system $\mathcal{M}_{\mathbf{a}}$ is

$$\kappa_{\mathbf{x}}(\mathbf{b};\,\mathbf{a}) = |2P_{\mathbf{a}}(Y = 1 \mid \mathbf{x}) - 1|\zeta_{\mathbf{x}}(\mathbf{b};\,\mathbf{a})P_{\mathbf{a}}(\mathbf{x}) \tag{1–105}$$

where $\zeta_{\mathbf{x}}(\mathbf{b};\,\mathbf{a})$ is an auxiliary function defined by

$$\zeta_{\mathbf{x}}(\mathbf{b};\,\mathbf{a}) = \begin{cases} 0, & \text{if } P_{\mathbf{b}}(Y = 1 \mid \mathbf{x}) \geqslant 0.5,\ P_{\mathbf{a}}(Y = 1 \mid \mathbf{x}) \geqslant 0.5 \\ 0, & \text{if } P_{\mathbf{b}}(Y = 1 \mid \mathbf{x}) < 0.5,\ P_{\mathbf{a}}(Y = 1 \mid \mathbf{x}) < 0.5 \\ 1, & \text{otherwise} \end{cases} \tag{1–106}$$

Robustness is obtained by summing $\kappa_{\mathbf{x}}(\mathbf{b};\,\mathbf{a})$ over all $\mathbf{x}$.

Depending on the model $\mathcal{M}_{\mathbf{a}}$, for any observation vector $\mathbf{x}$ there is a set of vectors $\mathcal{U}_{\mathbf{a},\mathbf{x}} = \{\mathbf{u}_1, \mathbf{u}_2, \ldots, \mathbf{u}_{m(\mathbf{x})}\}$ arising from realizations of $\mathbf{I_a}$ via intersections with W_z such that the center pixel is 1-valued and the transition $\mathbf{u}_i \to \mathbf{x}$ is possible with application of $\Xi_{\mathbf{a}}$. There also is a collection $\mathcal{V}_{\mathbf{a},\mathbf{x}} = \{\mathbf{v}_1, \mathbf{v}_2, \ldots, \mathbf{v}_{n(\mathbf{x})}\}$ such that the center pixel is 0-valued and the transition $\mathbf{v}_j \to \mathbf{x}$ can occur under $\Xi_{\mathbf{a}}$. Depending on the probability law for $\Xi_{\mathbf{a}}$, each transition has associated with it a conditional probability $P_{\mathbf{a}}(\mathbf{x} \mid \mathbf{u}_i)$ or $P_{\mathbf{a}}(\mathbf{x} \mid \mathbf{v}_j)$. These probabilities determine the value of the optimal filter when $\mathbf{x}$ is observed. Specifically, $P_{\mathbf{a}}(\mathbf{x})$ and $P_{\mathbf{a}}(Y = 1 \mid \mathbf{x})$ can be found from $P_{\mathbf{a}}(\mathbf{x} \mid \mathbf{u}_i)$, $P_{\mathbf{a}}(\mathbf{x} \mid \mathbf{v}_j)$, $P_{\mathbf{a}}(\mathbf{u}_i)$ and $P_{\mathbf{a}}(\mathbf{v}_j)$:

$$P_{\mathbf{a}}(\mathbf{x}) = \sum_{i=1}^{m(\mathbf{x})} P_{\mathbf{a}}(\mathbf{u}_i) P_{\mathbf{a}}(\mathbf{x} \mid \mathbf{u}_i) + \sum_{j=1}^{n(\mathbf{x})} P_{\mathbf{a}}(\mathbf{v}_j) P_{\mathbf{a}}(\mathbf{x} \mid \mathbf{v}_j) \tag{1–107}$$

$$P_{\mathbf{a}}(Y = 1 \mid \mathbf{x}) = \frac{P_{\mathbf{a}}(Y = 1, \mathbf{x})}{P_{\mathbf{a}}(\mathbf{x})} = \frac{\sum_{i=1}^{m(\mathbf{x})} P_{\mathbf{a}}(\mathbf{u}_i) P_{\mathbf{a}}(\mathbf{x} \mid \mathbf{u}_i)}{\sum_{i=1}^{m(\mathbf{x})} P_{\mathbf{a}}(\mathbf{u}_i) P_{\mathbf{a}}(\mathbf{x} \mid \mathbf{u}_i) + \sum_{j=1}^{n(\mathbf{x})} P_{\mathbf{a}}(\mathbf{v}_j) P_{\mathbf{a}}(\mathbf{x} \mid \mathbf{v}_j)} \tag{1–108}$$

thereby providing an analytic formulation for the optimal filter. According to Eq. 1–105,

$$\kappa_{\mathbf{x}}(\mathbf{b}; \mathbf{a}) = \left| \sum_{i=1}^{m(\mathbf{x})} P_{\mathbf{a}}(\mathbf{u}_i) P_{\mathbf{a}}(\mathbf{x} \mid \mathbf{u}_i) - \sum_{j=1}^{n(\mathbf{x})} P_{\mathbf{a}}(\mathbf{v}_j) P_{\mathbf{a}}(\mathbf{x} \mid \mathbf{v}_j) \right| \zeta_{\mathbf{x}}(\mathbf{b}; \mathbf{a}) \tag{1–109}$$

Tractable formulations are achievable with a sparse-noise constraint [33]. *Sparse noise* relative to W is degradation for which, if $\mathbf{u}$ and $\mathbf{x}$ are the ideal and observed vectors in W_z, then $\mathbf{u}$ and $\mathbf{x}$ differ at most in a single component. At most one component of $\mathbf{u}$ is switched by $\Xi_{\mathbf{t}}$. Now denote a vector by $\mathbf{x} = (x_0, x_1, x_2, \ldots, x_{n-1})$, where x_0 is the value at z and the others are observed by raster scanning the remainder of W_z. For sparse noise, let $\mathbf{x}_k = (k, x_1, x_2, \ldots, x_{n-1})$, $k = 0, 1$, and $\mathbf{x}_{k,i}$ be the same as $\mathbf{x}_k$ except that the ith component is switched. Then $\mathcal{U}_{\mathbf{a},\mathbf{x}_0} = \{\mathbf{x}_1\}$, $\mathcal{V}_{\mathbf{a},\mathbf{x}_1} = \{\mathbf{x}_0\}$,

$$\begin{aligned} \mathcal{U}_{\mathbf{a},\mathbf{x}_1} &= \{\mathbf{x}_1, \mathbf{x}_{1,1}, \mathbf{x}_{1,2}, \ldots, \mathbf{x}_{1,n-1}\} \\ \mathcal{V}_{\mathbf{a},\mathbf{x}_0} &= \{\mathbf{x}_0, \mathbf{x}_{0,1}, \mathbf{x}_{0,2}, \ldots, \mathbf{x}_{0,n-1}\} \end{aligned} \tag{1–110}$$

From Eq. 1–108 (and suppressing the subscript "**a**" in "$P_{\mathbf{a}}$" to ease notation),

$$P(Y = 1 \mid \mathbf{x}_1) = \frac{P(\mathbf{x}_1) P(\mathbf{x}_1 \mid \mathbf{x}_1) + \sum_{i=1}^{n-1} P(\mathbf{x}_{1,i}) P(\mathbf{x}_1 \mid \mathbf{x}_{1,i})}{P(\mathbf{x}_1) P(\mathbf{x}_1 \mid \mathbf{x}_1) + \sum_{i=1}^{n-1} P(\mathbf{x}_{1,i}) P(\mathbf{x}_1 \mid \mathbf{x}_{1,i}) + P(\mathbf{x}_0) P(\mathbf{x}_1 \mid \mathbf{x}_0)} \tag{1–111}$$

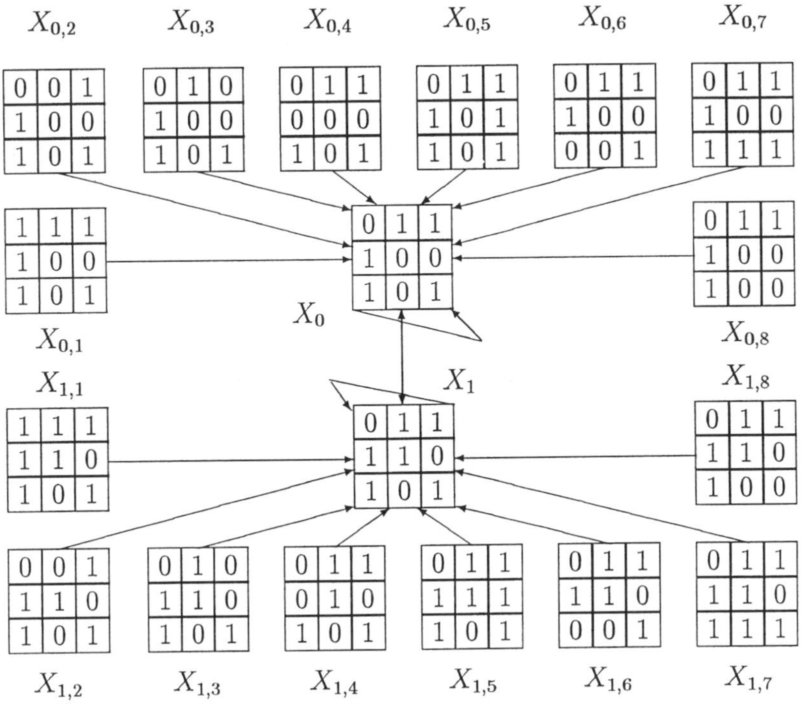

Figure 1–19. Possible transitions for sparse noise.

$$P(Y=1 \mid \mathbf{x}_0) = \frac{P(\mathbf{x}_1)P(\mathbf{x}_0 \mid \mathbf{x}_1)}{P(\mathbf{x}_1)P(\mathbf{x}_0 \mid \mathbf{x}_1) + P(\mathbf{x}_0)P(\mathbf{x}_0 \mid \mathbf{x}_0) + \sum_{i=1}^{n-1} P(\mathbf{x}_{0,i})P(\mathbf{x}_0 \mid \mathbf{x}_{0,i})} \tag{1–112}$$

thereby providing an analytic formulation of the optimal filter. Letting $\mathbf{x}^i$ denote the vector differing from $\mathbf{x}$ in component i, regardless of the center value, from Eq. 1–109,

$$\kappa_{\mathbf{x}}(\mathbf{b};\mathbf{a}) = \left| P(\mathbf{x})P(\mathbf{x} \mid \mathbf{x}) - P\left(\mathbf{x}^0\right)P\left(\mathbf{x} \mid \mathbf{x}^0\right) + \sum_{i=1}^{n-1} P\left(\mathbf{x}^i\right)P\left(\mathbf{x} \mid \mathbf{x}^i\right) \right| \zeta_{\mathbf{x}}(\mathbf{b};\mathbf{a}) \tag{1–113}$$

Figure 1–19 depicts the possible transitions for sparse noise when using a 3×3 window.

Now suppose the degradation operator is independent of the ideal image. Suppose $\mathbf{u}$ and $\mathbf{x}$ are identical except for component j, for which $u_j \neq x_j$, $0 \leqslant j \leqslant n-1$, and let γ denote complementation, so that $\gamma(x_j) = u_j$. The probability of the

Figure 1–20. Fonts: (a) triplex; (b) gothic; (c) noisy triplex, $p = 0.053$; (d) noisy gothic, $p = 0.053$.

transition $\mathbf{u} \to \mathbf{x}$ is

$$
\begin{aligned}
P(\mathbf{u} \to \mathbf{x}) &= P\big(\gamma(x_j) = u_j,\ x_i = u_i \text{ for } i \neq j\big) \\
&= P\big(\gamma(x_j) = u_j\big) P\big(x_i = u_i \text{ for } i \neq j \mid \gamma(x_j) = u_j\big) \\
&= P\big(\gamma(x_j) = u_j\big) \qquad (1\text{–}114)
\end{aligned}
$$

where the conditional probability in the second equality is 1 owing to noise sparseness. $P(\gamma(x_j) = u_j)$ is the probability that $\Xi_{\mathbf{a}}$ flips the jth component of $\mathbf{u}$, but this is independent of $\mathbf{u}$. Thus, it is the probability that $\Xi_{\mathbf{a}}$ flips a pixel value. By stationarity, this probability, p, called the *intensity* of the independent sparse noise,

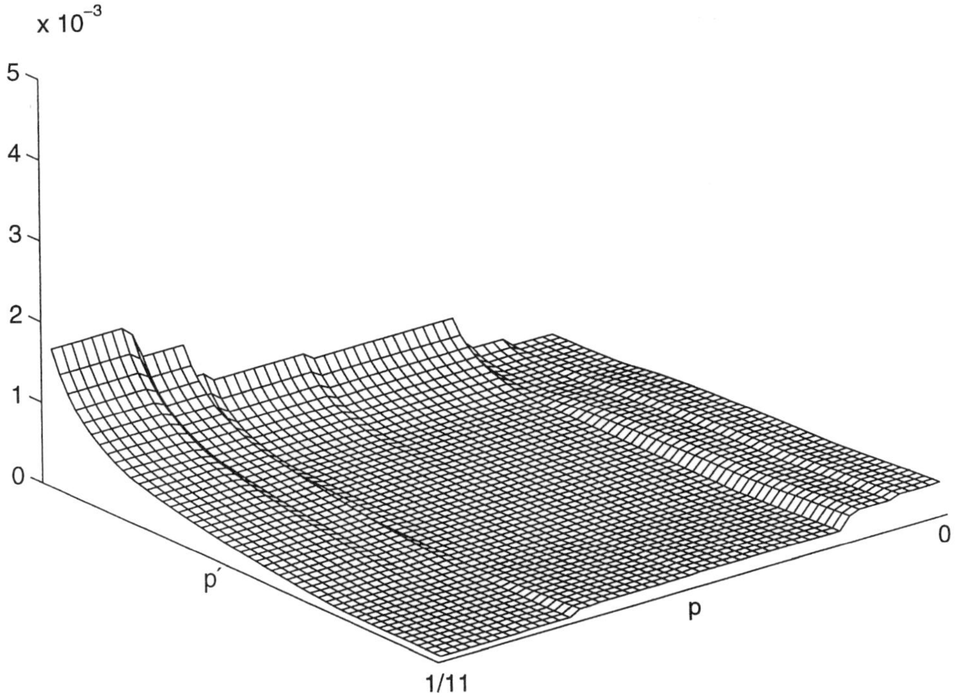

Figure 1–21. Robustness surface for triplex font.

is common across the entire domain. ($\Xi_{\mathbf{a}}$ does not necessarily act independently at each pixel, but the probability of flipping a value is common.) Consequently, for independent sparse noise, $P(\mathbf{u} \to \mathbf{x}) = p$ if $\mathbf{u} \neq \mathbf{x}$ and $P(\mathbf{u} \to \mathbf{u}) = 1 - np$. Note that $0 \leqslant p \leqslant 1/n$. Hence, $P(\mathbf{x}_k \mid \mathbf{x}_{k,i}) = p$ for $k = 0, 1$ and $i = 0, 1, \ldots, n-1$, and $P(\mathbf{x}_k \mid \mathbf{x}_k) = 1 - np$ for $k = 0, 1$. Define the ideal-image parameter

$$\lambda_{\mathbf{x}} = \sum_{i=1}^{n-1} P\left(\mathbf{x}^i\right) \tag{1–115}$$

To specify the window center in the parameter, $\lambda_{\mathbf{x},1}$ and $\lambda_{\mathbf{x},0}$ denote that $x_0 = 1$ and $x_0 = 0$, respectively. Using this parameter, Eqs. 1–111, 1–112, and 1–113 reduce to

$$P(Y = 1 \mid \mathbf{x}_1) = \frac{(1 - np)P(\mathbf{x}_1) + p\lambda_{\mathbf{x},1}}{(1 - np)P(\mathbf{x}_1) + p\lambda_{\mathbf{x},1} + pP(\mathbf{x}_0)} \tag{1–116}$$

$$P(Y = 1 \mid \mathbf{x}_0) = \frac{pP(\mathbf{x}_1)}{pP(\mathbf{x}_1) + (1 - np)P(\mathbf{x}_0) + p\lambda_{\mathbf{x},0}} \tag{1–117}$$

$$\kappa_{\mathbf{x}}(\mathbf{b}; \mathbf{a}) = \left|(1 - np)P(\mathbf{x}) + p\lambda_{\mathbf{x}} - pP\left(\mathbf{x}^0\right)\right|\zeta_{\mathbf{x}}(\mathbf{b}; \mathbf{a}) \tag{1–118}$$

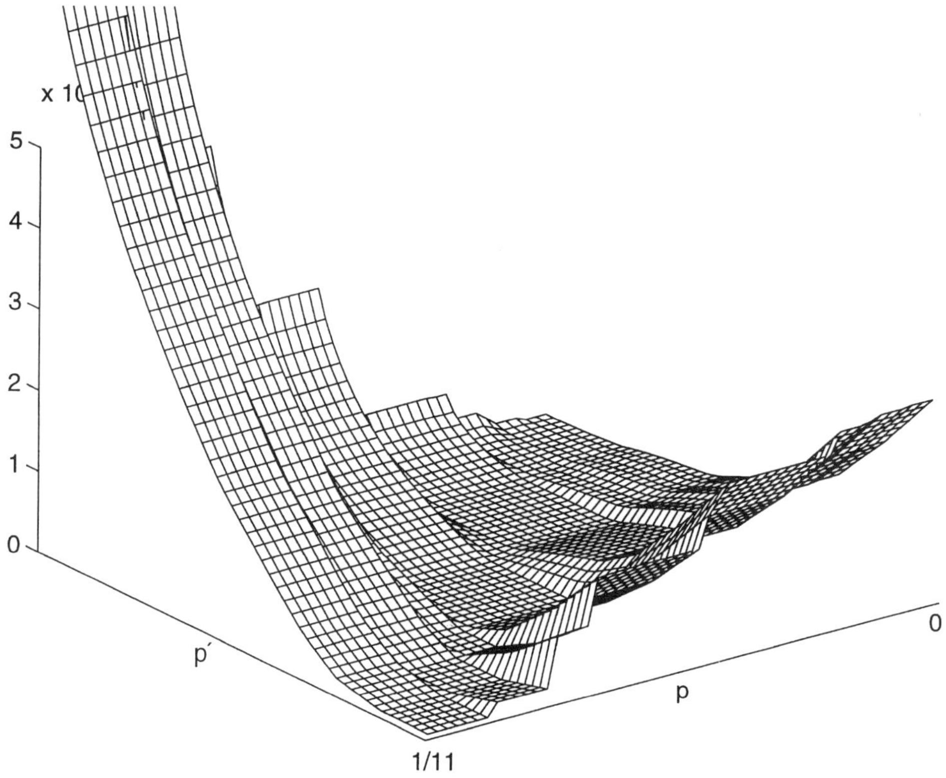

Figure 1–22. Robustness surface for gothic font.

The probabilities and robustness are analytically expressed via ideal-image parameters and p, without explicit reference to the observed process.

For independent sparse noise, robustness $\kappa(p'; p)$ is a function of two scalar variables and is geometrically represented by a *robustness surface* that is zero on the diagonal $p = p'$. For fixed p', $\kappa(p'; p)$ gives the robustness relative to design at p'. $\Psi_{p'}$ is qualitatively robust if the curve of $\kappa(p'; p)$ as a function of p is fairly flat. In terms of two variables, qualitative robustness relates to the flatness of the surface $\kappa(p'; p)$ about the diagonal. Since $\kappa(p'; p)$ only depends on ideal-image parameters and intensity, the actual degradation operator is inconsequential; only the noise intensity matters. There are various constrained random point processes that produce independent sparse noise.

EXAMPLE 1–10. We consider restoration of images degraded by independent sparse noise using a 3×3 window. Robustness for independent sparse noise is essentially analytic because it is computed from Eq. 1–118, in which only 512 probabilities $P(\mathbf{x})$ must be estimated and these can be estimated with great precision. Parts a and b of Fig. 1–20 show realizations of triplex and gothic fonts;

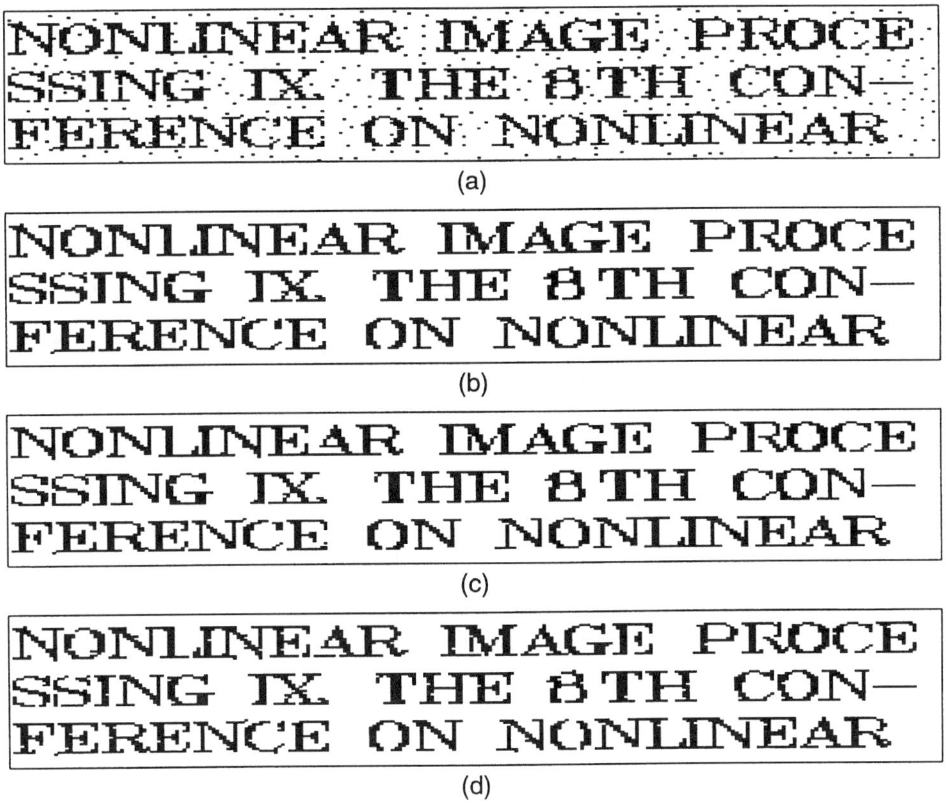

Figure 1–23. Restoration of noisy triplex font: (a) noisy triplex font, $p = 0.030$; (b) restoration by $\Psi_{0.03}$; (c) restoration by $\Psi_{0.06}$; restoration by $\Psi_{0.01}$.

parts c and d show sparse-noise-degraded versions of these realizations for intensity $p = 0.053$. Robustness surfaces for triplex and gothic fonts are shown in Figs. 1–21 and 1–22, respectively, where p values run from 1/11 out to 0 and the p' (design) axis is to the left. Figure 1–23 shows a noisy realization of triplex font for $p = 0.03$ and the results of filtering the realization by the optimal filters for $p' = p$, $p' = 0.06$, and $p' = 0.01$. Input MAE (part a) is 0.030 and MAEs for the filtered images (computed over large realizations) are 0.0123, 0.0129, and 0.0133 for $p' = p$, $p' = 0.06$, and $p' = 0.01$, respectively. Robustness values are

$$\kappa(0.03, 0.06) = 0.0006 \quad \text{and} \quad \kappa(0.03, 0.01) = 0.0010.$$

Figure 1–24 shows a noisy realization of gothic font for $p = 0.03$ and the results of filtering the realization by the optimal filters for $p' = p$, $p' = 0.06$, and $p' = 0.01$. Input MAE is 0.030 and MAEs for the filtered images are 0.0167, 0.0182, and 0.0189 for $p' = p$, $p' = 0.06$, and $p' = 0.01$, respectively,

$$\kappa(0.03, 0.06) = 0.0015 \quad \text{and} \quad \kappa(0.03, 0.01) = 0.0022.$$

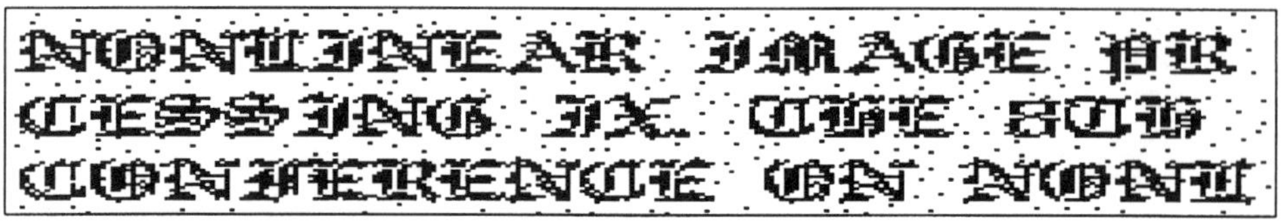

(a)

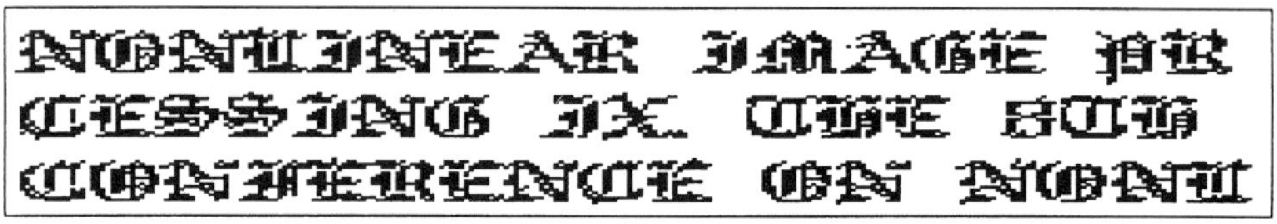

(b)

(c)

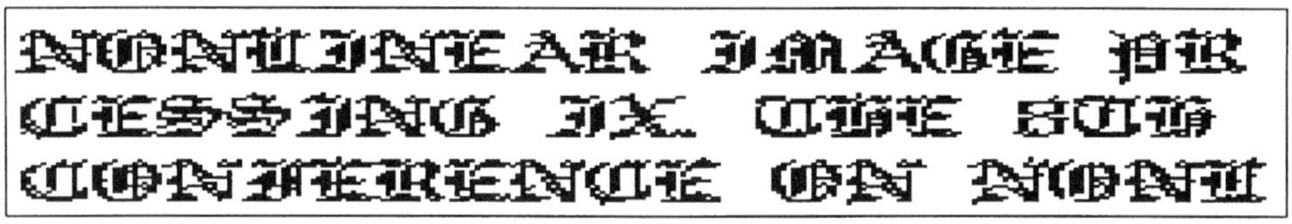

(d)

Figure 1–24. Restoration of noisy gothic font: (a) noisy triplex font, $p = 0.030$; (b) restoration by $\Psi_{0.03}$; (c) restoration by $\Psi_{0.06}$; restoration by $\Psi_{0.01}$.

REFERENCES

[1] Matheron, G., *Random Sets and Integral Geometry*, Wiley, New York, 1975.

[2] Maragos, P., and R. Schafer, "Morphological Filters – Part I: Their Set-Theoretic Analysis and Relations to Linear Shift-Invariant Filters," *IEEE Trans. on Acoustics, Speech, and Signal Processing*, **35**, 1987.

[3] Giardina, C. R., and E. R. Dougherty, *Morphological Methods in Image and Signal Processing*, Prentice-Hall, Englewood Cliffs, 1988.

[4] Dougherty, E. R., and R. M. Haralick, "Unification of Nonlinear Filtering in the Context of Binary Logical Calculus — Part I: Binary Filters," *Mathematical Imaging and Vision*, **2** (2), 1992.

[5] Serra, J., *Image Analysis and Mathematical Morphology*, Academic Press, London, 1982.

[6] Banon, G. J. F., and J. Barrera, "Minimal Representation for Translation Invariant Set Mappings by Mathematical Morphology," *SIAM J. Applied Mathematics*, **51**, 1991.

[7] Goutsias, J., "Morphological Analysis of Discrete Random Shapes," *Mathematical Imaging and Vision*, **2** (2/3), 1992.

[8] Barrera, J., Dougherty, E. R., and N. S. Tomita, "Automatic Programming of Binary Morphological Machines by Design of Statistically Optimal Operators in the Context of Computational Learning Theory," *Electronic Imaging*, **6** (1), 1997.

[9] Dougherty, E. R., and R. P. Loce, "Optimal Binary Differencing Filters: Design, Logic Complexity, Precision Analysis, and Application to Digital Document Processing," *Electronic Imaging*, **5** (1), 1996.

[10] Loce, R. P., and E. R. Dougherty, *Enhancement and Restoration of Digital Documents: Statistical Design of Nonlinear Algorithms*, SPIE Press, Bellingham, 1997.

[11] Dougherty, E. R., and R. P. Loce, "Optimal Mean-Absolute-Error Hit-or-Miss Filters: Morphological Representation and Estimation of the Binary Conditional Expectation," *Optical Engineering*, **32** (4), 1993.

[12] Dougherty, E. R., and R. P. Loce, "Precision of Morphological-Representation Estimators for Translation-Invariant Binary Filters: Increasing and Nonincreasing," *Signal Processing*, **40** (3), 1994.

[13] McCluskey, E. J., " Minimization of Boolean Functions," *Bell System Tech. J.*, **35** (5), 1956.

[14] Quine, W. V., "The Problem of Simplifying Truth Functions," *American Math. Monthly*, **59** (8), 1952.

[15] Barrera, J., Terada, R., Côrrea da Silva, F. S., and N. S. Tomita, "Automatic Programming of Morphological Machines for OCR," *Mathematical Morphology and its Applications to Image and Signal Processing*, Atlanta, 1996.

[16] Schmitt, M., *Des algorithmes morphologiques a l'a intelligence artificielle*, These present a l'Ecole Superieur des Mines de Paris, Fontainebleau, 1989.

[17] Dougherty, E. R., "Optimal Mean-Square *N*-Observation Digital Morphological Filters – Part I: Optimal Binary Filters," *CVGIP: Image Understanding*, **55** (1), 1992.

[18] Loce, R. P., and E. R. Dougherty, "Facilitation of Optimal Binary Morphological Filter Design Via Structuring-Element Libraries and Observation Constraints," *Optical Engineering*, **31** (5), 1992.

[19] Loce, R. P., and E. R. Dougherty, "Optimal Morphological Restoration: The Morphological Filter Mean-Absolute-Error Theorem," *Visual Communication and Image Representation*, **3** (4), 1992.

[20] Dougherty, E. R., and R. P. Loce, "Efficient Design Strategies for the Optimal Binary Digital Morphological Filter: Probabilities, Constraints, and Structuring-Element Libraries," in *Mathematical Morphology in Image Processing*, ed. E. R. Dougherty, Marcel Dekker, New York, 1993.

[21] Mathew, A. V., Dougherty, E. R., and V. Swarnakar, "Efficient Derivation of the Optimal Mean-Square Binary Morphological Filter from the Conditional Expectation via a Switching Algorithm for the Discrete Power-Set Lattice," *Circuits, Systems, and Signal Processing*, **12** (3) 1993.

[22] Han, C. C., and K. C. Fan, "A Greedy and Branch & Bound Searching Algorithm for Finding the Optimal Morphological Filter on Binary Images," *IEEE Signal Processing Letters*, **1**, 1994.

[23] Dougherty, E. R., Zhang, Y., and Y. Chen, "Optimal Iterative Increasing Binary Morphological Filters," *Optical Engineering*, **35** (12), 1996.

[24] Zhang, Y., Loce, R. P., and E. R. Dougherty, "Document Enhancememt Using Optimal Iterative and Paired Morphological Filters," *Proc. SPIE*, **3027**, 1997.

[25] Haussler, D., "Decision Theoretic Generalizations of the PAC Model for Neural Nets and Other Learning Applications," *Information and Computation*, **100**, 1992.

[26] Barrera, J., Tomita, N. S., and F. S. Côrrea da Silva, "Automatic Programming of Morphological Machines by PAC Learning," *Proc. SPIE*, **2568**, 1995.

[27] Valiant, L., "A Theory of the Learnable," *Comm. ACM*, **27**, 1984.

[28] Kong, T. Y., and A. Rosenfeld, "Digital Topology: Introduction and Survey," *CVGIP*, **48**, 1989.

[29] Crimmons, T., and W. Brown, "Image Algebra and Automatic Shape Recognition," *IEEE Trans. Aerospace and Electronic Systems*, **21**, 1985.

[30] Zhao, D., and D. Daut, "Morphological Hit-or-Miss Transformation for Shape Recognition," *Visual Communication and Image Representation*, **2** (3), 1991.

[31] Dougherty, E. R., and D. Zhao, "Model-Based Characterization of Statistically Optimal Design for Morphological Shape Recognition Algorithms via the Hit-or-Miss Transform," *Visual Communication and Image Representation*, **3** (2), 1992.

[32] Barrera, J., Dougherty, E. R., and N. S. Hirata, "Design of Optimal Morphological Operators Using Prior Filters," *Acta Stereologica*, **16** (3), 1998.

[33] Grigoryan, A. M., and E. R. Dougherty, "Robustness of Optimal Binary Filters," *Electronic Imaging*, **7** (1), January, 1998.

CHAPTER 2

COMPUTATIONAL GRAY-SCALE OPERATORS

Edward R. Dougherty
Texas A&M University
College Station, Texas

Junior Barrera
University of Sao Paulo
Sao Paulo, Brazil

This chapter treats representation and optimization of translation-invariant windowed gray-scale digital image operators. The binary operators treated in Chapter 1 form a subclass within the class of operators discussed in the present chapter; however, owing to the special nature of binary operators and their applications, we have chosen to discuss them separately. The operators are termed *computational* because each is defined via a window function with finite domain and range. Historically, both linear and nonlinear operators have typically been considered in the context of continous domain and range, or at least infinite range. Digitization has often been considered as some sort of coarse approximation for the necessities of computation. There is no other choice for linear operators when working with images whose gray values are in a finite subinterval of the natural numbers, since all operators on such images are nonlinear. However, for nonlinear operators one can deal directly with finite-range digital images and we will do so. In fact, we will make much use of the finite nature of the problem for the purposes of both representation and optimization. The computational nonlinear structure has been introduced precisely to achieve these purposes [1].

2.1 COMPUTATIONAL FUNCTIONS

Because windowed gray-scale image operators are defined via window functions, we begin by treating discrete-valued functions defined on vectors of discrete-valued components. For digital image processing with an l-valued input image and an m-valued output image, we consider functions of the form $y = \psi(x_1, x_2, \ldots, x_n)$, where $(x_1, x_2, \ldots, x_n) \in L^n$, $y \in M$, and the value sets are defined by $L = \{0, 1, \ldots, l\}$ and $M = \{0, 1, \ldots, m\}$ for $l, m \geqslant 1$. Letting L^n denote the lattice of n-vectors with values in L, these functions are of the form $\psi\colon L^n \to M$. We denote the space of all such functions by *Fun*$[L^n, M]$, refer to them as *computational functions*, and denote them by lower-case Greek letters.

ψ:

3	2	2	2	3
2	2	2	2	2
1	1	1	2	2
0	1	1	2	2
	0	1	2	3

Figure 2–1. A computational function ψ.

If $l = 1$, input vectors are binary; if $m = 1$, functions are binary-valued. There are $(l+1)^n$ vectors in L^n and $(m+1)^{(l+1)^n}$ functions in $Fun[L^n, M]$. In the original formulation of computational morphology, L and M were normalized so that all values were between 0 and 1 [1]. Figure 2–1 shows a computational function ψ defined from $\{0, 1, 2, 3\}^2$ to $\{0, 1, 2, 3\}$.

Our primary concern is representation of functions in $Fun[L^n, M]$, since it is these from which translation-invariant windowed image operators are constructed. To this end, various elementary operations must be defined. For any two vectors $\mathbf{x} = (x_1, x_2, \ldots, x_n)$, $\mathbf{y} = (y_1, y_2, \ldots, y_n) \in L^n$, partial order is defined by $\mathbf{x} \leqslant \mathbf{y}$ if and only if $x_i \leqslant y_i$ for $i = 1, 2, \ldots, n$. *Maximum* and *minimum* operations between vectors are defined componentwise by

$$\begin{aligned} \mathbf{x} \vee \mathbf{y} &= (x_1 \vee y_1, x_2 \vee y_2, \ldots, x_n \vee y_n) \\ \mathbf{x} \wedge \mathbf{y} &= (x_1 \wedge y_1, x_2 \wedge y_2, \ldots, x_n \wedge y_n) \end{aligned} \tag{2–1}$$

and *quasi-complementation* (also called *negation* or *involution*) is defined by

$$\mathbf{x}' = (x_1, x_2, \ldots, x_n)' = (l - x_1, l - x_2, \ldots, l - x_n) \tag{2–2}$$

For any nonempty subset $\mathcal{A} \subset L^n$, let $\mathcal{A}^-$ denote the set of minimal elements in $\mathcal{A}$. Since L^n is finite, $\mathcal{A}^-$ is nonempty.

For $y = 1, 2, \ldots, m$, function ψ: $L^n \to M$ induces the *level slices* $\mathcal{S}_0[\psi]$, $\mathcal{S}_1[\psi]$, $\ldots$, $\mathcal{S}_m[\psi]$ and the *kernel sets* $\mathcal{K}_1[\psi]$, $\mathcal{K}_2[\psi]$, $\ldots$, $\mathcal{K}_m[\psi]$ defined by

$$\begin{aligned} \mathcal{S}_y[\psi] &= \{\mathbf{x} \in L^n\colon \psi(\mathbf{x}) = y\} \\ \mathcal{K}_y[\psi] &= \{\mathbf{x} \in L^n\colon \psi(\mathbf{x}) \geqslant y\} \end{aligned} \tag{2–3}$$

$\mathcal{S}_0[\psi], \mathcal{S}_1[\psi], \ldots, \mathcal{S}_m[\psi]$ form a partition of L^n and $\mathcal{S}_y[\psi]$ and $\mathcal{K}_y[\psi]$ are related by

$$\mathcal{K}_y[\psi] = \bigcup_{j=y}^{m} \mathcal{S}_j[\psi] \tag{2–4}$$

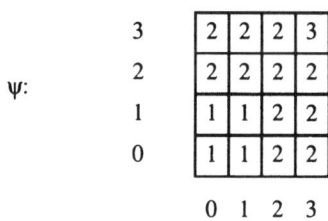

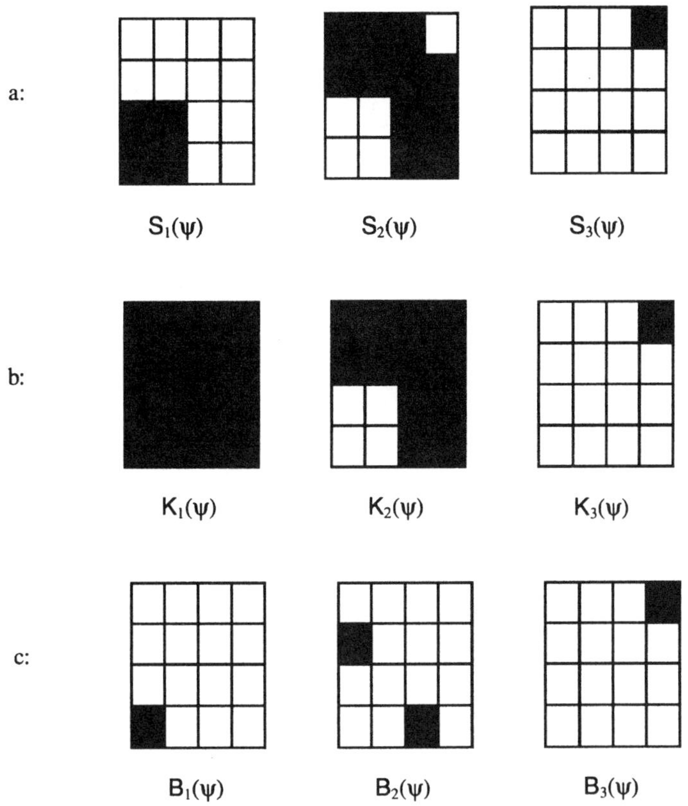

Figure 2–2. The slices (a), kernel (b), and increasing basis (c) of the computational function of Fig. 2–1.

Letting $\mathcal{B}_y[\psi] = \mathcal{K}_y[\psi]^-$ for $y = 1, 2, \ldots, m$,

$$\begin{aligned} \mathcal{K}[\psi] &= \{\mathcal{K}_1[\psi], \mathcal{K}_2[\psi], \ldots, \mathcal{K}_m[\psi]\} \\ \mathcal{B}[\psi] &= \{\mathcal{B}_1[\psi], \mathcal{B}_2[\psi], \ldots, \mathcal{B}_m[\psi]\} \end{aligned} \tag{2–5}$$

are called the *kernel* and *increasing basis* of ψ, respectively. The slices, kernel, and increasing basis of the computational function ψ of Fig. 2–1 are shown in Fig. 2–2.

The basic building block of gray-scale digital operators is *n-elemental erosion*, which is defined, for any $\mathbf{r} \in L^n$, by $\varepsilon_{\mathbf{r}}(\mathbf{x}) = 1$ if $\mathbf{r} \leqslant \mathbf{x}$ and $\varepsilon_{\mathbf{r}}(\mathbf{x}) = 0$ otherwise. With $\mathbf{r}$ assumed fixed and $\mathbf{x}$ varying over L^n, $\varepsilon_{\mathbf{r}}$ is a binary function on L^n with $\mathbf{r}$ playing the role of a *structuring element*. Architecturally, n-elemental erosion can be implemented by comparators. Consider the 1-elemental erosion (also called *elemental erosion*), which, for any $r \in L$, is defined by $\varepsilon_r(x) = 1$ if $r \leqslant x$ and $\varepsilon_r(x) = 0$ otherwise. With r assumed fixed, ε_r is a binary function on L and ε_r is

$\varepsilon_{\mathbf{r}}$

3	0	1	1	1
2	0	1	1	1
1	0	1	1	1
0	0	0	0	0
	0	1	2	3

Figure 2–3. The elemental erosion by the structuring element $\mathbf{r} = (1, 1)$.

logically a comparator. Letting $\mathbf{r} = (r_1, r_2, \ldots, r_n)$,

$$\varepsilon_{\mathbf{r}}(\mathbf{x}) = \bigwedge_{i=1}^{n} \varepsilon_{r_i}(x_i) \tag{2–6}$$

The representation theory of computational functions provides a comparator-based algebra in the context of mathematical morphology [2, 3]. In the mathematical literature, elemental erosion is usually referred to as the *zeta function*. Figure 2–3 shows the 2-elemental erosion $\varepsilon_{(1,1)}$ of the function ψ of Fig. 2–1.

Elemental erosion and, consequently, n-elemental erosion commute with minimum,

$$\varepsilon_r\left(\bigwedge_{i\in I} x_i\right) = \bigwedge_{i\in I} \varepsilon_r(x_i) \tag{2–7}$$

$$\varepsilon_{\mathbf{r}}\left(\bigwedge_{i\in I} \mathbf{x}_i\right) = \bigwedge_{i\in I} \varepsilon_{\mathbf{r}}(\mathbf{x}_i) \tag{2–8}$$

where the index set I can be infinite and in Eq. 2–8 the minimum on the left-hand side (LHS) operates on L-valued vectors and the minimum on the RHS operates on binary numbers. Since elemental erosion and n-elemental erosion commute with minimum and map the maximum element of the lattice domain to the maximum element of the range [$\varepsilon_r(l) = 1$ and $\varepsilon_{\mathbf{r}}(\mathbf{l}) = 1$, where $\mathbf{l} = (l, l, \ldots, l)$ is the maximum element in L^n], they are erosions in the algebraic sense of mathematical morphology on complete lattices [4].

Because there is a quasi-complement operation in L^n, we can define the *dual* of a function ψ: $L^n \to M$ by $\psi^*(\mathbf{x}) = \psi(\mathbf{x}')'$. The dual of n-elemental erosion is *n-elemental dilation*, $\delta_{\mathbf{r}}(\mathbf{x}) = \varepsilon_{\mathbf{r}}(\mathbf{x}')'$, the outer prime denoting Boolean complementation. $\delta_{\mathbf{r}}$ commutes with supremum and $\delta_{\mathbf{r}}(\mathbf{0}) = 0$, where $\mathbf{0} = (0, 0, \ldots, 0)$ is the minimum element in L^n. Hence, n-elemental dilation is a dilation in the sense of mathematical morphology on complete lattices [4]. From Eq. 2–6 and De

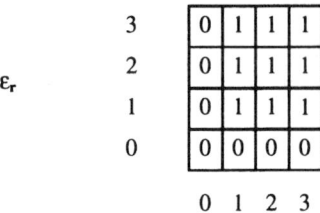

Morgan's law, n-elemental dilation possesses a computational representation as

$$\delta_{\mathbf{r}}(\mathbf{x}) = \bigvee_{i=1}^{n} \varepsilon_{r_i}\left(x_i'\right)' \tag{2–9}$$

2.2 REPRESENTATION OF INCREASING COMPUTATIONAL FUNCTIONS

For binary logical functions, classical logical sum-of-product representation leads to optimization paradigms for binary image operators. Computational representations serve the same role for gray-scale image operators. The computational representation theory can be placed into the framework of complete lattices, but we will not do so, leaving the more abstract setting for the literature [2, 3]. Nonetheless, so that the key role of order is made clear in the representation theory, we will base the theory on the following lemma, which holds in a more general lattice setting (but not for all complete lattices). The subsequent theorem is stated in terms of a summation, but this is a special form of a more general theorem that holds when the range is not totally ordered and that cannot be expressed via summation. Both the lemma and the theorem concern increasing computational functions: a function $\psi\colon L^n \to M$ is increasing if, for any $\mathbf{x}, \mathbf{y} \in L^n$, $\mathbf{x} \leqslant \mathbf{y}$ implies $\psi(\mathbf{x}) \leqslant \psi(\mathbf{y})$.

LEMMA 2–1. *If a function $\psi\colon L^n \to M$ is increasing, then it possesses the representation*

$$\psi(\mathbf{x}) = \bigvee \left\{ y\colon \bigvee_{\mathbf{r} \in \mathcal{K}_y[\psi]} \varepsilon_{\mathbf{r}}(\mathbf{x}) = 1 \right\} \tag{2–10}$$

PROOF. Let $\xi(\mathbf{x})$ denote the function defined by the RHS of Eq. 2–10 and suppose $\psi(\mathbf{x}) = z$. Since $\mathbf{x} \in \mathcal{K}_z[\psi]$ and $\varepsilon_{\mathbf{x}}(\mathbf{x}) = 1$, z is an element of the set over which the outer maximum is being taken and therefore $\xi(\mathbf{x}) \geqslant \psi(\mathbf{x})$. Suppose $\xi(\mathbf{x})$ is strictly greater than $\psi(\mathbf{x})$. Then there must exist y and $\mathbf{r} \in \mathcal{K}_y[\psi]$ such that $\mathbf{r} \leqslant \mathbf{x}$ and $y > \psi(\mathbf{x})$. But $\mathbf{r} \in \mathcal{K}_y[\psi]$ means that $\psi(\mathbf{r}) \geqslant y$. Thus, $\mathbf{r} \leqslant \mathbf{x}$ and $\psi(\mathbf{r}) \geqslant y > \psi(\mathbf{x})$, which contradicts the inreasingness of ψ. ■

THEOREM 2–1. *If $\psi\colon L^n \to M$ is increasing, then it possesses morphological representation*

$$\psi(\mathbf{x}) = \sum_{y=1}^{m} \bigvee_{\mathbf{r} \in \mathcal{K}_y[\psi]} \varepsilon_{\mathbf{r}}(\mathbf{x}) \tag{2–11}$$

PROOF. Let $\eta(\mathbf{x})$ denote the function defined by the RHS of Eq. 2–11 and A be the set of all y over which the outer maximum of Eq. 2–10 is taken. If $k \leqslant y$, then $y \in A$ implies $k \in A$, since $\mathcal{K}_y[\psi] \subset \mathcal{K}_k[\psi]$, and therefore $\eta(\mathbf{x}) \geqslant \psi(\mathbf{x})$. In the other direction, if $\eta(\mathbf{x}) = z$, then there exists $\mathbf{r} \in \mathcal{K}_z[\psi]$ such that $\varepsilon_{\mathbf{r}}(\mathbf{x}) = 1$, which means that $z \in A$ and $z \leqslant \psi(\mathbf{x})$. ∎

In terms of elemental erosions, Eq. 2–11 takes the form

$$\psi(\mathbf{x}) = \sum_{y=1}^{m} \bigvee_{\mathbf{r}\in\mathcal{K}_y[\psi]} \bigwedge_{i=1}^{n} \varepsilon_{r_i}(x_i) \tag{2–12}$$

For binary images, $l = 1$, $m = 1$, and Eq. 2–12 reduces to

$$\psi(\mathbf{x}) = \bigvee_{\mathbf{r}\in\mathcal{K}_1[\psi]} \bigwedge_{i=1}^{n} \varepsilon_{r_i}(x_i). \tag{2–13}$$

Let $\mathcal{K}_1[\psi] = \{\mathbf{r}_1, \mathbf{r}_2, \ldots, \mathbf{r}_p\}$ and for $\mathbf{r}_j \in \mathcal{K}_1[\psi]$ let its 1-valued components be labeled $r_{j1}, r_{j2}, \ldots, r_{jn(j)}$. Then the preceding equation can be rewritten as

$$\psi(\mathbf{x}) = \bigvee_{j=1}^{p} \bigwedge_{i=1}^{n(j)} \varepsilon_{r_{ji}}(x_{ji}) = \bigvee_{j=1}^{p} \bigwedge_{i=1}^{n(j)} x_{ji} \tag{2–14}$$

which is a positive logical (sum-of-product) expansion. Hence, the representations of Eqs. 2–11 and 2–12 are morphological and computational (comparator-based) extensions, respectively, of the positive logical representation of an increasing Boolean operator.

Relative to Eqs. 2–11 and 2–12, minimal representation results from maximally reducing the sets of vectors over which the maxima are taken to eliminate redundancy. For minimal representation, the maxima must be taken over sets of vectors that yield the same function and for which no further reduction is possible. As stated by the next theorem, the increasing basis provides minimal representation.

THEOREM 2–2. *If function ψ: $L^n \to M$ is increasing, then it has minimal representation*

$$\psi(\mathbf{x}) = \sum_{y=1}^{m} \bigvee_{\mathbf{r}\in\mathcal{B}_y[\psi]} \varepsilon_{\mathbf{r}}(\mathbf{x}) \tag{2–15}$$

PROOF. First we show that the reduced representation gives $\psi(\mathbf{x})$. Suppose the maximum of Eq. 2–11 is 1-valued for y. Then there exists $\mathbf{r} \in \mathcal{K}_y[\psi]$ such that

$$1: \quad \begin{array}{|c|c|c|c|}\hline 1&1&1&1\\\hline 1&1&1&1\\\hline 1&1&1&1\\\hline 1&1&1&1\\\hline\end{array}$$

$$+$$

$$2: \quad \begin{array}{|c|c|c|c|}\hline 1&1&1&1\\\hline 1&1&1&1\\\hline 0&0&1&1\\\hline 0&0&1&1\\\hline\end{array} = \begin{array}{|c|c|c|c|}\hline 1&1&1&1\\\hline 1&1&1&1\\\hline 0&0&0&0\\\hline 0&0&0&0\\\hline\end{array} \vee \begin{array}{|c|c|c|c|}\hline 0&0&1&1\\\hline 0&0&1&1\\\hline 0&0&1&1\\\hline 0&0&1&1\\\hline\end{array}$$

$$+$$

$$3: \quad \begin{array}{|c|c|c|c|}\hline 0&0&0&1\\\hline 0&0&0&0\\\hline 0&0&0&0\\\hline 0&0&0&0\\\hline\end{array}$$

$$=$$

$$\psi: \quad \begin{array}{|c|c|c|c|}\hline 2&2&2&3\\\hline 2&2&2&2\\\hline 1&1&2&2\\\hline 1&1&2&2\\\hline\end{array}$$

Figure 2–4. Minimal representation of the computational function ψ of Fig. 2–1.

$\varepsilon_{\mathbf{r}}(\mathbf{x}) = 1$. There exists $\mathbf{s} \in \mathcal{B}_y[\psi]$ such that $\mathbf{s} \leqslant \mathbf{r}$. Moreover, $\varepsilon_{\mathbf{s}}(\mathbf{x}) = 1$ since $\mathbf{s} \leqslant \mathbf{r} \leqslant \mathbf{x}$. Hence, the function defined by the RHS of Eq. 2–11 is bounded by the function defined by the RHS of Eq. 2–15. Since the maxima on the RHS of Eq. 2–15 are over subsets of the sets over which the maxima of Eq. 2–11 are taken, the reverse inequality must hold and the two RHS functions are equal. To show that further reduction is impossible, suppose $\mathcal{B}_y$ is a proper subset of $\mathcal{B}_y[\psi]$, $\mathbf{x} \in \mathcal{B}_y[\psi] - \mathcal{B}_y$, and $\mathcal{B}_y[\psi]$ is replaced by $\mathcal{B}_y$ in the maximum of Eq. 2–15 to form the function $\xi(\mathbf{x})$ on the RHS. Then $\xi(\mathbf{x}) < \psi(\mathbf{x})$ because, for all $\mathbf{r} \in \mathcal{B}_y$, $\varepsilon_{\mathbf{r}}(\mathbf{x}) = 0$ and thus

$$\bigvee_{\mathbf{r} \in \mathcal{B}_y} \varepsilon_{\mathbf{r}}(\mathbf{x}) = 0 \qquad \blacksquare \qquad (2\text{–}16)$$

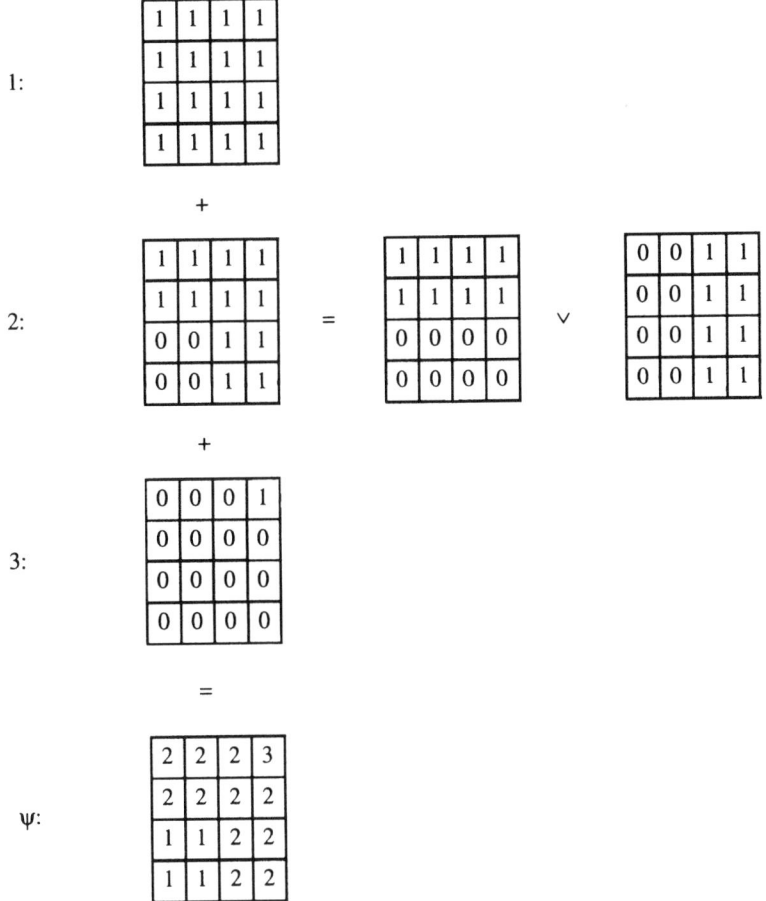

In terms of elemental erosions, Eq. 2–15 takes the form

$$\psi(\mathbf{x}) = \sum_{y=1}^{m} \bigvee_{\mathbf{r}\in\mathcal{B}_y[\psi]} \bigwedge_{i=1}^{n} \varepsilon_{r_i}(x_i) \tag{2–17}$$

In the binary case, Eq. 2–14 again results; however, this time it is the minimal positive Boolean expansion, so that the representations of Eqs. 2–15 and 2–17 are morpholgoical and computational extensions, respectively, of the minimal positive logical representation of an increasing Boolean operator. Figure 2–4 shows the minimal decomposition of the computational function ψ presented in Fig. 2–1.

Whereas Theorem 2–1 is a special case of a more general lattice theorem applying to functions $\psi\colon L^n \to M$, the minimal representation of Theorem 2–2 can also be derived from a more general lattice result, which is the minimal form of Lemma 2–1 given by

$$\psi(\mathbf{x}) = \bigvee \left\{ y\colon \bigvee_{\mathbf{r}\in\mathcal{B}_y[\psi]} \varepsilon_{\mathbf{r}}(\mathbf{x}) = 1 \right\} \tag{2–18}$$

Theorems 2–1 and 2–2 provide necessary representations for increasing computational functions. Their form provides a sufficient representation in the sense that for any class of sets $\mathcal{K}_1, \mathcal{K}_2, \ldots, \mathcal{K}_m \subset L^n$, the operator defined by

$$\psi(\mathbf{x}) = \sum_{y=1}^{m} \bigvee_{\mathbf{r}\in\mathcal{K}_y} \varepsilon_{\mathbf{r}}(\mathbf{x}) \tag{2–19}$$

is increasing because elemental erosion is increasing. However, a distinquishing fact of Eq. 2–11, and why it is equivalent to Eq. 2–10, is that unless $\psi(\mathbf{x}) = 0$, the maximum is 1-valued for $y = 1, 2, \ldots, j = \psi(\mathbf{x})$ and 0-valued thereafter. Such is not necessarily the case for Eq. 2–19. This situation can be remedied by requiring that $\mathcal{K}_1 \supset \mathcal{K}_2 \supset \cdots \supset \mathcal{K}_m$, but this does not really address the issue correctly since no such inclusion relation holds for the minimal representation of Eq. 2–15. Instead, we make the lesser requirement that $\mathcal{K}_1, \mathcal{K}_2, \ldots, \mathcal{K}_m$ have the following property: if $\mathbf{r} \in \mathcal{K}_y$ and $z \leqslant y$, then there exists $\mathbf{s} \in \mathcal{K}_z$ such that $\mathbf{s} \leqslant \mathbf{r}$. Under this condition, which we will term *consistency*, if the maximum is 1-valued for y, then it is 1-valued for all $z \leqslant y$.

Requiring $\mathcal{K}_1, \mathcal{K}_2, \ldots, \mathcal{K}_m$ to be consistent does not make them nonredundant. In fact, $\mathcal{K}_1, \mathcal{K}_2, \ldots, \mathcal{K}_m$ form an increasing basis for the operator they define via Eq. 2–19 if and only if they are consistent and each is *self-minimal*, meaning that $\mathcal{K}_y = \mathcal{K}_y^-$, or, equivalently, there does not exist y and $\mathbf{r}, \mathbf{s} \in \mathcal{K}_y$ such that $\mathbf{r}$ is strictly less than $\mathbf{s}$. Filter design involves finding a self-minimal, consistent class of sets that determines a good-performing filter.

2.3 FLAT COMPUTATIONAL FUNCTIONS

The most basic computational function is *flat erosion* (*minimum*). To define it, note that each vector in M^n is indexed over the set $N = \{1, 2, \ldots, n\}$. For any subset $B \subset N$, flat erosion with structuring element B is the function ε_B: $M^n \to M$ defined by

$$\varepsilon_B(x_1, x_2, \ldots, x_n) = \bigwedge_{i \in B} x_i \tag{2–20}$$

Its kernel sets are given by

$$\mathcal{K}_y[\varepsilon_B] = \{(x_1, x_2, \ldots, x_n) \in M^n\colon x_i \geqslant y \text{ for } i \in B\} \tag{2–21}$$

Define the indicator function for B by $I_B(i) = 1$ if $i \in B$ and $I_B(i) = 0$ if $i \notin B$, and the indicator vector for B by $\mathbf{I}_B = (I_B(1), I_B(2), \ldots, I_B(n))$. Then $\mathcal{K}_y[\varepsilon_B] = \{\mathbf{x} \in M^n\colon \mathbf{x} \geqslant y\mathbf{I}_B\}$ and $\mathcal{B}_y[\varepsilon_B] = \{y\mathbf{I}_B\}$. By Theorem 2–2, flat erosion possesses the representation

$$\varepsilon_B(\mathbf{x}) = \sum_{y=1}^{m} \varepsilon_{y\mathbf{I}_B}(\mathbf{x}) = \sum_{y=1}^{m} \bigwedge_{i=1}^{n} \varepsilon_{yI_B(i)}(x_i) \tag{2–22}$$

If $B_1, B_2, \ldots, B_p \subset N$, the *flat function* corresponding to the class $\mathcal{B} = \{B_1, B_2, \ldots, B_p\}$ is $\xi_{\mathcal{B}}$: $M^n \to M$ defined by

$$\xi_{\mathcal{B}}(\mathbf{x}) = \bigvee_{j=1}^{p} \varepsilon_{B_j}(\mathbf{x}) = \bigvee_{i=1}^{p} \bigwedge_{i \in B_j} x_i \tag{2–23}$$

If $B_j \subset B_k$, then B_k can be dropped from $\mathcal{B}$ without changing the flat function $\xi_{\mathcal{B}}$. On the other hand, if B_k does not contain a subset among the other sets in $\mathcal{B}$, then B_k cannot be dropped without changing $\xi_{\mathcal{B}}$. $\mathcal{B}$ provides minimal representation according to Eq. 2–23 for the flat function ξ if $\xi = \xi_{\mathcal{B}}$ and no two sets in $\mathcal{B}$ are properly related by subsethood. We will assume that flat-function representations are minimal.

The median function over M^n is an example of a flat function. Assuming n odd, minimal representation for the median $\mu_{\mathcal{B}}$ is given by the class $\mathcal{B}$ of subsets $B_1, B_2, \ldots, B_p \subset N$ of size $(n+1)/2$. To see this, one need only recognize two points: first, there exists a set B_j indexing the $(n+1)/2$ largest values among the components of any vector in M^n, so that the minimum value in B_j is the median; second, for any $k \neq j$, the minimum value over B_k is less than or equal to the minimum value over B_j. Figure 2–5 shows the increasing basis of the median defined on $\{0, 1, 2, 3\}^3$.

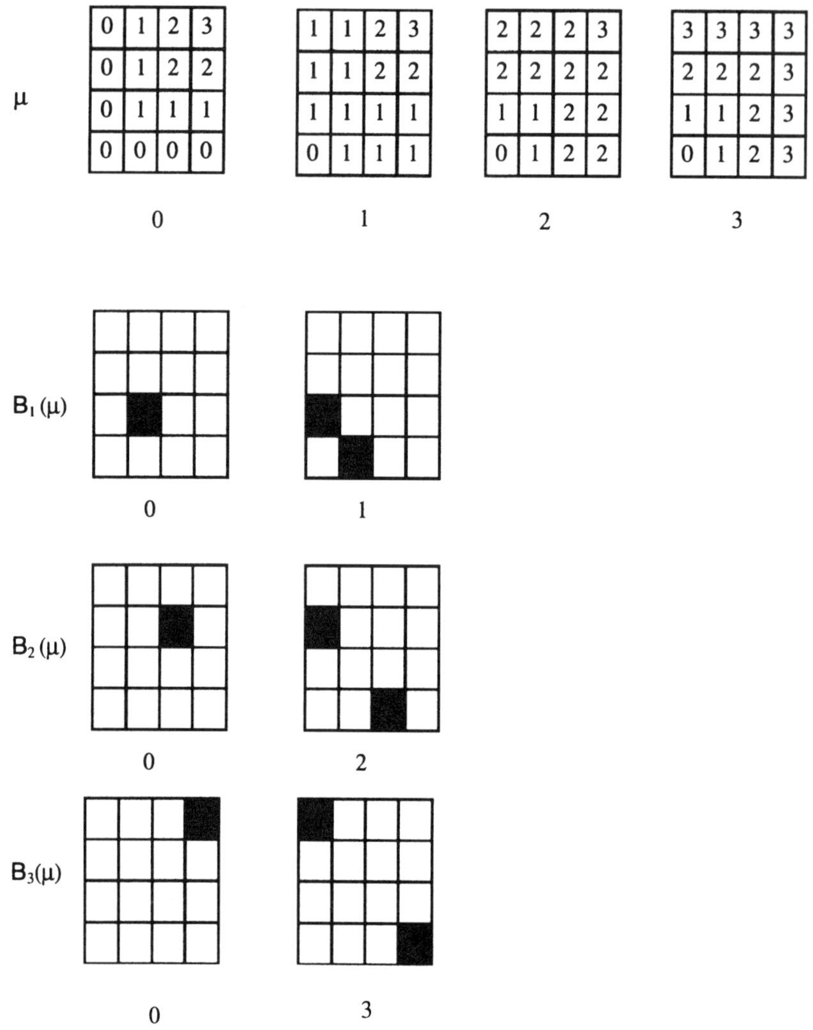

Figure 2–5. Increasing basis of the median on $\{0, 1, 2, 3\}^3$.

Returning to a general flat function $\xi_{\mathcal{B}}$, the basis sets for $\xi_{\mathcal{B}}$ are given by

$$\mathcal{B}_y[\xi_{\mathcal{B}}] = \{yI_{B_1}, yI_{B_2}, \ldots, yI_{B_p}\} \tag{2–24}$$

with each $\mathcal{B}_y[\xi_{\mathcal{B}}]$ composed of p scalar-multiplied indicator vectors. By Theorem 2–2,

$$\xi_{\mathcal{B}}(\mathbf{x}) = \sum_{y=1}^{m} \bigvee_{j=1}^{p} \varepsilon_{yI_{B_j}}(\mathbf{x}) = \sum_{y=1}^{m} \bigvee_{j=1}^{p} \bigwedge_{i=1}^{n} \varepsilon_{yI_{B_j}(i)}(x_i) \tag{2–25}$$

As seen from either the original formulation or the elemental erosion representation, a flat function is determined once the class $\mathcal{B}$ is specified.

Corresponding to flat erosion is *flat dilation*, defined for a subset $B \subset N$ by

$$\delta_B(\mathbf{x}) = \bigvee_{i \in B} x_i \tag{2–26}$$

The kernel sets, basis sets, and computational representation for flat dilation are given by

$$\mathcal{K}_y[\delta_B] = \{(x_1, x_2, \ldots, x_n) \in M^n\text{: there exists } i \text{ for which } x_i \geqslant y\} \tag{2–27}$$

$$\mathcal{B}_y[\delta_B] = \{(yI_B(1), 0, \ldots, 0), (0, yI_B(2), 0, \ldots, 0), \ldots, (0, \ldots, 0, yI_B(n))\} \tag{2–28}$$

$$\delta_B(\mathbf{x}) = \sum_{y=1}^{m} \bigvee_{i=1}^{n} \varepsilon_{yI_B(i)}(x_i) \tag{2–29}$$

2.4 INCREASING GRAY-TO-BINARY IMAGE OPERATORS

Each function ψ: $L^n \to M$ serves as a computational window function for an image operator Ψ. To formalize this relationship and apply the computational representations to image operators, let $Fun[Z^d, M]$ denote the space of images (functions) defined on Z^d, d-dimensional discrete Cartesian space, taking values in M. An image operator is a mapping Ψ: $Fun[Z^d, L] \to Fun[Z^d, M]$, the gray scales of the input and output images being L and M, respectively, and not necessarily equal. Ψ is defined by the computational function ψ over window $W = \{w_1, w_2, \ldots, w_n\}$ if, for any image $f \in Fun[Z^d, L]$, $\Psi(f)$ evaluated at pixel z is given by

$$\Psi(f)(z) = \psi[f(z + w_1), f(z + w_2), \ldots, f(z + w_n)] \tag{2–30}$$

In this case we say that Ψ is a *W-operator* with window (computational) function ψ. Because the same computational function is used at every pixel, Ψ is translation invariant. Characterization of ψ yields a characterization of Ψ, and conversely. We denote image operators by capital Greek letters Ψ, Φ, $\Lambda, \ldots$ and their corresponding window functions by the corresponding lower-case Greek letters $\psi, \phi, \lambda, \ldots$.

In the present section we restrict our attention to *W*-operators taking gray-scale images as inputs and having binary output images. Thus, Ψ: $Fun[Z^d, L] \to Fun[Z^d, \{0, 1\}]$. For such operators, definitions of the original binary morphological operators go over essentially unchanged and computational representations are straightforward.

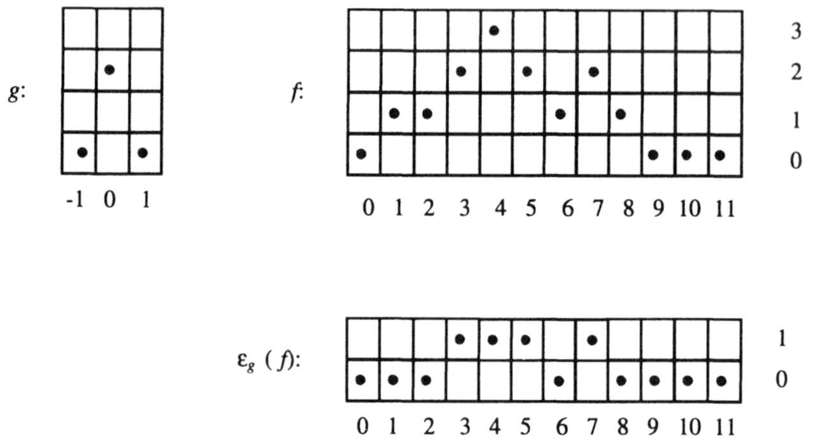

Figure 2–6. Gray to binary erosion by g.

For $g \in Fun[Z^d, L]$, *erosion* by *structuring element (function)* g is the image operator E_g: $Fun[Z^d, L] \to Fun[Z^d, \{0, 1\}]$ defined by

$$E_g(f)(z) = \begin{cases} 1, & \text{if } g_z \leqslant f \\ 0, & \text{otherwise} \end{cases} \tag{2–31}$$

where the translate g_z is defined by $g_z(y) = g(y - z)$. Because we desire all operations to be computational, we assume that g has finite support, supp(g), meaning that g is only nonzero on a finite subset of Z^d. As defined, erosion is a gray-to-binary image operator and its definition is fully analogous to binary-to-binary erosion. It is translation-invariant, increasing, and is a lattice erosion because it commutes with infimum and erosion by g of the maximal function in $Fun[Z^d, L]$, the function being identically l-valued, gives the maximal function in $Fun[Z^d, \{0, 1\}]$, the function being identically 1-valued.

If supp$(g) = \{v_1, v_2, \ldots, v_n\}$ and we let

$$\mathbf{g} = \big(g(v_1), g(v_2), \ldots, g(v_n)\big) \tag{2–32}$$
$$\mathbf{f}(z) = \big(f(z + v_1), f(z + v_2), \ldots, f(z + v_n)\big) \tag{2–33}$$

then, from Eq. 2–6, we see that $E_g(f)$ has the computational representation

$$E_g(f)(z) = \varepsilon_{\mathbf{g}}\big(\mathbf{f}(z)\big) = \bigwedge_{i=1}^{n} \varepsilon_{g(v_i)}\big(f(z + v_i)\big) \tag{2–34}$$

The minimum gives the expression for the computational function and the window is $W = \text{supp}(g)$. This erosion is illustrated for a one-dimensional signal in Fig. 2–6.

For $g \in Fun[Z^d, L]$ possessing finite support, *dilation* by *structuring element (function)* g is the gray-to-binary image operator

$$\Delta_g\colon Fun[Z^d, L] \to Fun[Z^d, \{0, 1\}]$$

defined by $\Delta_g(f) = \mathrm{E}_{\check{g}}(f')'$, where $f'(z) = l - f(z)$ and $\check{g}(z) = g(-z)$. Dilation is translation-invariant, increasing, and a lattice dilation [commutes with supremum and $\Delta_g(0) = 0$]. It has computational representation

$$\Delta_g(f)(z) = \bigvee_{i=1}^{n} \varepsilon_{g(-v_i)}\big(f(z - v_i)'\big)' \tag{2–35}$$

and the window is $W = \mathrm{supp}(\check{g})$.

In analogy to binary logical opening, for any $g \in Fun[Z^d, L]$ having finite support, *opening* by *structuring element* g is the image operator $\Gamma_g\colon Fun[Z^d, L] \to Fun[Z^d, \{0, 1\}]$ defined by $\Gamma_g = \Delta_g \mathrm{E}_g$. The next proposition gives the fitting characterization.

PROPOSITION 2–1. *Gray-to-binary opening possesses the structural characterization*

$$\Gamma_g(f)(z) = \begin{cases} 1, & \text{if there exists } u \text{ such that } z \in \mathrm{supp}(g_u) \text{ and } g_u \leqslant f \\ 0, & \text{otherwise} \end{cases} \tag{2–36}$$

PROOF. $\Gamma_g(f)(z) = 1$ if and only if $\mathrm{E}_{\check{g}}(\mathrm{E}_g(f)')(z) = 0$ if and only if $\check{g}_z$ is not less than or equal to $\mathrm{E}_g(f)'$ if and only if there exists $u \in \mathrm{supp}(\check{g})_z$ such that $\mathrm{E}_g(f)'(u) = 0$ if and only if there exists $u \in \mathrm{supp}(\check{g})_z$ such that $g_u \leqslant f$. The result follows by recognizing that $(\check{g})_z(u) = g_u(z)$ so that $u \in \mathrm{supp}(\check{g})_z$ if and only if $z \in \mathrm{supp}(g_u)$. ■

To obtain the computational representation of opening, let $\mathbf{g}$ be given by Eq. 2–32 and

$$\check{\mathbf{g}} = \big(g(-v_1), g(-v_2), \ldots, g(-v_n)\big) \tag{2–37}$$

Using the iterative definition of opening in conjunction with De Morgan's law yields

$$\begin{aligned} \Gamma_g(f)(z) &= \varepsilon_{\check{\mathbf{g}}}\big[\big(\mathrm{E}_g(f)(z+v_1)', \mathrm{E}_g(f)(z+v_2)', \ldots, \mathrm{E}_g(f)(z+v_n)'\big)\big]' \\ &= \left(\bigwedge_{i=1}^{n} \varepsilon_{g(-v_i)}\big(\mathrm{E}_g(f)(z+v_i)'\big)\right)' \end{aligned}$$

$$= \left(\bigwedge_{i=1}^{n} \varepsilon_{g(-v_i)} \left(\bigwedge_{j=1}^{n} \varepsilon_{g(v_j)}\big(f(z+v_i+v_j)\big) \right)' \right)'$$

$$= \bigvee_{i=1}^{n} \varepsilon_{g(-v_i)} \left(\bigvee_{j=1}^{n} \varepsilon_{g(v_j)}\big(f(z+v_i+v_j)\big)' \right)' \qquad (2\text{–}38)$$

The last expression is the computational function for opening and $W = \mathrm{supp}(g) \oplus \mathrm{supp}(g)$, where $\oplus$ denotes Minkowski addition.

The expression of Eq. 2–38 simplifies greatly if supp(g) is symmetric about the origin. Then $g(-v_i) > 0$ for $i = 1, 2, \ldots, n$ and the outer elemental erosion is 1-valued if and only if the inner elemental erosion is 0-valued for some j, which means that the complement of the outer elemental erosion is 1-valued if and only if the inner elemental erosion is 1-valued for all j. Hence, for symmetrically supported structuring elements,

$$\Gamma_g(f)(z) = \bigvee_{i=1}^{n} \bigwedge_{j=1}^{n} \varepsilon_{g(v_j)}\big(f(z+v_i+v_j)\big) \qquad (2\text{–}39)$$

The inner minimum is an erosion. Replacing it by its equivalent erosion yields

$$\Gamma_g(f) = \bigvee_{i=1}^{n} \mathrm{E}_g(f_{v_i}) = \bigvee_{i=1}^{n} \mathrm{E}_{g_{-v_i}}(f) = \bigvee_{\{g_v:\ 0 \in \mathrm{supp}(g_v)\}} \mathrm{E}_{g_v}(f) \qquad (2\text{–}40)$$

The opening is formed by eroding by all translates of the structuring element whose supports contain the origin and then taking the maximum.

2.5 Representation of Increasing Gray-to-Gray Image Operators

Elemental erosion, which yields a binary output, is crucial to representation of increasing computational functions of the form ψ: $L^n \to M$; gray-to-binary erosion plays a key role in representing increasing gray-to-gray W-operators Ψ: $Fun[Z^d, L] \to Fun[Z^d, M]$. For any image operator Ψ (not necessarily increasing) with window function ψ and for $y = 1, 2, \ldots, m$, we define the sets

$$\mathcal{K}_y[\Psi] = \big\{g \in Fun[Z^d, L]\text{: } \mathrm{supp}(g) \subset W \text{ and } \Psi(g)(0) \geqslant y\big\} \qquad (2\text{–}41)$$

$\mathcal{B}_y[\Psi] = \mathcal{K}_y[\Psi]^-$ is the set of minimal elements in $\mathcal{K}_y[\Psi]$. The *kernel* and *increasing basis* of Ψ are defined by

$$\begin{aligned} \mathcal{K}[\Psi] &= \big\{\mathcal{K}_1[\Psi], \mathcal{K}_2[\Psi], \ldots, \mathcal{K}_m[\Psi]\big\} \\ \mathcal{B}[\Psi] &= \big\{\mathcal{B}_1[\Psi], \mathcal{B}_2[\Psi], \ldots, \mathcal{B}_m[\Psi]\big\} \end{aligned} \qquad (2\text{–}42)$$

respectively. The relationship between $\mathcal{K}_y[\Psi]$ and $\mathcal{K}_y[\psi]$ is straightforward: $g \in \mathcal{K}_y[\Psi]$ if and only if $\mathbf{g} \in \mathcal{K}_y[\psi]$, where $\mathbf{g}$ is the vector composed of the values of g in W. More specifically, $\Psi(g)(0) = \psi(\mathbf{g})$.

We state the fundamental representation in minimal (basis) form; it holds also in kernel form.

THEOREM 2–3. *If* Ψ: *Fun*$[Z^d, L] \rightarrow$ *Fun*$[Z^d, M]$ *is an increasing* W*-operator, then*

$$\Psi(f) = \sum_{y=1}^{m} \bigvee_{g \in \mathcal{B}_y[\Psi]} \mathrm{E}_g(f) \tag{2–43}$$

PROOF. Let $W = \{w_1, w_2, \ldots, w_n\}$ and ψ be the computational function for Ψ. Then Eq. 2–30 in conjunction with Theorem 2–2 yields

$$\Psi(f)(z) = \sum_{y=1}^{m} \bigvee_{g \in \mathcal{B}_y[\psi]} \bigwedge_{i=1}^{n} \varepsilon_{g_i}\big(f(z + w_i)\big) \tag{2–44}$$

For each vector $\mathbf{g} \in L^n$, define the function g on W by $g(w_i) = g_i$. The relation $\mathbf{g} \leftrightarrow g$ is a one-to-one, order-preserving mapping between L^n and the images of *Fun*$[Z^d, L]$ whose supports lie in W. Hence,

$$\Psi(f)(z) = \sum_{y=1}^{m} \bigvee_{g \in \mathcal{B}_y[\Psi]} \bigwedge_{i=1}^{n} \varepsilon_{g(w_i)}\big(f(z + w_i)\big) \tag{2–45}$$

which, according to Eq. 2–34, is the representation of Eq. 2–43. ■

Figure 2–7 demonstrates the application of Theorem 2–3 to the operator Ψ defined by Fig. 2–1 to a signal f with $W = \{0, 1\}$.

According to Eq. 2–41, $g \in \mathcal{K}_y[\Psi]$ if and only if $\mathrm{supp}(g) \subset W$ and $\psi(\mathbf{g}) \geqslant 0$. A slightly different, but equivalent, view can be taken without requiring $\mathrm{supp}(g) \subset W$; rather, we can consider any function g: $Z^d \rightarrow L$ and define $g \in \mathcal{K}_y[\Psi]$ if and only if $\Psi(g|_W)(0) \geqslant 0$, where $g|_W$ is the restriction of g to W defined by $g|_W(x) = g(x)$ if $x \in W$ and $g|_W(x) = 0$ if $x \notin W$. For this interpretation, $\mathbf{g}$ is still the vector composed of the values of g in W and $\Psi(g|_W)(0) = \psi(\mathbf{g})$.

2.6 STACK FILTERS

The most applied nonlinear gray-scale operators are stack (flat) filters. The simplest stack filter is flat erosion [5]. For any subset $B \subset W$, *flat erosion* is the gray-to-gray

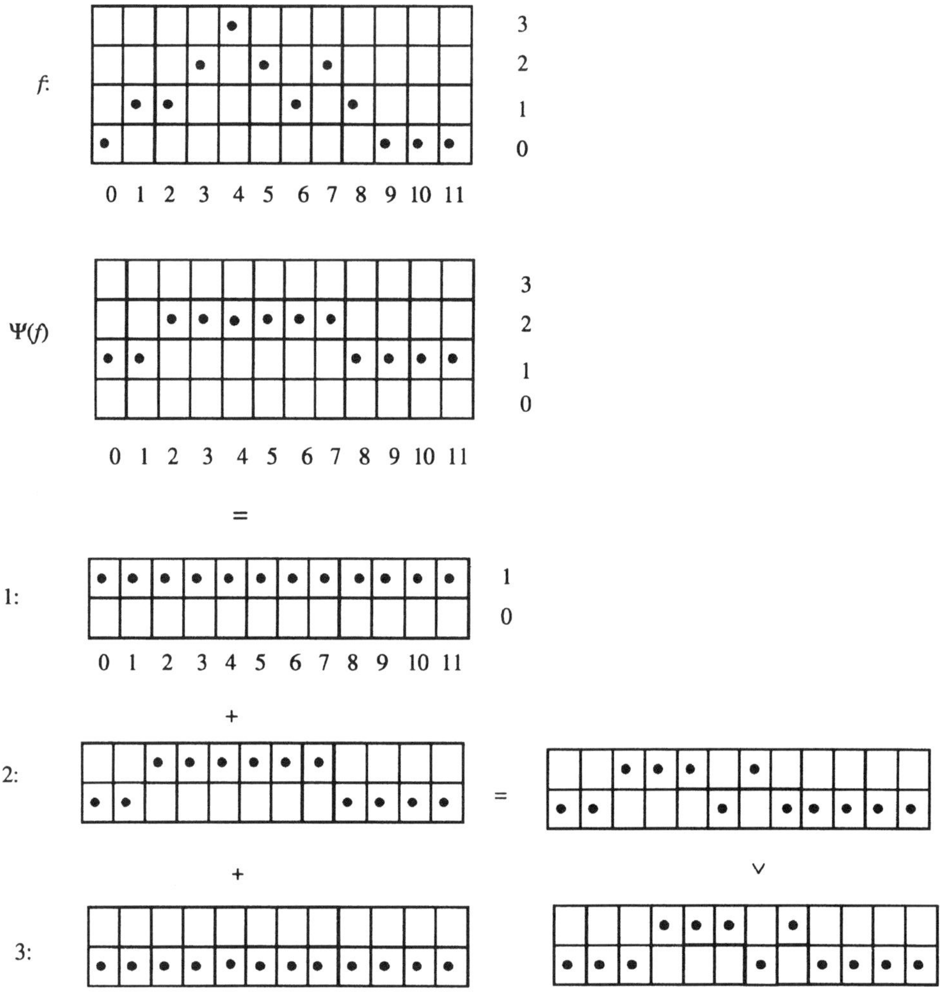

Figure 2–7. Representation of the increasing image operator Ψ of Fig. 2–1 with $W = \{0, 1\}$.

image operator E_B: $Fun[Z^d, M] \rightarrow Fun[Z^d, M]$ defined by the computational flat erosion ε_B:

$$\mathrm{E}_B(f)(z) = \varepsilon_B\big[f(z+w_1), f(z+w_2), \ldots, f(z+w_n)\big] = \bigwedge_{w_i \in B} f(z+w_i) \quad (2\text{–}46)$$

More generally, for $B_1, B_2, \ldots, B_p \subset W$, the *stack filter* corresponding to the class $\mathcal{B} = \{B_1, B_2, \ldots, B_p\}$ is defined by the computational stack filter $\xi_{\mathcal{B}}$: $M^n \rightarrow M$ of Eq. 2–23:

$$\Xi_{\mathcal{B}}(f)(z) = \xi_{\mathcal{B}}\big[f(z+w_1), f(z+w_2), \ldots, f(z+w_n)\big]$$

$$= \bigvee_{j=1}^{p} \bigwedge_{w_i \in B_j} f(z + w_i)$$

$$= \bigvee_{j=1}^{p} \mathrm{E}_{B_j}(f)(z) \tag{2–47}$$

where, for the sake of minimality, we assume that no set in $\mathcal{B}$ is a proper subset of a distinct set in $\mathcal{B}$ [6]. The class $\mathcal{B}$ is commonly called the "basis" of $\Xi_{\mathcal{B}}$; to avoid confusion with the computational basis, we call $\mathcal{B}$ the *set basis* of $\Xi_{\mathcal{B}}$. According to Eq. 2–47, a stack filter is a maximum of flat erosions, and for this reason $\Xi_{\mathcal{B}}$ is often called a *flat filter*. Because $\xi_{\mathcal{B}}$ is an increasing computational function, $\Xi_{\mathcal{B}}$ is an increasing image operator.

Using the computational form of $\xi_{\mathcal{B}}$ from Eq. 2–25 yields

$$\Xi_{\mathcal{B}}(f)(z) = \sum_{y=1}^{m} \bigvee_{j=1}^{p} \bigwedge_{i=1}^{n} \varepsilon_{yI_{B_j}(w_i)}\big(f(z + w_i)\big) \tag{2–48}$$

And from Eq. 2–34,

$$\Xi_{\mathcal{B}}(f)(z) = \sum_{y=1}^{m} \bigvee_{j=1}^{p} \mathrm{E}_{yI_{B_j}}(f)(z) \tag{2–49}$$

From Eq. 2–24 and the relationship between $\mathcal{K}_y[\Xi_{\mathcal{B}}]$ and $\mathcal{K}_y[\xi_{\mathcal{B}}]$,

$$\mathcal{B}_y[\Xi_{\mathcal{B}}] = \{yI_{B_1}, yI_{B_2}, \ldots, yI_{B_p}\} \tag{2–50}$$

so that the representation of $\Xi_{\mathcal{B}}$ in Eq. 2–49 is the representation ensured by Theorem 2–3.

The representation of Eq. 2–49 can be viewed differently by recognizing that $\mathrm{E}_{yI_{B_j}}(f)(z) = 1$ if and only if $f(x) \geqslant y$ for all $x \in (B_j)_z$, where the inner subscript j indexes the set and the outer subscript z is a translation. If, for each image $f \in Fun[Z^d, M]$ and $k = 1, 2, \ldots, m$, we define the binary *threshold image* $f^k \in Fun[Z^d, \{0, 1\}]$ by $f^k(z) = 1$ if $f(z) \geqslant k$ and $f^k(z) = 0$ if $f(z) < k$, then

$$\bigvee_{j=1}^{p} \mathrm{E}_{yI_{B_j}}(f)(z) = 1 \tag{2–51}$$

if and only if there exists at least one j for which $f^y(x) = 1$ for all $x \in (B_j)_z$. But this is equivalent to the minimum of f^y over $(B_j)_z$ being 1 for at least one j, which means that

$$\bigvee_{j=1}^{p} \mathrm{E}_{yI_{B_j}}(f)(z) = \bigvee_{j=1}^{p} \bigwedge_{w_i \in B_j} f^y(z + w_i) = \Xi_{\mathcal{B}}(f^y)(z) \tag{2–52}$$

because the inner expression is the stack filter $\Xi_{\mathcal{B}}$ applied to the threshold image f^y. Hence, from Eq. 2–49 we obtain the original formulation of a stack filter,

$$\Xi_{\mathcal{B}}(f) = \sum_{j=1}^{m} \Xi_{\mathcal{B}}(f^y) \tag{2–53}$$

The stack filter is applied to each of the threshold images and summed to arrive at the stack filter applied to the gray-scale image itself. The terminology arises from viewing each binary image as a set. Since $k \leqslant j$ implies $f^k \geqslant f^j$, the threshold images "stack" on top of one another, with each threshold set being a subset of the preceding one. Since $\Xi_{\mathcal{B}}$ is increasing, $\Xi_{\mathcal{B}}(f^k) \geqslant \Xi_{\mathcal{B}}(f^j)$, so that the outputs stack in the same order. According to Eq. 2–53, the gray-scale output is obtained by summing the stack values.

A key property of stack filters is that they commute with thresholding, meaning that

$$\Xi_{\mathcal{B}}(f^k) = \Xi_{\mathcal{B}}(f)^k \tag{2–54}$$

for any $k \in M$. Equation 2.54 results from the following equivalences: $\Xi_{\mathcal{B}}(f)^k(z) = 1 \Leftrightarrow \Xi_{\mathcal{B}}(f)(z) \geqslant k$ [by definition of threshold function] $\Leftrightarrow \Xi_{\mathcal{B}}(f^j)(z) = 1$ for all $j \leqslant k$ [by Eq. 2–53] $\Leftrightarrow \Xi_{\mathcal{B}}(f^k)(z) = 1$ [because $\Xi_{\mathcal{B}}$ is increasing].

For n odd, the moving median $\mathrm{M}_{\mathcal{B}}$ is the stack filter whose set basis $\mathcal{B}$ consists of all subsets $B_1, B_2, \ldots, B_p \subset W$ having $(n+1)/2$ pixels. $\mathrm{M}_{\mathcal{B}}$ is the maximum of flat erosions having structuring elements $B_1, B_2, \ldots, B_p$. Its window function is the computational median $\mu_{\mathcal{B}}$.

For subset $B \subset W$, *flat dilation* Δ_B is the stack filter with set basis $\mathcal{B} = \{\{z\}\colon z \in B\}$, the set of all singleton subsets of B. From Eq. 2–47,

$$\Delta_B(f)(z) = \bigvee_{x \in B} f(z + x) \tag{2–55}$$

so that Δ_B is the moving maximum over B and its window function is the computational flat dilation δ_B. Note that, to emphasize the structuring element, we are being slightly inconsistent with notation by writing Δ_B rather than $\Delta_{\mathcal{B}}$.

Table 2.1. Elemental erosion values.

y	$\varepsilon_y(f(-2))$	$\varepsilon_y(f(-1))$	$\varepsilon_y(f(0))$	$\varepsilon_y(f(1))$	$\varepsilon_y(f(2))$
1	1	1	1	1	1
2	1	1	1	0	1
3	1	0	0	0	1

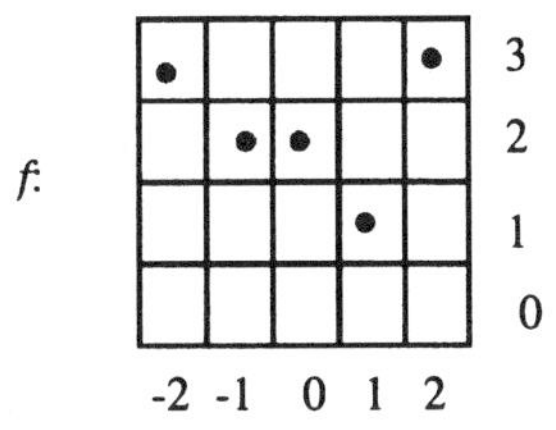

Figure 2–8. Image used to compute $\Gamma_B(f)(0)$.

An important stack filter is flat opening. For any set B, it is defined iteratively via erosion and dilation: $\Gamma_B = \Delta_B \mathrm{E}_B$. According to Eq. 2–55, it can be expressed using flat erosions:

$$\Gamma_B(f)(z) = \bigvee_{x \in B} \mathrm{E}_B(f)(z+x) = \bigvee_{x \in B} \mathrm{E}_{B_x}(f)(z) \tag{2–56}$$

Since there are erosions by B_x for all $x \in B$, the operator window is $B \oplus B$.

The computational representation of opening using the window $W = B \oplus B$ is given by

$$\Gamma_B(f)(z) = \sum_{y=1}^{m} \bigvee_{x \in B} \bigwedge_{i=1}^{n} \varepsilon_{y I_{B_x}(w_i)}\big(f(z+w_i)\big) \tag{2–57}$$

To illustrate the computational structure of opening, consider signal f of Fig. 2–8 and structuring element $B = \{-1, 0, 1\}$. Then $\mathrm{E}_B(f)(-1) = 2$, $\mathrm{E}_B(f)(0) = \mathrm{E}_B(f)(1) = 1$, and $\Gamma_B(f)(0) = 2$. The relevant values of the elemental erosion are given in Table 2.1. The min-max computations of Eq. 2–57 for Table 2.1 are

$$\begin{aligned} &(1 \wedge 1 \wedge 1) \vee (1 \wedge 1 \wedge 1) \vee (1 \wedge 1 \wedge 1) = 1 \\ &(1 \wedge 1 \wedge 1) \vee (1 \wedge 1 \wedge 0) \vee (1 \wedge 0 \wedge 1) = 1 \\ &(1 \wedge 0 \wedge 0) \vee (0 \wedge 0 \wedge 0) \vee (0 \wedge 0 \wedge 1) = 0 \end{aligned} \tag{2–58}$$

Summation gives $\Gamma_B(f)(0) = 2$.

2.7 NONFLAT EROSION

To represent classical gray-scale erosion in the computational structure, we begin by defining, for a fixed vector **b**, the computational *nonflat erosion* function $\hat{\varepsilon}_{\mathbf{b}}: M^n \to M$ by

$$\hat{\varepsilon}_{\mathbf{b}}(\mathbf{x}) = \bigwedge_{i=1}^{n} (x_i - b_i) \tag{2–59}$$

for any $\mathbf{x} = (x_1, x_2, \ldots, x_n) \in M^n$, where the hat notation is to distinguish nonflat erosion $\hat{\varepsilon}_{\mathbf{b}}$ from n-elemental erosion $\varepsilon_{\mathbf{b}}$. So that $\hat{\varepsilon}_{\mathbf{b}}$ is range-preserving, meaning that output vectors are M-valued when input vectors are M-valued, it is required that all components of **b** be nonpositive and at least one be 0-valued. If these two conditions are satisfied, then **b** is said to be *range-preserving*. In the classical approach to gray-scale morphology, no such restrictions are placed on the structuring element; however, it is assumed that the gray range is the set of integers together with plus and minus infinity (or, in the case of real-valued functions, the extended real line) [5, 7–11]. Because we require digitally implementable filters, we restrict **b**. If **b** is range-preserving, then, unless $\mathbf{b} = \mathbf{0}$, $\mathbf{b} \notin M^n$, meaning **b** is outside the domain space of the mapping it defines. The kernel of $\hat{\varepsilon}_{\mathbf{b}}$ consists of the sets

$$\mathcal{K}_y\big[\hat{\varepsilon}_{\mathbf{b}}\big] = \big\{(x_1, x_2, \ldots, x_n) \in M^n\colon\ x_i \geqslant b_i + y \text{ for all } i\big\} \tag{2–60}$$

The basis consists of the singleton basis classes

$$\mathcal{B}_y\big[\hat{\varepsilon}_{\mathbf{b}}\big] = \big\{(0 \vee (b_1 + y), 0 \vee (b_2 + y), \ldots, 0 \vee (b_n + y)\big\} \tag{2–61}$$

Computational representation is given by

$$\hat{\varepsilon}_{\mathbf{b}}(\mathbf{x}) = \sum_{y=1}^{m} \bigwedge_{i=1}^{n} \varepsilon_{0 \vee (b_i + y)}(x_i) \tag{2–62}$$

Generalization to maxima of computational erosions is similar to maxima of flat erosions.

Gray-to-gray image erosion is defined via the computational function $\hat{\varepsilon}_{\mathbf{b}}$. We need to consider a structuring function g, so that the vector **g** defined via Eq. 2–32 for the window W is range-preserving, in which case g is called range-preserving. There is no requirement that g have its values in M. If g is range-preserving, then gray-to-gray erosion with structuring function g is defined by

$$\begin{aligned}\widehat{\mathrm{E}}_g(f)(z) &= \hat{\varepsilon}_{\mathbf{b}}\big[f(z + w_1), f(z + w_2), \ldots, f(z + w_n)\big] \\ &= \bigwedge_{i=1}^{n} f(z + w_i) - g(w_i)\end{aligned} \tag{2–63}$$

Its computational representation is

$$\widehat{\mathrm{E}}_g(f)(z) = \sum_{y=1}^{m} \bigwedge_{i=1}^{n} \varepsilon_{0\vee(g(w_i)+y)}\big(f(z+w_i)\big) \tag{2–64}$$

If $g_1, g_2, \ldots, g_p$ are range-preserving, then the union-of-erosion filter for the class $\mathcal{G} = \{g_1, g_2, \ldots, g_p\}$ is defined by

$$\Psi_{\mathcal{G}}(f)(z) = \bigvee_{j=1}^{p} \bigwedge_{i=1}^{n} f(z+w_i) - g_j(w_i) \tag{2–65}$$

To avoid redundancy, it is assumed that there do not exist two functions in $\mathcal{G}$ properly related by function inequality. Under this assumption, $\mathcal{G}$ is called a *function basis*. The computational representation and representation according to Theorem 2–3 for $\Psi_{\mathcal{G}}$ are

$$\Psi_{\mathcal{G}}(f)(z) = \sum_{y=1}^{m} \bigvee_{j=1}^{n} \bigwedge_{i=1}^{n} \varepsilon_{0\vee(g_j(w_i)+y)}\big(f(z+w_i)\big) \tag{2–66}$$

$$\Psi_{\mathcal{G}}(f) = \sum_{y=1}^{m} \bigvee_{j=1}^{n} \mathrm{E}_{0\vee(g_j+y)}(f) \tag{2–67}$$

where, in Eq. 2–67, $0 \vee (g_j + y)$ is the maximum of the zero function and the function g_j offset by y.

2.8 REPRESENTATION OF GENERIC COMPUTATIONAL FUNCTIONS

So far we have focused on representation of increasing computational functions; the present section treats representation of generic (not necessarily increasing) computational functions. Representation of increasing functions has been based on elemental erosion; by introducing the notion of elemental anti-dilation, we can obtain similar representations for decreasing functions [ψ is decreasing if, for all $\mathbf{x}, \mathbf{y} \in L^n$, $\mathbf{x} \leqslant \mathbf{y}$ implies $\psi(\mathbf{x}) \geqslant \psi(\mathbf{y})$]. More generally, for representation of generic computational functions, the elemental building block will be a function formed by the minimum of an elemental erosion and an elemental anti-dilation.

For fixed $s \in L$, the 1-*elemental anti-dilation* (or simply, *elemental anti-dilation*) δ_s^a is defined, for any $x \in L$, by $\delta_s^a(x) = 1$ if $x \leqslant s$ and $\delta_s^a(x) = 0$ otherwise. δ_s^a is a binary function on L. For $\mathbf{s} = (s_1, s_2, \ldots, s_n) \in L^n$, the n-*elemental anti-dilation*

$\delta_{\mathbf{s}}^{a}$ is defined by

$$\delta_{\mathbf{s}}^{a}(\mathbf{x}) = \bigwedge_{i=1}^{n} \delta_{s_i}^{a}(x_i) \tag{2–68}$$

for any vector $\mathbf{x} = (x_1, x_2, \ldots, x_n) \in L^n$. $\delta_{\mathbf{s}}^{a}(\mathbf{x}) = 1$ if and only if $\mathbf{x} \leqslant \mathbf{s}$. Elemental and n-elemental anti-dilation satisfy

$$\delta_{s}^{a}\left(\bigvee_{i\in I} x_i\right) = \bigwedge_{i\in I} \delta_{s}^{a}(x_i) \tag{2–69}$$

$$\delta_{\mathbf{s}}^{a}\left(\bigvee_{i\in I} \mathbf{x}_i\right) = \bigwedge_{i\in I} \delta_{\mathbf{s}}^{a}(\mathbf{x}_i) \tag{2–70}$$

respectively, where the index set I can be infinite. In Eq. 2–70, the maximum on the LHS operates on L^n, whereas the minimum on the RHS operates on binary vectors. Since n-elemental anti-dilation satisfies Eq. 2–70 and maps the minimum element of lattice L^n to the maximum element of the range $[\delta_{\mathbf{s}}^{a}(\mathbf{0}) = 1]$, it is an anti-dilation in the algebraic sense of mathematical morphology on complete lattices.

For fixed $r, s \in L$ such that $r \leqslant s$, the 1-*elemental sup-generating function* (or, simply, *elemental sup-generating function*) $\lambda_{r,s}$ is defined for all $x \in L$ by $\lambda_{r,s}(x) = 1$ if $r \leqslant x \leqslant s$ and $\lambda_{r,s}(x) = 0$ otherwise. For $\mathbf{r} = (r_1, r_2, \ldots, r_n)$, $\mathbf{s} = (s_1, s_2, \ldots, s_n) \in L^n$ such that $\mathbf{r} \leqslant \mathbf{s}$, the *n-elemental sup-generating function* $\lambda_{\mathbf{r},\mathbf{s}}$ is defined for any vector $\mathbf{x} = (x_1, x_2, \ldots, x_n) \in L^n$ by

$$\lambda_{\mathbf{r},\mathbf{s}}(\mathbf{x}) = \bigwedge_{i=1}^{n} \lambda_{r_i,s_i}(x_i) \tag{2–71}$$

$\lambda_{\mathbf{r},\mathbf{s}}(\mathbf{x}) = 1$ if and only if $\mathbf{r} \leqslant \mathbf{x} \leqslant \mathbf{s}$. Equivalently,

$$\lambda_{r,s} = \varepsilon_r \wedge \delta_s^a \tag{2–72}$$

$$\lambda_{\mathbf{r},\mathbf{s}} = \varepsilon_{\mathbf{r}} \wedge \delta_{\mathbf{s}}^a \tag{2–73}$$

Since $\lambda_{r,s}$ and $\lambda_{\mathbf{r},\mathbf{s}}$ are each given as a minimum of an algebraic erosion and algebraic anti-dilation, they are sup-generating functions in the sense of mathematical morphology on complete lattices [12].

The functions $\lambda_{r,s}$ and $\lambda_{\mathbf{r},\mathbf{s}}$ can also be expressed as minima of erosions,

$$\lambda_{r,s}(x) = \varepsilon_r(x) \wedge \varepsilon_{s'}\left(x'\right) \tag{2–74}$$

$$\lambda_{\mathbf{r},\mathbf{s}}(\mathbf{x}) = \varepsilon_{\mathbf{r}}(\mathbf{x}) \wedge \varepsilon_{\mathbf{s}'}\left(\mathbf{x}'\right) \tag{2–75}$$

$$\underset{\lambda_{\mathbf{r},\mathbf{s}}}{\begin{array}{|c|c|c|c|}\hline 0&0&0&0\\\hline 0&1&1&0\\\hline 0&1&1&0\\\hline 0&0&0&0\\\hline\end{array}} = \underset{\varepsilon_{\mathbf{r}}}{\begin{array}{|c|c|c|c|}\hline 0&1&1&1\\\hline 0&1&1&1\\\hline 0&1&1&1\\\hline 0&0&0&0\\\hline\end{array}} \wedge \underset{\delta^{a}_{\mathbf{s}}}{\begin{array}{|c|c|c|c|}\hline 0&0&0&0\\\hline 1&1&1&0\\\hline 1&1&1&0\\\hline 1&1&1&0\\\hline\end{array}}$$

Figure 2–9. Sup-generating computational function as the minimum of an erosion and an anti-dilation.

These last expressions are hit-or-miss formulations of the computational sup-generating functions and, because of these formulations, $\lambda_{r,s}$ is also called the *elemental hit-or-miss function* with hit and miss parameters r and s, respectively, and $\lambda_{\mathbf{r},\mathbf{s}}$ the *n-elemental hit-or-miss function* with hit and miss parameters $\mathbf{r}$ and $\mathbf{s}$, respectively. Figure 2–9 shows a sup-generating function defined on $\{0, 1, 2, 3\}^2$.

The next proposition provides a generalization of Theorem 2–1 to generic computational functions by providing representations in terms of slices and kernel sets.

PROPOSITION 2–2. *If* $\psi\colon L^n \to M$, *then*

$$\psi(\mathbf{x}) = \sum_{y=1}^{m} y \bigvee_{\mathbf{r}\in\mathcal{S}_y[\psi]} \lambda_{\mathbf{r},\mathbf{r}}(\mathbf{x}) \tag{2–76}$$

or, in terms of the kernel of ψ,

$$\psi(\mathbf{x}) = \sum_{y=1}^{m} \bigvee_{\mathbf{r}\in\mathcal{K}_y[\psi]} \lambda_{\mathbf{r},\mathbf{r}}(\mathbf{x}) \tag{2–77}$$

PROOF. To demonstrate Eq. 2–76, let $\xi_y(\mathbf{x})$ be the function defined by the union. We have the following sequence of equivalences: $\xi_y(\mathbf{x}) = 1 \Leftrightarrow$ there exists $\mathbf{r} \in \mathcal{S}_y[\psi]$ such that $\lambda_{\mathbf{r},\mathbf{r}}(\mathbf{x}) = 1 \Leftrightarrow$ there exists $\mathbf{r} \in \mathcal{S}_y[\psi]$ such that $\mathbf{x} = \mathbf{r} \Leftrightarrow \mathbf{x} \in \mathcal{S}_y[\psi] \Leftrightarrow \psi(\mathbf{x}) = y$. Thus, there exists exactly one nonzero union in the sum and that is the yth term if and only if $\psi(\mathbf{x}) = y$. The kernel expansion follows from the first expansion since $\mathbf{r} \in \mathcal{S}_y[\psi]$ if and only if $\mathbf{r} \in \mathcal{K}_k[\psi]$ for $k = 1, 2, \ldots, y$ and $\mathbf{r} \notin \mathcal{K}_k[\psi]$ for $k > y$. ■

Theorem 2–2 provided a minimal representation relative to Theorem 2–1. To achieve minimal representation for nonincreasing operators it is necessary to introduce the concept of interval. An *interval* $[\mathbf{r}, \mathbf{s}]$ in L^n consists of the set of all $\mathbf{x} \in L^n$ such that $\mathbf{r} \leqslant \mathbf{x} \leqslant \mathbf{s}$. Figure 2–10 shows an interval in $\{0, 1, 2, 3\}^2$. Before

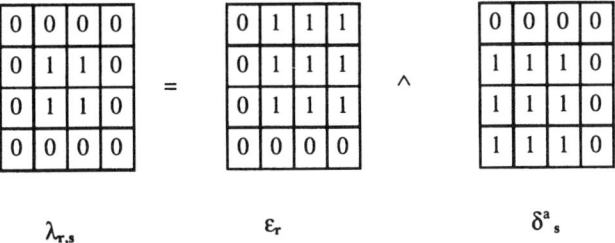

Figure 2–10. Interval [(0, 1), (3, 2)].

providing a minimal representation in the next theorem, we first show how the inclusions of the vector **r** in the slices and kernel sets in Proposition 2–2 can be replaced by inclusions of intervals [**r**, **s**].

PROPOSITION 2–3. *If* $\psi\colon L^n \to M$, *then*

$$\psi(\mathbf{x}) = \sum_{y=1}^{m} y \bigvee_{[\mathbf{r},\mathbf{s}] \subset \mathcal{S}_y[\psi]} \lambda_{\mathbf{r},\mathbf{s}}(\mathbf{x}) \tag{2–78}$$

or, in terms of kernel sets,

$$\psi(\mathbf{x}) = \sum_{y=1}^{m} \bigvee_{[\mathbf{r},\mathbf{s}] \subset \mathcal{K}_y[\psi]} \lambda_{\mathbf{r},\mathbf{s}}(\mathbf{x}) \tag{2–79}$$

PROOF. The proof of Eq. 2–78 is very similar to that of Eq. 2–76. Letting $\xi_y(\mathbf{x})$ be the function defined by the union, there is the following sequence of equivalences: $\xi_y(\mathbf{x}) = 1 \Leftrightarrow$ there exists $[\mathbf{r}, \mathbf{s}] \subset \mathcal{S}_y[\psi]$ such that $\lambda_{\mathbf{r},\mathbf{s}}(x) = 1 \Leftrightarrow$ there exists $[\mathbf{r}, \mathbf{s}] \subset \mathcal{S}_y[\psi]$ such that $\mathbf{r} \leqslant \mathbf{x} \leqslant \mathbf{s} \Leftrightarrow \mathbf{x} \in \mathcal{S}_y[\psi] \Leftrightarrow \psi(\mathbf{x}) = y$. The kernel expansion follows similarly: $[\mathbf{r}, \mathbf{s}] \subset \mathcal{S}_y[\psi]$ if and only if $[\mathbf{r}, \mathbf{s}] \subset \mathcal{K}_k[\psi]$ for $k = 1, 2, \ldots, y$ and $[\mathbf{r}, \mathbf{s}] \not\subset \mathcal{K}_k[\psi]$ for $k > y$. ■

An interval $[\mathbf{r}, \mathbf{s}]$ is said to be *maximal* among a collection of intervals if there does not exist a distinct interval $[\mathbf{u}, \mathbf{v}]$ in the collection such that $[\mathbf{u}, \mathbf{v}] \supset [\mathbf{r}, \mathbf{s}]$. A collection of intervals composed only of maximal intervals is said to be *maximal*. If $I_y[\psi]$ is the set of all intervals lying in $\mathcal{K}_y[\psi]$, then the set, $\mathcal{M}_y[\psi]$, of maximal intervals in $I_y[\psi]$ is called the *basis set* of ψ for level y and

$$\mathcal{M}[\psi] = \{\mathcal{M}_1[\psi], \mathcal{M}_2[\psi], \ldots, \mathcal{M}_m[\psi]\} \tag{2–80}$$

is called the *basis* of ψ. If $\mathcal{J}_y[\psi]$ is the set of all intervals lying in $\mathcal{S}_y[\psi]$, then the set, $\mathcal{N}_y[\psi]$, of maximal intervals in $\mathcal{J}_y[\psi]$ is called the *slice-basis set* of ψ for level y and

$$\mathcal{N}[\psi] = \{\mathcal{N}_1[\psi], \mathcal{N}_2[\psi], \ldots, \mathcal{N}_m[\psi]\} \tag{2–81}$$

is called the slice *basis* of ψ.

THEOREM 2–4. *If* ψ: $L^n \to M$, *then*

$$\psi(\mathbf{x}) = \sum_{y=1}^{m} y \bigvee_{[\mathbf{r},\mathbf{s}]\in\mathcal{N}_y[\psi]} \lambda_{\mathbf{r},\mathbf{s}}(\mathbf{x}) \tag{2–82}$$

or, in basis form,

$$\psi(\mathbf{x}) = \sum_{y=1}^{m} \bigvee_{[\mathbf{r},\mathbf{s}]\in\mathcal{M}_y[\psi]} \lambda_{\mathbf{r},\mathbf{s}}(\mathbf{x}) \tag{2–83}$$

PROOF. We prove Eq. 2–83. For all $y \in M$, $\mathcal{K}_y[\psi]$ is finite. Hence, for $[\mathbf{x}, \mathbf{y}] \subset \mathcal{K}_y[\psi]$, there exists $[\mathbf{u}, \mathbf{v}] \in \mathcal{M}_y[\psi]$ such that $[\mathbf{x}, \mathbf{y}] \subset [\mathbf{u}, \mathbf{v}]$. Furthermore, $\lambda_{\mathbf{x},\mathbf{y}} \leqslant \lambda_{\mathbf{u},\mathbf{v}}$, so that $\lambda_{\mathbf{x},\mathbf{y}}$ is redundant in the decomposition of Eq. 2–79 and the decomposition of Eq. 2–83 is sufficient to represent ψ. A similar argument applies to Eq. 2–82. ■

The representations of Eqs. 2–77 and 2–83 are extensions of the binary-logical disjunctive normal form and the minimal sum-of-products logical expansion. There exist more general extensions in the complete lattice framework, both for computational morphological representation [3] and for general lattices [12]. Figures 2–11 and 2–12 illustrate the basis and slice representations of a computational function, respectively.

2.9 REPRESENTATION OF GENERIC GRAY-TO-GRAY IMAGE OPERATORS

Representation of generic image operators is achieved via the gray-to-binary *hit-or-miss transform* $\Lambda_{g,h}$: $Fun[Z^d, L] \to Fun[Z^d, \{0, 1\}]$. If $\mathrm{supp}(h) \subset W$ and $g \leqslant h$, then $\Lambda_{g,h}$ is defined by

$$\Lambda_{g,h}(f) = \mathrm{E}_g(f) \wedge \Delta_h^a(f) \tag{2–84}$$

where Δ_h^a is the *anti-dilation* image operator defined by

$$\Delta_h^a(f)(z) = \delta_{\mathbf{h}}^a(\mathbf{f}(z)) \tag{2–85}$$

where $\mathbf{h}$ is the vector composed of the gray values of h on W. Computationally,

$$\Lambda_{g,h}(f)(z) = \lambda_{\mathbf{g},\mathbf{h}}(f(z+w_1), f(z+w_2), \ldots, f(z+w_n))$$

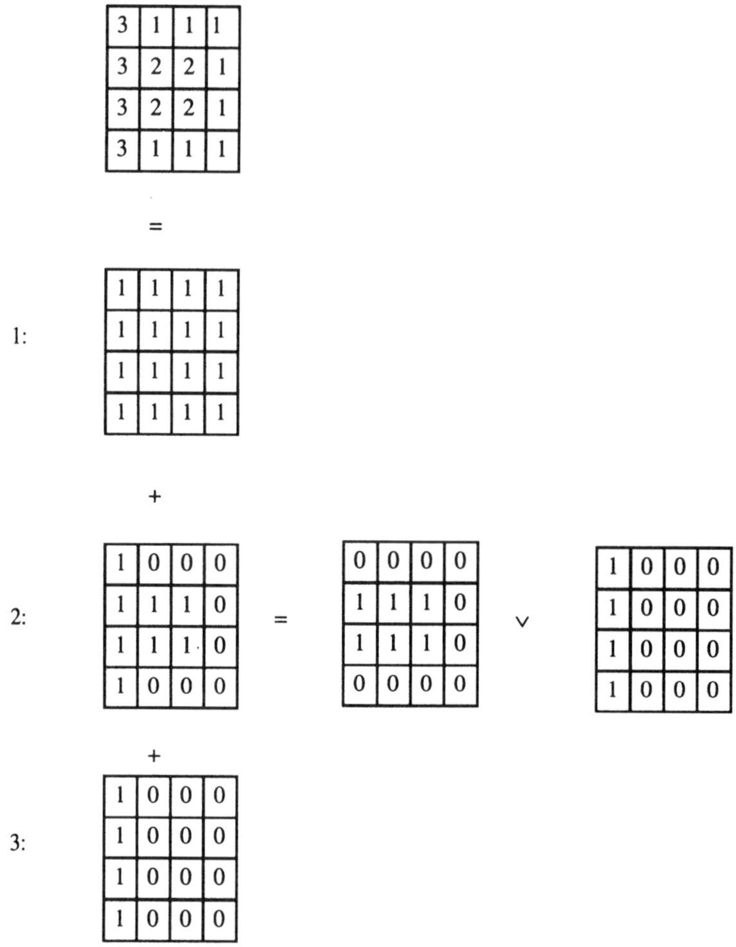

Figure 2–11. Basis decomposition of a nonincreasing computational function.

$$= \varepsilon_{\mathbf{g}}\big(f(z+w_1), \ldots, f(z+w_n)\big) \wedge \delta_{\mathbf{h}}^{a}\big(f(z+w_1), \ldots, f(z+w_n)\big)$$

$$= \bigwedge_{i=1}^{n} \varepsilon_{g(w_i)}\big(f(z+w_i)\big) \wedge \delta_{h(w_i)}^{a}\big(f(z+w_i)\big) \tag{2–86}$$

The gray-to-binary hit-or-miss transform can be used for template matching in the same way as the binary hit-or-miss transform. If we let g be an image patch with support in W, the patch may be detected at a pixel z by $\Lambda_{g,g}$, which means that the restriction of image f to the translated window W_z matches the patch:

$$\Lambda_{g,g}(f)(z) = \begin{cases} 1, & \text{if } f|W_z \equiv g_z \\ 0, & \text{otherwise} \end{cases} \tag{2–87}$$

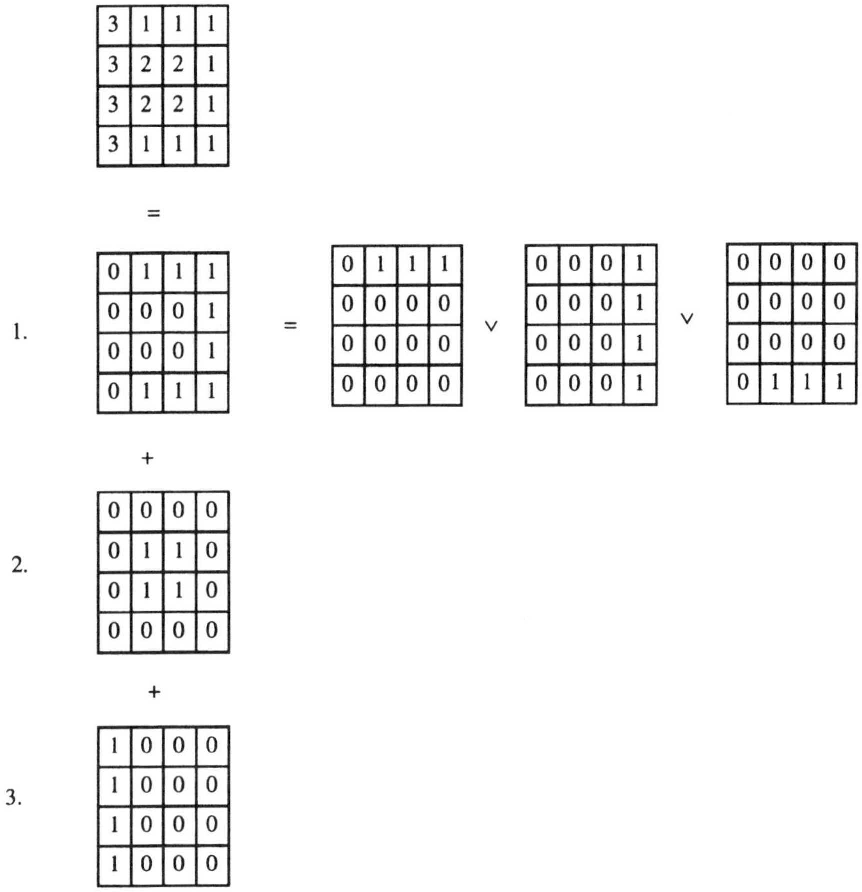

Figure 2–12. Representation of the computational function of Fig. 2–1 by slices.

The exactness of the match is softened by using $\Lambda_{g,h}$.

Representation of generic image operators follows directly from representation of generic computational functions. The *basis* $\mathcal{M}[\Psi]$ and *slice basis* $\mathcal{N}[\Psi]$ of the generic operator Ψ with computational function ψ are defined by the basis sets and slice-basis sets $\mathcal{M}_y[\Psi] = \{g\colon \mathbf{g} \in \mathcal{M}_y[\psi]\}$ and $\mathcal{N}_y[\Psi] = \{g\colon \mathbf{g} \in \mathcal{N}_y[\psi]\}$, respectively, where $\operatorname{supp}(g) \subset W$. Interval $[g, h] \subset \mathcal{M}_y[\Psi]$ if and only if $\operatorname{supp}(h) \subset W$ and, for any function q such that $g \leqslant q \leqslant h$, $\mathbf{q} \in \mathcal{M}_y[\psi]$. The following representation theorem corresponds to Theorem 2–4.

THEOREM 2–5. *If* Ψ: $Fun[Z^d, L] \to Fun[Z^d, M]$ *is a* W*-operator, then, in slice-basis form,*

$$\Psi(f) = \sum_{y=1}^{m} y \bigvee_{[g,h]\in\mathcal{N}_y[\Psi]} \Lambda_{g,h}(f) \tag{2–88}$$

In basis form,

$$\Psi(f) = \sum_{y=1}^{m} \bigvee_{[g,h]\in\mathcal{M}_y[\Psi]} \Lambda_{g,h}(f) \tag{2–89}$$

PROOF. We demonstrate Eq. 2–89:

$$\begin{aligned}\sum_{y=1}^{m} \bigvee_{[g,h]\in\mathcal{M}_y[\Psi]} \Lambda_{g,h}(f)(z) &= \sum_{y=1}^{m} \bigvee_{[\mathbf{g},\mathbf{h}]\in\mathcal{M}_y[\psi]} \lambda_{\mathbf{g},\mathbf{h}}\big(f(z+w_1),\ldots,f(z+w_n)\big)\\ &= \psi\big(f(z+w_1), f(z+w_2),\ldots,f(z+w_n)\big)\end{aligned} \tag{2–90}$$

where the second inequality is a direct application of Eq. 2–83. ■

2.10 OPTIMAL GRAY-SCALE COMPUTATIONAL FILTERS

For a gray-scale computational function $\psi \in Fun[L^n, M]$, the *mean-absolute error* of ψ as an estimator $\psi(X_1, X_2, \ldots, X_n)$ of a random variable Y is given by

$$\begin{aligned}MAE\langle\psi\rangle &= E\big[|Y - \psi(X_1, X_2, \ldots, X_n)|\big]\\ &= \sum_{(x_1,x_2,\ldots,x_n)\in L^n,\ y\in M} \big|y - \psi(x_1, x_2, \ldots, x_n)\big|\\ &\quad \times f(x_1, x_2, \ldots, x_n, y)\end{aligned} \tag{2–91}$$

where $f(x_1, x_2, \ldots, x_n, y)$ is the joint density of the random variables $X_1, X_2, \ldots, X_n, Y$. For generic filter optimization, one needs to choose the optimal filter ψ_{opt} from among all filters by selecting the filter with minimim MAE. In the present section we will restrict our attention to increasing computational functions.

Finding an optimal increasing computational filter is equivalent to finding a consistent collection of self-minimal sets $\mathcal{K}_1, \mathcal{K}_2, \ldots, \mathcal{K}_m$ of vectors in L^n for which the function defined via Eq. 2–19 is an optimal estimator of Y over the class of all increasing functions. If ψ is defined via $\mathcal{K}_1, \mathcal{K}_2, \ldots, \mathcal{K}_m$, then they form the basis of ψ, so that we can relabel the vector classes by $\mathcal{B}_1[\psi], \mathcal{B}_2[\psi], \ldots, \mathcal{B}_m[\psi]$ and have the representation of Eq. 2–15. If we let

$$\psi_y(\mathbf{x}) = \bigvee_{\mathbf{r}\in\mathcal{B}_y[\psi]} \varepsilon_{\mathbf{r}}(\mathbf{x}) \tag{2–92}$$

then we can rewrite the representation of Eq. 2–15 as

$$\psi(\mathbf{x}) = \sum_{y=1}^{m} \psi_y(\mathbf{x}) \tag{2–93}$$

ψ_y is a binary-valued computational function possessing increasing basis $\mathcal{B}_y[\psi]$. Because $\psi_y(\mathbf{x})$ can be equivalently expressed as

$$\psi_y(\mathbf{x}) = \bigvee_{\mathbf{r} \in \mathcal{K}_y[\psi]} \varepsilon_{\mathbf{r}}(\mathbf{x}) \tag{2–94}$$

we see that $\psi_1(\mathbf{x}) \geqslant \psi_2(\mathbf{x}) \geqslant \cdots \geqslant \psi_m(\mathbf{x})$. In fact, from Eq. 2–94 we see that $\psi_y(\mathbf{x}) = 1$ if and only if there exists $\mathbf{r}$ such that $\psi(\mathbf{r}) \geqslant y$ and $\mathbf{r} \leqslant \mathbf{x}$, which by the increasing nature of ψ means that $\psi(\mathbf{x}) \geqslant y$. The functions $\psi_y(\mathbf{x})$ are called *level functions*.

As in the binary setting, there is an MAE theorem for increasing W-operators, which we state here in the context of computational functions [13]. For ψ: $L^n \to M$ having increasing basis composed of the basis sets

$$\mathcal{B}_y[\psi] = \{\mathbf{r}_{y,1}, \mathbf{r}_{y,2}, \ldots, \mathbf{r}_{y,n(y)}\} \tag{2–95}$$

for $y = 1, 2, \ldots, m$, the MAE for ψ as an estimator of Y has the recursive representation

$$MAE\langle\psi\rangle = \sum_{y=1}^{m} MAE\langle\psi_y^{n(y)-1}\rangle + MAE\langle\mathbf{r}_{y,n(y)}\rangle - MAE\langle\phi_y^{n(y)-1}\rangle \tag{2–96}$$

where $MAE\langle\mathbf{r}_{y,n(y)}\rangle$ is the MAE associated with the n-elemental erosion with structuring element $\mathbf{r}_{y,n(y)}$, and $\psi_y^{n(y)-1}$ and $\phi_y^{n(y)-1}$ are functions with increasing bases

$$\begin{aligned} \mathcal{B}[\psi_y^{n(y)-1}] &= \{\mathbf{r}_{y,1}, \mathbf{r}_{y,2}, \ldots, \mathbf{r}_{y,n(y)-1}\} \\ \mathcal{B}[\phi_y^{n(y)-1}] &= \{\mathbf{r}_{y,1} \vee \mathbf{r}_{y,n(y)}, \mathbf{r}_{y,2} \vee \mathbf{r}_{y,n(y)}, \ldots, \mathbf{r}_{y,n(y)-1} \vee \mathbf{r}_{y,n(y)}\} \end{aligned} \tag{2–97}$$

Note that

$$\mathcal{B}[\psi_y] = \{\mathbf{r}_{y,1}, \mathbf{r}_{y,2}, \ldots, \mathbf{r}_{y,n(y)}\} = \mathcal{B}[\psi_y^{n(y)-1}] \cup \{\mathbf{r}_{y,n(y)}\} \tag{2–98}$$

To appreciate how the recursive error representation can be employed in the design of an optimal filter, consider the following idealized search strategy. Suppose optimization is to be accomplished by selecting bases from m structuring element collections $\mathcal{C}_1, \mathcal{C}_2, \ldots, \mathcal{C}_m$, where

$$\mathcal{C}_y = \{\mathbf{r}_{y,1}, \mathbf{r}_{y,2}, \ldots, \mathbf{r}_{y,t(y)}\} \tag{2–99}$$

For $s = 1, 2, \ldots, t(y)$, let $\mathcal{C}_{y,s}$ denote the closure of $\mathcal{C}_y$ under maxima of s elements in $\mathcal{C}_y$. Then $\mathcal{C}_{y,1} = \mathcal{C}_y$ and $\mathcal{C}_{y,1} \subset \mathcal{C}_{y,2} \subset \cdots \subset \mathcal{C}_{y,t(y)}$. Suppose we know the MAE for structuring elements in the closures. We can proceed in the following recursive manner. For any 2-erosion-per-level filter ψ^2 with

$$\mathcal{B}[\psi^2] = \{\{\mathbf{r}_{1,i_{11}}, \mathbf{r}_{1,i_{12}}\}, \{\mathbf{r}_{2,i_{21}}, \mathbf{r}_{2,i_{22}}\}, \ldots, \{\mathbf{r}_{m,i_{m1}}, \mathbf{r}_{m,i_{m2}}\}\} \tag{2–100}$$

MAE is given in terms of single-structuring element erosions by

$$MAE\langle\psi^2\rangle = \sum_{y=1}^{m} MAE\langle\mathbf{r}_{y,i_{y1}}\rangle + MAE\langle\mathbf{r}_{y,i_{y2}}\rangle - MAE\langle\mathbf{r}_{y,i_{y1}} \vee \mathbf{r}_{y,i_{y2}}\rangle \tag{2–101}$$

Now suppose ψ^3 is a 3-erosion-per-level filter with

$$\begin{aligned}\mathcal{B}[\psi^3] = \{&\{\mathbf{r}_{1,i_{11}}, \mathbf{r}_{1,i_{12}}, \mathbf{r}_{1,i_{13}}\}, \{\mathbf{r}_{2,i_{21}}, \mathbf{r}_{2,i_{22}}, \mathbf{r}_{2,i_{23}}\}, \ldots, \\ &\{\mathbf{r}_{m,i_{m1}}, \mathbf{r}_{m,i_{m2}}, \mathbf{r}_{m,i_{m3}}\}\}\end{aligned} \tag{2–102}$$

Then the MAE theorem gives

$$\begin{aligned}MAE\langle\psi^3\rangle = \sum_{y=1}^{m} &MAE\langle\mathbf{r}_{y,i_{y1}}, \mathbf{r}_{y,i_{y2}}\rangle + MAE\langle\mathbf{r}_{y,i_{y3}}\rangle \\ &- MAE\langle\mathbf{r}_{y,i_{y1}} \vee \mathbf{r}_{y,i_{y3}}, \mathbf{r}_{y,i_{y2}} \vee \mathbf{r}_{y,i_{y3}}\rangle\end{aligned} \tag{2–103}$$

All terms on the RHS of the equation can be obtained from the previous stage of the iteration and therefore can be used on the third stage. The error-decomposition process can be continued recursively.

As given, the search procedure is idealized. In practice, other considerations need to be taken into account, for instance, utilizing a different number of structuring elements for each level and ensuring the necessary order relations so that the resulting structuring-element collections form a basis.

2.11 Application: Quantization Range Conversion

Computational increasing-filter representation, in particular the mean-absolute error theorem, has been used in practice to design filters for quantization range conversion in digital documents involving multiple bits per pixel [14]. In this application, an image is mapped from 1 to N-bits/pixel while maintaining constant spatial sampling resolution. Quantization range conversion takes advantage of greater gray-scale resolution for the printer than for the digitally stored document. Character edges are enhanced because some edge pixels are mapped to intermediate gray levels that lessen the jagged appearance of curved and angled strokes. A windowed portion of the input digital document is observed and, based on the observed values in the window, the center pixel is mapped to a value within the output quantization range. The computational window function, ψ: $\{0, 1\}^n \rightarrow \{0, 1, \ldots, 2^N - 1\}$, which performs this task, defines a quantization-conversion filter Ψ.

Design of ψ is performed via computer search, where the search is performed over combinations of structuring elements to find a combination yielding minimum MAE. Single-erosion MAE values are input to the search and their combinations employed as bases are evaluated rapidly for MAE through the recursion of Eq. 2–96. Owing to the combinatoric nature of the procedure, constraints are placed on filter design to limit the search: a small window is used and the number of basis sets is limited. Because the enhancement primarily concerns nontrivial estimation only about character edges, acceptable results can be obtained using a small window.

For a concrete application, consider 1-bit-to-2-bit quantization range conversion with a 5-pixel cross window. Assuming an increasing filter, the function ψ is determined by three basis sets: $\mathcal{B}_1[\psi]$, $\mathcal{B}_2[\psi]$, $\mathcal{B}_3[\psi]$. Using statistics derived from image realizations and assuming that each basis set contains four structuring elements (each of which is binary because the domain of ψ is $\{0, 1\}^5$), the basis sets of Fig. 2–13 were obtained. Figure 2–14 shows the effect of applying the filter. Parts a and b show the binary input and the 2-bit output, respectively. The visual difference between the two digital images when they are printed is shown by the corresponding print simulations of Fig. 2–15. (For a detailed account of the manner in which image statistics were gathered and the images of Fig. 2–15 simulated, see Ref. [14].)

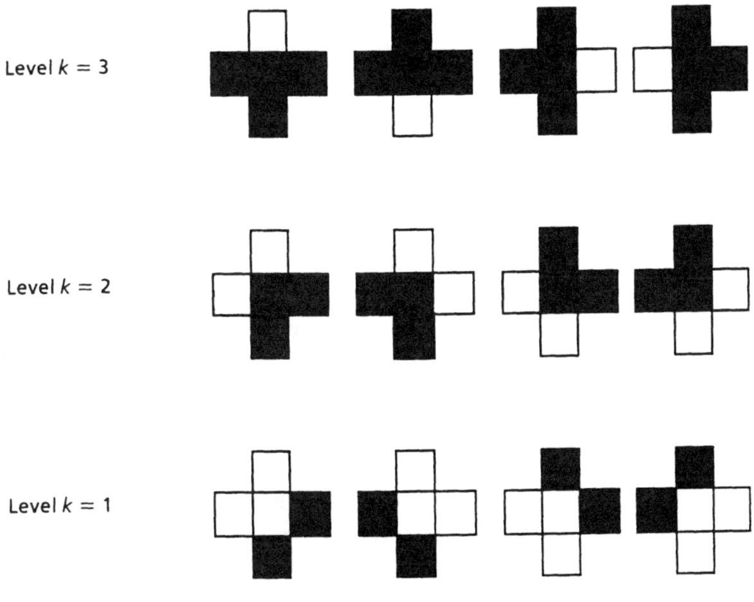

Figure 2–13. Optimal 4-erosion basis.

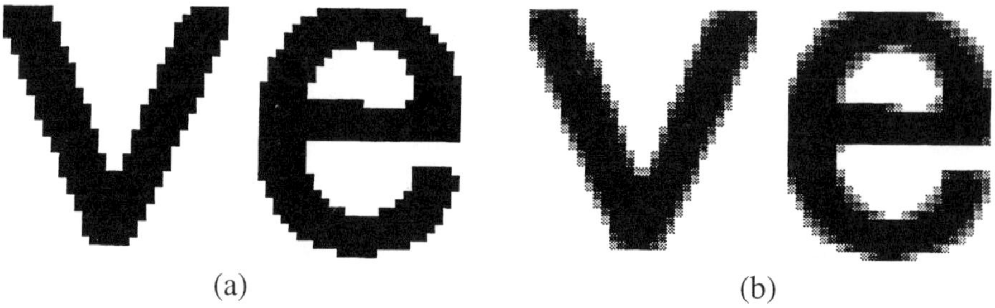

Figure 2–14. Character image showing jaggie reduction: (a) before filtering – binary; (b) after filtering – quaternary.

Figure 2–15. Print simulations showing jaggie reduction: (a) unfiltered; (b) filtered image.

2.12 Comments on Gray-Scale Morphology

The computational framework permits the representation of all translation-invariant windowed operators via maxima of elemental hit-or-miss operations and in the special case of increasing operators provides representation via maxima of elemental erosions. These representations can be used for filter optimization by finding the optimal bases of structuring elements. Because the elementary operations of erosion, anti-dilation, and sup-generating functions satisfy the corresponding lattice properties, the computational structure provides a representation and optimization theory in the context of mathematical morphology and is therefore termed *computational gray-scale morphology*. We have chosen to separate the binary theory from the gray-scale theory for two reasons: (1) there are special methods applicable to the binary theory and (2) the binary theory is basically a morphological formulation of classical Boolean algebra, whereas the gray-scale theory provides an extension of Boolean algebra appropriate for digital image processing. In the present section we wish to discuss a number of points regarding the latter reason, in particular, the evolution of gray-scale morphology along different paths.

Mathematical morphology has evolved from a strictly binary theory [15] to a gray-scale function theory [7, 16]. to an operator theory within the context of complete lattices [4, 17, 18]. As a theory concerning lattice operators, special focus has been placed on increasing operators; as set or function theory, the focus has been on increasing, translation-invariant operators, translation invariance relating to the additive group structure in the plane for sets and in both the domain and range for functions (gray-scale images).

For the binary theory, Matheron's original representation of increasing, translation-invariant set operators as unions of set erosions has played a central role [15]. This representation is an extension of the classical Boolean representation of a positive Boolean function as a sum of products void of complementation (Eq. 1–5), which in the operator framework becomes the morphological representation of Eq. 1–13. The representation (and subsequent refinements) will be studied in detail in Chapter 3; however, for the sake of completeness and current discussion we state it here without proof. For two sets S and B in the Euclidean (Cartesian) plane, the *erosion* of S by structuring element B is defined by

$$\mathrm{E}_B(S) = \{z\colon\ B_z \subset S\} \tag{2–104}$$

If Ψ is a translation-invariant set operator, then the *kernel* of Ψ is defined by

$$\mathcal{K}[\Psi] = \{A\colon\ 0 \in \Psi(A)\} \tag{2–105}$$

The original erosion representation for increasing, translation-invariant set operators stated that every such operator Ψ can be expressed as

$$\Psi(S) = \bigcup_{B \in \mathcal{K}[\Psi]} \mathrm{E}_B(S) \tag{2–106}$$

(Theorem 3–1). This representation reduces to a sum of products for windowed operators. Equation 2–106 can be reduced to a minimal basis representation that it extends the minimal logical representation of Eq. 2–15 by taking the basis of Ψ to be the class of minimal elements in the kernel [9, 11]; however, in the general set framework the kernel is infinite, existence of a basis is not ensured, and proving the existence of a basis requires transfinite induction using Zorn's lemma (Theorem 3–2).

The disjunctive normal form for a (not necessarily increasing) logical operator (Eq. 1–1) and its reduction to minimal form (Eq. 1–2) are morphologically expressed by the hit-or-miss union of Eq. 1–18. Representation of translation-invariant set operators is attained by extending hit-or-miss representation to set operators (Theorem 3–4) [19].

Our main interest in the current section is gray-scale morphological representation. As originally formulated by Serra, gray-scale erosion and dilation were a moving minimum (Eq. 2–46) and moving maximum, respectively [16]. Subsequently, Sternberg [7, 20] extended the definition of gray-scale erosion according to Eq. 2–63 (see comments of Serra [21]). Geometrically, nonflat erosion is found by translating the structuring element to the appropriate point and finding the maximum amount one can "push up" the structuring element and have it still remain beneath the graph of the image. As defined here, nonflat erosion requires that the structuring element be range-preserving; no such requirement was made originally. If one makes the assumption that an image is a function whose range consists of the extended real line (real line together with plus and minus infinity), then there is no need for range preservation. The question of range preservation has not often been raised since the vast majority of gray-scale applications use flat erosion.

More generally, we can consider gray-scale operators that are suprema of flat erosions,

$$\Psi(f) = \bigvee_{B \in \mathcal{B}} \mathrm{E}_B(f) \tag{2–107}$$

or suprema of nonflat erosions,

$$\Psi(f) = \bigvee_{g \in \mathcal{B}} \widehat{\mathrm{E}}_g(f) \tag{2–108}$$

where in the first case $\mathcal{B}$ is a class of sets and in the second case a class of functions. Both expressions take the form of the original set-theoretic representation; both yield increasing, translation-invariant operators; and the second expression includes the first as a special case because letting g be the indicator function of a set produces the flat erosion by that set. Wendt *et al.* [6] referred to filters of the form given in Eq. 2–107 as stack filters. Clearly, the class of stack filters does not encompass the class of increasing, translation-invariant function operators because the class of stack filters is a proper subclass of the operator class represented by Eq. 2–108. Moreover, stack filters are range-preserving.

Maragos and Schafer [9] showed that, under the assumption that extended-real-valued functions are upper semicontinuous, all increasing, translation-invariant operators possess nonflat erosion expansions according to Eq. 2–108, and Giardina and Dougherty [11] showed a similar result without an upper-semicontinuity restriction. For image ranges, the key point here is that translation invariance is with respect to both domain and range: Ψ is translation invariant if $\Psi(f_x + y) = \Psi(f)_x + y$ for any point x and any numerical value y. No problems arise with respect to range translation so long as the range is the extended real line. This strictly function-space theory was superseded by Serra's recognition that the proper algebraic framework for morphological representation is the complete lattice [4, 23] and that increasing function operators possess morphological representation on account of their lattice properties [23] (see also Matheron [24]). The role of lattice theory was further explicated by Heijmans and Ronse [17]. Finally, from a general algebraic perspective, representation for lattice operators (not necessarily increasing) was completely characterized in the context of mathematical morphology by Banon and Barrera [12].

While there are no theoretical problems with assuming that functions have values in the extended real line, there is a practical problem because digital processing involves a finite discrete range. Recognizing this difficulty, Ronse has proposed truncation in the definitions of gray-scale morphological operations to maintain the original range [25]. A second truncation approach has been taken by Hsueh, who defines morphological operations via l-images [26]. By fuzzifying binary morphology, Sinha and Dougherty arrive at (gray-scale-like) morphological operators on fuzzy membership functions that are range-preserving [27]. For fuzzy morphology, there is a weak form of Eq. 2–108 that reduces to the classical erosion representation when sets are crisp. In an effort to reformulate gray-scale morphology for finite ranges, while essentially preserving the orginal fitting characterization of nonflat erosion, Heijmans [28] has redefined the morphological operations so that the maximum gray value "absorbs" the structuring element as it is pushed up from beneath the signal whenever the signal takes on the maximum gray value. Intuitively, erosion is still found by fitting, but the maximum gray level acts somewhat like infinity in that it does not act as a surface point in the signal graph. The approach results in an erosion operation that preserves much of the original intuition, is range-preserving, and leads to a restricted form of the representation of Eq. 2–108.

Computational morphology arose from a desire to extend the logical interpretations of the binary erosion and hit-or-miss representations, and to provide a range-preserving, gray-scale morphology in which binary morphology becomes a special case of gray-scale morphology with binary range $\{0, 1\}$. The main motivation was the latter and it developed out of a practical need having to do with the design of statistically optimal nonlinear filters.

If we wish to use the representation of Eq. 2–108 for finding an optimal increasing, translation-invariant filter, the problem is to find a basis $\mathcal{B}$ of structuring functions, where in this case the minimality requirement for a basis means that there do not exist two functions $g_1, g_2 \in \mathcal{B}$ such that $g_1 \leqslant g_2$. While the general theory does not require the filter to be windowed (the structuring elements to have finite supports within a common window), practical filter implementation does require windowing. Two anomalies arise. Because binary optimization involves $\{0, 1\}$-valued functions and binary morphology is algebraically embedded into Sternberg's classical gray-scale morphology via $\{-\infty, 0\}$-valued functions, the binary optimization theory [29] does not fall out of the gray-scale optimization theory [30] as a special case. Second, the classical gray-scale optimization theory assumes finite-valued random variables, which means that, as in real-world digital images, signals possess a finite range, say $[0, m]$. There does exist a minimal finite *fundamental set* of nonflat structuring elements from which to select the optimal basis, but because the gray-scale optimization theory is based on the classical fitting definition of erosion, the range of the filtered signals is $[-m, 2m]$. Therefore some sort of truncation must be employed to bring filtered signals back into the original gray range. The entire problem manifests itself in the fact that the MAE theorems for representation of binary filters and gray-scale filters possess different proofs [31]. Indeed, in the latter case the representation involves summation over $[-m, 2m]$. In the case of computational morphology, there is no distinct representation or optimization theory for binary filters.

Computational morphology provides a unified framework for representation and optimization of finite-windowed, translation-invariant digital filters. For the most part, full optimzation is computationally impractical and constraints must be applied to achieve practical ends. Although constraints may result in specialized representations (such as representations for increasing and stack filters), these representations themselves can be expressed within the framework of computational morphology. A key aspect of future research is the formalization of constraints that are suitable for real-world problems and that lead to tractable optimization procedures, whether via direct error minimization or via adaptive methods such as gradient-based searches.

REFERENCES

[1] Dougherty, E. R., and D. Sinha, "Computational Mathematical Morphology," *Signal Processing*, **38**, 1994.

[2] Dougherty, E. R., and Sinha, D., "Computational Gray-Scale Morphology on Lattices (A Comparator-Based Image Algebra) Part I: Architecture," *Real-Time Imaging*, **1** (1), 1995.

[3] Dougherty, E. R., and Sinha, D., "Computational Gray-Scale Morphology on Lattices (A Comparator-Based Image Algebra) Part II: Image Operators," *Real-Time Imaging*, **1** (4), 1995.

[4] Serra, J., "Mathematical Morphology for Complete Lattices," in *Image Analysis and Mathematical Morphology, Vol. 2: Theoretical Advances*, ed. J. Serra, Academic Press, New York, 1988.

[5] Serra, J., ed., *Image Analysis and Mathematical Morphology*, Vol. 2, Academic Press, New York, 1988.

[6] Wendt, P. D., Coyle, E. J., and N. C. Gallagher, "Stack Filters," *IEEE Trans. on Acoustics, Speech, and Signal Processing*, **34**, 1986.

[7] Sternberg, S., "Grayscale Morphology," *Computer Vision, Graphics, and Image Processing*, **35** (3), September, 1986.

[8] Haralick, R., Sternberg, S., and X. Zhuang, "Image Analysis Using Mathematical Morphology," *IEEE Trans. on Pattern Analysis and Machine Intelligence*, **9** (4), 1987.

[9] Maragos, P., and R. W. Schafer, "Morphological Filters — Part I: Their Set-Theoretic Analysis and Relations to Linear Shift-Invariant Filters," *IEEE Trans. on Acoustics, Speech, and Signal Processing*, **35**, 1987.
[10] Maragos, P., and R. W. Schafer, "Morphological Filters - Part II: Their Relations to Median, Order-Statistics, and Stack Filters," *IEEE Trans. on Acoustics, Speech, and Signal Processing*, **35**, 1987.
[11] Giardina, C. R., and E. R. Dougherty, *Morphological Methods in Image and Signal Processing*, Prentice-Hall, Englewood Cliffs, 1988.
[12] Banon, G. J. F., and J. Barrera, "Decomposition of Mappings Between Complete Lattices by Mathematical Morphology, Part I. General Lattices," *Signal Processing*, **30**, 1993.
[13] Loce, R. P., and E. R. Dougherty, "Mean-Absolute-Error Representation and Optimization of Computational-Morphological Filters," *Computer Vision, Graphics, and Image Processing: Image Understanding*, **57** (1), 1995.
[14] Loce, R. P., and E. R. Dougherty, *Enhancement and Restoration of Digital Documents: Statistical Design of Nonlinear Algorithms*, SPIE Press, Bellingham, 1997.
[15] Matheron, G., *Random Sets and Integral Geometry*, John Wiley, New York, 1975.
[16] Serra, J., *Image Analysis and Mathematical Morphology*, Academic Press, New York, 1982.
[17] Heijmans, H. J., and C. Ronse, "The Algebraic Basis of Mathematical Morphology, I. Dilations and Erosions," *Computer Vision, Graphics, and Image Processing*, **50**, 1990.
[18] Ronse, C., and H. J. Heijmans, "The Algebraic Basis of Mathematical Morphology, II. Openings and Closings," *Computer Vision, Graphics, and Image Processing*, **54**, 1991.
[19] Banon, G. J. F., and J. Barrera, "Minimal Representation of Translation-Invariant Set Mappings by Mathematical Morphology," *SIAM Journal on Applied Mathematics*, **51** (6), 1991.
[20] Sternberg, S., "Cellular Computers and Biomedical Image Processing," in *Biomedical Images and Computers*, eds. J. Sklansky and J. C. Bisconte, Lecture Notes in Medical Informatics, 17, Springer-Verlag, Berlin, 1982.
[21] Serra, J., "Anamorphoses and Function Lattices," in *Mathematical Morphology in Image Processing*, ed. E. R. Dougherty, Marcel Dekker, New York, 1993.
[22] Serra, J., "Introduction to Morphological Filters," in *Image Analysis and Mathematical Morphology, Vol. 2: Theoretical Advances*, ed. J. Serra, Academic Press, New York, 1988.
[23] Serra, J., "Dilation and Filtering for Numerical Functions," in *Image Analysis and Mathematical Morphology, Vol. 2: Theoretical Advances*, ed. J. Serra, Academic Press, New York, 1988.
[24] Matheron, G., "Filters and Lattices," in *Image Analysis and Mathematical Morphology, Vol. 2: Theoretical Advances*, ed. J. Serra, Academic Press, New York, 1988.

[25] Ronse, C., "Why Mathematical Morphology Needs Complete Lattices," *Signal Processing*, **21**, 1990.

[26] Hsueh, Y.-C., "Mathematical Morphology on l-images," *Signal Processing*, **26**, 1992.

[27] Sinha, D., and E. R. Dougherty, "A General Axiomatic Theory of Intrinsically Fuzzy Mathematical Morphologies," *IEEE Trans. on Fuzzy Systems*, **3** (4), 1995.

[28] Heijmans, H. J., "Theoretical Aspects of Gray-Scale Morphology," *IEEE Trans. on Pattern Analysis and Machine Intelligence*, **13**, 1991.

[29] Dougherty, E. R., "Optimal Mean-Square N-Observation Digital Morphological Filters – Part I: Optimal Binary Filters," *Computer Vision, Graphics, and Image Processing: Image Understanding*, **55** (1), 1992.

[30] Dougherty, E. R., "Optimal Mean-Square N-Observation Digital Morphological Filters – Part II: Optimal Gray-Scale Filters," *Computer Vision, Graphics, and Image Processing: Image Understanding*, **55** (1), 1992.

[31] Loce, R. P., and E. R. Dougherty, "Optimal Morphological Restoration: The Morphological Filter Mean-Absolute-Error Theorem," *Visual Communication and Image Representation*, **3** (4), 1992.

[32] Dougherty, E. R., and E. Kraus, "Shape Analysis and Reduction of the Morphological Basis for Digital Moving Averages," *SIAM Journal on Applied Mathematics*, **51** (6), 1991.

CHAPTER 3

TRANSLATION-INVARIANT SET OPERATORS

Edward R. Dougherty
Texas A&M University
College Station, Texas

Having expressed the representation theory for windowed operators in the context of classical binary logical calculus, we discuss the general theory of translation-invariant operators on n-dimensional Euclidean space R^n. While the theory is developed for R^n, the algeraic apsects apply to n-dimensional discrete space. It can be viewed as a generalization of the logical theory, which itself is the special case of the general theory as it applies to window operators on discrete space. Historically, the theory fits into the framework of mathematical morphology as developed by Matheron [1] and Serra [2] from roots in Minkowski [3] and Hadwiger [4, 5].

3.1 TRANSLATION-INVARIANT OPERATORS

We begin with some basic definitions and related propositions concerning translation-invariant operators. Let $\mathcal{P}$ denote the power set (set of all subsets) of R^n. For $n = 2$, $\mathcal{P}$ is the class of continuous binary images. A mapping $\Psi\colon \mathcal{P} \to \mathcal{P}$ is said to be *translation invariant* if $\Psi(S_x) = \Psi(S)_x$ for all $S \in \mathcal{P}$ and $x \in R^n$, where $S_x = \{x + z\colon\ z \in S\}$. Matheron calls Ψ a *τ-mapping*. The *kernel* of Ψ is defined by

$$\mathcal{K}[\Psi] = \{S\colon\ 0 \in \Psi(S)\} \tag{3–1}$$

Point $z \in \Psi(S)$ if and only if $S \in \mathcal{K}[\Psi]_z$.

PROPOSITION 3–1. *Let Ψ_1 and Ψ_2 be translation-invariant mappings. Then $\mathcal{K}[\Psi_1] \subset \mathcal{K}[\Psi_2]$ if and only if $\Psi_1 \subset \Psi_2$. In particular, $\Psi_1 = \Psi_2$ if and only if $\mathcal{K}[\Psi_1] = \mathcal{K}[\Psi_2]$.*

PROOF. Suppose $\mathcal{K}[\Psi_1] \subset \mathcal{K}[\Psi_2]$. Let S be an arbitrary set and $z \in \Psi_1(S)$. Then $0 \in \Psi_1(S)_{-z} = \Psi_1(S_{-z})$ and $S_{-z} \in \mathcal{K}[\Psi_1] \subset \mathcal{K}[\Psi_2]$. But $S_{-z} \in \mathcal{K}[\Psi_2]$ implies (by similar reasoning) that $z \in \Psi_2(S)$, and we conclude that $\Psi_1 \subset \Psi_2$. Conversely, suppose $\Psi_1 \subset \Psi_2$. Then $S \in \mathcal{K}[\Psi_1]$ implies $0 \in \Psi_1(S) \subset \Psi_2(S)$, which implies $S \in \mathcal{K}[\Psi_2]$, and we conclude $\mathcal{K}[\Psi_1] \subset \mathcal{K}[\Psi_2]$. ■

Mapping Ψ is *increasing* if it preserves ordering within $\mathcal{P}$, namely, $S_1 \subset S_2$ implies $\Psi(S_1) \subset \Psi(S_2)$. Subset partial ordering in $\mathcal{P}$ induces a partial order relation among mappings: $\Psi_1 \subset \Psi_2$ if and only if $\Psi_1(S) \subset \Psi_2(S)$ for all $S \in \mathcal{P}$.

PROPOSITION 3–2. *A union of translation-invariant mappings is translation invariant and a union of increasing mappings is increasing. In the case of translation invariance, the kernel of the union is equal to the union of the kernels.*

PROOF. Let $\Psi = \bigcup_{i \in I} \Psi_i$. If Ψ_i is translation invariant for each $i \in I$, because union commutes with translation by a point,

$$\Psi(S_z) = \bigcup_{i \in I} \Psi_i(S_z) = \bigcup_{i \in I} \Psi_i(S)_z = \left(\bigcup_{i \in I} \Psi_i(S)\right)_z = \Psi(S)_z \tag{3–2}$$

Moreover, if $S \subset T$, then $\Psi_i(S) \subset \Psi_i(T)$ for all $i \in I$, so that

$$\Psi(S) = \bigcup_{i \in I} \Psi_i(S) \subset \bigcup_{i \in I} \Psi_i(T) = \Psi(T) \tag{3–3}$$

Finally, $0 \in \Psi(S)$ iff $0 \in \Psi_i(S)$ for some $i \in I$. Hence, $S \in \mathcal{K}[\Psi]$ if and only if there exists $i \in I$ such that $S \in \mathcal{K}[\Psi_i]$. ■

The *dual mapping* of a mapping Ψ is $\Psi^*\colon \mathcal{P} \to \mathcal{P}$ defined by $\Psi^*(S) = \Psi(S^c)^c$. Since $S^{cc} = S$, $\Psi^{**} = \Psi$. The dual of the dual is the original mapping.

PROPOSITION 3–3. *If Ψ is translation invariant, so is Ψ^* and*

$$\mathcal{K}[\Psi^*] = \{S\colon S^c \notin \mathcal{K}[\Psi]\} \tag{3–4}$$

If Ψ is increasing, so is Ψ^*.

PROOF. Translation invariance of the dual follows from translation invariance of Ψ and the fact that complementation commutes with translation $[(S_z)^c = (S^c)_z]$. Specifically,

$$\Psi^*(S_z) = \Psi\left((S_z)^c\right)^c = \Psi\left((S^c)_z\right)^c = \left[\Psi(S^c)_z\right]^c = \left[\Psi(S^c)^c\right]_z = \Psi^*(S)_z \tag{3–5}$$

Increasingness of Ψ^* follows from increasingness of Ψ because complementaion reverses subset ordering [$S \subset T$ iff $S^c \supset T^c$]. Specifically, if $S \subset T$, then, by the increasingness of Ψ, $\Psi(T^c) \subset \Psi(S^c)$. Taking complements yields $\Psi^*(S) \subset \Psi^*(T)$. As for the kernel,

$$\mathcal{K}[\Psi^*] = \{S\colon 0 \in \Psi^*(S)\} = \{S\colon 0 \notin \Psi(S^c)\} = \{S\colon S^c \notin \mathcal{K}[\Psi]\} \;\blacksquare \tag{3–6}$$

Operator Ψ is said to *commute with intersection* if

$$\Psi\left(\bigcap_{i\in I} S_i\right) = \bigcap_{i\in I} \Psi(S_i) \tag{3–7}$$

for any collection of sets $\{S_i\}$ over an arbitrary index set I. It *commutes with union* if

$$\Psi\left(\bigcup_{i\in I} S_i\right) = \bigcup_{i\in I} \Psi(S_i) \tag{3–8}$$

PROPOSITION 3–4. *Ψ commutes with intersection if and only if Ψ^* commutes with union.*

PROOF. If Ψ commutes over intersection then, by De Morgan's law,

$$\Psi^*\left(\bigcup_{i\in I} S_i\right) = \Psi\left(\bigcap_{i\in I} S_i^c\right)^c = \left(\bigcap_{i\in I} \Psi(S_i^c)\right)^c = \bigcup_{i\in I} \Psi^*(S_i) \tag{3–9}$$

The converse is demonstrated similarliy. ■

A mapping Ψ: $\mathcal{P} \to \mathcal{P}$ is *antiextensive* if $\Psi(S) \subset S$ for any $S \in \mathcal{P}$. Ψ is *extensive* if $\Psi(S) \supset S$ for any $S \in \mathcal{P}$. If Ψ is antiextensive, then $\Psi(S^c) \subset S^c$, so that $\Psi^*(S) = \Psi(S^c)^c \supset S^{cc} = S$. Since the argument can be reversed, Ψ is antiextensive if and only if Ψ^* is extensive. For translation-invariant operators, antiextensivity and extensivity can be characterized via the kernel.

PROPOSITION 3–5. *Let $\mathcal{P}_0$ be the subset of all sets in $\mathcal{P}$ that contain the origin and let Ψ be translation-invariant. Then* (i) *Ψ is antiextensive if and only if $\mathcal{P}_0 \supset \mathcal{K}[\Psi]$ and* (ii) *Ψ is extensive if and only if $\mathcal{P}_0 \subset \mathcal{K}[\Psi]$.*

PROOF. We show (i). Suppose Ψ is antiextensive and $S \in \mathcal{K}[\Psi]$. Since $\Psi(S) \subset S$ and $0 \in \Psi(S)$, $0 \in S$, so that $S \in \mathcal{P}_0$. Conversely, suppose $\mathcal{P}_0 \supset \mathcal{K}[\Psi]$, but that Ψ is not antiextensive. Then there exists set S such that $\Psi(S) \not\subset S$, which means there exists a point $z \in \Psi(S) - S$. Since Ψ is translation invariant, $0 \in \Psi(S)_z - S_z = \Psi(S_z) - S_z$. Hence, $S_z \in \mathcal{K}[\Psi]$ but $S_z \notin \mathcal{P}_0$, which contradicts the supposition $\mathcal{P}_0 \supset \mathcal{K}[\Psi]$. ■

The basic building blocks of increasing, translation-invariant mappings are erosions. The definition extends directly from the windowed setting with *erosion* of set S by set (structuring element) B being defined by

$$\mathrm{E}_B(S) = \{x\colon\ B_x \subset S\} \tag{3–10}$$

Fixing B and letting S vary over $\mathcal{P}$ yields an operator $\Psi(S) = \mathrm{E}_B(S)$. The next proposition summarizes the most basic properties of erosion.

PROPOSITION 3–6. *Erosion satisfies the following properties*:

(i) $\mathrm{E}_B(S_z) = \mathrm{E}_B(S)_z$ [*translation invariant*]

(ii) $S_1 \subset S_2$ *implies* $\mathrm{E}_B(S_1) \subset \mathrm{E}_B(S_2)$ [*increasing*]

(iii) $\mathrm{E}_{B_z}(S) = \mathrm{E}_B(S)_{-z}$

(iv) $\mathrm{E}_B(S_1 \cap S_2) = \mathrm{E}_B(S_1) \cap \mathrm{E}_B(S_2)$

(v) $\mathrm{E}_{B_1 \cup B_2}(S) = \mathrm{E}_{B_1}(S) \cap E_{B_2}(S)$

PROOF. Part (i) is established by the relation

$$\{x\colon B_x \subset S_z\} = \{x\colon B_{x-z} \subset S\} = \{y + z\colon B_y \subset S\} = \{y\colon B_y \subset S\}_z \qquad (3\text{–}11)$$

Since $S_1 \subset S_2$, part (ii) follows because $B_x \subset S_1$ implies $B_x \subset S_2$. Part (iii) is established by

$$\{x\colon (B_z)_x \subset S\} = \{x\colon B_{z+x} \subset S\} = \{y - z\colon B_y \subset S\} = \{y\colon B_y \subset S\}_{-z} \qquad (3\text{–}12)$$

As for part (iv), $B_x \subset S_1 \cap S_2$ if and only if $B_x \subset S_1$ and $B_x \subset S_2$; and part (v) holds because $(B_1 \cup B_2)_x \subset S$ if and only if $(B_1)_x \subset S$ and $(B_2)_x \subset S$. ■

Property (iv) holds for an erosion of any union, countable or uncountable, so that erosion commutes with intersection: for any index set I,

$$\mathrm{E}_B\left(\bigcap_{i\in I} S_i\right) = \bigcap_{i\in I} \mathrm{E}_B(S_i) \qquad (3\text{–}13)$$

In the algebraic formulation of mathematical morphology on complete lattices [6], any operator that commutes with infimum (intersection) and for which the maximal element is invariant [for $\mathcal{P}$, meaning that $\Psi(R^n) = R^n$] is called an *erosion*. The erosion operation defined in Eq. 3–10 via the structuring element B is often called a *structural erosion* [7].

Property (v) also holds for any union:

$$\mathrm{E}_{\bigcup_{i\in I} B_i}(S) = \bigcap_{i\in I} \mathrm{E}_{B_i}(S) \qquad (3\text{–}14)$$

Since, for any single point x, $\mathrm{E}_{\{x\}}(S) = S_{-x}$, erosion can be expressed via intersection:

$$\mathrm{E}_B(S) = \mathrm{E}_{\bigcup_{x\in B}\{x\}}(S) = \bigcap_{x\in B} \mathrm{E}_{\{x\}}(S) = \bigcap_{x\in B} S_{-x} = \bigcap_{x\in -B} S_x \qquad (3\text{–}15)$$

where $-B = \{-x\colon x \in B\}$.

The intersection on the right-hand side defines the classical *Minkowski subtraction*, so we see that $\mathrm{E}_B(S)$ is equal to the Minkowski subtraction of S by $-B$.

The dual mapping for erosion by B is *dilation* by $-B$. Dilation of S by (structuring element) B is defined according to duality by

$$\Delta_B(S) = \mathrm{E}_{-B}\left(S^c\right)^c \tag{3–16}$$

Because the dual of the dual is the original operator,

$$\mathrm{E}_B(S) = \Delta_{-B}\left(S^c\right)^c \tag{3–17}$$

It follows from Propositions 3–3 and 3–4 that Δ_B is translation invariant and increasing. Since erosion commutes with intersection, its dual dilation commutes with union:

$$\Delta_B\left(\bigcup_{i\in I} S_i\right) = \bigcup_{i\in I} \Delta_B(S_i) \tag{3–18}$$

In the algebraic formulation of mathematical morphology on complete lattices [6], any operator that commutes with supremum (union) and for which the minimal element is invariant [which for $\mathcal{P}$ means that $\Psi(\emptyset) = \emptyset$] is called a *dilation*. Thus, the dilation defined in Eq. 3–16 is often called a *structural dilation* [7].

Whereas erosion can be directly expressed as an intersection of translates of the set, dilation can be expressed as a union of translates:

$$\Delta_B(S) = \mathrm{E}_{-B}\left(S^c\right)^c = \left(\bigcap_{x\in -B} S^c_{-x}\right)^c = \bigcup_{x\in B} S_x \tag{3–19}$$

Since $S_x = \{y + x\colon\ y \in S\}$, the preceding union can be expanded to yield

$$\Delta_B(S) = \bigcup_{y\in S, x\in B} \{y + x\} \tag{3–20}$$

The union on the right-hand side of Eq. 3–20 is symmetric in S and B. This union is called the *Minkowski addition* of S and B, and is denoted by $S \oplus B$. Thus, $\Delta_B(S) = S \oplus B$. The symmetry of the union implies (from the corresponding properties of ordinary addition) that Minkowski addition is both commutative and associative:

$$S \oplus T = T \oplus S \tag{3–21}$$

$$(S \oplus T) \oplus U = S \oplus (T \oplus U) \tag{3–22}$$

According to Eq. 3–10, erosion is characterized by translates of the structuring element fitting within the set. Dilation, the dual of erosion, can be characterized by hitting.

PROPOSITION 3–7. *Dilation possesses the formulation*

$$\Delta_B(S) = \{x\colon (-B)_x \cap S \neq \emptyset\} \tag{3–23}$$

PROOF. The proposition follows from Eq. 3–19 and commutativity:

$$\begin{aligned}\{x\colon (-B)_x \cap S \neq \emptyset\} &= \{x\colon \exists z \in S \text{ such that } z \in (-B)_x\} \\ &= \{x\colon \exists z \in S \text{ such that } z - x \in -B\} \\ &= \{x\colon \exists z \in S \text{ such that } x \in B_z\} \\ &= \bigcup_{z \in S} B_z = B \oplus S = \Delta_B(S) \qquad \blacksquare\end{aligned} \tag{3–24}$$

Because $0 \in \mathrm{E}_B(S)$ if and only if $B = B_0 \subset S$, erosion by set B has the kernel

$$\mathcal{K}[\mathrm{E}_B] = \{S\colon S \supset B\} \tag{3–25}$$

Since dilation by B is the dual of erosion by $-B$,

$$\begin{aligned}\mathcal{K}[\Delta_B] &= \{S\colon S^c \notin \mathcal{K}[\mathrm{E}_{-B}]\} \\ &= \{S\colon -B \not\subset S^c\} \\ &= \{S\colon S \cap (-B) \neq \emptyset\}\end{aligned} \tag{3–26}$$

3.2 REPRESENTATION OF INCREASING TRANSLATION-INVARIANT OPERATORS

In a finite-logical windowed setting, the principal representation regarding increasing mappings is their represention as a sum of products in which there are no complements, which was then expressed as a union of erosions. A fundamental theorem of Matheron is that the erosion expansion applies to operators on subsets of R^n.

THEOREM 3–1 [1]. *If Ψ is increasing and translation invariant, then it has the representation*

$$\Psi(S) = \bigcup_{B \in \mathcal{K}[\Psi]} \mathrm{E}_B(S) \tag{3–27}$$

PROOF. Because Ψ is increasing, if $B \in \mathcal{K}[\Psi]$ and $A \supset B$, then $0 \in \Psi(B) \subset \Psi(A)$ and $A \in \mathcal{K}[\Psi]$. Thus $\{A\colon A \supset B\} \subset \mathcal{K}[\Psi]$. More generally,

$$\bigcup_{B \in \mathcal{K}[\Psi]} \{A\colon A \supset B\} \subset \mathcal{K}[\Psi] \tag{3–28}$$

Since the reverse inclusion is trivial,

$$\mathcal{K}[\Psi] = \bigcup_{B \in \mathcal{K}[\Psi]} \{A\colon A \supset B\} \tag{3–29}$$

Hence, by Proposition 3–2,

$$\mathcal{K}[\Psi] = \bigcup_{B \in \mathcal{K}[\Psi]} \mathcal{K}[\mathrm{E}_B] = \mathcal{K}\left[\bigcup_{B \in \mathcal{K}[\Psi]} \mathrm{E}_B\right] \tag{3–30}$$

and the representation follows by Proposition 3–1. ■

COROLLARY 3–1. *If Ψ is increasing and translation invariant, then*

$$\Psi(S) = \bigcap_{B \in \mathcal{K}[\Psi^*]} \Delta_{-B}(S) \tag{3–31}$$

PROOF. Because the dual of Ψ is also translation invariant and increasing, by the theorem,

$$\Psi^*(S^c) = \bigcup_{B \in \mathcal{K}[\Psi^*]} \mathrm{E}_B(S^c) \tag{3–32}$$

Taking complements, applying duality, and recognizing that $\Psi^*(S^c)^c = \Psi(S)$ yields

$$\Psi(S) = \bigcap_{B \in \mathcal{K}[\Psi^*]} \mathrm{E}_B(S^c)^c \tag{3–33}$$

The result follows since dilation by $-B$ is the dual of erosion by B. ■

The erosion representation of an increasing, translation-invariant operator over the kernel is highly redundant. Redundancy can often be eliminated as in the finite-logical setting; that is, by deleting from the union expansion all erosions whose structuring elements contain structuring elements of other erosions. In the infinite setting this reduction is a bit problematic because of the existence of infinite totally ordered families of sets, a family $\{S_i\}_{i \in I}$ (I countable or uncountable) being totally ordered if for any $i, i' \in I$, either $S_i \subset S_{i'}$ or $S_{i'} \subset S_i$. The problem is one of transfinite induction because no finite search through the family need reach a lower bound for the family, a lower bound being a set S such that $S \subset S_i$ for all i. We can define the concept of basis and show uniqueness of a basis should one exist in a manner analogous to the finite-logical setting; however, to provide a sufficient condition for the existence of a basis (which always must exist in the finite-logical setting), we will need to employ Zorn's lemma.

A subset $\mathcal{B}[\Psi]$ of $\mathcal{K}[\Psi]$ is called a *(increasing) basis* if (1) no set in $\mathcal{B}[\Psi]$ is a proper subset of a distinct set in $\mathcal{B}[\Psi]$ and (2) for any $T \in \mathcal{K}[\Psi]$, there exists $B \in \mathcal{B}[\Psi]$ such that $B \subset T$. The notation "$\mathcal{B}[\Psi]$" is permitted since there can be at most one basis. To see uniqueness, suppose $\mathcal{B}_1$ and $\mathcal{B}_2$ are subsets of $\mathcal{K}[\Psi]$ satisfying the two basis conditions and $B \in \mathcal{B}_1$. Since $B \in \mathcal{K}[\Psi]$, there exists $B' \in \mathcal{B}_2$ such that $B' \subset B$ and, since $B' \in \mathcal{K}[\Psi]$, there exists $B'' \in \mathcal{B}_1$ such that $B'' \subset B' \subset B$. Since $B'' \subset B$ and $B'', B \in \mathcal{B}_1$, it must be that $B'' = B$. Hence, $B = B' \in \mathcal{B}_2$ and $\mathcal{B}_1 \subset \mathcal{B}_2$. By symmetry, $\mathcal{B}_2 \subset \mathcal{B}_1$ and $\mathcal{B}_1 = \mathcal{B}_2$.

THEOREM 3–2 [9]. *If Ψ is increasing, translation invariant, and possesses a basis, then*

$$\Psi(S) = \bigcup_{B \in \mathcal{B}[\Psi]} \mathrm{E}_B(S) \tag{3–34}$$

Moreover, if Ψ commutes with intersection, then it possesses a basis.

PROOF. We first show the union representation. Since $\mathcal{B}[\Psi] \subset \mathcal{K}[\Psi]$, the union over $\mathcal{B}[\Psi]$ is a subset of the union of $\mathcal{K}[\Psi]$. On the other hand, if $B \in \mathcal{K}[\Psi]$, there exists $B' \in \mathcal{B}[\Psi]$ such that $B' \subset B$. The latter inclusion implies $\mathrm{E}_B(S) \subset \mathrm{E}_{B'}(S)$. Thus, the union over $\mathcal{K}[\Psi]$ is a subset of the union over $\mathcal{B}[\Psi]$ and Eq. 3–34 is verified. Next, suppose Ψ commutes with intersection. Let $\mathcal{B}$ denote the collection of all sets in $\mathcal{K}[\Psi]$ that do not have a proper subset in $\mathcal{K}[\Psi]$. We show that $\mathcal{B}$ is a basis. The first condition is trivially fulfilled. As for the second condition, let $S \in \mathcal{K}[\Psi]$ and let $\mathcal{S}$ denote the class of all subsets of S in $\mathcal{K}[\Psi]$. $\mathcal{S}$ is partially ordered by containment. Suppose $\{U_i\}_{i \in I}$ is a totally ordered subset of $\mathcal{S}$. Let $U_0 = \bigcap_{i \in I} U_i$. Since $U_i \in \mathcal{K}[\Psi]$ for all i, $0 \in \Psi(U_i)$ for all i. Since Ψ commutes with intersection, $0 \in \bigcap_{i \in I} \Psi(U_i) = \Psi(U_0)$. Hence $U_0 \in \mathcal{K}[\Psi]$ and $U_0 \in \mathcal{S}$. Moreover, U_0 is a lower bound for $\{U_i\}$. Consequently, every totally ordered subclass of $\mathcal{S}$ possesses a minimal element in $\mathcal{S}$. By Zorn's lemma, $\mathcal{S}$ possesses a minimal element, say S_0. We claim $S_0 \in \mathcal{B}$. If not, then S_0 has a proper subset $S_1 \in \mathcal{S}$, but this would mean that S_0 is not minimal. Hence, $S_0 \subset S$ and $S_0 \in \mathcal{B}$, verifying that $\mathcal{B}$ is a basis for Ψ. ■

PROPOSITION 3–8 [10]. *If Ψ is increasing, translation invariant, and every set in $\mathcal{K}[\Psi]$ possesses a finite subset in $\mathcal{K}[\Psi]$, then Ψ has a basis and every basis set is finite.*

PROOF. Again let $\mathcal{B}$ denote the collection of all sets in $\mathcal{K}[\Psi]$ that do not have a proper subset in $\mathcal{K}[\Psi]$. To show the second condition for a basis, let $S \in \mathcal{K}[\Psi]$. There exists a finite set $S_1 \in \mathcal{K}[\Psi]$ such that $S_1 \subset S$. If $S_1 \in \mathcal{B}$, we are done; if not, there exists a proper subset S_2 of S_1 such that $S_2 \in \mathcal{K}[\Psi]$. Continue the process. Since S_1 is finite, the process must come to an end with some $S_m \in \mathcal{B}$. ■

3.3 REPRESENTATION OF NONINCREASING TRANSLATION-INVARIANT OPERATORS

Representation of a nonincreasing, translation-invariant mapping depends on the *hit-or-miss transform*, which for disjoint structuring elements (sets) E and F is defined by

$$\Lambda_{E,F}(S) = \mathrm{E}_E(S) \cap \mathrm{E}_F(S^c) \tag{3–35}$$

For representation, we require the notion of the *closed interval* $[A, B]$ determined by two sets A and B, which is simply the family of all sets $T \in \mathcal{P}$ such that $A \subset T \subset B$. If $\mathcal{C}$ is a collection of sets, then $\mathcal{C}$ is equal to the union of closed intervals contained within it:

$$\mathcal{C} = \bigcup_{[A,B]\subset\mathcal{C}} [A, B] \tag{3–36}$$

THEOREM 3–3 [11]. *If Ψ is translation invariant, then it possesses the representation*

$$\Psi(S) = \bigcup_{[E,F^c]\subset\mathcal{K}[\Psi]} \Lambda_{E,F}(S) \tag{3–37}$$

PROOF. Because the intervals within the kernel cover the kernel,

$$\mathcal{K}[\Psi] = \bigcup_{[E,F^c]\subset\mathcal{K}[\Psi]} [E, F^c] \tag{3–38}$$

It is immediately apparent from the definitions of the kernel and the hit-or-miss transform that the kernel of the hit-or-miss transform with structuring pair (E, F) is $[E, F^c]$. Consequently, the theorem follows, as did Theorem 3–1: the kernel of the mapping defined by the union is equal to the union of the kernels, this latter union is $\mathcal{K}[\Psi]$, and two mappings possessing the same kernel are identical. ■

Dual representation is achieved via the dual of the hit-or-miss transform. Application of De Morgan's laws in conjunction with the duality relation between erosion and dilation yields the dual, $\Lambda^*_{E,F}$, of the hit-or-miss transform $\Lambda_{E,F}$:

$$\Lambda^*_{E,F}(S) = \Delta_{-E}(S) \cup \Delta_{-F}(S^c) \tag{3–39}$$

COROLLARY 3–2. *If Ψ is translation invariant, then it possesses the representation*

$$\Psi(S) = \bigcup_{[E,F^c]\subset\mathcal{K}[\Psi^*]} \Lambda^*_{E,F}(S) \tag{3–40}$$

PROOF. Apply the theorem and De Morgan's law to Ψ expressed as $\Psi(S) = \Psi^*(S^c)^c$. ■

As with finite-logic representation (disjunctive normal form), reduction is possible. A set $\mathcal{B}$ of closed intervals in $\mathcal{K}[\Psi]$ is said to satisfy the *representation condition* for Ψ if and only if for any closed interval $[A, B] \subset \mathcal{K}[\Psi]$ there exists a closed interval $[A', B'] \in \mathcal{B}$ such that $[A, B] \subset [A', B']$.

THEOREM 3–4 [11]. *If Ψ is translation invariant and $\mathcal{B}$ is a set of closed intervals contained in $\mathcal{K}[\Psi]$ satisfying the representation condition for Ψ, then*

$$\Psi(S) = \bigcup_{[E,F^c]\in\mathcal{B}} \Lambda_{E,F}(S) \tag{3–41}$$

PROOF. Let Φ be the mapping defined by the union. We need to show that $\Phi = \Psi$. Since each interval in $\mathcal{B}$ is a subset of $\mathcal{K}[\Psi]$, $\Phi \subset \Psi$ by Theorem 3–3. In the other direction, for each $[E, F^c] \subset \mathcal{K}[\Psi]$, there exists $[E', F'^c] \in \mathcal{B}$ such that $[E, F^c] \subset [E', F'^c]$. Hence,

$$\Psi(S) = \bigcup_{[E,F^c]\subset\mathcal{K}[\Psi]} \Lambda_{E,F}(S) \subset \bigcup_{[E,F^c]\subset\mathcal{K}[\Psi]} \Lambda_{E',F'}(S) \subset \Phi(S) \ \blacksquare \tag{3–42}$$

As for minimal representation, the situation is somewhat problematic. In the present setting the *basis*, $\underline{\mathcal{B}}[\Psi]$, is defined to be the collection of all maximal closed intervals contained in $\mathcal{K}[\Psi]$, where by a *maximal closed interval* in $\mathcal{K}[\Psi]$ we mean a closed interval in $\mathcal{K}[\Psi]$ that is not properly contained in any other closed interval contained in $\mathcal{K}[\Psi]$. According to Theorem 3–4, if $\underline{\mathcal{B}}[\Psi]$ satisfies the representation condition, then the expansion can be taken over $\underline{\mathcal{B}}[\Psi]$:

$$\Psi(S) = \bigcup_{[E,F^c]\in\underline{\mathcal{B}}[\Psi]} \Lambda_{E,F}(S) \tag{3–43}$$

Moreover, if $\mathcal{B}$ is a collection of closed intervals satisfying the representation condition for Ψ, then $\underline{\mathcal{B}}[\Psi] \subset \mathcal{B}$ (were there a closed interval in $\underline{\mathcal{B}}[\Psi] - \mathcal{B}$, there would be a distinct closed interval in $\mathcal{B}$ containing it, thereby contradicting the $\underline{\mathcal{B}}[\Psi]$ maximality condition). Thus, in the sense that we are considering expansions over collections of closed intervals satisfying the representation condition, expansion over $\underline{\mathcal{B}}[\Psi]$ is minimal. In fact, there may still be redundancy because it is possible for a union of closed intervals in $\underline{\mathcal{B}}[\Psi]$ to contain a distinct closed interval in $\underline{\mathcal{B}}[\Psi]$ that can be deleted from the union expansion.

The minimization problem can be seen in the finite-logic setting. There, each minterm corresponds to a kernel element and redundancy is decreased via logic reduction. In the reduction method of Quine and McCluskey, the prime implicants of the Boolean function ψ correspond to the maximal closed intervals in the kernel, thereby leading to a basis expansion. If, however, an optimal set of prime implicants is found, then the products in the expansion correspond to maximal closed intervals contained in $\mathcal{K}[\Psi]$, and forming the hit-or-miss union expansion using these maximal closed intervals can result in a simpler expansion than the one over $\underline{\mathcal{B}}[\Psi]$.

Besides the general problem of minimization, there is another key difference between the representations of increasing and nonincreasing mappings. For increasing mappings, the basis need not exist; for nonincreasing mappings the basis exists by definition. But for nonincreasing mappings, existence of the basis by definition does not ensure that it satisfies the representation condition; whereas, for increasing mappings, existence of the basis *ipso facto* provides representation. Hence, the corresponding problem to existence of the basis for an increasing mapping is satisfaction of the representation condition for nonincreasing mappings. While a sufficient condition for existence in the increasing case can be framed in terms of commutation with intersection, a corresponding condition for satisfaction of the representation condition in the nonincreasing case involves formulation in terms of the hit-or-miss topologyand we shall leave that to the literature.

3.4 OPENINGS AND CLOSINGS

We have thus far considered representation relative to translation invariance and increasingness. Now we consider the effect of antiextensivity and idempotence. An operator Ψ is *idempotent* if $\Psi\Psi = \Psi$. The study of these properties involves openings: the *opening* of S by structuring element B is denoted by $\Gamma_B(S)$ [or $S \circ B$] and defined by the operator composition

$$\Gamma_B = \Delta_B \mathrm{E}_B \tag{3–44}$$

The next proposition gives a key property of opening for image processing: the opening is equal to the union of all translations of the structuring element that are subsets of the input image. It follows that opening is independent of the position of the structuring element: $\Gamma_B(S) = \Gamma_{B_y}(S)$ for any point y.

PROPOSITION 3–9. *Opening is given by*

$$\Gamma_B(S) = \bigcup_{B_y \subset S} B_y \tag{3–45}$$

PROOF. $z \in \Gamma_B(S)$ if and only if there exists $b \in B$ such that $z \in \mathrm{E}_B(S)_b$ if and only if there exists $b \in B$ and $w \in \mathrm{E}_B(S)$ such that $w = -b + z$ if and only if there exists w such that $w \in (-B)_z \cap \mathrm{E}_B(S)$ if and only if there exists w such $z \in B_w$ and $B_w \subset S$ if and only if z lies in the union of Eq. 3–45. ■

PROPOSITION 3–10. *Opening satisfies the following properties*:

(i) $\Gamma_B(S_z) = \Gamma_B(S)_z$ [*translation invariant*]

(ii) $S \subset T$ *implies* $\Gamma_B(S) \subset \Gamma_B(T)$ [*increasing*]

(iii) $\Gamma_B(S) \subset S$ [*antiextensive*]

(iv) $\Gamma_B\Gamma_B = \Gamma_B$ [*idempotent*]

PROOF. (i) As an iteration of translation-invariant mappings, opening is translation invariant. (ii) $B_y \subset S$ implies $B_y \subset T$, so that increasingness follows from Proposition 3–9. (iii) Antiextensivity follows at once from Proposition 3–9. (iv) $\Gamma_B(\Gamma_B(S)) \subset \Gamma_B(S)$ by antiextensivity. For the reverse inclusion, $z \in \Gamma_B(S)$ implies there exists y such that $z \in B_y \subset S$. $B_y \subset S$ implies $B_y \subset \Gamma_B(S)$, so that $z \in B_y \subset \Gamma_B(S)$. By Proposition 3–9, $z \in \Gamma_B(\Gamma_B(S))$. ■

Consider the signal-union-noise model, in which there is an underlying uncorrupted image S, a noise image N, and a corrupted image $S \cup N$. As a filter, opening is translation invariant, and because it is increasing and antiextensive,

$$\Gamma_B(S) \subset \Gamma_B(S \cup N) \subset S \cup N \tag{3–46}$$

Figure 3–1 shows how opening by a disk can be used to filter small background clutter from an image. The small clutter is removed because no translate of the structuring element fits into the clutter components. For this example, $\Gamma_B(S) = \Gamma_B(S \cup N)$.

A key property is respective openness. S is *open with respect to* B if $\Gamma_B(S) = S$. We say that S is B-*open*.

PROPOSITION 3–11. *If A is B-open, then $\Gamma_A(S) \subset \Gamma_B(S)$.*

PROOF. $z \in \Gamma_A(S)$ implies there exists y such that $z \in A_y \subset S$. Let $z = w + y \in S$, with $w \in A$. Since A is B-open, for each $a \in A$ there exists point $b(a)$ such that $a \in B_{b(a)} \subset A$. Hence, $z \in (B_{b(w)})_y \subset A_y \subset S$, so that $z \in \Gamma_B(S)$. ■

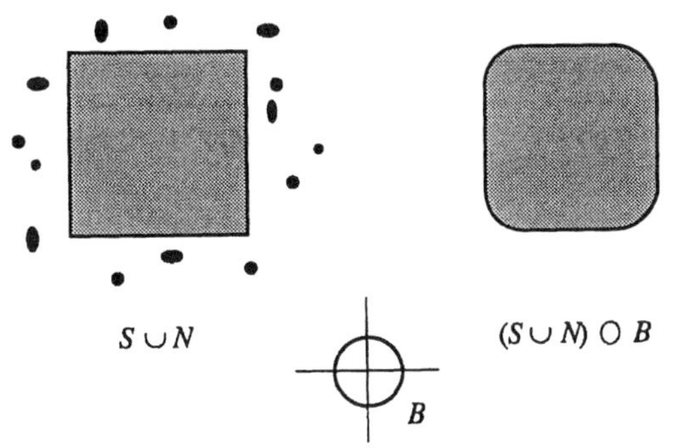

Figure 3–1. Suppression of clutter by opening.

PROPOSITION 3–12. *S is B-open if and only if there exists a set A such that* $S = \Delta_B(A)$.

PROOF. If S is B-open, then $S = \Delta_B(\mathrm{E}_B(S))$, so the implication holds with $A = \mathrm{E}_B(S)$. Conversely, suppose $S = \Delta_B(A)$ and $z \in S$. Then there exists $a \in A$ such that $z \in B_a$. By Eq. 3–19, $B_a \subset S$. Hence, $z \in \Gamma_B(S)$ and $S \subset \Gamma_B(S)$. By antiextensivity, $S = \Gamma_B(S)$. ■

A special case of Proposition 3–12 occurs when A and B are disks of radii r and s, respectively. Then $\Delta_B(A)$ is a disk of radius $r + s$. Hence, a disk is open with respect to any disk possessing a smaller radius.

PROPOSITION 3–13. *If A is B-open, then, for any image S,*

$$\Gamma_A\big(\Gamma_B(S)\big) = \Gamma_B\big(\Gamma_A(S)\big) = \Gamma_A(S) \tag{3–47}$$

PROOF. By antiextensivity and increasingness, $\Gamma_A(\Gamma_B(S)) \subset \Gamma_A(S)$. Since A is B-open, $\Gamma_A(S) \subset \Gamma_B(S)$. Idempotence implies

$$\Gamma_A(S) = \Gamma_A\big(\Gamma_A(S)\big) \subset \Gamma_A\big(\Gamma_B(S)\big) \tag{3–48}$$

and we conclude that $\Gamma_A(\Gamma_B(S)) = \Gamma_A(S)$. To prove the other identity, suppose $x \in \Gamma_A(S)$. Then there exists y such that $x \in A_y \subset S$. By B-openness, $x \in \Gamma_B(A)_y \subset S$. Thus, there exists a point w such that $x \in (B_w)_y \subset A_y \subset S$. By Proposition 3–9, $x \in (B_w)_y \subset \Gamma_A(S)$, so that $x \in \Gamma_B(\Gamma_A(S))$. Thus, $\Gamma_A(S) \subset \Gamma_B(\Gamma_A(S))$. The reverse inclusion follows by antiextensivity. ■

As an increasing, translation-invariant mapping, opening must possess an erosion representation. According to Proposition 3–9, $0 \in \Gamma_B(S)$ if and only if there exists

$y \in B$ such that $0 \in B_y \subset S$, which means that $y \in -B$ and $B_y \subset S$. The matter can be expressed as there exists $z \in B$ such that $B_{-z} \subset S$. Hence,

$$\mathcal{K}[\Gamma_B] = \bigcup_{z \in B} \{S\colon B_{-z} \subset S\} \tag{3–49}$$

It follows at once that $\{B_{-z}\}_{z \in B}$ satisfies the second condition of a basis, but it is not necessarily true that the first condition regarding no proper inclusions among basis sets is satisfied. However, if B is finite, then each translation possesses equal finite cardinality and there can be no proper subset relation among them, so that for finite structuring elements, $\mathcal{B}[\Gamma_B]$ consists of all translates of B that contain the origin. The finite assumption applies at once to window operators in the discrete setting.

Consider the digital setting and let B be a 2×2 structuring element. Opening by B can be obtained by unioning four erosions having the following four structuring elements:

$$\begin{pmatrix} 1 & 1 & 0 \\ 1 & \mathbf{1} & 0 \\ 0 & 0 & 0 \end{pmatrix} \quad \begin{pmatrix} 0 & 1 & 1 \\ 0 & \mathbf{1} & 1 \\ 0 & 0 & 0 \end{pmatrix} \quad \begin{pmatrix} 0 & 0 & 0 \\ 1 & \mathbf{1} & 0 \\ 1 & 1 & 0 \end{pmatrix} \quad \begin{pmatrix} 0 & 0 & 0 \\ 0 & \mathbf{1} & 1 \\ 0 & 1 & 1 \end{pmatrix}$$

Figure 3–2 illustrates opening of a set A using the basis expansion. Using the variables of Fig. 1–3, the corresponding Boolean function is

$$\gamma_B(x_1, x_2, \ldots, x_9) = x_1x_2x_4x_5 + x_2x_3x_5x_6 + x_4x_5x_7x_8 + x_5x_6x_8x_9 \tag{3–50}$$

Architecturally, the combinational logic circuit is shown in Fig. 3–3.

The dual of opening is called *closing*, is denoted by $\Phi_B(S)$ [or $S \bullet B$], and defined by

$$\Phi_B(S) = \Gamma_B\left(S^c\right)^c \tag{3–51}$$

Straightforward algebra yields

$$\Phi_B = \mathrm{E}_{-B}\Delta_{-B} \tag{3–52}$$

By duality and Proposition 3–10, closing is translation invariant, increasing, extensive, and idempotent. S is said to be B-*closed* if $\Phi_B(S) = S$. By duality, S is B-open if and only if S^c is B-closed. By duality in conjunction with Proposition 3–12, S is B-closed if and only if there exists a set A such that $S = \mathrm{E}_{-B}(A)$. Figure 3–4 shows a text image, the image degraded by salt noise, and the result of filtering the salt-degraded image by a 2×2 closing.

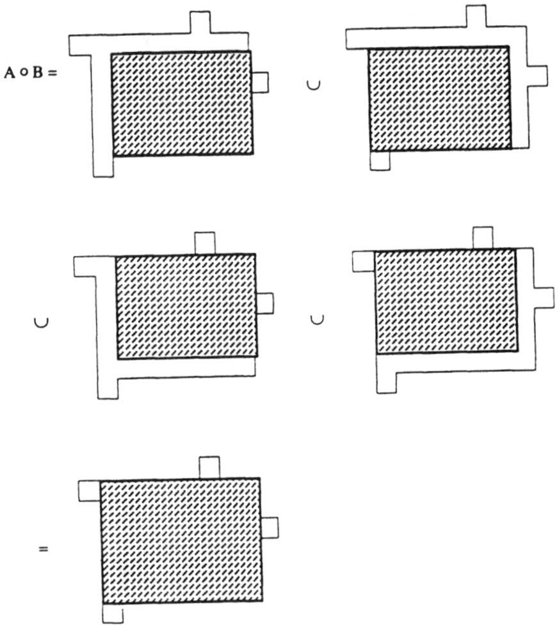

Figure 3–2. Erosion representation of 2 × 2 opening.

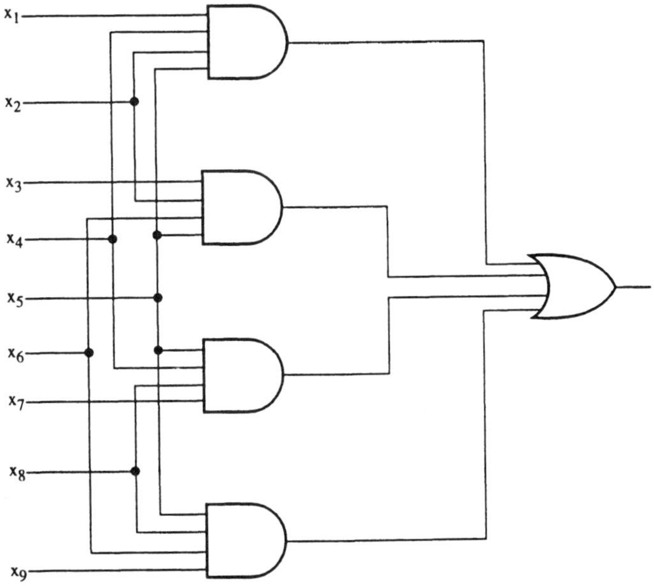

Figure 3–3. Combinational logic for 2 × 2 opening.

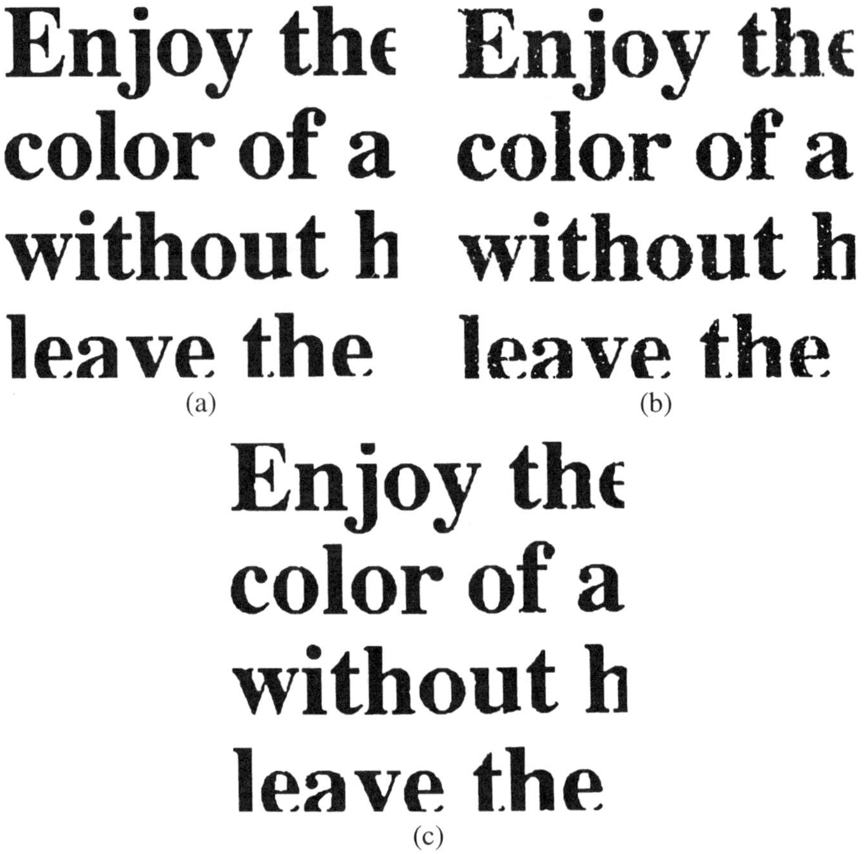

Figure 3–4. Restoration by 2 × 2 closing: (a) original image; (b) image degraded by salt noise; (c) closed degraded image.

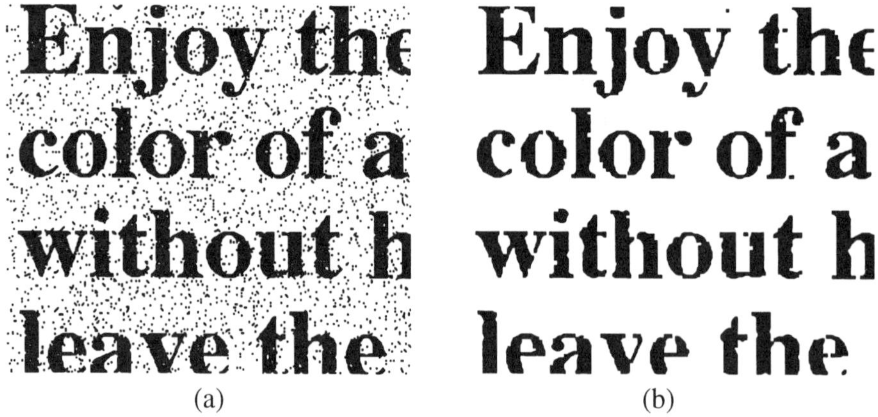

Figure 3–5. Restoration by 2 × 2 open-close: (a) image degraded by salt-and-pepper noise; (b) open-closed degraded image.

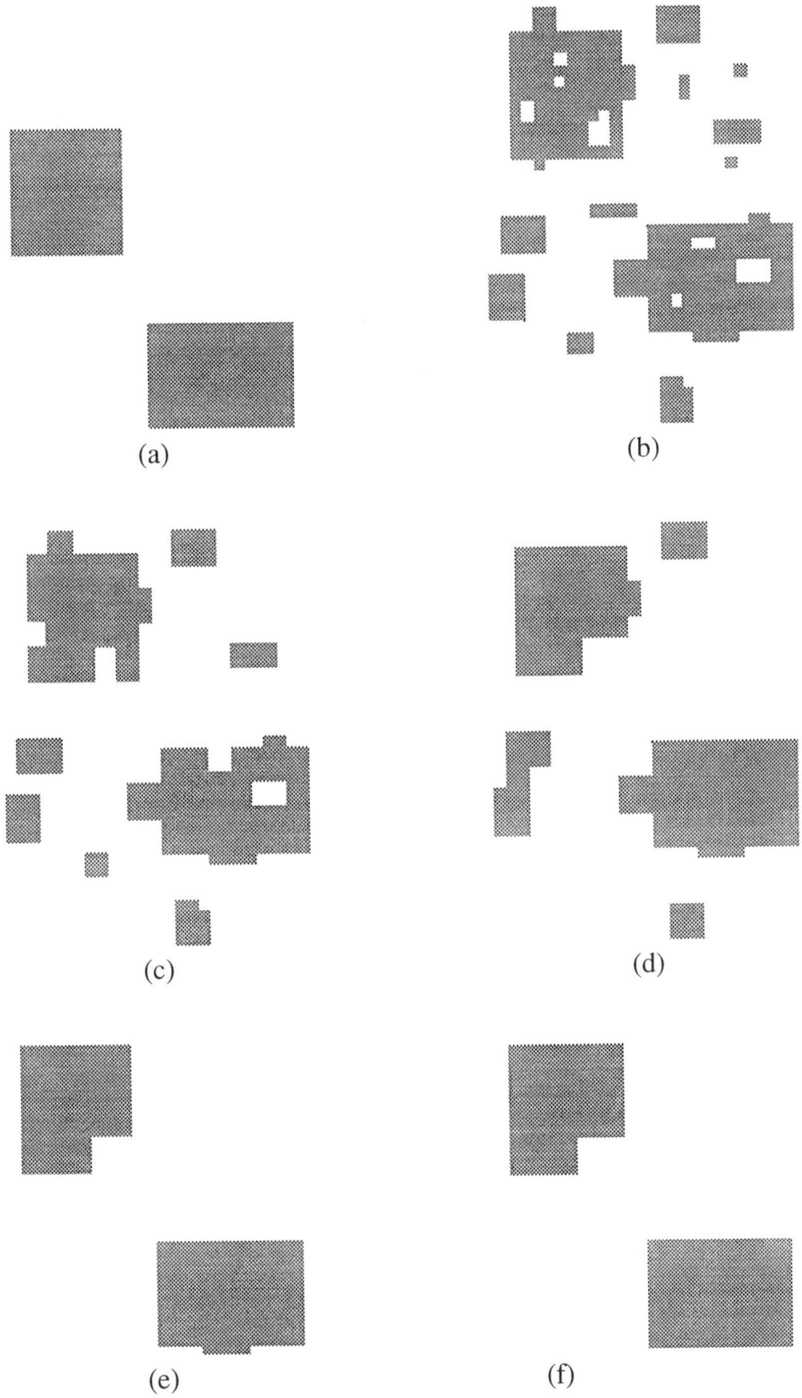

Figure 3–6. Restoration by alternating sequential filter: (a) original image; (b) image corrupted by both union and subtractive noise; (c) result of single-stage ASF; (d) result of two-stage ASF; (e) result of three-stage ASF; (f) result of four-stage ASF.

Opening can be used to filter pepper noise; closing can be used to filter salt noise. When there is both union and subtractive noise, one strategy is to open to eliminate union noise in the background and then close to fill subtractive noise in the foreground. The resulting filter is called an *open-close*. By duality, one can also close and then open, the filter then being called a *close-open*. Open-close is illustrated in Fig. 3–5, which shows a salt-and-pepper degraded realization of the original text image of Fig. 3–4 and the result of open-close with a 2×2 structuring element applied to the noisy image.

A potential pitfall of the open-close strategy occurs when large noise components need to be eliminated but a direct attempt to do so will destroy too much of the original image. One way around this problem is to employ an alternating sequential filter (ASF) [12]. Open-close filters are performed iteratively, beginning with a very small structuring element and then proceeding with ever-increasing structuring elements. The strategy is to eliminate small salt and pepper components, thereby allowing the larger structuring elements to more likely fit when they are eventually applied in the process. Figure 3–6 shows an uncorrupted image; the image corrupted by both union and subtractive noise; and the results of single-stage, two-stage, three-stage, and four-stage alternating-sequential filters.

3.5 REPRESENTATION OF OPENINGS AND CLOSINGS

Opening can be generalized in accordance with the basic properties of Proposition 3–10. A mapping $\Psi\colon \mathcal{P} \to \mathcal{P}$ is called an *algebraic opening* if it is increasing, antiextensive, and idempotent. Further, Ψ is called a *τ-opening* if it is translation invariant. Ψ is called an *algebraic closing* if it is increasing, extensive, and idempotent. It is called a *τ-closing* if it also translation invariant. Ψ is a τ-opening if and only if Ψ^* is a τ-closing.

The *invariant class* of any mapping Ψ, to be denoted by $Inv[\Psi]$, is the collection of all sets S for which $\Psi(S) = S$. Owing to idempotence, for any algebraic opening Ψ and any set S, $\Psi(S) \in Inv[\Psi]$. If Ψ is a τ-opening, then $S \in Inv[\Psi]$ if and only if $S_y \in Inv[\Psi]$ for all y, which means that $Inv[\Psi]$ is closed under translation. In fact, an algebraic opening Ψ is a τ-opening if and only if its invariant class is closed under translation.

A class $\mathcal{B}$ is called a *base* for Ψ, or a base for $Inv[\Psi]$, if $Inv[\Psi]$ is the class generated by $\mathcal{B}$ under translations and unions. This means that $S \in Inv[\Psi]$ if and only if there exists a subfamily $\{B_i\}_{i \in I}$ of $\mathcal{B}$ and points y_i such that

$$S = \bigcup_{i \in I} (B_i)_{y_i} \tag{3–53}$$

Bases for τ-openings are not unique. Indeed, $Inv[\Psi]$ is a base for itself. To see this, we need only show that $Inv[\Psi]$ is closed under unions since we already know it is closed under translation. If $S_i \in Inv[\Psi]$ for all $i \in I$, then

$$\Psi\left(\bigcup_{i\in I} S_i\right) \supset \Psi(S_i) = S_i \tag{3–54}$$

for all i and therefore

$$\Psi\left(\bigcup_{i\in I} S_i\right) \supset \bigcup_{i\in I} S_i \tag{3–55}$$

The reverse inclusion follows from antiextensivity, so that $\bigcup_{i\in I} S_i \in Inv[\Psi]$. The intent is to find small bases that determine filter behavior. A theorem of Matheron shows that openings by structuring elements are primary for representation of τ-openings.

THEOREM 3–5 [1]. *A mapping Ψ: $\mathcal{P} \to \mathcal{P}$ is a τ-opening if and only if there exists a class of sets $\mathcal{B}$ such that*

$$\Psi(S) = \bigcup_{B\in\mathcal{B}} \Gamma_B(S) \tag{3–56}$$

Moreover, $\mathcal{B}$ is a base for Ψ.

PROOF. First suppose Ψ is given by the union. As a union of increasing, translation invariant mappings, Ψ is increasing and translation invariant. As a union of antiextensive mappings, Ψ is antiextensive. As for idempotence, because of antiextensivity we need only show that $\Psi(S) \subset \Psi(\Psi(S))$. Now, $z \in \Psi(S)$ implies there exists $B \in \mathcal{B}$ and a point y such that $z \in B_y \subset \Gamma_B(S) \subset \Psi(S)$. Since $B_y \subset \Psi(S)$, application of Ψ a second time will leave $B_y \subset \Psi(\Psi(S))$ because $\Psi(B_y) = B_y$. Conversely, suppose Ψ is a τ-opening. Let Ξ be defined by the union with $\mathcal{B} = Inv[\Psi]$. Since we already know that Ξ is a τ-opening we need only show that $\Psi = \Xi$. According to Proposition 3–9,

$$\Xi(S) = \bigcup_{B\in Inv[\Psi],\, B_y\subset S} B_y \tag{3–57}$$

By idempotence, $\Psi(S) \in Inv[\Psi]$. Hence, $\Psi(S) \subset \Xi(S)$. Now, suppose $z \in \Xi(S)$. Then there exists $B \in Inv[\Psi]$ and a point y such that $z \in B_y \subset S$. Since Ψ is increasing and translation invariant, $\Psi(S) \supset \Psi(B_y) = \Psi(B)_y = B_y$. Hence, $z \in \Psi(S)$ and $\Psi = \Xi$. The claim that the union is over a base holds because $Inv[\Psi]$ is a base for itself. ■

COROLLARY 3–3. *If Ψ is a τ-opening, then its dual (a τ-closing) has the representation*

$$\Psi^*(S) = \bigcap_{B \in \mathcal{B}} \Gamma_B(S) \tag{3–58}$$

where $\mathcal{B}$ is a base for Ψ.

The representation of Eq. 3–56 provides a filter-design paradigm. If an image is composed of a disjoint union of grains (connected components), then unwanted grains can be eliminated according to their sizes relative to the structuring elements in the base $\mathcal{B}$. A key to good filtering is selection of appropriately sized structuring elements, since we wish to minimize elimination of signal grains and maximize elimination of noise grains. Optimally and adaptively designed τ-openings will be discussed in Chapter 4 in the context of granulometries.

Since each opening in the expansion of Eq. 3–56 can, according to Theorem 3–1, be represented as a union of erosions, substituting the erosion representation of each opening into Eq. 3–56 *ipso facto* produces an erosion representation for Ψ. However, even if a basis expansion is used for each opening, there is redundancy in the resulting expansion. In the case of a finite number of finite digital structuring elements, there exists a procedure to produce a minimal erosion expansion from the opening representation [13].

As applied to a grain image according to Eq. 3–56, a τ-opening passes certain components while eliminating others, but affects the passing grains. Corresponding to each τ-opening Ψ is an induced *reconstructive* τ-opening $\mathfrak{R}(\Psi)$ defined in the following manner: a connected component is passed in full by $\mathfrak{R}(\Psi)$ if it is not eliminated by Ψ; a connected component is eliminated by $\mathfrak{R}(\Psi)$ if it is eliminated by Ψ.

3.6 CONVEXITY

Convex sets play a significant role in the construction of Euclidean set operators. A set S is *convex* if, for any two points $x, y \in S$, the line segment $\lambda(x, y)$ between x and y is a subset of S. The line segment is defined by

$$\lambda(x, y) = \{ax + by\colon\ a \geqslant 0,\ b \geqslant 0,\ a + b = 1\} \tag{3–59}$$

For any $t > 0$, tS is convex if S is convex. Since the intersection of convex sets is convex, Eq. 3–15 yields the convexity of $\mathrm{E}_B(S)$ whenever S is convex.

PROPOSITION 3–14. *If A and B are convex sets, so are the dilation, erosion, opening, and closing of A by B.*

PROOF. Suppose $z,\ w \in \Delta_B(A)$, $r + s = 1$, $r \geqslant 0$, and $s \geqslant 0$. According to Eq. 3–20, there exist $a, a' \in A$ and $b, b' \in B$ such that $z = a + b$ and $w = a' + b'$. Owing to convexity,

$$rz + sw = \left(ra + sa'\right) + \left(rb + sb'\right) \in \Delta_B(A) \tag{3–60}$$

establishing the convexity of $\Delta_B(A)$. Convexity of opening and closing follow, since each is an iteration of a dilation and an erosion. ■

It is generally true that $(r + s)A \subset rA \oplus sA$, but the reverse inclusion does not always hold. However, if A is convex, then we have the identity

$$(r + s)A = rA \oplus sA \tag{3–61}$$

PROPOSITION 3–15. *If $r \geqslant s > 0$ and B is convex, then $\Gamma_{rB}(S) \subset \Gamma_{sB}(S)$ for any set S.*

PROOF. From the property of Eq. 3–61,

$$rB = sB \oplus (r - s)B \tag{3–62}$$

By Proposition 3–12, rB is sB-open, and the conclusion follows from Proposition 3–11. ■

If $t \geqslant 1$ and we replace r, s, and B in Eq. 3–62 by t, 1, and A, respectively, then Proposition 3–12 establishes the following proposition: if A is convex, tA is A-open for any $t \geqslant 1$. The converse is not generally valid; however, a significant theorem of Matheron states that it is valid under the assumption that A is compact. The proof is quite involved and we state the theorem without proof.

THEOREM 3–6 [1]. *Let A be a compact set. Then tA is A-open for any $t \geqslant 1$ if and only if A is convex.*

REFERENCES

[1] Matheron, G., *Random Sets and Integral Geometry*, John Wiley, New York, 1975.

[2] Serra, J., *Image Analysis and Mathematical Morphology*, Academic Press, New York, 1983.

[3] Minkowski, H., "Volumen and Oberflache," *Mathematical Annals*, Vol. 57, 1903.

[4] Hadwiger, H., *Altes und Neues Uber Konvexe Korper*, Birkhauser-Verlag, Basel, 1955.

[5] Hadwiger, H., *Vorslesungen Uber Inhalt, Oberflache and Isoperimetrie*, Springer-Verlag, Berlin, 1957.

[6] Serra, J., "Mathematical Morphology for Complete Lattices," in *Image Analysis and Mathematical Morphology, Vol. 2: Theoretical Advances*, ed. J. Serra, Academic Press, New York, 1988.

[7] Heijmans, H. J., and C. Ronse, "The Algebraic Basis of Mathematical Morphology, I. Dilations and Erosions," *Computer Vision, Graphics, and Image Processing*, **50**, 1990.

[8] Ronse, C., and H. J. Heijmans, "The Algebraic Basis of Mathematical Morphology, II. Openings and Closings," *Computer Vision, Graphics, and Image Processing*, **54**, 1991.

[9] Maragos, P., and R. Schafer, "Morphological Filters — Part I: Their Set-Theoretic Analysis and Relations to Linear Shift-Invariant Filters," *IEEE Trans. on Acoustics, Speech, and Signal Processing*, **35**, 1987.

[10] Giardina, C., R., and E. R. Dougherty, *Morphological Methods in Image and Signal Processing*, Prentice-Hall, Englewood Cliffs, 1988.

[11] Banon, G. J. F., and J. Barrera, "Minimal Representation of Translation-Invariant Set Mappings by Mathematical Morphology," *SIAM Journal on Applied Mathematics*, **51** (6), 1991.

[12] Sternberg, S., "Grayscale Morphology," *Computer Vision, Graphics, and Image Image Processing*, **35** (3), 1986.

[13] Dougherty, E. R., "Minimal Representation of τ-Openings via Pattern Bases," *Pattern Recognition Letters*, **14** (3), 1994.

CHAPTER 4

GRANULOMETRIC FILTERS

Edward R. Dougherty
Texas A&M University
College Station, Texas

Yidong Chen
National Institutes of Health
Bethesda, Maryland

Granulometries were introduced by Matheron to model parameterized sieving processes operating on random sets [1]. If an opening is applied to a binary image composed of a collection of disjoint grains, then some grains are passed (perhaps with diminution) and some are eliminated. If the structuring element is decreased or increased in size, then grains are more or less likely to pass. From the perspective of filtering, a parameterized class of filters is created and this permits us to try to find an optimal filter in the class. The situation is akin to optimal linear filtering, where the task is to find an optimal set of weights. As with linear filtering, we can consider optimal, adaptive, and bandpass filters. In addition to image-to-image filtering, a parameterized opening filter can be viewed in terms of its diminishing effect on image volume as the structuring element(s) increase in size. The resulting size distributions are powerful image descriptors, especially for classification of random textures.

4.1 GRANULOMETRIES

To motivate the definition of a granulometry, consider a set S decomposed as a disjount union

$$S = \bigcup_{i=1}^{n} S_i \tag{4–1}$$

Imagine that the components are passed over a sieve of mesh size $t > 0$ and that a parameterized filter Ψ_t is defined componentwise according to whether a component does or does not pass through the sieve: $\Psi_t(S_i) = S_i$ if S_i does not fall through the sieve; $\Psi_t(S_i) = \emptyset$ if S_i falls through the sieve. For the overall set,

$$\Psi_t(S) = \bigcup_{i=1}^{n} \Psi_t(S_i) \tag{4–2}$$

Since the components of $\Psi_t(S)$ form a subcollection of the components of S, Ψ_t is antiextensive. If $T \supset S$, then each component of T must contain a component of S, so that $\Psi_t(T) \supset \Psi_t(S)$ and Ψ_t is increasing. If the components are sieved iteratively with two different mesh sizes, then the output after both iterations depends only on the larger of the mesh sizes.

In accordance with these remarks and with $\mathcal{P}$ being the power set of R^n, an *algebraic granulometry* is defined as a family of operators Ψ_t: $\mathcal{P} \to \mathcal{P}$, $t > 0$, satisfying three properties:

(a) Ψ_t is antiextensive;

(b) Ψ_t is increasing;

(c) $\Psi_r \Psi_s = \Psi_s \Psi_r = \Psi_{\max\{r,s\}}$ for $r, s > 0$ [*mesh property*].

If $\{\Psi_t\}$ is an algebraic granulometry and $r \geqslant s$, then $\Psi_r = \Psi_s \Psi_r \subset \Psi_s$, where the equality follows from the mesh property and the inclusion from the antiextensivity of Ψ_r and the increasingness of Ψ_s. The granulometric axioms are equivalent to two conditions.

PROPOSITION 4–1. *$\{\Psi_t\}$ is an algebraic granulometry if and only if*

(i) *for any $t > 0$, Ψ_t is an algebraic opening*;

(ii) *$r \geqslant s > 0$ implies $Inv[\Psi_r] \subset Inv[\Psi_s]$* [invariance ordering].

PROOF. Assuming $\{\Psi_t\}$ is an algebraic granulometry, we need to show idempotence and the invariance ordering of (ii). For idempotence, $\Psi_t \Psi_t = \Psi_{\max\{t,t\}} = \Psi_t$. For invariance ordering, suppose $S \in Inv[\Psi_r]$. Then

$$\Psi_s(S) = \Psi_s\big(\Psi_r(S)\big) = \Psi_{\max\{s,r\}}(S) = \Psi_r(S) = S \tag{4–3}$$

To prove the converse, we need only show condition (c), since conditions (a) and (b) hold by virtue of Ψ_t being an algebraic opening. Suppose $r \geqslant s > 0$. By idempotence and condition (ii), $\Psi_r(S) \in Inv[\Psi_r] \subset Inv[\Psi_s]$. Hence, $\Psi_s \Psi_r = \Psi_r$. Consequently,

$$\Psi_r = \Psi_r \Psi_r = \Psi_r \Psi_s \Psi_r \subset \Psi_r \Psi_s \subset \Psi_r \tag{4–4}$$

where the two inclusions hold because Ψ_s is antiextensive and Ψ_r increasing. It follows that $\Psi_r \subset \Psi_r\Psi_s \subset \Psi_r$ and $\Psi_r\Psi_s = \Psi_r = \Psi_{\max\{s,r\}}$. In a way similar to Eq. 4–4,

$$\Psi_r = \Psi_r\Psi_r = \Psi_s\Psi_r\Psi_r \subset \Psi_s\Psi_r \subset \Psi_r \tag{4–5}$$

and it follows that $\Psi_s\Psi_r = \Psi_r = \Psi_{\max\{s,r\}}$, so that condition (c) is satisfied. ■

If Ψ_t is translation invariant for all $t > 0$, then $\{\Psi_t\}$ is called a *granulometry*. For a granulometry, condition (i) of Proposition 4–1 is changed to say that Ψ_t is a τ-opening. If, for any $t > 0$, Ψ_t satisfies the *Euclidean property*,

$$\Psi_t(S) = t\Psi_1(S/t) \tag{4–6}$$

then $\{\Psi_t\}$ is called a *Euclidean granulometry*. We are slightly changing the original terminology of Matheron. He did not distinguish between an algebraic granulometry and a granulometry. If $\{\Psi_t\}$ satisfied the two conditions of Proposition 4–1, he simply called it a "granulometry" and he placed translation invariance with the Euclidean property in defining a Euclidean granulometry. Our teminology stems from our desire to separate the algebraic granulometric conditions, translation invariance, and the Euclidean property.

In terms of sieving, translation invariance means that the sieve mesh is uniform throughout the space. The Euclidean condition means that scaling a set by $1/t$, sieving by Ψ_1, and then rescaling by t is the same as sieving by Ψ_t. We call Ψ_1 the *unit* of the granulometry. The simplest Euclidean granulometry is an opening by a parameterized structuring element, Γ_{tB}. For it, the Euclidean property states that

$$\Gamma_{tB}(S) = t\Gamma_B(S/t) \tag{4–7}$$

Through Proposition 3–9: $x \in \Gamma_{tB}(S)$ if and only if there exists y such that $x \in (tB)_y \subset S$; but $(tB)_y = tB_{y/t}$, so that $x \in (tB)_y \subset S$ if and only if $x/t \in B_{y/t} \subset S/t$, which means that $x/t \in \Gamma_B(S/t)$.

4.2 REPRESENTATION OF EUCLIDEAN GRANULOMETRIES

There is a general representation for Euclidean granulometries; prior to giving it, we develop some preliminaries. A crucial point to be established is that not just any class of sets can serve as the invariant class of a granulometric unit.

PROPOSITION 4–2. *If $\{\Psi_t\}$ is a granulometry, then it satisfies the Euclidean condition if and only if Inv$[\Psi_t] = t$Inv$[\Psi_1]$, which means that $S \in$ Inv$[\Psi_t]$ if and only if $S/t \in$ Inv$[\Psi_1]$.*

PROOF. Suppose the Euclidean condition is satisfied and $S \in Inv[\Psi_t]$. Then $\Psi_1(S/t) = \Psi_t(S)/t = S/t$, so that $S/t \in Inv[\Psi_1]$. Now suppose $S/t \in Inv[\Psi_1]$. Then $\Psi_t(S) = t\Psi_1(S/t) = t(S/t) = S$, so that $S \in Inv[\Psi_t]$. To show the converse, let $\Xi_t(S) = t\Psi_1(S/t)$. We claim that Ξ_t is a τ-opening. Anitextensivity, increasingness, and translation invariance follow at once from the corresponding properties of Ψ_t. For idempotence,

$$\Xi_t\left[\Xi_t(S)\right] = t\Psi_1\left[t\Psi_1(S/t)/t\right] = t\Psi_1\left[\Psi_1(S/t)\right] = t\Psi_1(S/t) = \Xi_t(S) \qquad (4\text{–}8)$$

Next, $S \in Inv[\Xi_t]$ if and only if $S/t \in Inv[\Psi_1]$, which according to the hypothesis means that $S \in Inv[\Psi_t]$. Thus, $Inv[\Xi_t] = Inv[\Psi_t]$ and it follows from Theorem 3–5 that $\Psi_t = \Xi_t$, and thus Ψ_t satisfies the Euclidean condition. ■

THEOREM 4–1 [1]. *Let I be a class of subsets of R^n. There exists a Euclidean granulometry for which I is the invariant class of the unit if and only if I is closed under union, translation, and scalar multiplication by all $t \geqslant 1$. Moreover, for such a class I, the corresponding Euclidean granulometry possesses the representation*

$$\Psi_t(S) = \bigcup_{B \in I} \Gamma_{tB}(S) \qquad (4\text{–}9)$$

where $Inv[\Psi_1] = I$.

PROOF. First suppose there exists a Euclidean granulometry $\{\Psi_t\}$ for which $I = Inv[\Psi_1]$. To prove closure under unions, consider a collection of sets $S_i \in Inv[\Psi_1]$ and let S be the union of the S_i. Since Ψ_1 is increasing,

$$S = \bigcup_i S_i = \bigcup_i \Psi_1(S_i) \subset \Psi_1\left(\bigcup_i S_i\right) = \Psi_1(S) \qquad (4\text{–}10)$$

Since Ψ_1 is antiextensive, the reverse inclusion holds, $S \in Inv[\Psi_1]$, and there is closure under unions. From Chapter 3, an algebraic opening is a τ-opening if and only if its invariant class is invariant under translation, so that $Inv[\Psi_1]$ is closed under translation. Now suppose $S \in Inv[\Psi_1]$ and $t \geqslant 1$. By the Euclidean condition, $tS \in Inv[\Psi_t]$ and

$$\Psi_1(tS) = \Psi_1\left[\Psi_t(tS)\right] = \Psi_{\max\{1,t\}}(tS) = \Psi_t(tS) = tS \qquad (4\text{–}11)$$

Therefore $tS \in Inv[\Psi_1]$ and $Inv[\Psi_1]$ is closed under scalar multiplication by $t \geqslant 1$. For the converse of the proposition, we need to find a granulometry for which I is the invariant class of the unit. According to Theorem 3–5, Ψ_t defined by Eq. 4–9 is a τ-opening with base tI. Since I is closed under unions and translations, $Inv[\Psi_t] = tI$ and $I = Inv[\Psi_1]$. To show $\{\Psi_t\}$ is a Euclidean granulometry, we must demonstrate invariance ordering and the Euclidean condition. Suppose $r \geqslant s > 0$

and $S \in Inv[\Psi_r]$. Then $S = rB$ for some $B \in I$. Since $r/s \geqslant 1$, $(r/s)B \in I$, which implies that $S = rB = sC$ for some $C \in I$, which implies that $S \in sI = Inv[\Psi_s]$. Finally, according to Proposition 4–2 the Euclidean condition holds since, by construction, $Inv[\Psi_t] = tInv[\Psi_1]$. ■

Taken together with Proposition 4–2, which states that invariant classes of Euclidean granulometries are determined by the invariant class of the unit, Theorem 4–1 characterizes the form and invariant classes of Euclidean granulometries. Nevertheless, as it stands, it does not provide a useful framework for filter design because one must construct invariant classes of units and we need a methodology for construction.

Suppose I is a class of sets closed under union, translation, and scalar multiplication by $t \geqslant 1$. A class $\mathcal{G}$ of sets is called a *generator* of I if the class closed under union, translation, and scalar multiplication by scalars $t \geqslant 1$ that is generated by $\mathcal{G}$ is I. If $\{\Psi_t\}$ is the Euclidean granulometry with $Inv[\Psi_1] = I$, then $\mathcal{G}$ is called a *generator* of $\{\Psi_t\}$. The next theorem provides a more constructive characterization of Euclidean granulometries.

THEOREM 4–2 [1]. *An operator family* $\{\Psi_t\}$, $t > 0$, *is a Euclidean granulometry if and only if there exists a class of images* $\mathcal{G}$ *such that*

$$\Psi_t(S) = \bigcup_{B \in \mathcal{G}} \bigcup_{r \geqslant t} \Gamma_{rB}(S) \tag{4–12}$$

Moreover, $\mathcal{G}$ *is a generator of* $\{\Psi_t\}$.

PROOF. We first show that Eq. 4–12 yields a Euclidean granulometry for any class $\mathcal{G}$. According to Theorem 3–5, Ψ_t is a τ-opening with base $\mathcal{B}_t = \{rB\colon B \in \mathcal{G},\ r \geqslant t\}$. If $u \geqslant t$, then $\mathcal{B}_u \subset \mathcal{B}_t$, which implies $Inv[\Psi_u] \subset Inv[\Psi_t]$. To show that $\{\Psi_t\}$ is a Euclidean granulometry, we apply Proposition 4–2 and Eq. 4–7. Indeed, $S/t \in Inv[\Psi_1]$ if and only if

$$\frac{S}{t} = \bigcup_{B \in \mathcal{G}} \bigcup_{s \geqslant 1} \Gamma_{sB}(S/t) = \frac{1}{t} \bigcup_{B \in \mathcal{G}} \bigcup_{s \geqslant 1} \Gamma_{tsB}(S) = \frac{1}{t} \bigcup_{B \in \mathcal{G}} \bigcup_{r \geqslant t} \Gamma_{rB}(S) \tag{4–13}$$

which, upon cancelling the $1/t$, says that $S/t \in Inv[\Psi_1]$ if and only if $S \in Inv[\Psi_t]$. Since $\mathcal{B}_1$ is a base for Ψ_1, the form of $\mathcal{B}_1$ shows that $\mathcal{G}$ is a generator of $\{\Psi_t\}$. As for the converse, since $\{\Psi_t\}$ is a Euclidean granulometry, Ψ_t has the representation of Eq. 4–9 and, since $I = Inv[\Psi_1]$ is a generator of itself,

$$\Psi_1(S) = \bigcup_{B \in I} \bigcup_{r \geqslant 1} \Gamma_{rB}(S) \tag{4–14}$$

By the Euclidean condition and Eq. 4–7,

$$\Psi_t(S) = t\bigcup_{B\in I}\bigcup_{r\geqslant 1}\Gamma_{rB}(S/t) = \bigcup_{B\in I}\bigcup_{r\geqslant 1}\Gamma_{rtB}(S) = \bigcup_{B\in I}\bigcup_{u\geqslant t}\Gamma_{uB}(S) \tag{4–15}$$

so that Ψ_t possesses a representation of the desired form. ■

Theorem 4–2 provides a methodology for constructing Euclidean granulometries: select a generator $\mathcal{G}$ and apply Eq. 4–12; however, such an approach is problematic in practice since it involves, for each t, a union over all $r \geqslant t$. To see the problem, suppose we choose a singleton generator $\mathcal{G} = \{B\}$. Then Eq. 4–12 yields the representation

$$\Psi_t(S) = \bigcup_{r\geqslant t}\Gamma_{rB}(S) \tag{4–16}$$

which is an uncountable union. According to Theorem 3–6, if B is compact and convex, then rB is tB-open, so that $\Gamma_{rB}(S) \subset \Gamma_{tB}(S)$ and the union reduces to the single opening $\Psi_t(S) = \Gamma_{tB}(S)$. Since Theorem 3–6 is an equivalence, for compact B, we require the convexity of B to obtain the reduction.

This reasoning extends to an arbitrary generator: for a generator composed of compact sets, the double union of Eq. 4–12 reduces to the single outer union over $\mathcal{G}$,

$$\Psi_t(S) = \bigcup_{B\in\mathcal{G}}\Gamma_{tB}(S) \tag{4–17}$$

if and only if $\mathcal{G}$ is composed of convex sets, in which case we say the granulometry is *convex*. The single union represents a parameterized τ-opening.

The generator sets of a convex granulometry are convex, and therefore connected. Hence, if $S_1, S_2, \ldots$ are mutually disjoint compact sets, then

$$\Psi_t\left(\bigcup_{i=1}^{\infty}S_i\right) = \bigcup_{i=1}^{\infty}\Psi_t(S_i) \tag{4–18}$$

That is, a convex granulometry is *distributive* and it can be viewed componentwise.

Although we have restricted our development to binary granulometries, as conceived by Matheron, the theory can be extended to gray-scale images [2, 3] and the algebraic theory to the framework of complete lattices [4].

4.3 RECONSTRUCTIVE GRANULOMETRIES

The representation of Eq. 4–17 can be generalized by separately parameterizing each structuring element, rather than simply scaling each by a common parameter. To avoid cumbersome subscripts, we will now switch to the infix notation $S \circ B$ for the opening of S by B. Assuming a finite number of convex structuring elements, individual structuring-element parameterization yields a family $\{\Psi_{\mathbf{r}}\}$ of multiparameter τ-openings of the form

$$\Psi_{\mathbf{r}}(S) = \bigcup_{k=1}^{n} S \circ B_k[\mathbf{r}_k] \tag{4–19}$$

where $\mathbf{r}_1, \mathbf{r}_2, \ldots, \mathbf{r}_n$ are parameter vectors governing the convex, compact structuring elements $B_1[\mathbf{r}_1], B_2[\mathbf{r}_2], \ldots, B_n[\mathbf{r}_n]$ composing the base of $\Psi_{\mathbf{r}}$ and $\mathbf{r} = (\mathbf{r}_1, \mathbf{r}_2, \ldots, \mathbf{r}_n)$. To keep the notion of sizing, we require (here and subsequently) the *sizing condition* that $\mathbf{r}_k \leqslant \mathbf{s}_k$ implies $B_k[\mathbf{r}_k] \subset B_k[\mathbf{s}_k]$ for $k = 1, 2, \ldots, n$, where vector order is defined by $(t_1, t_2, \ldots, t_m) \leqslant (u_1, u_2, \ldots, u_m)$ if and only if $t_j \leqslant u_j$ for $j = 1, 2, \ldots, m$.

$\Psi_{\mathbf{r}}$ is a τ-opening because any union of openings is a τ-opening; however, since the parameter is now a vector, the second condition of Proposition 4–1 does not apply as stated. We generalize the condition by employing componentwise ordering in the vector lattice, so that condtion (ii) of Proposition 4–1 becomes (ii′) $\mathbf{r} \geqslant \mathbf{s} > \mathbf{0} \Rightarrow Inv[\Psi_{\mathbf{r}}] \subset Inv[\Psi_{\mathbf{s}}]$. Condition (ii′) states that the mapping $\mathbf{r} \rightarrow Inv[\Psi_{\mathbf{r}}]$ is order reversing and we say that any family $\{\Psi_{\mathbf{r}}\}$ for which it holds is *invariance ordered*. If $\Psi_{\mathbf{r}}$ is a τ-opening for any $\mathbf{r}$ and a family $\{\Psi_{\mathbf{r}}\}$ is invariance ordered, then we call $\{\Psi_{\mathbf{r}}\}$ a *granulometry*. The family $\{\Psi_{\mathbf{r}}\}$ defined by Eq. 4–19 is not necessarily a granulometry because it need not be invariance ordered. As it stands, $\{\Psi_{\mathbf{r}}\}$ is simply a collection of τ-openings over a parameter space.

The failure of the family of Eq. 4–19 and other operator families defined via unions and intersections of parameterized openings to be granulometries is overcome by openingwise reconstruction and leads to the class of logical granulometries [5, 6]. Regarding Eq. 4–19, the induced reconstructive family $\{\Lambda_{\mathbf{r}}\}$, defined by

$$\Lambda_{\mathbf{r}}(S) = \bigcup_{k=1}^{n} \mathfrak{R}(S \circ B_k[\mathbf{r}_k]) = \mathfrak{R}\left(\bigcup_{k=1}^{n} S \circ B_k[\mathbf{r}_k]\right) \tag{4–20}$$

is a granulometry (since it is invariance ordered). As shown in Eq. 4–20, reconstruction can be performed openingwise or on the union. $\{\Lambda_{\mathbf{r}}\}$ is called a *disjunctive granulometry*.

Although Eq. 4–19 does not generally yield a granulometry without reconstruction, a salient special case occurs when each structuring element is of the form $t_i B_i$. In this case, for any n-vector $\mathbf{t} = (t_1, t_2, \ldots, t_n)$, $t_i > 0$, for $i = 1, 2, \ldots, n$, the filter takes the form

$$\Psi_{\mathbf{t}}(S) = \bigcup_{i=1}^{n} S \circ t_i B_i \tag{4–21}$$

To avoid useless redundancy, we assume that no set in the base is open with respect to another set in the base, meaning that for $i \neq j$, $B_i \circ B_j \neq B_i$. For any $\mathbf{t} = (t_1, t_2, \ldots, t_n)$ for which there exists $t_i = 0$, we define $\Psi_{\mathbf{t}}(S) = S$. $\{\Psi_{\mathbf{t}}\}$ is a *multivariate granulometry* (it being a granulometry without reconstruction) [7].

If the union of Eq. 4–19 is changed to an intersection and all conditions qualifying Eq. 4–19 are maintained, then the result is a family of multiparameter operators of the form

$$\Psi_{\mathbf{r}}(S) = \bigcap_{k=1}^{n} S \circ B_k[\mathbf{r}_k] \tag{4–22}$$

Each operator $\Psi_{\mathbf{r}}$ is translation invariant, increasing, and antiextensive but, unless $n = 1$, $\Psi_{\mathbf{r}}$ need not be idempotent. Hence $\Psi_{\mathbf{r}}$ is not generally a τ-opening and the family $\{\Psi_{\mathbf{r}}\}$ is not a granulometry. Each induced reconstruction $\mathfrak{R}(\Psi_{\mathbf{r}})$ is a τ-opening (is idempotent) but the family $\{\mathfrak{R}(\Psi_{\mathbf{r}})\}$ is not a granulometry because it is not invariance ordered. However, if reconstruction is performed openingwise, then the resulting intersection of reconstructions is invariance ordered and a granulometry. The family of operators

$$\Lambda_{\mathbf{r}}(S) = \bigcap_{k=1}^{n} \mathfrak{R}\big(S \circ B_k[\mathbf{r}_k]\big) \tag{4–23}$$

is called a *conjunctive granulometry*. In the conjunctive case, the equality of Eq. 4–21 is softened to an inequality: the reconstruction of the intersection is a subset of the intersection of the reconstructions.

Conjunction and disjunction can be combined to form a more general form of reconstructive granulometry:

$$\Lambda_{\mathbf{r}}(S) = \bigcup_{k=1}^{n} \bigcap_{j=1}^{m_k} \mathfrak{R}\big(S \circ B_{k,j}[\mathbf{r}_{k,j}]\big) \tag{4–24}$$

If S_i is a component of S and $x_{i,k,j}$ and y_i are the logical variables determined by the truth values of the equations $S_i \circ B_{k,j}[\mathbf{r}_{k,j}] \neq \emptyset$ and $\Lambda_{\mathbf{r}}(S_i) \neq \emptyset$ [or, equivalently, $\mathfrak{R}(S_i \circ B_{k,j}[\mathbf{r}_{k,j}]) = S_i$ and $\Lambda_{\mathbf{r}}(S_i) = S_i$], respectively, then y possesses the

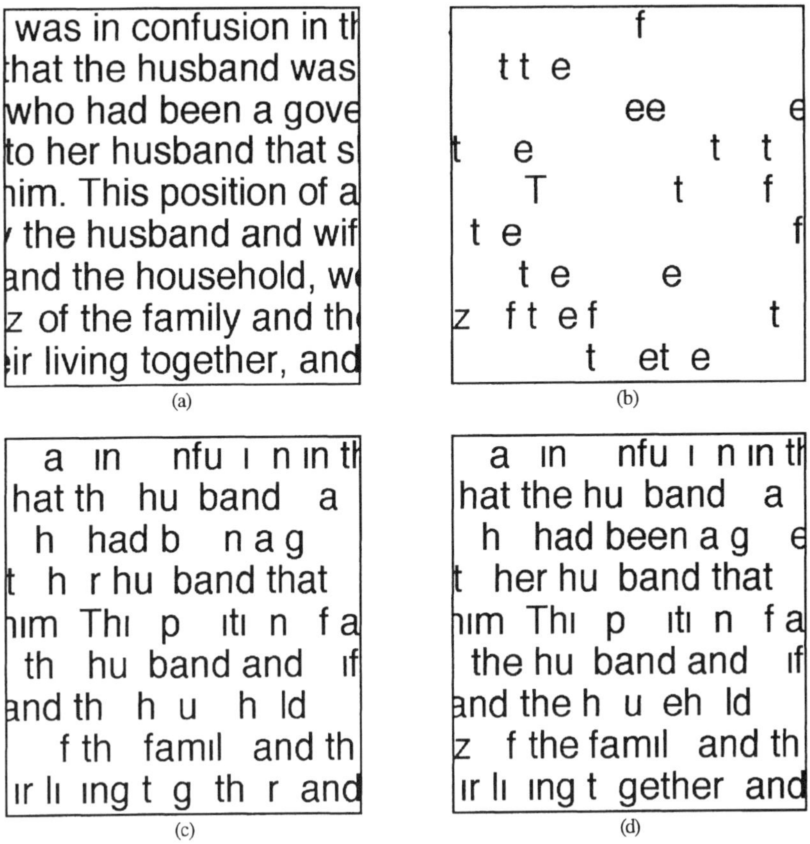
was in confusion in th
that the husband was
who had been a gove
to her husband that s
him. This position of a
the husband and wif
and the household, w
z of the family and th
ir living together, and
(a)
(b)
(c)
(d)

Figure 4–1. Disjunctive granulometry: (a) original text in Helvetica font; (b) opening by short horizontal line; (c) opening by vertical line; (d) union of parts (b) and (c) giving disjunctive granulometry. Note that Fig. 4–1(d) contains all characters containing either horizontal or vertical structuring elements.

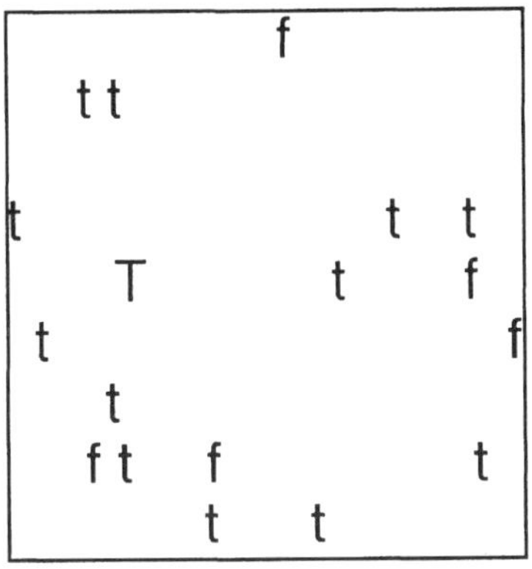

Figure 4–2. Conjunctive granulometry of the original image in Fig. 4–1(a) obtained by intersecting Figs. 4–1(b) and 4–1(c). Note that the image contains characters containing both horizontal or vertical structuring elements.

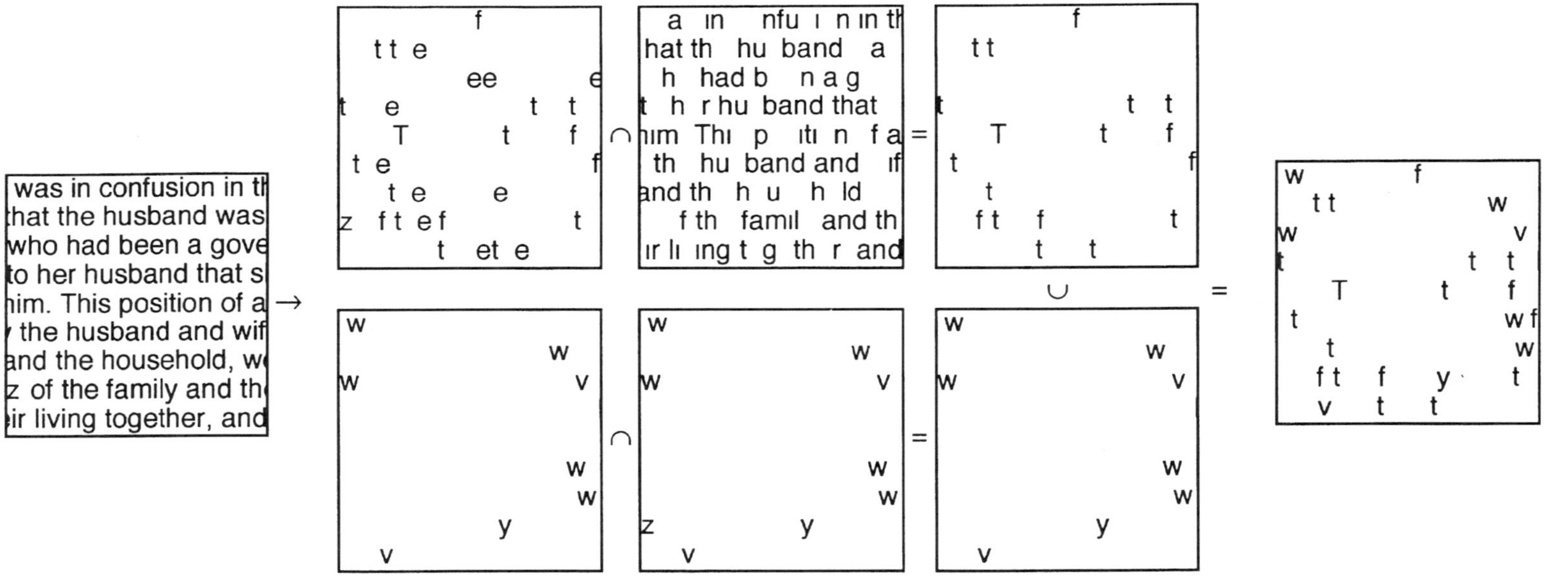
was in confusion in t
hat the husband was
who had been a gove
to her husband that s
him. This position of a
the husband and wif
and the household, w
z of the family and th
ir living together, and

Figure 4–3. Logical granulometry of the original image in Fig. 4–1(a). The upper row shows a conjunctive granulometry with horizontal and vertical structuring elements similar to Fig. 4–2 and the lower row shows a conjunctive granulometry using diagonal structuring elements. The two conjunctive granulometries are unioned to form a logical granulometry.

logical representation

$$y_i = \sum_{k=1}^{n} \prod_{j=1}^{m_k} x_{i,k,j} \tag{4–25}$$

We call $\{\Lambda_{\mathbf{r}}\}$ a *logical granulometry*. Component S_i is passed if and only if there exists k such that, for $j = 1, 2, \ldots, m_k$, there exists a translate of $B_{k,j}[\mathbf{r}_{k,j}]$ that is a subset of S_i. Disjunctive and conjunctive granulometries are special cases of logical granulometries and the latter compose a class of sieving filters that locate targets among clutter based on the size and shape of the target and clutter structural components. For fixed $\mathbf{r}$, we refer to $\Lambda_{\mathbf{r}}$ as a disjunctive, conjunctive, or logical opening, based on the type of reconstructive granulometry from which it arises. Figures 4–1 through 4–3 illustrate disjunctive, conjunctive, and logical openings, respectively. Logical openings form a subclass of a more general class of reconstructive sieving filters called *logical structural filters* [6]. These are not granulometric in the sense of Matheron; they need not be increasing.

4.4 OPTIMAL FILTERING BY RECONSTRUCTIVE GRANULOMETRIES

A basic problem of binary filtering is to remove background clutter (noise) to reveal a desired target (signal). In its primary form, the problem consists of a signal random set S, a noise random set N, an observed random set $S \cup N$, and a filter Ψ for whch $\Psi(S \cup N)$ provides an esimate of S, the goodness of the estimate being measured by a probabilistic error criterion $Er[\Psi(S \cup N), S]$. Assuming that signal grains are probabilistically larger than noise grains, the problem has been morphologically treated by forming Ψ as an opening, or union of openings. Such an approach naturally fits into the theory of granulometries. If we focus on the decision to pass or not pass a grain (connected component) in the observed image, then it is appropriate to consider optimization of reconstructive granulometric filters.

Consider the granular signal-union-noise model $S \cup N$, where

$$\begin{aligned} S &= \bigcup_{i=1}^{I} C[\mathbf{s}_i] + x_i \\ N &= \bigcup_{j=1}^{J} D[\mathbf{n}_j] + y_j \end{aligned} \tag{4–26}$$

I and J are random natural numbers; $\mathbf{s}_1, \mathbf{s}_2, \ldots, \mathbf{s}_I$ and $\mathbf{n}_1, \mathbf{n}_2, \ldots, \mathbf{n}_J$ are identically distributed to random vectors $\mathbf{s} = (s_1, s_2, \ldots, s_m)$ and $\mathbf{n} = (n_1, n_2, \ldots, n_m)$, respectively; $C[\mathbf{s}_i]$ and $D[\mathbf{n}_j]$ are random compact grains (connected components)

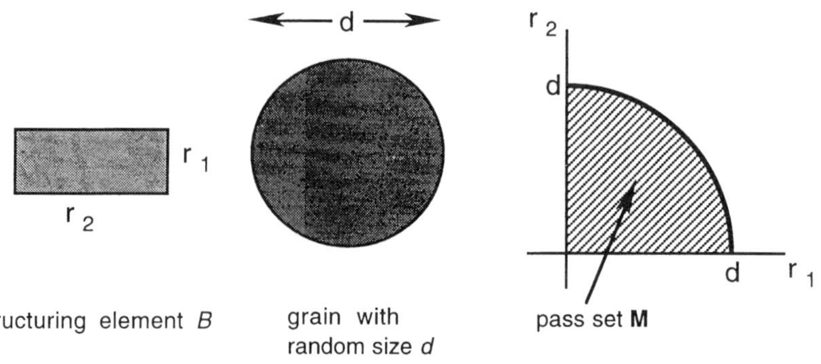

Figure 4–4. Pass set for a reconstructive opening, $\Lambda_{\mathbf{r}} = \mathfrak{R}(S \circ B[\mathbf{r}])$, with a rectangular structuring element.

governed by $\mathbf{s}_i$ and $\mathbf{n}_j$, respectively, and identically distributed with the *primary grains* $C[\mathbf{s}]$ and $D[\mathbf{n}]$; and x_i and y_j are random points governing grain locations (translations) under the constraint of grain disjointness. Error is from signal grains erroneously removed and noise grains erroneously passed. Optimization with respect to a family $\{\Lambda_{\mathbf{r}}\}$ of reconstructive granulometries is achieved by finding $\mathbf{r}$ to minimize the expected error $E[v[\Lambda_{\mathbf{r}}(S \cup N)\Delta S]]$, v and Δ denoting volume and symmetric difference. By distributivity, $\Lambda_{\mathbf{r}}(S \cup N) = \Lambda_{\mathbf{r}}(S) \cup \Lambda_{\mathbf{r}}(N)$.

Because $C[\mathbf{s}]$ and $D[\mathbf{n}]$ are random sets depending on the multivariate distributions of the random vectors $\mathbf{s}$ and $\mathbf{n}$, respectively, the parameter sets

$$\begin{aligned} \mathbf{M}_{C[\mathbf{s}]} &= \{\mathbf{r}\colon\ \Lambda_{\mathbf{r}}(C[\mathbf{s}]) = C[\mathbf{s}]\} \\ \mathbf{M}_{D[\mathbf{n}]} &= \{\mathbf{r}\colon\ \Lambda_{\mathbf{r}}(D[\mathbf{n}]) = D[\mathbf{n}]\} \end{aligned} \tag{4–27}$$

are random sets composed of vectors $\mathbf{r}$ for which $\Lambda_{\mathbf{r}}$ passes the primary grains $C[\mathbf{s}]$ and $D[\mathbf{n}]$, respectively. $\mathbf{M}_{C[\mathbf{s}]}$ and $\mathbf{M}_{D[\mathbf{n}]}$ are the regions in the parameter space where signal and noise grains, respectively, are passed. $\mathbf{M}_{C[\mathbf{s}]}$ and $\mathbf{M}_{D[\mathbf{n}]}$ are called the *signal* and *noise pass sets*. We often write $\mathbf{M}_{C[s]}$ and $\mathbf{M}_{D[\mathbf{n}]}$ as $\mathbf{M}_S$ and $\mathbf{M}_N$, respectively. As functions of $\mathbf{s}$ and $\mathbf{n}$, $\mathbf{M}_S = \mathbf{M}_S(s_1, s_2, \ldots, s_m)$ and $\mathbf{M}_N = \mathbf{M}_N(n_1, n_2, \ldots, n_m)$. Figure 4–4 shows a pass set.

Filter error corresponding to the parameter $\mathbf{r}$ is given by

$$\begin{aligned} e[\mathbf{r}] = E[I] \underset{\{\mathbf{s}\colon\ \mathbf{r} \notin \mathbf{M}_{C[\mathbf{s}]}\}}{\int \cdots \int} v\big[[C[\mathbf{s}]\big] f_S(s_1, s_2, \ldots, s_m) ds_1 ds_2 \cdots ds_m \\ + E[J] \underset{\{\mathbf{n}\colon\ \mathbf{r} \in \mathbf{M}_{D[\mathbf{n}]}\}}{\int \cdots \int} v\big[D[\mathbf{n}]\big] f_N(n_1, n_2, \ldots, n_m) dn_1 dn_2 \cdots dn_m \end{aligned} \tag{4–28}$$

where $E[I]$ and $E[J]$ are the expected numbers of signal and noise grains, respectively, and f_S and f_N are the multivariate densities for the random vectors $\mathbf{s}$ and $\mathbf{n}$, respectively [8]. In general, minimization of $e[\mathbf{r}]$ to find the optimal filter is mathematically prohibitive owing to the problematic nature of the domains of integration.

A special situation occurs when $\mathbf{M}_S$ and $\mathbf{M}_N$ are characterized by half-line inclusions of the component parameters: by which we mean that $\mathbf{r}$, $\mathbf{s}$, and $\mathbf{n}$ are of the same dimensions; $\mathbf{r} \in \mathbf{M}_S$ if and only $r_1 \leqslant M_{S,1}(s_1)$, $r_2 \leqslant M_{S,2}(s_2)$, $\ldots$, $r_m \leqslant M_{S,m}(s_m)$; and $\mathbf{r} \in \mathbf{M}_N$ if and only if $r_1 \leqslant M_{N,1}(n_1), r_2 \leqslant M_{N,2}(n_2)$, $\ldots$, $r_m \leqslant M_{N,m}(n_m)$. In this case we say the model is *separable* and Eq. 4–28 reduces. For $m = 2$, Eq. 4–28 becomes

$$e[\mathbf{r}] = E[I] \int_{-\infty}^{M_{S,1}^{-1}(r_1)} \int_{-\infty}^{M_{S,2}^{-1}(r_2)} v\big[C[s_1, s_2]\big] f_s(s_1, s_2) ds_2 ds_1 + E[J] \int_{M_{N,1}^{-1}(r_1)}^{\infty} \int_{M_{N,2}^{-1}(r_2)}^{\infty} v\big[D[n_1, n_2]\big] f_N(n_1, n_2) dn_2 dn_1 \quad (4\text{–}29)$$

If we restrict our attention to a single-parameter homothetic disjunctive granulometry

$$\Lambda_r(S) = \bigcup_{i=1}^{n} \mathfrak{R}(S \circ rB_i) \quad (4\text{–}30)$$

then the random sets of Eq. 4–27 can be replaced by the *granulometric sizes*

$$M_S = \sup\{r\colon r \in \mathbf{M}_S\} = \sup\{r\colon \Lambda_r(C[\mathbf{s}]) = C[\mathbf{s}]\} \quad (4\text{–}31)$$

$$M_N = \sup\{r\colon r \in \mathbf{M}_N\} = \sup\{r\colon \Lambda_r(D[\mathbf{n}]) = D[\mathbf{n}]\} \quad (4\text{–}32)$$

and the domains of integration in Eq. 4–27 become the regions $r > M_S$ and $r \leqslant M_N$.

For a mathematically straightforward example, consider the model of Eq. 4–30 and let $C[\mathbf{s}]$ be a randomly rotated ellipse with random axis lengths $2u$ and $2y$, $D[\mathbf{n}]$ be a randomly rotated rectangle with random sides of length $2w$ and $2z$, and the filter be $\Lambda_r(S) = \mathfrak{R}(S \circ rB)$, where B is the unit disk. Grains are disjoint. Then $M_S = \min\{u, y\}$, $M_N = \min\{w, z\}$, $v[C[u, y]] = \pi uy$, and $v[D[w, z]] = 4wz$. With f denoting probability densities,

$$e[r] = \pi E[I] \left[\int_0^r \int_0^\infty uyf(u, y) du dy + \int_r^\infty \int_0^r uyf(u, y) du dy \right]$$

$$+4E[J]\int_r^{\infty}\int_r^{\infty} wzf(w,z)dwdz \tag{4–33}$$

Minimization of Eq. 4–33 yields an optimal filter of the form $\Lambda_r(S) = \mathfrak{R}(S \circ rB)$. Assuming the four sizing variables are independent, Eq. 4–33 becomes

$$e[r] = \pi E[I]\left[\int_0^r\int_0^{\infty} uvf(u)f(v)dudv + \int_r^{\infty}\int_0^r uvf(u)f(v)dudv\right]$$
$$+4E[J]\int_r^{\infty}\int_r^{\infty} wzf(w)f(z)dwdz \tag{4–34}$$

Suppose u and v are gamma distributed with parameters α and β, and w and z are exponentially distributed with parameter b. For model parameters $\alpha = 12$, $\beta = 1$, $b = 0.2$, and $E[I] = E[J] = 20$, minimization of Eq. 4–34 occurs for $r = 5.95$. Figure 4–5 shows a realization of the union process and the realization filtered by Λ_6. One noise grain has erroneously passed and one signal grain has erroneously been eliminated.

For a single-parameter homothetic conjunctive example, let the primary signal grain $C[\mathbf{s}]$ be a nonrotated cross with each bar of width 1 and random length $2w \geqslant 1$ and the primary noise grain $D[\mathbf{n}]$ be a nonrotated cross with each bar of width 1, one bar of length $z \geqslant 1$, and the other bar of length $2z$. Let grain placement be constrained by disjointness and define the filter by $\Lambda_r(S) = \mathfrak{R}(S \circ rE) \cap \mathfrak{R}(S \circ rF)$, where E and F are unit-length vertical and horizontal lines, respectively. Then

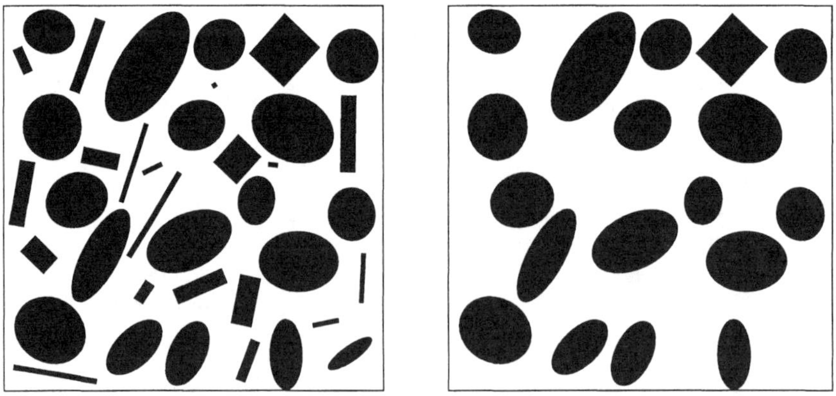

(a) (b)

Figure 4–5. Optimal disjunctive filtering: realization of the union process and the realization filtered by Λ_6.

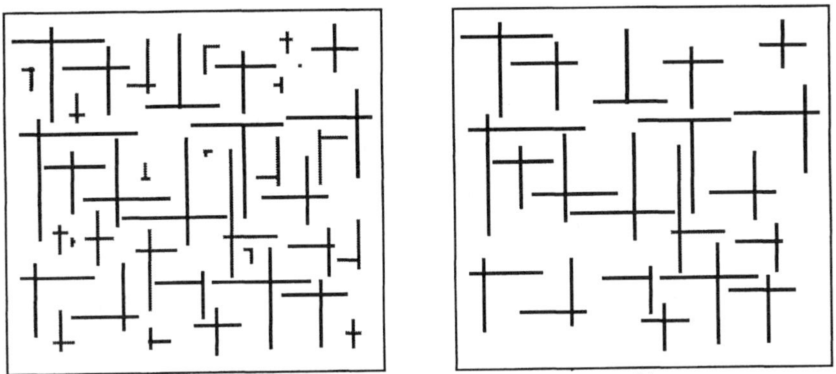

(a) (b)

Figure 4–6. Optimal conjunctive filtering: realization of the union process and the realization filtered by Λ_{14}.

$M_S = 2w,\ M_N = z,\ v[C[w]] = 4w - 1,\ v[D[z]] = 3z - 1$, and

$$e[r] = E[I] \int_0^{r/2} (4w - 1) f(w) dw + E[J] \int_r^{\infty} (3z - 1) f(z) dz \qquad (4\text{–}35)$$

Suppose w is gamma distributed with $\alpha = 12$ and $\beta = 1$, and z is exponentially distributed with $b = 0.2$. Minimization of Eq. 4–35 occurs at $r = 13.58$. Figure 4–6 shows a realization of the union process (signal black, noise gray) and the realization filtered by Λ_{14}.

Under disjointness, Eq. 4–28 is applied directly in terms of the probability models governing signal and noise; if grains are not disjoint, then segmentation must be applied before application of the error representation, and the modeling assumptions of Eq. 4–26 are relative to the segmentation procedure adopted. If segmentation is accomplished by the morphological watershed transformation [9], which we denote by Θ, then we need to find the probabilistic descriptions of the random outputs $\Theta(C[\mathbf{s}])$ and $\Theta(D[\mathbf{n}])$. Finding an output random set for the watershed is generally very difficult and involves statistical modeling of grain overlapping [8]. For many granular images (gels, toner, etc.), when there is overlapping it is often very modest, with the probability of increased overlapping diminishing rapidly. The watershed produces a segmentation line between grains and its precise geometric effect depends on the random geometry of the grains and the degree of overlapping, which is itself random. Even when input grain geometry is very simple, output geometry can be very complicated (as well as dependent on overlap statistics).

4.5 Adaptive Disjunctive Granulometric Filters

Owing to the mathematical obstacles in deriving an optimal filter and the difficulty of obtaining process statistics via estimators, adaptive approaches are used to obtain a filter that is (it is hoped) close to optimal. In adaptive design, a sequence of observations $T_1, T_2, T_3, \ldots$ is made and the filter is applied to each observation. Based on some criterion of goodness relating $\Lambda_{\mathbf{r}}(T_n)$ and S, the vector $\mathbf{r}$ is adapted. Adaptations yield a random-vector time series $\mathbf{r}_0, \mathbf{r}_1, \mathbf{r}_2, \mathbf{r}_3, \ldots$ resulting from transitions $\mathbf{r}_n \to \mathbf{r}_{n+1}$, where $\mathbf{r}_n$ is the state of the process at time n and $\mathbf{r}_0$ is the initial state vector. There are various sets of conditions on the scanning process, the form of the filter, and the adaptation protocol that result in the parameter process $\mathbf{r}_n$ forming a Markov chain whose state space is the parameter space of $\mathbf{r}$. When this is so, adaptive filtering is characterized via the behavior of the Markov chain $\mathbf{r}_n$, which can be assumed to possess a single irreducible class. Convergence of the adaptive filter means existence of a steady-state distribution and characteristics of filter behavior are the stationary characteristics of the Markov chain (mean function, covariance function, etc.) in the steady state. Our adaptive estimate of the optimal filter depends on the steady-state distribution of $\mathbf{r}_n$. For instance, we might use the filter $\Lambda_{\bar{\mathbf{r}}}$, where $\bar{\mathbf{r}}$ is the mean vector of $\mathbf{r}_n$ in the steady state. The mean vector can be estimated from a single realization (from a single sequence of observations $T_1, T_2, T_3, \ldots$) owing to ergodicity in the steady state. The size of the time interval over which $\mathbf{r}_n$ needs to be averaged in the steady state for a desired degree of precision can be found from the steady-state variance of $\mathbf{r}_n$.

Adaptation protocols vary for disjunctive granulometries and their complexity grows with the number of parameters [10]. Moreover, adaptation strategies exist for the class of logical granulometries and the more general class of logical structural filters [6]. Here we only consider adaptation of the single-parameter disjunctive granulometry of Eq. 4–30 [11]. We initialize Λ_r and scan $S \cup N$ to successively encounter grains. The adaptive filter will be of the form $\Lambda_{r(n)}$, where n corresponds to the nth grain encountered. When a grain G "arrives," there are four possibilities:

$$\begin{aligned} &\text{a. } G \text{ is a noise grain and } \Lambda_{r(n)}(G) = G,\\ &\text{b. } G \text{ is a signal grain and } \Lambda_{r(n)}(G) = \emptyset,\\ &\text{c. } G \text{ is a noise grain and } \Lambda_{r(n)}(G) = \emptyset,\\ &\text{d. } G \text{ is a signal grain and } \Lambda_{r(n)}(G) = G. \end{aligned} \tag{4–36}$$

In the latter two cases, the filter has acted as desired; in either of the first two it has not. Consequently, we employ the following adaptation rule:

$$\begin{aligned} &\text{i. } r \to r+1 && \text{if condition a occurs.}\\ &\text{ii. } r \to r-1 && \text{if condition b occurs.}\\ &\text{iii. } r \to r && \text{if conditions c or d occur.} \end{aligned} \tag{4–37}$$

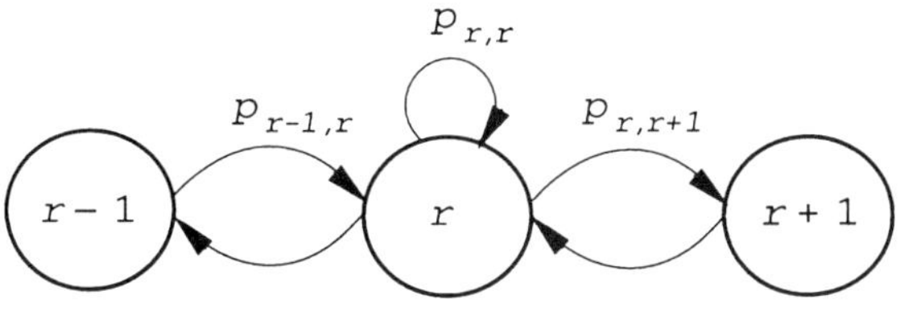

Figure 4–7. Transition probabilities.

Each arriving grain determines a step and we treat $r(n)$ as the state of the system at step n. Since all grain sizes are independent and there is no grain overlapping, $r(n)$ determines a discrete state-space Markov chain over a discrete parameter space. Three positive stationary transition probabilities are associated with each state r:

$$\begin{aligned} &\text{i. } p_{r,r+1} = P(N)P\big(\Lambda_r(D[\mathbf{n}]) = D[\mathbf{n}]\big) \\ &\text{ii. } p_{r,r-1} = P(S)P\big(\Lambda_r(C[\mathbf{s}]) = \emptyset\big) \\ &\text{iii. } p_{r,r} = P(S)P(\Lambda_r(C[\mathbf{s}]) = C[\mathbf{s}]) + P(N)P\big(\Lambda_r(D[\mathbf{n}]) = \emptyset\big) \end{aligned} \tag{4–38}$$

where $P(S)$ and $P(N)$ are the probabilities of a signal and noise grain arriving, respectively (Fig. 4–7). $P(S)$ and $P(N)$ depend on the protocol for selecting grains in the images. A number of these, together with the corresponding probabilities, have been employed [11]. In *weighted random point selection*, points in the image frame are randomly selected until a point in $S \cup N$ is chosen and the grain containing the point is considered. In *unweighted random point selection*, each grain in $S \cup N$ is labeled and labels are uniformly randomly selected with replacement. In *horizontal scanning*, the image is horizontally scanned at randomly chosen points along the side of the image frame, a grain is encountered if and only if it is cut by the scan line, and the scan line traverses the entire width of the image frame.

The transition probabilities can be expressed in terms of granulometric measure:

$$\begin{aligned} &\text{i. } p_{r,r+1} = P(N)P(M_N \geqslant r) \\ &\text{ii. } p_{r,r-1} = P(S)P(M_S < r) \\ &\text{iii. } p_{r,r} = P(S)P(M_S \geqslant r) + P(N)P(M_N < r) \end{aligned} \tag{4–39}$$

We let r be a nonnegative integer and transitions be plus or minus one; in fact, r need not be an integer and transitions could be of the form $r \to r + \rho$ and $r \to r - \rho$, where ρ is some positive constant.

Equivalence classes of the Markov chain are determined by the distributions of M_S and M_N. To avoid trivial anomalies, we assume that distribution supports are intervals with endpoints $a_S < b_S$ and $a_N < b_N$, where $0 \leqslant a_S$, $0 \leqslant a_N$, and it may be that $b_S = \infty$ or $b_N = \infty$. We assume $a_N \leqslant a_S < b_N \leqslant b_S$. Nonnull intersection of the supports ensures that the adpative filter does not trivially converge to an optimal filter that totally restores S. There are four cases regarding state communication.

Suppose $a_S < 1$ and $b_N = \infty$: then the Markov chain is irreducible since all states communicate (each state can be reached from every other state in a finite number of steps). Suppose $1 \leqslant a_S$ and $b_N = \infty$: then, for each state $r \leqslant a_S$, r is accessible from state s if $s < r$, but s is not accessible from r; on the other hand, all states $r \geqslant a_S$ communicate and form a single equivalence class. Suppose $a_S < 1$ and $b_N < \infty$: then, for each state $r \geqslant b_N$, r is accessible from state s if $s > r$, but s is not accessible from r; on the other hand, all states $r \leqslant b_N$ communicate and form a single equivalence class. Suppose $1 \leqslant a_S < b_N < \infty$: then states below a_S are accessible from states below themselves, but not conversely; states above b_N are accessible from states above themselves, but not conversely; and all states r such that $a_S \leqslant r \leqslant b_N$ communicate and form a single equivalence class. In sum, the states between a_S and b_N form an irreducible equivalence class C of the state space and each state outside C is transient. With certainty, the chain will eventually enter C and once inside C will not leave. Thus, we focus on C. In C the chain is irreducible and aperiodic. If it is also positive recurrent, then it will be ergodic and possess a steady-state distribution.

To find the steady-state distribution, let $p_r(n)$ be the probability that the system is in state r at step n and

$$\begin{aligned} \lambda_\kappa &= P(N)P(M_N \geqslant k) \\ \mu_k &= P(S)P(M_S < k). \end{aligned} \tag{4–40}$$

The Chapman–Kolmogorov equation yields

$$\begin{aligned} &p_r(n+1) - p_r(n) \\ &\quad = P(N)P(M_N \geqslant r-1)p_{r-1}(n) + P(S)P(M_S < r+1)p_{r+1}(n) \\ &\qquad - \big(P(S)P(M_S < r) + P(N)P(M_N \geqslant r)\big)p_r(n) \\ &\quad = \lambda_{r-1}p_{r-1}(n) + \mu_{r+1}p_{r+1}(n) - (\lambda_r + \mu_r)p_r(n) \end{aligned} \tag{4–41}$$

for $r \geqslant 1$. For $r = 0$, $p_{-1}(n) = 0$ and $\mu_0 = 0$ yield the initial state equation

$$p_0(n+1) - p_0(n) = \mu_1 p_1(n) - \lambda_0 p_0(n) \tag{4–42}$$

In the steady state, these equations form the system

$$\begin{cases} 0 = \mu_1 p_1 - \lambda_0 p_0 \\ 0 = \lambda_{r-1}p_{r-1} + \mu_{r+1}p_{r+1} - (\lambda_r + \mu_r)p_r \quad (r \geqslant 1) \end{cases} \tag{4–43}$$

and the solution is

$$\begin{cases} p_1 = \dfrac{\lambda_0}{\mu_1} p_0 \\ p_r = p_0 \displaystyle\prod_{k=1}^{r} \frac{\lambda_{k-1}}{\mu_k} \quad (r \geqslant 1) \end{cases} \tag{4–44}$$

where

$$p_0 = \frac{1}{1 + \sum_{r=1}^{\infty} \prod_{k=1}^{r} \frac{\lambda_{k-1}}{\mu_k}} \tag{4–45}$$

and it can be shown that the sum in the denominator of p_0 is finite [11], so that $p_0 > 0$.

Given convergence of Λ_r, in the sense that r reaches a steady state, key characteristics are the steady-state mean and variance. In the steady state, r is a random variable (and we slightly abuse notation by writing both it and its values as r). Its mean and variance are

$$\mu_r = \sum_{r=0}^{\infty} r p_r = \sum_{r=1}^{\infty} \left[r p_0 \prod_{k=1}^{r} \frac{\lambda_{k-1}}{\mu_k} \right] \tag{4–46}$$

$$\sigma_r^2 = \sum_{r=1}^{\infty} \left[r^2 p_0 \prod_{k=1}^{r} \frac{\lambda_{k-1}}{\mu_k} \right] - \mu_r^2 \tag{4–47}$$

Both mean and variance exist. In fact, r has finite moments of all orders [11].

For optimization, the key error measure is $E[e[r]]$. If the random primary grains $C[s]$ and $D[n]$ are governed by random variables s and n, respectively, and M_S and M_N are strictly increasing functions of s and n, respectively, then

$$e[r] = E[I] \int_{0}^{M_S^{-1}(r)} v\big[C[s]\big] f_S(s) ds + E[J] \int_{M_N^{-1}(r)}^{\infty} v\big[D[n]\big] f_N(n) dn \tag{4–48}$$

where $f_S(s)$ and $f_N(n)$ are the probability densities for s and n, respectively. Averaging over r in the steady state yields

$$E\big[e[r]\big] = \sum_{r=0}^{\infty} e[r] p_r \tag{4–49}$$

The optimal value of r, say $\hat{r}$, is found by minimizing Eq. 4–48. The expected cost of adaptivity can be measured by $E[e[r]] - e[\hat{r}]$, which must be nonnegative.

For an illustration of the adaptive method, consider a signal process consisting of randomly rotated ellipses having random major and minor axes A_S and B_S, respectively, where $A_S/B_S = 3$ and A_S is normally distributed with mean 20 and variance 2; and consider a noise process consisting of randomly rotated ellipses having

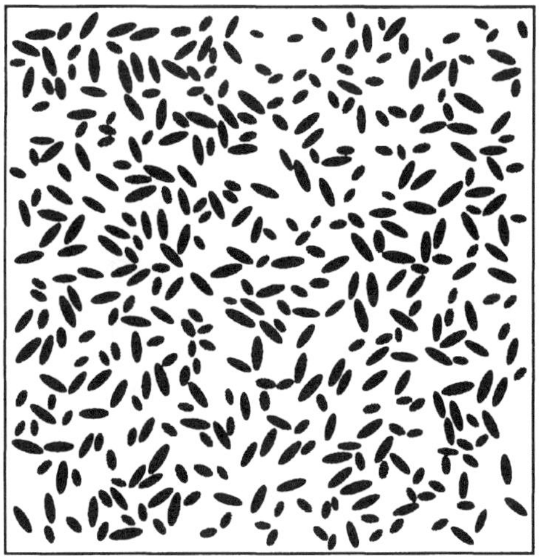

Figure 4–8. Realization of signal-union-noise process of randomly rotated and sized ellipses.

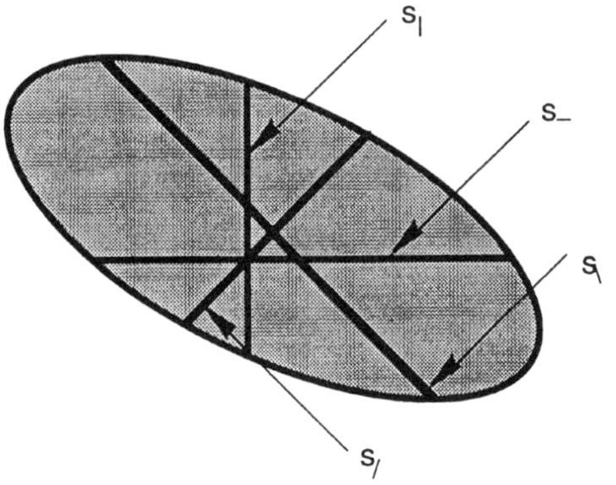

Figure 4–9. Granulometric measure of elliptical grain using four linear structuring elements.

random major and minor axes A_N and B_N, respectively, where $A_N/B_N = 2$ and A_N is normally distributed with mean 18 and variance 2. For both processes, the expected number of grains is 200. Figure 4–8 shows a realization of the union process (signal black, noise gray). The granulometric measure of an elliptical grain is shown in Fig. 4–9 for four linear structuring elements: vertical [|], horizontal [−], 45° diagonal [/], and −45° diagonal [\]. The maximum values of the sizing parameter for openings with single structuring elements are $s_|$, s_-, $s_/$, and $s_\backslash$, and the granulometric measure of the grain is $\max\{s_|, s_-, s_/, s_\backslash\}$. Figure 4–10 shows the empirical signal and noise granulometric-measure distributions, along with the steady-state distribution (over five realizations), for the τ-opening parameter. The mean of the parameter in the steady state is 17.2. The output of the disjunctive filter with $r = 18$ operating on the realization of Fig. 4–8 is shown in Fig. 4–11.

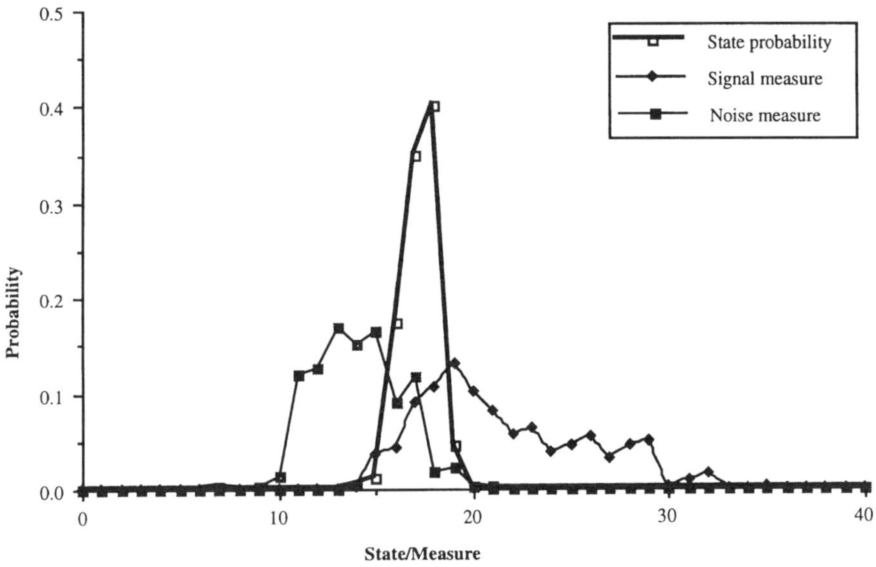

Figure 4–10. Empirical signal and noise granulometric measures, and steady-state distribution.

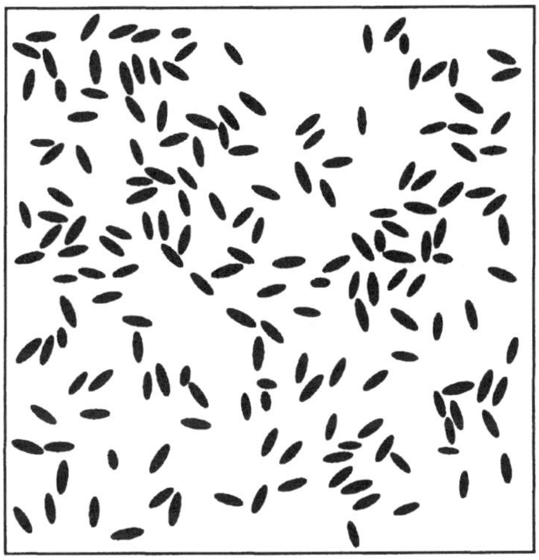

Figure 4–11. Filtered output of realization of Fig. 4–8.

4.6 SIZE DISTRIBUTIONS

For increasing t, a granulometry causes increasing diminution of a set. The rate of diminution is a powerful image descriptor. Consider a finite-generator convex Euclidean granulometry $\{\Psi_t\}$ of the form given in Eq. 4–17 with compact generating sets containing more than a single point. (More generally, we could consider upper semicontinuous Euclidean granulometries [1].) Fixing a compact set S of positive measure and treating t as a variable, we define the following *size distributions*:

$$\Omega(t) = v\big[\Psi_t(S)\big],$$

$$\Omega^\bullet(t) = v[S] - v\big[\Psi_t(S)\big], \tag{4–50}$$

$$\Phi(t) = \frac{\Omega^\bullet(t)}{v[S]} = 1 - \frac{\Omega(t)}{\Omega(0)}.$$

Ω gives the area remaining upon application of Ψ_t and is a decreasing function for which $\Omega(0) = v[s]$ and $\Omega(t) = 0$ for sufficiently large t. $\Omega^\bullet$ measures the area removed by Ψ_t and is an increasing function for which $\Omega^\bullet(0) = 0$ and $\Omega^\bullet(t) = v[S]$ for sufficiently large t. When necessary to terminologically distinguish Ω and $\Omega^\bullet$, we will refer to them as the *decreasing* and *increasing* size distributions, respectively. Φ is a probability distribution function. Φ and its derivative $\Phi' = d\Phi/dt$, which is a probability density, are called the *pattern spectrum* of S. For a random set, $\Omega(t)$, $\Omega^\bullet(t)$, and $\Phi(t)$ are random functions whose realizations are characteristics of the corresponding realizations of the random set. Taking the expectation of $\Omega^\bullet(t)$ gives the *mean size distribution* (*MSD*), $\mathrm{M}(t) = E[\Omega^\bullet(t)]$. The (generalized) derivative, $\mathrm{H}(t) = \mathrm{M}'(t)$, of the mean size distribution is called the *granulometric size density* (*GSD*). The MSD is not a probability distribution function since $\mathrm{M}(t) \to E[v[S]]$ as $t \to \infty$, and therefore the GSD is not a probability density. The expectation $E[\Phi(t)]$ is a probability distribution function, and we call its derivative, $\Sigma(t) = dE[\Phi(t)]/dt$, the *pattern-spectrum density* (*PSD*). Σ is a probability density and, under nonrestrictive regularity conditions, $\Sigma(t) = E[\Phi'(t)]$. The MSD, GSD, and PSD serve as partial descriptors of a random set in much the same way as the power spectral density partially describes a wide-stationary random function, and they play a role analogous to the power spectral density in designing optimal filters, but for granulometric, not linear, filters [12–15]. Size distributions are also used for texture and shape classification [16–21]. Various properties of size distributions have been studied: asymptotic behavior (relative to grain count) of the pattern-spectrum moments [22–24], effects of noise [25], continuous-to-discrete sampling [26], and estimation [27, 28]. Granulometric classification has also been applied to gray-scale textures [29–31].

Treating S as a random image, $\Phi(t)$ is a random function and its moments are random variables. Since the moments of $\Phi(t)$ are used for classification, three basic problems arise: (1) find expressions for the pattern-spectrum moments; (2) find expressions for the moments of the pattern-spectrum moments; (3) describe the probability distributions of the pattern-spectrum moments. We address some of these problems for the granulometry defined by $\Psi_t(S) = \{S \circ tB\}$ and a granular random set S whose realizations are disjoint unions of scalar multiples of compact primitives $A_1, A_2, \ldots, A_m$:

$$S = \bigcup_{i=1}^{m} \bigcup_{j=1}^{m_i} (r_{ij} A_i + x_{ij}) \tag{4–51}$$

S is random owing to the randomness of the scalar multiples r_{ij}, the locations x_{ij}, and the number $N = m \times m_i$ of grains. For this model, there exists a representation for the pattern-spectrum moments of S in terms of the pattern-spectrum moments of the primitives.

THEOREM 4–3 [22]. *If B is compact and convex, S satisfies Eq. 4–51, and $\mu^{(k)}(A_i)$ is the kth pattern-spectrum moment for the granulometry $\{A_i \circ tB\}$, then the kth pattern-spectrum moment for the granulometry $\{S \circ tB\}$ is given by*

$$\mu^{(k)}(S) = \frac{\sum_{i=1}^{m} v[A_i]\mu^{(k)}(A_i) \sum_{j=1}^{m_i} r_{ij}^{k+2}}{\sum_{i=1}^{m} v[A_i] \sum_{j=1}^{m_i} r_{ij}^{2}} \tag{4–52}$$

PROOF. Note that $\mu^{(k)}(S)$ is a random variable and $\mu^{(k)}(A_i)$ is a constant for $i = 1, 2, \ldots, m$. Let Ω, Ω_i, Φ, and Φ_i denote the size distribution for $\{S \circ tB\}$, the size distribution for $\{A_i \circ tB\}$, the normalized size distribution for Ω, and the normalized size distribution for Ω_i, respectively. From Eq. 4–7,

$$r_{ij} A_i \circ tB = r_{ij} \left[A_i \circ (t/r_{ij}) B \right] \tag{4–53}$$

Distributivity of the granulometry yields

$$\begin{aligned} \Omega(t) &= \sum_{i=1}^{m} \sum_{j=1}^{m_i} v[r_{ij} A_i \circ tB] \\ &= \sum_{i=1}^{m} \sum_{j=1}^{m_i} v\left[r_{ij} (A_i \circ (t/r_{ij}) B) \right] \\ &= \sum_{i=1}^{m} \sum_{j=1}^{m_i} r_{ij}^2 \Omega_i (t/r_{ij}) \end{aligned} \tag{4–54}$$

Normalization yields

$$\Phi(t) = 1 - \frac{\sum_{i=1}^{m} \sum_{j=1}^{m_i} r_{ij}^2 \Omega_i(t/r_{ij})}{\sum_{i=1}^{m} v[A_i] \sum_{j=1}^{m_i} r_{ij}^2} \tag{4–55}$$

Since

$$\Omega_i'(t/r_{ij}) = -v[A_i]\Phi_i'(t/r_{ij}) \tag{4–56}$$

differentiation gives

$$\Phi'(t) = \frac{\sum_{i=1}^{m} v[A_i] \sum_{j=1}^{m_i} r_{ij} \Phi_i'(t/r_{ij})}{\sum_{i=1}^{m} v[A_i] \sum_{j=1}^{m_i} r_{ij}^2} \tag{4–57}$$

Hence,

$$\begin{aligned} \mu^{(k)}(S) &= \frac{\sum_{i=1}^{m} v[A_i] \sum_{j=1}^{m_i} r_{ij} \int_0^{\infty} t^k \Phi_i'(t/r_{ij}) dt}{\sum_{i=1}^{m} v[A_i] \sum_{j=1}^{m_i} r_{ij}^2} \\ &= \frac{\sum_{i=1}^{m} v[A_i] \sum_{j=1}^{m_i} r_{ij}^{k+2} \int_0^{\infty} u^k \Phi_i'(u) du}{\sum_{i=1}^{m} v[A_i] \sum_{j=1}^{m_i} r_{ij}^2} \end{aligned} \tag{4–58}$$

which reduces to Eq. 4–52 since

$$\mu^{(k)}(A_i) = \int_0^{\infty} u^k \Phi_i'(u) du \quad \blacksquare \tag{4–59}$$

An interesting reduction occurs for the case of a single primitive A, where S becomes

$$S = \bigcup_{j=1}^{m} r_j A + x_j \tag{4–60}$$

Define the random variable Z_p for $p = 3, 4, \ldots$ by

$$Z_p = \frac{\sum_{i=1}^{m} r_i^p}{\sum_{i=1}^{m} r_i^2} \tag{4–61}$$

and assume that the scalar multiples r_j are independent and identically distributed, their common distribution being known as a *sizing distribution*. It can be shown that Z_p is asymptotically normal as $m \to \infty$ and there exist asymptotic expressions for the moments of Z_p [23]. For instance, if μ_i denotes the ith central moment of the sizing distribution, then

$$E[Z_3] = \frac{\mu_3 + 3\mu_1\mu_2 + \mu_1^3}{\mu_2 + \mu_1^2} + O\left(m^{-1}\right) \tag{4–62}$$

For the single-primitive model of Eq. 4–60, the representation of Eq. 4–52 reduces to

$$\mu^{(k)}(S) = \mu^{(k)}(A) Z_{k+2} \tag{4–63}$$

The kth granulometric moment of S is a constant times Z_{k+2}. Thus, the granulometric moments are asymptotically normal (as $m \to \infty$). Since there exist asymptotic expressions for the moments of Z_{k+2}, there exist asymptotic expressions for the moments of the granulometric moments. For instance, for the model of Eq. 4–60,

$$E\left[\mu^{(k)}(S)\right] = \mu^{(k)}(A) E[Z_{k+2}] \tag{4–64}$$

If the sizing distribution is gamma distributed with parameters α and β, then, according to Eq. 4–62,

$$E[Z_3] = (\alpha + 2)\beta + O\left(m^{-1}\right) \tag{4–65}$$

$$E\left[\mu^{(1)}(S)\right] = \mu^{(1)}(A)(\alpha + 2)\beta + O\left(m^{-1}\right) \tag{4–66}$$

4.7 Granulometric Classification

Before discussing the specifics of granulometric classification, we briefly review the classification paradigm. Suppose $\{S_1, S_2, \ldots, S_I\}$ is a class of random binary images and with each image S_i is associated a random m-vector $\mathbf{F}_i = (F_{i1}, F_{i2}, \ldots, F_{im})$. The vectors $\mathbf{F}_1, \mathbf{F}_2, \ldots, \mathbf{F}_I$ are to serve as descriptors of the images, so that F_{ij} is a *feature* of image S_i. The association between an image and its feature vector is defined by a mapping $\Gamma(S_i) = \mathbf{F}_i$. Feature-based classification can be characterized in the following manner: given an image realization, apply the mapping Γ to the realization to compute a feature-vector realization and, by comparing the feature-vector realization to $\mathbf{F}_1, \mathbf{F}_2, \ldots, \mathbf{F}_I$, decide to which of the given I random image processes the realization belongs. Comparison is accomplished by some statistical measure of difference.

In the case of granulometries, given a collection of convex, compact sets $B_1, B_2, \ldots, B_J$, there exist granulometric moments for each granulometry $\{S \circ tB_j\}$, $j = 1, 2, \ldots, J$. If we take the first q moments of each granulometry, then $m = qJ$ descriptors, $\mu^{(k)}(S_i; B_j)$, are generated for each S_i, thereby yielding, for each S_i, a feature vector

$$\mathbf{F}_i = \begin{pmatrix} \mu^{(1)}(S_i; B_1) \\ \mu^{(2)}(S_i; B_1) \\ \vdots \\ \mu^{(q)}(S_i; B_1) \\ \mu^{(1)}(S_i; B_2) \\ \vdots \\ \mu^{(q)}(S_i; B_J) \end{pmatrix} \tag{4–67}$$

For classification of individual pixels, size distributions are applied locally. The granulometries are applied to the whole image but a size distribution is computed at each pixel by taking pixel counts in a window about each pixel. The window should be kept as small as possible to make the classifier sensitive; however, it must be kept large enough to avoid misclassification owing to variability.

We demonstrate the method by thresholding the gray-scale texture images of Fig. 4–12 to obtain the binarized texture images of Fig. 4–13 and then applying granulometric classification to the binary images [32]. Five structuring elements are employed: vertical line, horizontal line, 45° line, −45° line, and circular disk. Three pattern-spectrum moments are used: mean, variance, and skewness. Granulometries are applied to both the foreground (black) and background (white), so that each structuring element yields six features, there being 30 features in all at each pixel. We apply Gaussian maximum-likelihood classification in which the

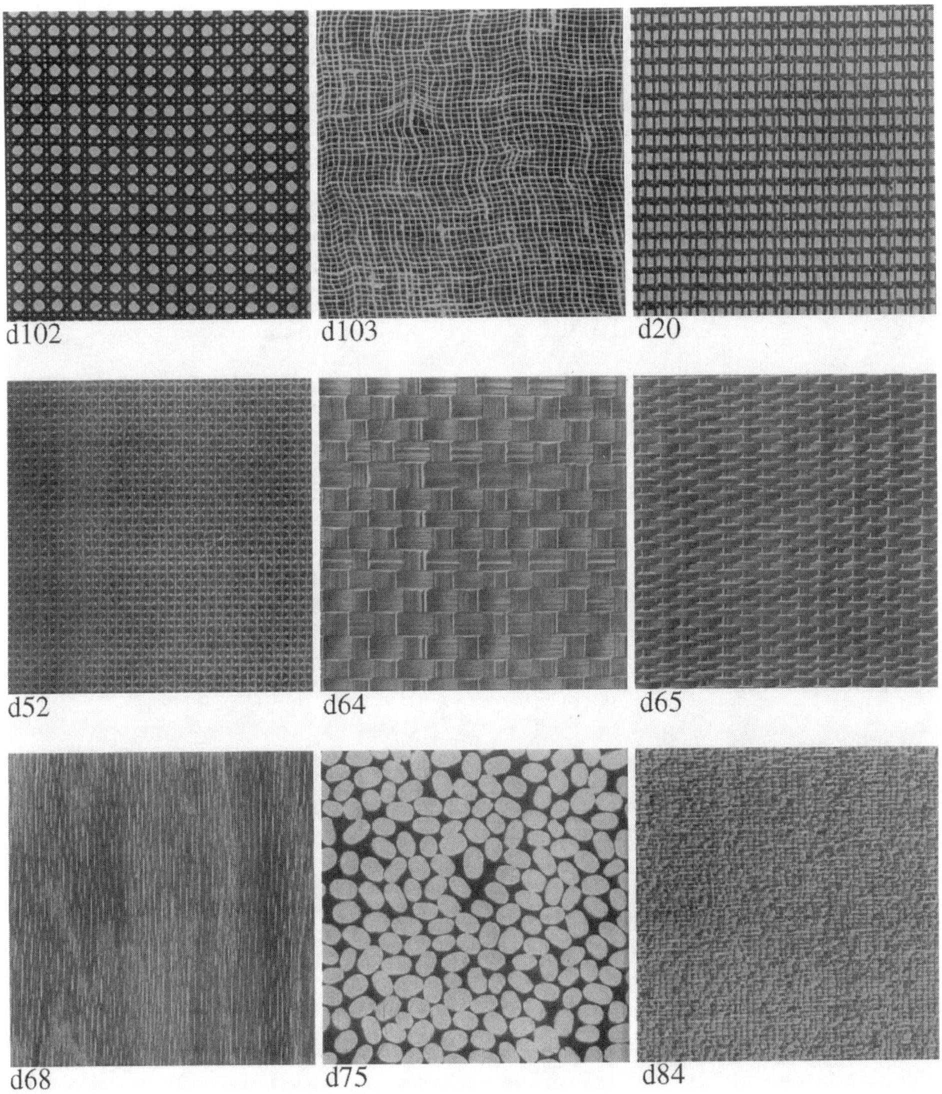

Figure 4–12. Gray-scale texture images.

mean vector and covariance matrix for each feature vector is used (not pooled). The left side of each image is used for training. Thus, classification on dependent and independent data is achieved by applying the classifier to the left and right sides of the images, respectively. Table 4.1 gives classification results for dependent data, overall classification accuracy being 99.96%. Table 4.2 gives results for independent data, ovarall accuracy falling to 95.5%. Three points should be noted regarding diminished accuracy: the major confusion occurs between textures d64 and d65, and d64 is actually an enlarged version of d65; the local features are obtained from using a 21×21 window, which is too small to capture the texture

Figure 4–13. Binary texture images.

Table 4.1. Classification accuracies for dependent data.

		Classified as								
		d102	d103	d20	d52	d64	d65	d68	d75	d84
Input	d102	100.00%	0.00%	0.00%	0.00%	0.00%	0.00%	0.00%	0.00%	0.00%
	d103	0.00%	99.98%	0.00%	0.00%	0.00%	0.00%	0.00%	0.00%	0.02%
	d20	0.00%	0.00%	100.00%	0.00%	0.00%	0.00%	0.00%	0.00%	0.00%
	d52	0.00%	0.10%	0.00%	99.90%	0.00%	0.00%	0.00%	0.00%	0.00%
	d64	0.00%	0.00%	0.00%	0.00%	100.00%	0.00%	0.00%	0.00%	0.00%
	d65	0.00%	0.00%	0.00%	0.00%	0.04%	99.96%	0.00%	0.00%	0.00%
	d68	0.00%	0.00%	0.00%	0.00%	0.17%	0.00%	99.83%	0.00%	0.00%
	d75	0.00%	0.00%	0.00%	0.00%	0.00%	0.00%	0.00%	100.00%	0.00%
	d84	0.00%	0.00%	0.00%	0.00%	0.00%	0.00%	0.00%	0.00%	100.00%

Table 4.2. Classification accuracies for independent data.

		Classified as								
		d102	d103	d20	d52	d64	d65	d68	d75	d84
Input	d102	99.8%	0.0%	0.0%	0.0%	0.1%	0.0%	0.0%	0.1%	0.0%
	d103	0.0%	90.9%	0.0%	0.0%	5.7%	0.0%	0.0%	0.0%	3.4%
	d20	0.0%	0.0%	99.7%	0.0%	0.3%	0.0%	0.0%	0.0%	0.0%
	d52	0.0%	0.6%	0.0%	98.7%	0.6%	0.0%	0.1%	0.0%	0.0%
	d64	0.0%	0.0%	0.0%	0.0%	99.6%	4.0%	0.0%	0.1%	0.0%
	d65	0.0%	0.0%	0.0%	0.0%	27.1%	72.9%	0.0%	0.0%	0.0%
	d68	0.0%	0.0%	0.0%	0.0%	1.3%	0.0%	98.7%	0.0%	0.0%
	d75	0.0%	0.0%	0.0%	0.0%	0.0%	0.0%	0.0%	100.00%	0.0%
	d84	0.0%	0.0%	0.0%	0.0%	0.6%	0.4%	0.0%	0.0%	99.0%

Table 4.3. Classification accuracies for independent data with d64 removed.

		Classified as							
		d102	d103	d20	d62	d65	d68	d75	d84
Input	d102	99.80%	0.00%	0.00%	0.00%	0.00%	0.00%	0.20%	0.00%
	d103	0.00%	93.94%	0.00%	0.00%	0.65%	0.00%	0.00%	5.41%
	d20	0.00%	0.00%	100.00%	0.00%	0.00%	0.00%	0.00%	0.00%
	d62	0.00%	0.70%	0.00%	99.17%	0.00%	0.14%	0.00%	0.00%
	d65	0.00%	0.00%	0.00%	0.00%	93.76%	0.00%	6.24%	0.00%
	d68	0.01%	0.00%	0.00%	0.00%	0.00%	99.99%	0.00%	0.00%
	d75	0.00%	0.00%	0.00%	0.00%	0.00%	0.00%	100.00%	0.00%
	d84	0.00%	0.00%	0.00%	0.00%	0.47%	0.00%	0.00%	99.53%

characteristics of d64, thereby causing large variation; there are slight illumination gradients in the binarized versions of d64 and d65, and these are inversely related, dark to light in d64 and light to dark in d65 (viewing from left to right). If d64 is eliminated from the texture set, and training and testing are done on the reduced set, then overall accuracies on dependent and independent data are 100% and 98.3%, respectively. Table 4.3 shows the individual results for classification on independent data.

The granulometric-feature set is typically reduced by using either the Karhunen-Loeve transform or sequential feature selection [29, 33]. Figure 4–14 shows classification accuracies for different numbers of features using reduction via feature selection and the Karhunen-Loeve transform on both dependent and independent data. The Karhunen-Loeve transform is performed using the pooled covariance matrix from all 9 image samples. Full accuracy is achieved using 15 features, and very little accuracy is lost using 9 features.

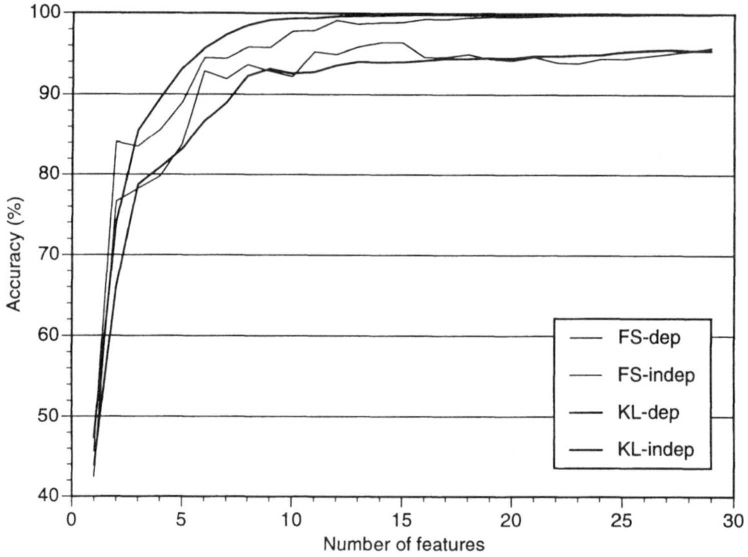

Figure 4–14. Classification accuracies for different numbers of features.

4.8 Discrete Granulometric Bandpass Filters

If $\{\Psi_t\}$ is a granulometry and $T = \{t_k\}$ is a sequence of points such that $0 = t_0 < t_1 < t_2 < \cdots$ and $t_k \to \infty$ as $k \to \infty$, then every set S is partitioned according to

$$S_k = \Psi_{t_k}(S) - \Psi_{t_{k+1}}(S) \tag{4–68}$$

for $k = 0, 1, 2, \ldots$. Specifically, $S = \bigcup_k S_k$ and $S_k \cap S_{k'}$ for $k \neq k'$. The family $\{S_k\}$ is called the *discrete granulometric spectrum* of S relative to $\{\Psi_t\}$ and T, and S_k the kth *spectral band* relative to $\{\Psi_t\}$ and T. For digital processing, we assume $T = \{0, 1, 2, \ldots\}$ and just call $\{S_k\}$ the discrete granulometric spectrum (relative to $\{\Psi_t\}$). The discrete pattern and granulometric spectra are related by $\Phi'_S(k) = v[S_k]/v[S]$.

A *discrete granulometric bandpass filter* Λ_Π is defined by choosing a *pass set* of orders $\Pi \subset T$ and defining

$$\Lambda_\Pi(S) = \bigcup_{t_k \in \Pi} S_k \tag{4–69}$$

Π^c is called the *fail set*. Let Σ denote the mapping $S \to \{S_k\}$, which takes compact sets into set families via the discrete granulometric spectrum, and Θ_Π be the selection function defined by the pass set. Then the discrete bandpass filter is defined by the composition

$$\Lambda_\Pi = \Sigma^{-1}\Theta_\Pi\Sigma \tag{4–70}$$

which shows the spectral nature of a granulometric bandpass filter: Λ_Π is defined by selecting spectral bands to pass (whereas a linear bandpass filter selects frequency bands).

To find an optimal granulometric bandpass filter in the signal-union-noise ($S \cup N$) model, we need to express the estimation error associated with Λ_Π. If $S \cap N = \emptyset$ and the granulometry is distributive, then

$$\begin{aligned}\Lambda_\Pi(S \cup N)\Delta S &= \left(\bigcup_{t_k \in \Pi}(S \cup N)_{t_k}\right)\Delta\left(\bigcup_{k=0}^{\infty} S_{t_k}\right)\\ &= \left(\bigcup_{t_k \in \Pi} N_{t_k}\right)\Delta\left(\bigcup_{t_k \in \Pi^c} S_{t_k}\right)\end{aligned} \tag{4–71}$$

Taking the measure and expectation in Eq. 4–71 yields the error for Λ_Π:

$$e[\Lambda_\Pi] = E\big[v[\Lambda_\Pi(S \cup N)\Delta S]\big] = \sum_{t_k \in \Pi} E\big[v[N_{t_k}]\big] + \sum_{t_k \in \Pi^c} E\big[v[S_{t_k}]\big] \tag{4–72}$$

Error is minimized by the decision rule $t_k \in \Pi$ if and only if

$$E\big[v[N_{t_k}]\big] \leqslant E\big[v[S_{t_k}]\big] \tag{4–73}$$

The optimal pass and fail sets are denoted by $\Pi\langle\Psi\rangle$ and $\Pi\langle\Psi\rangle^c$, respectively, and they determine the optimal filter Λ_Ψ relative to the spectrum associated with $\{\Psi_t\}$ and T.

Equation 4–73 can be reformulated in terms of the increasing size distribution $\Omega^\bullet$ of S relative to $\{\Psi_t\}$ and T. Using the relationship

$$v[S_{t_k}] = \Omega_S(t_k) - \Omega_S(t_{k+1}) \tag{4–74}$$

the error equation of Eq. 4–72 takes the form

$$\begin{aligned}e[\Lambda_\Pi] &= \sum_{t_k \in \Pi} E\big[v[\Omega_N(t_k) - \Omega_N(t_{k+1})]\big]\\ &\quad + \sum_{t_k \in \Pi^c} E\big[v[\Omega_S(t_k) - \Omega_S(t_{k+1})]\big]\end{aligned} \tag{4–75}$$

The decision rule for the induced granulometric bandpass filter is $t_k \in \Pi\langle\Psi\rangle$ if and only if

$$E[\Omega_N(t_k) - \Omega_N(t_{k+1})] \leqslant E\big[\Omega_S(t_k) - \Omega_S(t_{k+1})\big] \tag{4–76}$$

Rewriting the inequality in terms of $\Omega^\bullet$ and dividing both sides by $t_{k+1} - t_k$ yields

$$E\left[\frac{\Omega_N^\bullet(t_{k+1}) - \Omega_N^\bullet(t_k)}{t_{k+1} - t_k}\right] \leqslant E\left[\frac{\Omega_S^\bullet(t_{k+1}) - \Omega_S^\bullet(t_k)}{t_{k+1} - t_k}\right] \tag{4–77}$$

In terms of the mean size distributions,

$$\frac{\mathrm{M}_N(t_{k+1}) - \mathrm{M}_N(t_k)}{t_{k+1} - t_k} \leqslant \frac{\mathrm{M}_S(t_{k+1}) - \mathrm{M}_S(t_k)}{t_{k+1} - t_k} \tag{4–78}$$

The decision rule for t_k being in the optimal pass set involves difference quotients of M and holds no matter how fine the sampling (no matter how small $t_{k+1} - t_k$).

Were M differentiable everywhere (in some suitable sense), then letting $t_{k+1} - t_k \to 0$ would yield a rule involving the derivatives of the granulometric size densities, namely, it would reduce to $t_k \in \Pi\langle\Psi\rangle$ if and only if $\mathrm{H}_N(t_k) \leqslant \mathrm{H}_S(t_k)$. Since $t_{k+1} - t_k$ can be made arbitrarily small, one might conjecture that $t_{k+1} - t_k$ could be sent to 0 to obtain a decision rule for continuous granulometric spectra, namely, $t \in \Pi\langle\Psi\rangle$ if and only if $\mathrm{H}_N(t) \leqslant \mathrm{H}_S(t)$. The matter is not so simple because $\Omega_S^\bullet$ and $\Omega_N^\bullet$ are random functions and neither the realizations of $\Omega^\bullet$ nor M need be differentiable everywhere.

4.9 Continuous Granulometric Bandpass Filters

We turn now to the continuous granulometric spectrum. For theoretical reasons we assume that $\{\Psi_t\}$ is a distributive Euclidean granulometry that is upper semicontinuous. We will leave the details on upper semicontinuity to other sources, noting simply that if $\{\Psi_t\}$ has a finite generator of compact, convex sets, then $\{\Psi_t\}$ is upper semicontinuous. The *continuous granulometric spectrum* of a set S relative to $\{\Psi_t\}$ is defined, for $t \geqslant 0$, by

$$S_t = \bigcap_{h>0} \Psi_t(S) - \Psi_{t+h}(S) \tag{4–79}$$

Equivalently,

$$S_t = \Psi_t(S) - \bigcup_{h>0} \Psi_{t+h}(S) \tag{4–80}$$

Each S_t is called a *spectral component* of S.

The spectral components partition S; that is, they are mutually disjoint and their union equals S. To demonstrate disjointness, if $r > t$, then, for $h = r - t$,

$$S_t \subset \Psi_t(S) - \Psi_{t+h}(S) = \Psi_t(S) - \Psi_r(S) \subset \Psi_t(S) - S_r \tag{4–81}$$

Thus, $S_t \cap S_r = \emptyset$. To see that $S = \bigcup_t S_t$, clearly $S_t \subset S$ for all t, so that $\bigcup_t S_t \subset S$. On the other hand, suppose $x \in S$ and let $t(x) = \sup\{t\colon x \in \Psi_t(S)\}$. Because the granulometry is upper semicontinuous, $x \in \limsup_{t>t(x)} \Psi_t(S) \subset \Psi_{t(x)}(S)$. Since $x \notin \Psi_{t(x)+h}(S)$ for $h > 0$, we conclude that $x \in S_{t(x)}$. Hence, $S \subset \bigcup_t S_t$, and we conclude that $\{S_t\}$ forms a partition of S.

For any t-interval I, the *spectral band* $G_S(I)$ of the image S determined by $\{\Psi_t\}$ is the union of the spectral components over I, namely,

$$G_S(I) = \bigcup_{t \in I} S_t \tag{4–82}$$

The measure of the band is $M_S(I) = v[G_S(I)]$.

A continuous granulometric bandpass filter is defined via the notion of a *countable-interval subset* of $[0, \infty)$, which is a subset $\Pi \subset [0, \infty)$ that can be expressed as a countable union of disjoint intervals Π_i, where singleton point sets are treated as intervals of length zero. Without loss of generality we assume that $i < j$ implies, for $t \in \Pi_i$ and $r \in \Pi_j$, that $t < r$ and there exists $s \notin \Pi$ such that $t < s < r$. This assumption means that Π_i is to the left of Π_j and that Π_i and Π_j are separated by Π^c, which we denote by Χ. If $\{\Psi_t\}$ is an upper semicontinuous distributive Euclidean granulometry and Π is a countable-union subset, then the *continuous granulometric bandpass filter* Λ_Π is defined by

$$\Lambda_\Pi(S) = \bigcup_{t \in \Pi} S_t = \bigcup_{i=1}^{\infty} G_S(\Pi_i) \tag{4–83}$$

where the second union may be finite. Π and Χ are the pass and fail sets for Λ_Π. Χ has a decomposition similar to that for Π: $\mathrm{X} = \bigcup_j \mathrm{X}_j$.

We consider the optimal continuous granulometric bandpass filter in the disjoint, compact signal-union-noise model. The observed image is $S \cup N$, signal and noise are compact, and $S \cap N = \emptyset$. The error associated with the filter Λ_Π is

$$e[\Lambda_\Pi] = E\big[v[S \Delta \Lambda_\Pi(S \cup N)]\big] \tag{4–84}$$

A key proposition (not proven here) expresses $e[\Lambda_\Pi]$ in terms of the MSD.

THEOREM 4–4 [14]. *For the disjoint, compact signal-union-noise model $S \cup N$,*

$$e[\Lambda_\Pi] = \sum_{i=1}^{\infty} E\big[M_N(\Pi_i)\big] + \sum_{j=1}^{\infty} E\big[M_S(\mathrm{X}_j)\big] \tag{4–85}$$

Moreover, expectations for the intervals $[a, b)$, (a, b), $(a, b]$, *and* $[a, b]$, $a \leqslant b$, *are*

$$\begin{aligned} E\big[M_S([a,b))\big] &= \mathrm{M}_S(b) - \mathrm{M}_S(a) \\ E\big[M_S((a,b))\big] &= \mathrm{M}_S(b) - \mathrm{M}_S(a^+) \\ E\big[M_S((a,b])\big] &= \mathrm{M}_S(b^+) - \mathrm{M}_S(a^+) \\ E\big[M_S([a,b])\big] &= \mathrm{M}_S(b^+) - \mathrm{M}_S(a) \end{aligned} \tag{4–86}$$

where $+$ *denotes a one-sided limit from the right and analogous expressions hold for* N.

In general, M is an increasing function of the variable t and $\mathrm{M}(0) = 0$. We will assume that M is of bounded variation and continuous from the left. For practical images, M will be of bounded variation; however (as we will see in an example), it need not be. Under these assumptions, the following properties hold: M is continuous except on at most a countable set, M is differentiable a.e. with respect to Lebesgue measure, the GSD $\mathrm{H} = \mathrm{M}'$ is defined a.e. and is integrable, and M possesses the Lebesgue decomposition

$$\mathrm{M}(t) = K(t) + A(t) \tag{4–87}$$

where K is increasing, $K'(t) = 0$ a.e., A is increasing, A is absolutely continuous, A is differentiable a.e., and

$$A(t) = \int_0^t \mathrm{H}(u)du \tag{4–88}$$

K is known as the *singular part* of the decomposition and we assume it is a step function with a countable number of steps (which places no practical constraint on the decomposition). We denote the jump at t by $J(t)$.

THEOREM 4–5 [14]. *If countable-union pass and fail sets are decomposed into sets of intervals* $\langle a_1, b_1\rangle$, $\langle a_2, b_2\rangle, \ldots$ *and* $\langle c_1, d_1\rangle$, $\langle c_2, d_2\rangle, \ldots$, *respectively (where angle brackets indicate that it does not matter whether the endpoints are included in the interval), then*

$$e[\Lambda_\Pi] = \sum_{i=1}^{\infty} \int_{a_i}^{b_i} \mathrm{H}_N(u)du + \sum_{j=1}^{\infty} \int_{c_j}^{d_j} \mathrm{H}_S(u)du + \sum_{t \in \Pi} J_N(t) + \sum_{t \in \mathrm{X}} J_S(t) \tag{4–89}$$

PROOF. We first show that, relative to the Lebesgue decomposition of M_S, the expected spectral measure corresponding to any interval I with left and right endpoints a and b is

$$E\big[M_S(I)\big] = A_S(b) - A_S(a) + \sum_{t \in I} J_S(t) \tag{4–90}$$

Using the decomposition of Eq. 4–87 in conjunction with Eq. 4–86 yields

$$\begin{aligned}
E\big[M_S([a,b))\big] &= K_S(b) - K_S(a) + A_S(b) - A_S(a) \\
E\big[M_S((a,b))\big] &= K_S(b) - K_S(a^+) + A_S(b) - A_S(a) \\
E\big[M_S((a,b])\big] &= K_S(b^+) - K_S(a^+) + A_S(b) - A_S(a) \\
E\big[M_S([a,b])\big] &= K_S(b^+) - K_S(a) + A_S(b) - A_S(a)
\end{aligned} \tag{4–91}$$

Using the assumption that K_S is continuous from the left, analysis of the relationship between the jumps and the right-hand limits yields Eq. 4–90. $E[M_N(I)]$ possesses an analogous representation. Combining Eqs. 4–85 and 4–90 yields the desired result. ■

It needs to be recognized that in Eq. 4–89 the derivatives exist a.e. and, in general, the integrals need to be taken in the Lebesgue sense, ignoring the sets over which the MSDs are not differentiable.

It remains to construct an optimal filter (pass set) for a given granulometry. Let $D[S]$ and $D[N]$ denote the sets on which M_S and M_N are differentiable, respectively. The point sets at which A_S and A_N are differentiable include $D[S]$ and $D[N]$, respectively. K_S and K_N each have a countable number of jumps, their jump sets to be denoted by $J[S]$ and $J[N]$, respectively. For the upper semicontinuous distributive Euclidean granulometry $\{\Psi_t\}$ and the disjoint, compact signal-union noise model $S \cup N$, we define the *$\{\Psi_t\}$-induced pass set* $\Pi\langle\Psi\rangle$ by $t \in \Pi\langle\Psi\rangle$ if and only if one of the following three conditions is satisfied: (i) $t \in D[S] \cap D[N]$ and $A'_S(t) \geqslant A'_N(t)$; (ii) $t \in J[S] - J[N]$; (iii) $t \in J[S] \cap J[N]$ and $J_S(t) \geqslant J_N(t)$. The corresponding fail set is denoted by $\mathrm{X}\langle\Psi\rangle = \Pi\langle\Psi\rangle^c$. The corresponding filter is denoted by Λ_Ψ and is called the *$\{\Psi_t\}$-induced filter*.

The next theorem states that Λ_Ψ is optimal in the sense that it possesses minimum error among all granulometric bandpass filters induced by countable-union subsets. For this theorem to make sense, $\Pi\langle\Psi\rangle$ must be a countable-union subset. Since the MSD and its absolutely continuous part need only be differentiable a.e., some care must be taken; that is, we need some sufficient condition for $\Pi\langle\Psi\rangle$ being a countable-union subset that does not hinder application of the theory. It can be shown that, if M_S and M_N possess continuous derivatives except on sets without limit points, then $\Pi\langle\Psi\rangle$ is a countable-union subset. An added bonus when this assumption holds is that the integral of Eq. 4–88 is Riemann. We can now state the basic optimality theorem for granulometric bandpass filters. It follows immediately from the definition of the induced pass set and the error formulation of Theorem 4–5.

THEOREM 4–6 [14]. *If* $\Pi\langle\Psi\rangle$ *is a countable-union subset, then* $e[\Lambda_\Psi] \leqslant e[\Lambda_\Pi]$ *for any other countable-union subset* Π.

It is common for the MSD to have no singular part. In this case, the error representation of Theorem 4–5 reduces to the first two summands of Eq. 4–89 and the induced pass set is defined by

$$\Pi\langle\Psi\rangle = \{t\colon \mathrm{H}_S(t) \geqslant \mathrm{H}_N(t)\} \tag{4–92}$$

If either derivative fails to exist or the derivatives are equal at t, then the choice of pass or fail set for t is irrelevant. Indeed, any set of measure zero can be switched between the pass and fail sets without changing the error, and therefore the optimality. Furthermore, if

$$\Pi\langle\Psi\rangle = \bigcup_{i=1}^{\infty}\langle a_i, b_i\rangle \tag{4–93}$$

where $b_i < a_{i+1}$ for $i = 1, 2, \ldots$, then the induced optimal filter is given by

$$\begin{aligned}\Lambda_\Psi(S) &= \bigcup_{i=1}^{\infty} G_S(\langle a_i, b_i\rangle)\\ &= \bigcup_{i=1}^{\infty}\ \bigcup_{a_i<t\leqslant b_i}\left(\Psi_t(S) - \bigcup_{h>0}\Psi_{t+h}(S)\right)\\ &= \bigcup_{i=1}^{\infty}\Psi_{b_i} - \Psi_{a_i}\end{aligned} \tag{4–94}$$

where, owing to zero singular parts, we have replaced each interval $\langle a_i, b_i\rangle$ by $(a_i, b_i]$. The bandpass character of the filter is evident in Eq. 4–94.

To see why it is common for M to have no singular part, first note that, for each realization of the random set, $\Omega^\bullet$ is an increasing function of the variable t and $\Omega^\bullet(0) = 0$, and $\Omega^\bullet$ is continuous from the left. If $\Omega^\bullet$ is also of bounded variation, then it possesses a decomposition of the form given in Eq. 4–87. Suppose, for the moment, that $\Omega^\bullet$ is singular and has steps of size $g(r_i)$ at y identically distributed random points $r_1, r_2, \ldots, r_y$, governed by the continuous probability density $f(r)$. Then

$$\Omega^\bullet(t) = \sum_{i=1}^{y} g(r_i)T[t; r_i] \tag{4–95}$$

where $T[t; r_i] = 1$ if $t > r_i$ and $T[t; r_i] = 0$ if $t \leqslant r_i$. Taking the expectation,

$$M(t) = \sum_{i=1}^{y} E\big[g(r_i)T[t; r_i]\big] = y \int_0^t g(r)f(r)dr \tag{4–96}$$

If the number of jumps is a random variable Y with mean μ_Y, then

$$M(t) = \mu_Y \int_0^t g(r)f(r)dr \tag{4–97}$$

M is therefore a continuous function, the key assumption being that the distribution $f(r)$ of the jump points is a continuous density. Since M is increasing, it is of bounded variation if and only if the product gf forms an integrable function over $[0, \infty)$. If we further assume that g and f are continuous functions, then M is differentiable with derivative

$$H(t) = \mu_Y g(t)f(t) \tag{4–98}$$

At a minimum, Eq. 4–98 holds for all points at which both g and f are continuous.

We consider two examples. The first illustrates the effect of the sizing distributions. Let B be a ball of unit radius and C be a square of unit radius (edge length 2). Let S and N be disjoint unions of randomly sized homothetics of B and C governed by continuous sizing densities $f_S(r)$ and $f_N(r)$, respectively, and suppose S and N are disjoint. Let the expected number of signal and noise grains be μ_S and μ_N, respectively. Then

$$\begin{aligned} S &= \bigcup_{i=1}^{Y} r_i B + x_i \\ N &= \bigcup_{j=1}^{Z} r_j C + y_j \end{aligned} \tag{4–99}$$

with $E[Y] = \mu_S$ and $E[Z] = \mu_N$. Let

$$\Psi_t(S) = (S \circ tB) \cup (S \circ tC) \tag{4–100}$$

Then the realizations of both $\Omega_S^\bullet$ and $\Omega_N^\bullet$ are step functions, so that Eqs. 4–97 and 4–98 apply. $H_S(t) = \mu_S \pi t^2 f_S(t)$ and $H_N(t) = \mu_N 4t^2 f_N(t)$. Thus, $t \in \Pi\langle\Psi\rangle$ is and only if

$$\mu_S f_S(t) \geqslant (4/\pi)\mu_N f_N(t) \tag{4–101}$$

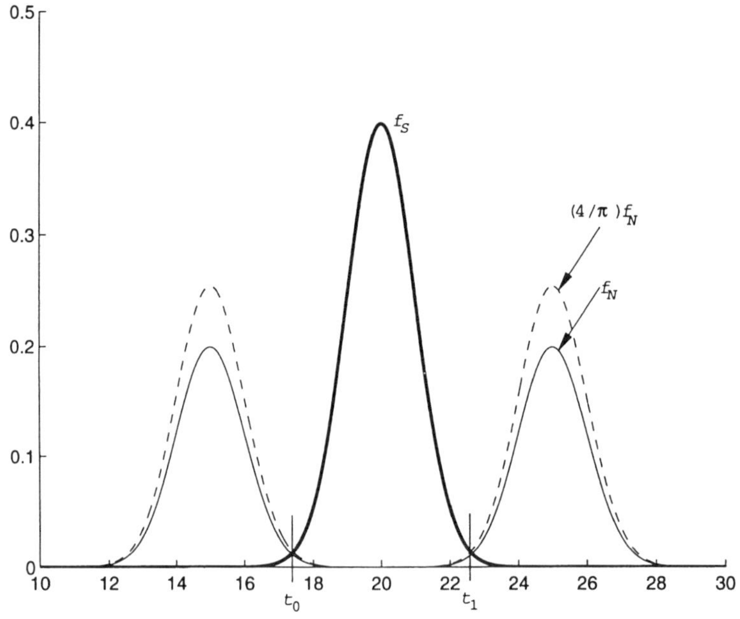

Figure 4–15. Signal and noise sizing distributions for unimodal signal and bimodal noise.

Suppose the signal and noise sizing densities are unimodal and bimodal, respectively, as shown in Fig. 4–15 (in which it is assumed that $\mu_S = \mu_N$). Then there exist points t_0 and t_1 such that $t \in \Pi\langle\Psi\rangle$ if and only if $t_0 \leqslant t \leqslant t_1$. Equations 4–89 and 4–94 apply. Thus, $\Lambda_\Psi = \Psi_{t_0} - \Psi_{t_1}$ and

$$e[\Lambda_\Pi] = 4\mu_N \int_{t_0}^{t_1} r^2 f_N(r)dr + \pi\mu_S\left(\int_0^{t_0} r^2 f_S(r)dr + \int_{t_1}^{\infty} r^2 f_S(r)dr\right) \quad (4\text{–}102)$$

This second example illustrates effects on the spectral bands. Consider the same setting as the previous example, except let $\Psi_t(S) = S \circ tB$. $\Omega_S^\bullet$ and H_S are unchanged and

$$\Omega_N^\bullet(t) = \sum_{j=1}^{Z}(4-\pi)t^2(1 - T[t;r_j]) + 4r_j^2 T[t;r_j] \quad (4\text{–}103)$$

$$\mathrm{H}_N(t) = \mu_N\left[\pi t^2 f_N(t) + (4-\pi)2t(1 - F_N(t))\right] \quad (4\text{–}104)$$

where Z is the random number of squares in an image and $F_N(t)$ is the probability distribution function for $f_N(t)$. The absolutely continuous part A_N does not vanish and results from the continuous removal of spectral bands from squares by disks of increasing radii. From Eq. 4–104, $t \in \Pi\langle\Psi\rangle$ if and only if

$$\mu_S f_S(t) \geqslant \mu_N\left[f_N(t) + 2(4-\pi)\pi^{-1}t^{-1}(1 - F_N(t))\right] \quad (4\text{–}105)$$

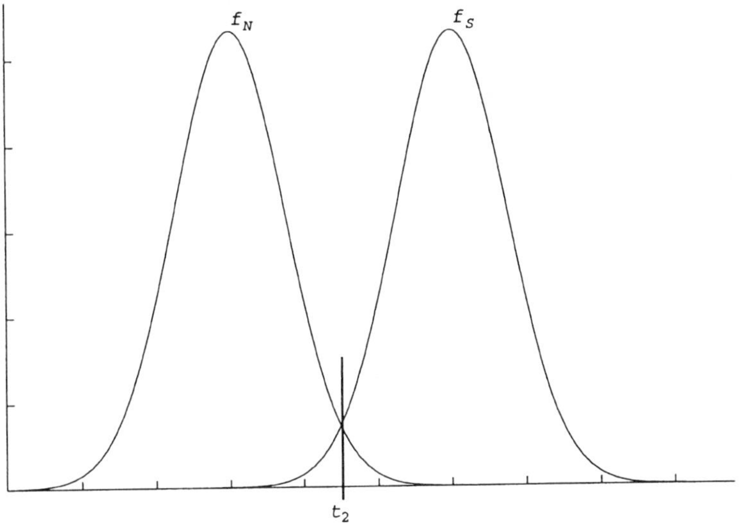

Figure 4–16. Signal and noise sizing distributions for unimodal signal and noise.

For simplicity, assume $\mu_S = \mu_N$. If the sizing distributions are of the sort shown in Fig. 4–16 and f_S is sufficiently far to the right so that the second summand on the right only adds slightly to $f_N(t)$ in the region where the distributions overlap, then $t \in \Pi\langle\Psi\rangle$ if and only if $t \geqslant t_2$, where t_2 is a point slightly to the right of where the densities intersect in Fig. 4–16, and the optimal filter is $\Lambda_\Psi(S) = S \circ t_2 B$. Some noise squares have radii exceeding t_2. These will pass; however, opening by $t_2 B$ will round a good portion of their corners so that they will have a fairly circular appearance. Thus, $\Lambda_\Psi(S \cup N)$ will consist of signal disks and rounded noise grains with radii exceeding t_2, the key point being that the optimal filter removes the small spectral bands in the noise, thereby making the shape of passing noise more like the signal. Because the structuring element is isotropic, squares can be randomly rotated without affecting the outcome. Figure 4–17(a) shows a realization of $S \cup N$ for which f_S and f_N are standard normal densities with unit variances and means 22 and 16, respectively, and $\mu_S = \mu_N = 20$. In this case, solution of Eq. 4–105 yields the cut-off $t_2 = 19.0015$. Figure 4–17(b) shows the realization of Fig. 4–17(a) opened by a disk of radius 19. The arrow in Fig. 4–17(b) points to a noise grain that has been erroneously passed but, owing to deleted spectral bands, is rounded off at the corners.

If the signal and noise MSDs possess null singular parts, then the optimal pass set can be interpreted in terms of the differential inequality of Eq. 4–92. More can be said if we interpret the derivatives as generalized functions. To wit, assume M_S and M_N are continuously differentiable except on sets without limit points. By definition of $\Pi\langle\Psi\rangle$, we can ignore points outide

$$D_{S,N} = \big(D[S] \cap D[N]\big) \cup J[S] \cup J[N] \tag{4–106}$$

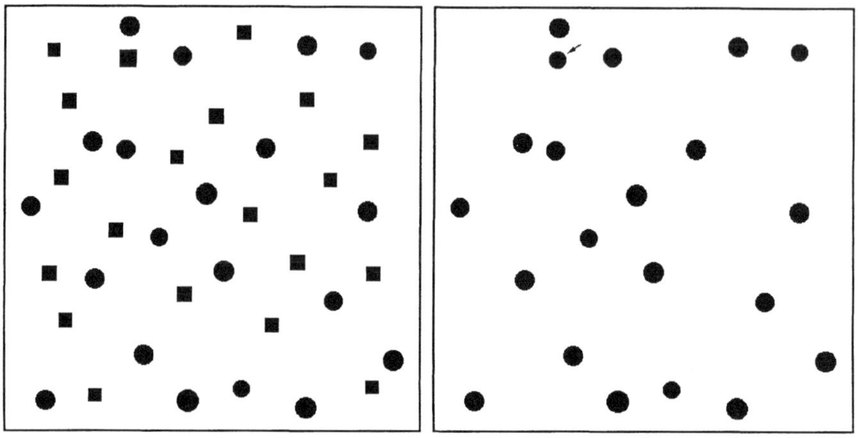

Figure 4–17. Filter showing spectral effects: (a) signal union noise; (b) filtered image (arrow pointing to passing noise grain).

At such points, M_S and M_N are continuous but do not possess derivatives. Under the hypothesis, all such points are isolated and their inclusion or lack of inclusion in $\Pi\langle\Psi\rangle$ has no effect on the error. Moreover, the jump points of K_S and K_N are isolated and therefore K_S and K_N can be represented as generalized functions. Hence, on $D_{S,N}$,

$$\begin{aligned} H_S(t) &= \sum_{i=1}^{\infty} J_S[t_{s,i}]\delta(t - t_{s,i}) + A'_S(t) \\ H_N(t) &= \sum_{j=1}^{\infty} J_N[t_{N,j}]\delta(t - t_{N,j}) + A'_N(t) \end{aligned} \tag{4–107}$$

where $J[S] = \{t_{S,1}, t_{S,2}, \ldots\}$ and $J[N] = \{t_{N,1}, t_{N,2}, \ldots\}$. Under the usual interpretation of impulse functions and their intensities, $\Pi\langle\Psi\rangle$ can be interpreted as $t \in \Pi\langle\Psi\rangle$ if and only if $H_S(t) \geqslant H_N(t)$.

REFERENCES

[1] Matheron, *Random Sets and Integral Geometry*, John Wiley, New York, 1975.
[2] Dougherty, E. R., "Euclidean Gray-Scale Granulometries: Representation and Umbra Inducement," *Mathematical Imaging and Vision*, **1** (1),1992.
[3] Kraus, E., Heijmans, H. J., and E. R. Dougherty, "Gray-Scale Morphological Granulometries Compatible with Spatial Scaling," *Signal Processing*, **34**, 1993.
[4] Heijmans, H. J., *Morphological Operators*, Academic Press, New York, 1995.

[5] Dougherty, E. R., and Y. Chen, "Logical Granulometric Filtering in the Signal-Union-Clutter Model," in *Random Sets: Theory and Applications*, eds. J. Goutsias, R. Mahler, and C. Nguyen, Springer-Verlag, New York, 1997.

[6] Dougherty, E. R., and Y. Chen, "Logical Structural Filters," *Optical Engineering*, **37** (6), 1998.

[7] Batman, S., and E. R. Dougherty, "Size Distributions for Multivariate Morphological Granulometries: Texture Classification and Statistical Properties," *Optical Engineering*, **36** (5), 1997.

[8] Dougherty, E. R., and C. Cuciurean-Zapan, "Optimal Reconstructive τ-Openings for Disjoint and Statistically Modeled Nondisjoint Grains," *Signal Processing*, **56**, 1997.

[9] Meyer, F., and S. Beucher, "Morphological Segmentation," *Visual Communication and Image Representation*, **1** (1), 1990.

[10] Chen, Y., and E. R. Dougherty, "Markovian Analysis of Adaptive Reconstructive Multiparameter τ-Openings," *Mathematical Imaging and Vision*, **9** (1), 1999.

[11] Chen, Y., and E. R. Dougherty, "Adaptive Reconstructive τ-Openings: Convergence and the Steady-State Distribution," *Electronic Imaging*, **5** (3), 1996.

[12] Dougherty, E. R., Haralick, R. M., Chen, Y., Agerskov, C., Jacobi, U., and P. H. Sloth, "Estimation of Optimal τ-Opening Parameters Based on Independent Observation of Signal and Noise Pattern Spectra," *Signal Processing*, **29**, 1992.

[13] Haralick, R. M., Katz, P. L., and E. R. Dougherty, "Model-Based Morphology: The Opening Spectrum," *Computer Vision, Grpahics, and Image Processing: Graphical Models and Image Processing*, **57** (1), 1995.

[14] Dougherty, E. R., "Optimal Binary Morphological Bandpass Filters Induced by Granulometric Spectral Representation," *Mathematical Imaging and Vision*, **7** (2), 1997.

[15] Schonfeld, D., and J. Goutsias, "Optimal Morphological Pattern Restoration from Noisy Binary Images," *IEEE Trans. on Pattern Analysis and Machine Intelligence*, **13**, 1991.

[16] Maragos, P., "Pattern Spectrum and Multiscale Shape Representation," *IEEE Trans. on Pattern Analysis and Machine Intelligence*, **11**, 1989.

[17] Dougherty, E. R., and J. Pelz, "Morphological Granulometric Analysis of Electrographic Images — Size Distribution Statistics for Process Control," *Optical Engineering*, **30** (4), 1991.

[18] Dougherty, E. R., Pelz, J., Sand, F., and A. Lent, "Morphological Segmentation by Local Granulometric Size Distributions," *Electronic Imaging*, **1** (1), 1992.

[19] Dougherty, E. R., Newell, J. T., and J. B. Pelz, "Morphological Texture-Based Maximum-Likelihood Pixel Classification Based on Local Granulometric Moments," *Pattern Recognition*, **25** (10), 1992.

[20] Chakravarthy, B., Grivas, D., and M. Skolnick, "Morphological Analysis of Pavement Surface Condition," in *Mathematical Morphology in Image Processing*, ed. E. Dougherty, Marcel Dekker, New York, 1993.

[21] Vincent, L., and E. R. Dougherty, "Morphological Segmentation for Textures and Particles," in *Digital Image Processing Methods*, ed. E. Dougherty, Marcel Dekker, New York, 1994.
[22] Dougherty, E. R., and F. Sand, "Representation of Linear Granulometric Moments for Deterministic and Random Binary Euclidean Images," *Visual Communication and Image Representation*, **6** (1), 1995.
[23] Sand, F., and E. R. Dougherty, "Asymptotic Normality of the Morphological Pattern-Spectrum Moments and Orthogonal Granulometric Generators," *Visual Communication and Image Representation*, **3** (2), 1992.
[24] Sand, F., and E. R. Dougherty, "Asymptotic Granulometric Mixing Theorem: Morphological Estimation of Sizing Parameters and Mixture Proportions," *Pattern Recognition*, **31** (1), 1998.
[25] Bettoli, B., and E. R. Dougherty, "Linear Granulometric Moments of Noisy Binary Images," *Mathematical Imaging and Vision*, **3** (3), 1993.
[26] Dougherty, E. R., and C. R. Giardina, "Error Bounds for Mophologically Derived Measurements," *SIAM J. on Applied Mathematics*, **47** (2), 1987.
[27] Sivakumar, K., and J. Goutsias, "Monte Carlo Estimation of Morphological Granulometric Discrete Size Distributions," in *Mathematical Morphology and its Applications to Image Processing*, eds. J. Serra and P. Soille, Kluwer Academic Publishers, Boston, 1994.
[28] Sivakumar, K., and J. Goutsias, "Discrete Morphological Size Distributions and Densities: Estimation Techniques and Applications," *Electronic Imaging*, **6** (1), 1997.
[29] Chen, Y., and E. R. Dougherty, "Gray-scale Morphological Granulometric Texture Classification," *Optical Engineering*, **33** (8), 1994.
[30] Chen, Y., Dougherty, E. R., Totterman, S., and J. Hornak, "Classification of Trabecular Structure in Magnetic Resonance Images Based on Morphological Granulometries," *Magnetic Resonance Medicine*, **29** (3), 1993.
[31] Baeg, S., Batman, S., Dougherty, E. R., Kamat, V., Kehtarnavaz, N. D., Kim, S., Popov, A., Sivakumar, K., and R. Shah, "Unsupervised Morphological Granulometric Texture Segmentation of Digital Mammograms," *Electronic Imaging*, **8** (1), 1999.
[32] Brodatz, P., *Textures: A Photographic Album for Artists and Designers*, Dover, New York, 1966.
[33] Young, T. Y., and K.-S. Fu, *Handbook of Pattern Recognition and Image Processing*, Academic Press, Orlando, 1986.

CHAPTER 5

EASY RECIPES FOR MORPHOLOGICAL FILTERS[1]

Henk J. A. M. Heijmans
CWI
Amsterdam, The Netherlands

5.1 INTRODUCTION

A *(morphological) filter* is an increasing operator ψ on a complete lattice $\mathcal{L}$ which is idempotent:

$$\psi^2 = \psi.$$

For some basic literature on morphological filters the reader may refer to the second book edited by Serra [23] (in particular Chap. 6 by Matheron and Chap. 8 by Serra), the tutorial paper by Serra and Vincent [24], to our book [5, Chaps. 12–13], and to [2, 7, 17].

This paper discusses various classes of morphological filters such as openings and closings, annular filters, alternating sequential filters (or AS-filters), and self-dual filters. A large part of the mathematical theory for morphological filters holds on arbitrary complete lattices. However, when we give concrete examples, we often restrict ourselves to the case of subsets of a Euclidean or discrete space.

This paper aims to be mathematically rigorous in the sense that it gives precise definitions and propositions. As a general rule, proofs will be included when they provide additional insight. However, if a proof is rather technical, or when it requires substantial preparations, it will be omitted; in such cases appropriate references will be given.

In the remainder of this section we summarize the contents of this paper. Section 5.2 contains a brief discussion of the complete lattice framework for morphology. In Section 5.3 we discuss two elementary classes of filters, namely, openings and closings. A simple and general (but not the only) way to get openings and closings is by composing dilations and erosions that form adjunctions; a generalization

[1] This paper is a revised and extended version of our lecture "Morphological Filters," which was presented at the summer school, Morphological Image and Signal Processing, September 27–30, 1995, Zakopane, Poland.

of this idea leads to a class of filters which is relatively unknown, the *adjunctional filters*. In Section 5.4, annular filters will be treated. Such filters are given by very simple mathematical expressions. By composing openings and closings, one obtains a class of filters which has proved its importance in practice: the *alternating sequential filters*. These filters are discussed in Section 5.5. In Section 5.6, we introduce operators which are closely related to filters, namely *overfilters, underfilters, inf-overfilters*, and *sup-underfilters*, and describe several methods for their construction. In this section we also treat the *rank-max opening* and the *rank-min closing*, and introduce a new AS-filter obtained by composition of such openings and closings. The class of AS-filters introduced in Section 5.5 can be generalized by composing overfilters and underfilters instead of openings and closings. This generalization is the topic of Section 5.7; we also present some interesting examples there. A general way to construct idempotent operators is by iteration of operators which are not idempotent. In Section 5.8 it is explained that pointwise convergence of the iterates ψ^n of an operator ψ to a limit operator ψ^∞ (along with the continuity of ψ) guarantees that ψ^∞ is idempotent. An important class of operators which satisfy this pointwise convergence criterion are the operators which are *activity-extensive*. These operators are introduced in Section 5.9. There we also discuss the *center operator*. Section 5.10 deals with *self-dual filters*. Such filters treat foreground and background identically (unlike openings, closings, and AS-filters), and as such they are of great importance. The simplest self-dual filter is a special case of the annular filter treated in Section 5.4. For this filter we can give an explicit expression. All other (nontrivial) self-dual filters we know of cannot be expressed by an explicit formula, but their action is described in terms of an iteration procedure. Such a procedure uses a self-dual operator which is activity-extensive. We give detailed results (using the center operator) on the construction of such operators, and some explicit examples.

The filters discussed in this paper are all applied to the same binary input image, the right image in Fig. 5–1. It is obtained from the left image by "adding" salt-and

Figure 5–1. The undistorted image (left) and our test image (right). The size of these images is 128×128 pixels. The black pixels represent the foreground, the white pixels the background.

pepper noise; approximately 15% of the pixels have been affected by this noise. In all images shown in this paper, the black pixels represent the foreground and the white pixels the background.

Although many of our examples concern binary images (our test image is also binary), we can apply them to gray-scale images, too. For that goal we have to consider the usual flat operator extension discussed in [5, 22].

We conclude with some remarks about notation. If E is a set, then we denote by $\mathcal{P}(E)$ the power set of E comprising all subsets of E. When we write $\mathbb{E}^d$, we mean the d-dimensional product of $\mathbb{E}$, where $\mathbb{E}$ is a group; in practical cases $\mathbb{E} = \mathbb{Z}$ or $\mathbb{R}$ with the additive group structure. Using the notation $\mathbb{E}^d$ enables us to treat the discrete case $\mathbb{Z}^d$ and the continuous case $\mathbb{R}^d$ simultaneously.

5.2 MORPHOLOGY ON COMPLETE LATTICES

By its very nature, mathematical morphology is set-oriented, and as such directed toward binary images. However, from the early days of morphology onward, there has been a need for a more general theory covering different object spaces, in particular gray-scale images. Matheron and Serra [23] were the first to observe that a general framework for morphology can be achieved if one starts from the assumption that the object space is a complete lattice. This idea has been carried further by various people, in particular Heijmans and Ronse [9, 19] and Roerdink [13, 14]. A comprehensive account of the complete lattice framework can be found in [5].

5.2.1 BASIC THEORY

In this subsection we give some basic results. In the next section some of them will be applied to gray-scale images.

DEFINITION 5–1. A *complete lattice* is a set $\mathcal{L}$ with a partial ordering '$\leqslant$' such that every subset $\mathcal{H}$ of $\mathcal{L}$ has an infimum (greatest lower bound) and a supremum (least upper bound). The least element of $\mathcal{L}$ is denoted by O, the greatest element by I.

See [1] or [5] for further details. Throughout this section we assume that $\mathcal{L}$ is a complete lattice. A simple example is the family $\mathcal{P}(E)$ ordered by inclusion.

A well-known principle in the theory of partially ordered sets is the *duality principle*. This principle originates from the (trivial) fact that if $\mathcal{L}$ is a partial ordered set, then $\mathcal{L}$ with the dual partial ordering $\leqslant'$ defined by "$X \leqslant' Y$ if and only if $Y \leqslant X$" is a partial ordered set, too. As a result, to every definition or statement referring to $\leqslant$ there corresponds a dual one referring to $\leqslant'$.

Let $\mathcal{L}, \mathcal{M}$ be complete lattices and let $\psi : \mathcal{L} \to \mathcal{M}$; by this notation we mean that ψ is an operator from $\mathcal{L}$ to $\mathcal{M}$. We say that ψ is *increasing* if $X \leqslant X'$ implies that $\psi(X) \leqslant \psi(X')$. It is *decreasing* if $X \leqslant X'$ implies that $\psi(X) \geqslant \psi(X')$. On the collection of operators from $\mathcal{L}$ to $\mathcal{M}$ one can define a partial ordering as follows: $\psi \leqslant \psi'$ if $\psi(X) \leqslant \psi'(X)$, for every $X \in \mathcal{L}$. The set of operators, as well as the set of increasing operators, constitutes a complete lattice under this partial ordering.

An operator $\psi : \mathcal{L} \to \mathcal{L}$ is called an *automorphism* if it is increasing and bijective. One can easily show that every automorphism satisfies $\psi(\bigvee_{i \in I} X_i) = \bigvee_{i \in I} \psi(X_i)$ and $\psi(\bigwedge_{i \in I} X_i) = \bigwedge_{i \in I} \psi(X_i)$, for every family of sets X_i. An operator ψ is called a *negation* if it is decreasing, bijective, and satisfies $\psi^2 = \mathsf{id}$. Here id, or $\mathsf{id}_{\mathcal{L}}$, denotes the *identity operator* given by $\mathsf{id}(X) = X$, for every $X \in \mathcal{L}$. A negation satisfies $\psi(\bigvee_{i \in I} X_i) = \bigwedge_{i \in I} \psi(X_i)$ and $\psi(\bigwedge_{i \in I} X_i) = \bigvee_{i \in I} \psi(X_i)$. If ψ is a negation, then we call $X^* = \psi(X)$ the *negative* of X. (Although our notation may suggest otherwise, negations are not unique in general.) On the Boolean lattice $\mathcal{P}(E)$, the complement operator $X \mapsto X^c$ defines a negation.

Let $\psi : \mathcal{L} \to \mathcal{M}$ and suppose that both complete lattices possess a negation, then the *negative operator* $\psi^* : \mathcal{L} \to \mathcal{M}$ is defined by

$$\psi^*(X) = \left[\psi(X^*)\right]^*.$$

An operator $\psi : \mathcal{L} \to \mathcal{L}$, where $\mathcal{L}$ possesses a negation, is called *self-dual* if

$$\psi^* = \psi.$$

The key notion in mathematical morphology is that of an adjunction.

DEFINITION 5–2. Let $\mathcal{L}, \mathcal{M}$ be complete lattices, let $\varepsilon : \mathcal{L} \to \mathcal{M}$ and $\delta : \mathcal{M} \to \mathcal{L}$. The pair (ε, δ) is called an *adjunction* between $\mathcal{L}$ and $\mathcal{M}$ if

$$\delta(Y) \leqslant X \iff Y \leqslant \varepsilon(X), \tag{5–1}$$

for $X \in \mathcal{L}$ and $Y \in \mathcal{M}$.

PROPOSITION 5–1. *If (ε, δ) is an adjunction between $\mathcal{L}$ and $\mathcal{M}$, then*

(a) $\varepsilon(\bigwedge_{i \in I} X_i) = \bigwedge_{i \in I} \varepsilon(X_i)$ *for every collection X_i in $\mathcal{L}$; in particular, ε is an increasing operator.*

(b) $\delta(\bigvee_{i \in I} Y_i) = \bigvee_{i \in I} \delta(Y_i)$ *for every collection Y_i in $\mathcal{M}$; in particular, δ is an increasing operator.*

(c) $\varepsilon\delta \geqslant \mathsf{id}_{\mathcal{M}}$ *and* $\delta\varepsilon \leqslant \mathsf{id}_{\mathcal{L}}$.

(d) $\varepsilon\delta\varepsilon = \varepsilon$ *and* $\delta\varepsilon\delta = \delta$.

PROOF. (a) Suppose that (ε, δ) is an adjunction between $\mathcal{L}$ and $\mathcal{M}$; we show that ε is an erosion. Suppose $X_i \in \mathcal{L}$ for $i \in I$; given $Y \in \mathcal{M}$, it holds that $\delta(Y) \leqslant \bigwedge_{i\in I} X_i$ if and only if $\delta(Y) \leqslant X_i$ for every $i \in I$. This, however, is equivalent to $Y \leqslant \varepsilon(X_i)$ for every $i \in I$; that is, $Y \leqslant \bigwedge_{i\in I} \varepsilon(X_i)$. On the other hand, by the adjunction relation, $\delta(Y) \leqslant \bigwedge_{i\in I} X_i$ if and only if $Y \leqslant \varepsilon(\bigwedge_{i\in I} X_i)$. But this implies $\varepsilon(\bigwedge_{i\in I} X_i) = \bigwedge_{i\in I} \varepsilon(X_i)$.

(b) Dual statement of (a).

(c) Choosing $X = \delta(Y)$ in Eq. 5–1, we get $Y \leqslant \varepsilon\delta(Y)$, which proves the first relation. The second follows by duality.

(d) From (c) and the increasingness of ε and δ we get that $\varepsilon\delta\varepsilon \geqslant \varepsilon$ and $\varepsilon\delta\varepsilon \leqslant \varepsilon$; therefore, equality holds. Similarly, it follows that $\delta\varepsilon\delta = \delta$. ■

An operator ε which satisfies the relation given under (a) is called *erosion*. An operator satisfying relation (b) is called *dilation*. Thus an adjunction is formed by a dilation and an erosion satisfying the adjunction relation 5–1.

PROPOSITION 5–2. *With every erosion $\varepsilon : \mathcal{L} \to \mathcal{M}$ there corresponds a unique dilation $\delta : \mathcal{M} \to \mathcal{L}$ such that (ε, δ) is an adjunction. This dilation is given by*

$$\delta(Y) = \bigwedge \{X \in \mathcal{L} \mid Y \leqslant \varepsilon(X)\}.$$

Dually, with every dilation $\delta : \mathcal{M} \to \mathcal{L}$ there corresponds a unique erosion $\varepsilon : \mathcal{L} \to \mathcal{M}$ such that (ε, δ) is an adjunction. This erosion is given by

$$\varepsilon(X) = \bigvee \{Y \in \mathcal{M} \mid \delta(Y) \leqslant X\}.$$

PROOF. Suppose that ε is an erosion, and let δ be given by the expression above. We show that Eq. 5–1 holds. First, if $\delta(Y) \leqslant X$, then by applying ε to both sides and using the fact that it distributes over infima, we get

$$\varepsilon\delta(Y) = \bigwedge \{\varepsilon(X') \mid Y \leqslant \varepsilon(X')\} \leqslant \varepsilon(X),$$

and therefore $Y \leqslant \varepsilon(X)$. On the other hand, if $Y \leqslant \varepsilon(X)$, then by definition $\delta(Y) \leqslant X$. This proves Eq. 5–1.

It remains to us to prove uniqueness of δ. Suppose δ' is another operator such that ε, δ' satisfy Eq. 5–1. Then

$$\delta'(Y) \leqslant X \iff Y \leqslant \varepsilon(X) \iff \delta(Y) \leqslant X.$$

But this yields immediately that $\delta(Y) = \delta'(Y)$. The second part of the statement follows by the duality principle. ∎

The adjunction relation, though mathematically very simple, provides pairs of operators ε, δ with special properties as illustrated by Propositions 5–1 and 5–2. Also, the proof of our next result can be easily established by using this adjunction relation.

PROPOSITION 5–3.

(a) *Let* (ε, δ) *and* (ε', δ') *be adjunctions between* $\mathcal{L}$ *and* $\mathcal{M}$. *Then* $\varepsilon' \leqslant \varepsilon$ *if and only if* $\delta' \geqslant \delta$.

(b) *Let* $(\varepsilon_i, \delta_i)$ *be an adjunction between* $\mathcal{L}$ *and* $\mathcal{M}$ *for every* $i \in I$. *Then* $(\bigwedge_{i \in I} \varepsilon_i, \bigvee_{i \in I} \delta_i)$ *is an adjunction between* $\mathcal{L}$ *and* $\mathcal{M}$ *as well.*

(c) *Let* (ε, δ) *be an adjunction between* $\mathcal{L}$ *and* $\mathcal{M}$, *and let* (ε', δ') *be an adjunction between* $\mathcal{M}$ *and* $\mathcal{N}$. *Then* $(\varepsilon'\varepsilon, \delta\delta')$ *is an adjunction between* $\mathcal{L}$ *and* $\mathcal{N}$.

If both $\mathcal{L}$ and $\mathcal{M}$ possess a negation and (ε, δ) is an adjunction between $\mathcal{L}$ and $\mathcal{M}$, then $(\delta^*, \varepsilon^*)$ is an adjunction between $\mathcal{M}$ and $\mathcal{L}$.

The operators ε_A, δ_A on $\mathcal{P}(\mathbb{E}^d)$ given by

$$\varepsilon_A(X) = X \ominus A = \{h \in \mathbb{E}^d \mid A_h \subseteq X\} = \bigcap_{a \in A} X_{-a}$$

$$\delta_A(X) = X \oplus A = \bigcup_{a \in A} X_a$$

define an adjunction. Here X_h denotes the translate of X over a vector h, i.e., $X_h = \{x + h \mid x \in X\}$. Furthermore, A is a given subset of $\mathbb{E}^d$, called a *structuring element*.

5.2.2 APPLICATION TO GRAY-SCALE FUNCTIONS

We start with some notation and terminology. We represent gray-scale images mathematically as functions $F : \mathbb{E}^d \to \mathcal{T}$, where $\mathcal{T}$ is the gray-value set. Depending on the application at hand we may choose for $\mathcal{T}$ the set $\overline{\mathbb{R}} = \mathbb{R} \cup \{-\infty, +\infty\}$, $\overline{\mathbb{R}}_+ = \mathbb{R}_+ \cup \{+\infty\}$, $\overline{\mathbb{Z}} = \mathbb{Z} \cup \{-\infty, +\infty\}$, $\overline{\mathbb{Z}}_+ = \mathbb{Z}_+ \cup \{+\infty\}$, the bounded interval $[0, 1]$, or the finite set $\{0, 1, \ldots, N\}$. All these examples have in common a complete lattice structure.

By $\text{Fun}(\mathbb{E}^d, \mathcal{T})$ we represent the set of all functions $F : \mathbb{E}^d \to \mathcal{T}$. If $\mathcal{T} = \overline{\mathbb{R}}$, we shall simply write $\text{Fun}(\mathbb{E}^d)$; in all other cases we will include the gray-value set in our notation.

We assume throughout this subsection that $\mathcal{T} = \overline{\mathbb{R}}$. Most of its content, however, carries over to the case $\mathcal{T} = \overline{\mathbb{Z}}$.

For $h \in \mathbb{E}^d$ and $F \in \mathrm{Fun}(\mathbb{E}^d)$, the *horizontal translate* F_h is defined by

$$F_h(x) = F(x-h), \quad x \in \mathbb{E}^d.$$

The *vertical translate* $F + v$, where $v \in \mathbb{R}$, is defined by

$$(F+v)(x) = F(x) + v, \quad x \in \mathbb{E}^d.$$

Given a function $G \in \mathrm{Fun}(\mathbb{E}^d)$, we define the operators

$$\Delta_G(F) = F \oplus G, \quad \mathcal{E}_G(F) = F \ominus G,$$

respectively given by

$$(F \oplus G)(x) = \bigvee_{h \in \mathbf{E}^d} \big[F(x-h) + G(h)\big], \tag{5–2}$$

$$(F \ominus G)(x) = \bigwedge_{h \in \mathbf{E}^d} \big[F(x+h) - G(h)\big]. \tag{5–3}$$

We call G an *additive structuring function*. In the case of ambiguous expressions we use the convention that $s + t = -\infty$ if $s = -\infty$ or $t = -\infty$, and $s - t = +\infty$ if $s = +\infty$ or $t = -\infty$. The following result is easily proved.

PROPOSITION 5–4. *The pair* $(\mathcal{E}_G, \Delta_G)$ *defines an adjunction on* $\mathrm{Fun}(\mathbb{E}^d)$.

Both Δ_G and $\mathcal{E}_G$ are translation invariant with respect to horizontal as well as vertical translations, i.e., both operators have the following property:

$$\Psi(F_h + v) = \big[\Psi(F)_h\big] + v,$$

for $h \in \mathbb{E}^d$ and $v \in \mathbb{R}$. An operator with this property is called a *T-operator*. If Ψ is only invariant under horizontal translations, i.e.,

$$\Psi(F_h) = \big[\Psi(F)_h\big],$$

then it is called an *H-operator*.

The mapping $F \to -F$, where $(-F)(x) = -F(x)$ defines a negation on $\mathrm{Fun}(\mathbb{E}^d)$. Writing $F^* = -F$, we have the following duality relations:

$$(F \oplus G)^* = F^* \ominus \check{G}, \quad (F \ominus G)^* = F^* \oplus \check{G},$$

where $\check{G}$ is the reflection of G with respect to the origin, that is, $\check{G}(x) = G(-x)$.

A general way to construct T-operators is by using (extensions of) Boolean functions. If b is an increasing Boolean function of n variables, then we extend b to a function mapping $\overline{\mathbb{R}}^n$ into $\overline{\mathbb{R}}$ as follows: products are replaced by infima, sums by suprema, 0 by $-\infty$ and 1 by $+\infty$. For example, if $b(u_1, u_2, u_3) = u_1 + u_2 u_3$, then $b(t_1, t_2, t_3) = t_1 \vee (t_2 \wedge t_3)$.

Let $A = \{a_1, \ldots, a_n\}$ be a finite structuring element and b an increasing Boolean function of n variables, define the increasing T-operator Ψ_b on $\text{Fun}(\mathbb{E}^d)$ by

$$\Psi_b(F)(x) = b\big(F(x + a_1), \ldots, F(x + a_n)\big).$$

In fact, gray-scale operators obtained using Boolean functions all belong to the class of so-called *flat operators*.

Evidently, if $\mathcal{T}$ is a complete lattice, then $\text{Fun}(E, \mathcal{T})$ is also a complete lattice with partial ordering

$$F \leqslant F' \text{ iff } F(x) \leqslant F'(x), \text{ for } x \in E.$$

Heijmans and Ronse [9] (see also [5, Sect. 5.1]) have given a complete description of adjunctions on $\text{Fun}(E, \mathcal{T})$, where E is an arbitrary set and $\mathcal{T}$ a complete lattice.

PROPOSITION 5–5. *The pair $(\mathcal{E}, \Delta)$ is an adjunction on* $\text{Fun}(E, \mathcal{T})$ *if and only if for every $x, y \in E$ there exists an adjunction $(e_{y,x}, d_{x,y})$ on $\mathcal{T}$ such that*

$$\Delta(F)(y) = \bigvee_{x \in E} d_{x,y}\big(F(x)\big)$$
$$\mathcal{E}(F)(x) = \bigwedge_{y \in E} e_{y,x}\big(F(y)\big).$$

We take $E = \mathbb{E}^d$ and focus on adjunctions in which both the dilation and the erosion are H-operators; such adjunctions are called *H-adjunctions*. The next proposition follows easily from the previous one.

PROPOSITION 5–6. *The pair $(\mathcal{E}, \Delta)$ is an H-adjunction on* $\text{Fun}(\mathbb{E}^d, \mathcal{T})$ *if and only if for every $h \in \mathbb{E}^d$ there exists an adjunction (e_h, d_h) on $\mathcal{T}$ such that*

$$\Delta(F)(y) = \bigvee_{h \in \mathbb{E}^d} d_h\big(F(y - h)\big)$$
$$\mathcal{E}(F)(x) = \bigwedge_{h \in \mathbb{E}^d} e_h\big(F(x + h)\big).$$

If $\mathcal{T} = \overline{\mathbb{R}}$, we can obtain the adjunction $(\mathcal{E}_G, \Delta_G)$ using the additive structuring function G if we take $d_h(t) = t + G(h)$ and $e_h(t) = t - G(h)$. In [5, Sect. 11–5] we discuss some other H-adjunctions. Below we apply Proposition 5–6 to the case where the gray-value set $\mathcal{T}$ is finite.

If $\mathcal{T} = \{0, 1, \ldots, N\}$, the adjunction given by Eqs. 5–2 and 5–3 becomes meaningless, since $\mathcal{T}$ is not closed under addition and subtraction. If one tries to overcome this problem by truncating values below 0 and above N, one does not get adjunctions: see [4] or [5, Sect. 11–9]. It turns out that we can use the characterization of H-adjunctions given in Proposition 5–6. This characterization utilizes adjunctions on $\mathcal{T}$, in the present case $\{0, 1, \ldots, N\}$.

Define, for $v \in \mathbb{Z}$, the operation $t \mapsto t \dot{+} v$ on $\{0, 1, \ldots, N\}$ given by

$$\begin{cases} 0 \dot{+} v = 0, & \\ t \dot{+} v = 0, & \text{if } t > 0 \text{ and } t + v \leqslant 0, \\ t \dot{+} v = t + v, & \text{if } t > 0 \text{ and } 0 \leqslant t + v \leqslant N, \\ t \dot{+} v = N, & \text{if } t > 0 \text{ and } t + v > N, \end{cases}$$

and the operation $t \mapsto t \dot{-} v$ by

$$\begin{cases} t \dot{-} v = 0, & \text{if } t < N \text{ and } t - v \leqslant 0, \\ t \dot{-} v = t - v, & \text{if } t < N \text{ and } 0 \leqslant t - v \leqslant N, \\ t \dot{-} v = N, & \text{if } t < N \text{ and } t - v > N, \\ N \dot{-} v = N. & \end{cases}$$

Let, for example, $N = 10$. Then $(6 \dot{+} 5) \dot{-} 4 = 10$ and $6 \dot{+} (5 \dot{-} 4) = 7$. The operation $\dot{+}$ is neither commutative $(0 \dot{+} 1 \neq 1 \dot{+} 0)$ nor associative $((3 \dot{+} 0) \dot{+} 5 = 3 \dot{+} 5 = 8 \neq 3 = 3 \dot{+} 0 = 3 \dot{+} (0 \dot{+} 5))$.

A simple computation shows that the pair $e(t) = t \dot{-} v$, $d(t) = t \dot{+} v$ defines an adjunction on $\{0, 1, \ldots, N\}$ for every $v \in \mathbb{Z}$. For an illustration, see Fig. 5–2.

In combination with Proposition 5–6, this yields an interesting class of H-adjunctions with dilation and erosion, respectively, given by

$$(F \dot{\oplus} G)(x) = \bigvee_{h \in \operatorname{dom} G} \big(F(x - h) \dot{+} G(h)\big),$$

$$(F \dot{\ominus} G)(x) = \bigwedge_{h \in \operatorname{dom} G} \big(F(x + h) \dot{-} G(h)\big).$$

Here G is a function with domain $\operatorname{dom}(G)$ and values in $\mathbb{Z}$. In fact, one takes $d_h(t) = t \dot{+} G(h)$, $e_h(t) = t \dot{-} G(h)$, for $h \in \operatorname{dom}(G)$ and $d_h \equiv 0$, $e_h \equiv N$ for $h \notin \operatorname{dom}(G)$.

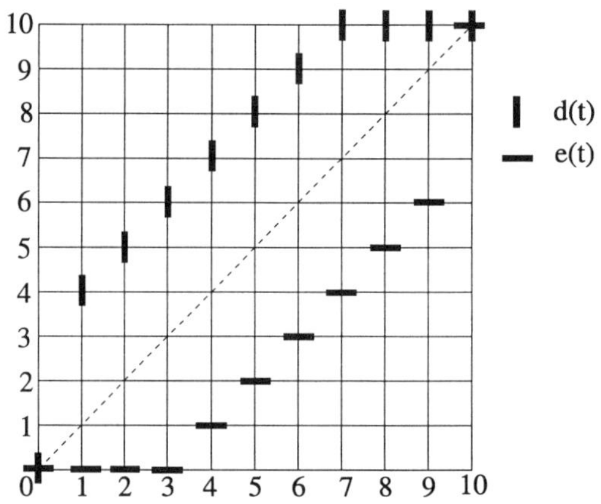

Figure 5–2. The pair $e(t) = t \dot{-} 3,\ d(t) = t \dot{+} 3$ forms an adjunction on $\mathcal{T} = \{0, 1, \ldots, 10\}$.

It is easy to verify that

$$(F \dot{+} v) \dot{\oplus} G = (F \dot{\oplus} G) \dot{+} v,$$
$$(F \dot{-} v) \dot{\ominus} G = (F \dot{\ominus} G) \dot{-} v,$$

if $v \geqslant 0$. More results can be found in [5, Sect. 11–9].

5.3 Openings and Closings

This section contains a brief description of some basic properties of openings and closings, and introduces adjunctional filters.

5.3.1 Basic Facts

DEFINITION 5–3. An *opening* α is an operator on a complete lattice $\mathcal{L}$ which is increasing, idempotent, and anti-extensive ($\alpha(X) \leqslant X$ for every $X \in \mathcal{L}$). Dually, a *closing* β is an operator which is increasing, idempotent and extensive ($\beta(X) \geqslant X$ for every X).

The results presented in this section are mostly concerned with openings; analogous statements for closings follow from the duality principle [5].

If ψ is an operator on the complete lattice $\mathcal{L}$, then the *invariance domain* of ψ is

$$\mathrm{Inv}(\psi) = \{X \in \mathcal{L} \mid \psi(X) = X\}.$$

Elements of $\mathrm{Inv}(\psi)$ are sometimes called *fixpoints* or *roots*. From (c)–(d) in Proposition 5–1 the following result is clear:

PROPOSITION 5–7. *If* (ε, δ) *is an adjunction between* $\mathcal{L}$ *and* $\mathcal{M}$, *then* $\delta\varepsilon$ *is an opening on* $\mathcal{L}$ *and* $\varepsilon\delta$ *is a closing on* $\mathcal{M}$.

The opening resulting from Minkowski subtraction followed by Minkowski addition, i.e.,

$$X \circ A = (X \ominus A) \oplus A = \bigcup \{A_h \mid h \in \mathbb{E}^d \text{ and } A_h \subseteq X\}$$

is called a *structural opening*. The invariance domain of the opening $\delta\varepsilon$ is $\mathrm{Ran}(\delta)$. The previous result can be extended as follows.

PROPOSITION 5–8. *Let* α *be an opening on the complete lattice* $\mathcal{M}$ *and let* (ε, δ) *be an adjunction between* $\mathcal{L}$ *and* $\mathcal{M}$, *then* $\delta\alpha\varepsilon$ *is an opening on* $\mathcal{L}$ *with invariance domain* $\{\delta(Y) \mid Y \in \mathrm{Inv}(\alpha)\}$.

PROOF. It is evident that $\alpha' = \delta\alpha\varepsilon$ is increasing. Furthermore, $\delta\alpha\varepsilon \leqslant \delta \mathrm{id} \varepsilon = \delta\varepsilon \leqslant \mathrm{id}$, hence α' is anti-extensive. It remains to prove that $\alpha'^2 \geqslant \alpha'$ (note that the reverse inequality is trivial by the anti-extensivity of α'). Now

$$\alpha'^2 = \delta\alpha\varepsilon\delta\alpha\varepsilon \geqslant \delta\alpha^2\varepsilon = \delta\alpha\varepsilon = \alpha',$$

where we have used that $\varepsilon\delta \geqslant \mathrm{id}$.

Every fixpoint of α' is of the form $\delta(Y)$, where $Y \in \mathrm{Inv}(\alpha)$. To prove the converse, take $Y \in \mathrm{Inv}(\alpha)$ and consider $\alpha'\delta(Y)$. Since α' is anti-extensive, we have $\alpha'\delta(Y) \leqslant \delta(Y)$. On the other hand, since $\varepsilon\delta \geqslant \mathrm{id}$,

$$\alpha'\delta(Y) = \delta\alpha\varepsilon\delta(Y) \geqslant \delta\alpha(Y) = \delta(Y),$$

where we used that $\alpha(Y) = Y$. This concludes the proof. ■

PROPOSITION 5–9. *Let* α_1, α_2 *be openings on the complete lattice* $\mathcal{L}$. *The following assertions are equivalent*:

(i) $\alpha_1 \leqslant \alpha_2$;

(ii) $\alpha_1\alpha_2 = \alpha_2\alpha_1 = \alpha_1$;

(iii) $\mathrm{Inv}(\alpha_1) \subseteq \mathrm{Inv}(\alpha_2)$.

In particular, $\alpha_1 = \alpha_2$ *if and only if* $\mathrm{Inv}(\alpha_1) = \mathrm{Inv}(\alpha_2)$.

PROOF. Let α_1, α_2 be openings.

(i) $\Rightarrow$ (ii): If $\alpha_1 \leqslant \alpha_2$, then $\alpha_1\alpha_2 \geqslant \alpha_1\alpha_1 = \alpha_1$. Since the reverse inequality is trivially satisfied, one gets $\alpha_1\alpha_2 = \alpha_1$. The identity $\alpha_2\alpha_1 = \alpha_1$ is proved in a similar way.

(ii) ⇒ (iii): Let $X \in \mathrm{Inv}(\alpha_1)$, that is, $\alpha_1(X) = X$. Then $\alpha_2(X) = \alpha_2\alpha_1(X) = \alpha_1(X) = X$, and therefore $X \in \mathrm{Inv}(\alpha_2)$.

(iii) ⇒ (i): As $\alpha_1(X) \in \mathrm{Inv}(\alpha_1) \subseteq \mathrm{Inv}(\alpha_2)$, one gets $\alpha_1(X) = \alpha_2\alpha_1(X) \leqslant \alpha_2(X)$. ■

If $\alpha_1(X) = X \circ A$ and $\alpha_2(X) = X \circ B$, then these equivalent conditions hold if A is B-open, i.e., $A \circ B = A$. We recall the following result.

PROPOSITION 5–10. *If α_i, $i \in I$, are openings, then $\bigvee_{i\in I} \alpha_i$ is an opening, too.*

PROOF. Let α_i, $i \in I$, be openings, and put $\alpha = \bigvee_{i\in I} \alpha_i$. It is evident that α is increasing and anti-extensive. We show that it is idempotent. By the anti-extensivity, it follows immediately that $\alpha^2 \leqslant \alpha$. The converse inequality also holds, since

$$\alpha^2 = \bigvee_{i\in I} \alpha_i\alpha \geqslant \bigvee_{i\in I} \alpha_i\alpha_i = \bigvee_{i\in I} \alpha_i = \alpha.$$

This proves the result. ■

However, it is easy to construct examples which show that neither the infimum nor the composition of two openings is an opening in general.

5.3.2 ANNULAR OPENING

An opening on $\mathcal{P}(\mathbb{E}^d)$ which is given by a simple expression, but which is not of structural type, is the *annular opening* given by $X \mapsto X \cap X \oplus A$, where $A \subseteq \mathbb{E}^d$ is a structuring element which is symmetric, i.e., $A = \check{A}$. Here $\check{A}$ is the reflected structuring element: $\check{A} = \{-a \mid a \in A\}$. The proof that this operator defines an opening indeed, is not very difficult; see e.g., [23] or [5, Prop. 4–27]. An illustration of the effect of the annular opening can be found in Fig. 5–3.

In [5] we discuss various manifestations of the annular opening; see also [19]. In the next section we discuss a generalization of the annular opening, called an *annular filter*.

5.3.3 ADJUNCTIONAL FILTERS

If (ε, δ) is an adjunction on the complete lattice $\mathcal{L}$, then $\varepsilon\delta$ is a closing and $\delta\varepsilon$ an opening; see Proposition 5–7. More generally, if $k \geqslant 1$ then $\varepsilon^k\delta^k$ is a closing and $\delta^k\varepsilon^k$ is an opening. This follows easily from the observation that $(\varepsilon^k, \delta^k)$ is an adjunction, too. The composition $\varepsilon\delta^2\varepsilon$ is a filter, for

$$\varepsilon\delta^2\varepsilon\varepsilon\delta^2\varepsilon = \varepsilon(\delta^2\varepsilon^2\delta^2)\varepsilon = \varepsilon\delta^2\varepsilon,$$

where we have used that $\delta^2\varepsilon^2\delta^2 = \delta^2$ since $(\varepsilon^2, \delta^2)$ is an adjunction [5].

In [3] we have established the following general result:

PROPOSITION 5–11. *Let* (ε, δ) *be an adjunction on* $\mathcal{L}$ *and let* ψ *be of the form*

$$\psi = \varepsilon^{e_n}\delta^{d_n}\cdots\varepsilon^{e_2}\delta^{d_2}\varepsilon^{e_1}\delta^{d_1},$$

where $e_i, d_i \geqslant 0$ *are integers and* $\sum_{i=1}^{n} e_i = \sum_{i=1}^{n} d_i$. *Then* ψ *is a filter.*

For example, $\varepsilon^3\delta^2\varepsilon\delta^2$ is a filter. The filters given by Proposition 5–11 are called *adjunctional filters*; refer to [3] for additional results.

5.4 ANNULAR FILTERS

In what follows, δ_A and ε_A denote *dilation* and *erosion* on $\mathcal{P}(\mathbb{E}^d)$ by the structuring element A, respectively; that is

$$\delta_A(X) = X \oplus A \text{ and } \varepsilon_A(X) = X \ominus A.$$

In Section 5.3.2 we introduced annular openings. Such openings have the form $\alpha = \mathrm{id} \wedge \delta_A$. Dually, annular closings are given by $\beta = \mathrm{id} \vee \varepsilon_B$. Here A, B are symmetric structuring elements which do not contain the origin. Annular filters, which were introduced by the author in [8] and investigated in great detail in [10, 20], are a combination of both operators.

5.4.1 ANNULAR FILTERS FOR BINARY IMAGES

Let A, B be structuring elements in $\mathbb{E}^d$ which are symmetric and which do not contain the origin. Consider the operator

$$\omega = (\mathrm{id} \wedge \delta_A) \vee \varepsilon_B. \tag{5–4}$$

Throughout this subsection we assume that

$$A \cap B \neq \emptyset. \tag{5–5}$$

As a result we have

$$\varepsilon_B \leqslant \delta_A,$$

and thus we find that ω can alternatively be written as $\omega = (\mathrm{id} \vee \varepsilon_B) \wedge \delta_A$. In the sequel we write

$$\omega = \delta_A \wedge \mathrm{id} \vee \varepsilon_B,$$

showing that the expression is independent of the order in which the infimum and supremum are computed. In [10] the following result has been established:

PROPOSITION 5–12. *Let A, B be as before; the operator $\omega = \delta_A \wedge \text{id} \vee \varepsilon_B$ is a filter if and only if*

$$A \cap B \cap (A \oplus B) \neq \emptyset. \tag{5-6}$$

It is self-dual if and only if $A = B$.

We mention four examples for $\mathbb{Z}^2$ where Eqs. 5–5 and 5–6 hold. Observe that in all examples one may interchange A and B.

$$A = B = \begin{array}{|ccc|}\hline \bullet & \bullet & \bullet \\ \bullet & \cdot & \bullet \\ \bullet & \bullet & \bullet \\ \hline\end{array}$$

$$A = \begin{array}{|ccc|}\hline \bullet & \bullet & \bullet \\ \bullet & \cdot & \bullet \\ \bullet & \bullet & \bullet \\ \hline\end{array} \quad \text{and} \quad B = \begin{array}{|ccc|}\hline & \bullet & \\ \bullet & \cdot & \bullet \\ & \bullet & \\ \hline\end{array}$$

$$A = \begin{array}{|ccccc|}\hline & & \bullet & & \\ \bullet & \bullet & \cdot & \bullet & \bullet \\ & & \bullet & & \\ \hline\end{array} \quad \text{and} \quad B = \begin{array}{|ccc|}\hline & \bullet & \\ \bullet & \cdot & \bullet \\ & \bullet & \\ \hline\end{array}$$

$$A = \begin{array}{|ccc|}\hline \bullet & \cdot & \bullet \\ \bullet & \cdot & \bullet \\ \bullet & \cdot & \bullet \\ \hline\end{array} \quad \text{and} \quad B = \begin{array}{|ccc|}\hline \bullet & \bullet & \bullet \\ \cdot & \cdot & \cdot \\ \bullet & \bullet & \bullet \\ \hline\end{array}$$

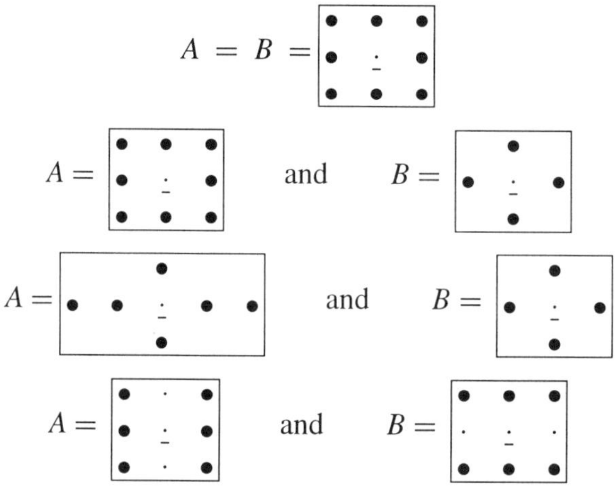

Considering A and B as sets which determine foreground and background adjacency, respectively, one obtains an interesting geometric interpretation of the annular filter. Thus, saying that two points $x, y \in \mathbb{Z}^2$ are foreground adjacent if $x - y \in A$ and background adjacent if $x - y \in B$, the operator ω given by Eq. 5–4 removes points in X which have no foreground neighbors in X, and it adds points from X^c which have no background neighbors in X^c.

The first example, with $A = B$, yields a self-dual annular filter. For this structuring element we apply the annular opening and the annular filter to our test image; see

Figure 5–3. Annular opening (left) and annular filter (right).

Fig. 5–3. Observe that the annular opening removes only isolated noise pixels from the foreground (black pixels), whereas the annular filter removes all isolated noise pixels, that is, from the foreground as well as from the background.

5.4.2 ANNULAR FILTERS FOR GRAY-SCALE IMAGES

Consider the complete lattice of gray-scale images modeled by $\mathrm{Fun}(\mathbb{E}^d)$. For a structuring function A with domain $\mathrm{dom}(A) \subseteq \mathbb{E}^d$ and range in $\overline{\mathbb{R}}$, the gray-scale dilation Δ_A and gray-scale erosion $\mathcal{E}_A$ are given by (see Eqs. 5–2, 5–3):

$$\Delta_A(F)(x) = \bigvee_{h \in \mathrm{dom}(A)} \big(F(x-h) + A(h)\big)$$

$$\mathcal{E}_A(F)(x) = \bigwedge_{h \in \mathrm{dom}(A)} \big(F(x+h) - A(h)\big).$$

In [19] it was shown that the operator $\mathrm{id} \wedge \Delta_A$ is an opening, the annular opening for gray-scale images, if

(i) $\mathrm{dom}(A)$ is symmetric

(ii) $A(x) + A(-x) \geqslant 0$ for $x \in \mathrm{dom}(A)$.

In order that $\mathrm{id} \wedge \Delta_A$ is not the identity mapping, one has to assume that

$$0 \notin \mathrm{dom}(A).$$

Given two symmetric structuring functions A and B, we define the structuring function $A \sqcap B$ as follows:

$$\mathrm{dom}(A \sqcap B) = \big\{x \in \mathrm{dom}(A) \cap \mathrm{dom}(B) \mid \min(A(x), B(x)) + \min(A(-x), B(-x)) \geqslant 0\big\},$$

and

$$(A \sqcap B)(x) = \min(A(x), B(x)) \ \text{ if } x \in \mathrm{dom}(A \sqcap B).$$

The next result can be found in [20].

PROPOSITION 5–13. *Let A, B be two structuring functions which satisfy* (i)–(ii) *above, as well as*

$$\big[A \oplus B \oplus (A \sqcap B)\big](0) \geqslant 0;$$

then $\Omega = \Delta_A \wedge \mathrm{id} \vee \mathcal{E}_B$ is a morphological filter.

In [20] some examples are given.

5.5 AS-FILTERS

Perhaps the most interesting class of morphological filters is obtained by composing openings and closings.

PROPOSITION 5–14. *Let $\alpha_1 \geqslant \alpha_2 \geqslant \cdots \geqslant \alpha_N$ be openings and let $\beta_1 \leqslant \beta_2 \leqslant \cdots \leqslant \beta_N$ be closings. Every composition of operators of these two sequences is a filter.*

The result in this form was first stated by Schonfeld and Goutsias [21]. We introduce the following notation: if $\psi_1, \psi_2, \ldots$ are operators, then

$$(\psi)_n = \psi_n \psi_{n-1} \cdots \psi_1, \quad n \geqslant 1.$$

More generally, if $\phi_1, \phi_2, \ldots$ is another sequence of operators, then

$$\begin{aligned}(\psi\phi)_n &= \psi_n\phi_n\psi_{n-1}\phi_{n-1}\cdots\psi_1\phi_1,\\ (\psi\phi\psi)_n &= \psi_n\phi_n\psi_n\psi_{n-1}\phi_{n-1}\psi_{n-1}\cdots\psi_1\phi_1\psi_1.\end{aligned}$$

In what follows we fix a sequence of openings

$$\alpha_1 \geqslant \alpha_2 \geqslant \cdots \geqslant \alpha_N,$$

and a sequence of closings

$$\beta_1 \leqslant \beta_2 \leqslant \cdots \leqslant \beta_N.$$

From Proposition 5–14 we get that the compositions $(\alpha\beta)_n, (\beta\alpha)_n, (\alpha\beta\alpha)_n, (\beta\alpha\beta)_n$ are filters. These filters are called *alternating sequential filters* or *AS-filters* [5, 23, 24]. Furthermore, the sequences $(\alpha\beta)_n, (\beta\alpha)_n, (\alpha\beta\alpha)_n, (\beta\alpha\beta)_n$ are absorbing in the following sense: a sequence of operators $\psi_1, \psi_2, \ldots$ is said to be *absorbing* if

$$\psi_n\psi_m = \psi_n, \quad n \geqslant m.$$

In fact, one can easily show that for any family of increasing operators ψ_n for which $(\psi)_n$ is a filter, one automatically gets that the sequence $(\psi)_n$ is absorbing.

The following inequalities are easily established:

$$(\alpha\beta\alpha)_n \leqslant \left\{ \begin{matrix} (\alpha\beta)_n \\ (\beta\alpha)_n \end{matrix} \right\} \leqslant (\beta\alpha\beta)_n.$$

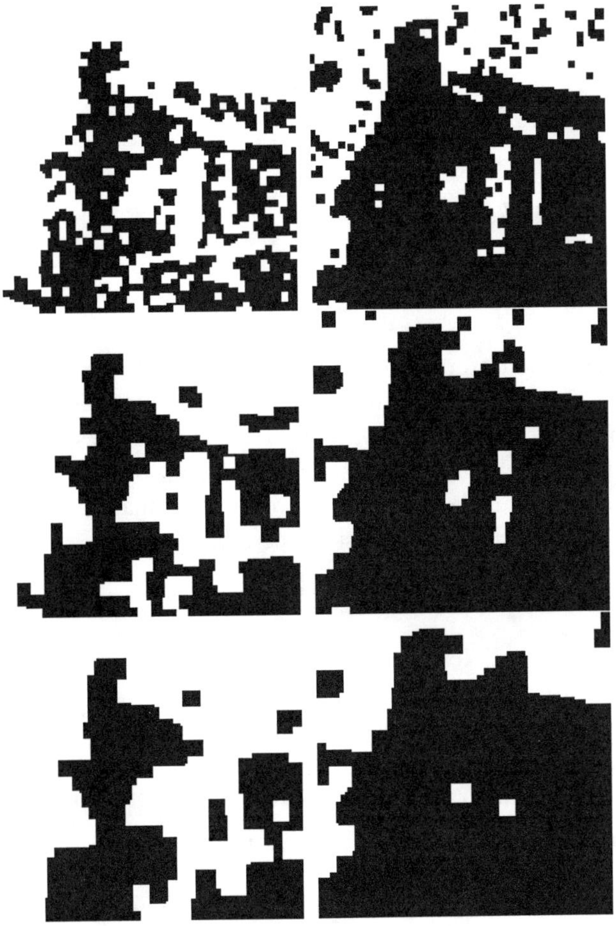

Figure 5–4. $(\beta\alpha)_n(X)$ (left) and $(\alpha\beta)_n(X)$ (right). $n = 1, 2, 3$ in the first, second, and third row, respectively.

In practice (discrete case) one usually constructs AS-filters starting from a structuring element A, and defining $\alpha_n(X) = X \circ nA$ and $\beta_n(X) = X \bullet nA$, where $nA = A \oplus \cdots \oplus A$ (n terms). In Fig. 5–4 we apply $(\alpha\beta)_n$ and $(\beta\alpha)_n$ to our test image of Fig. 5–1; we choose for A the 3×3 square.

5.6 OVERFILTERS AND INF-OVERFILTERS

This section discusses overfilters and inf-overfilters. From the duality principle [5] we know that analogous results hold for the dual concepts, underfilters and sup-underfilters.

5.6.1 DEFINITIONS AND BASIC PROPERTIES

DEFINITION 5–4. An increasing operator ψ is called

(a) an *overfilter* if $\psi^2 \geqslant \psi$;

(b) an *inf-overfilter* if $\psi(\mathrm{id} \wedge \psi) = \psi$;

(c) an *underfilter* if $\psi^2 \leqslant \psi$;

(d) a *sup-underfilter* if $\psi(\mathrm{id} \vee \psi) = \psi$.

We make some simple observations. It is clear that overfilters and underfilters are dual in the sense of the duality principle. The same is true for inf-overfilters and sup-underfilters. In this section we mostly restrict ourselves to (inf-) overfilters. Since $\psi(\mathrm{id} \wedge \psi) \leqslant \psi^2$, every inf-overfilter is also an overfilter. To prove that an operator ψ is an inf-overfilter, we only have to show that $\psi(\mathrm{id} \wedge \psi) \geqslant \psi$, as the reverse inequality is trivial for increasing operators.

PROPOSITION 5–15. *The family of (inf-) overfilters is closed under suprema.*

PROOF. Assume that ψ_i, $i \in I$, are inf-overfilters, and put $\psi = \bigvee_{i \in I} \psi_i$. Thus $\psi(\mathrm{id} \wedge \psi) \geqslant \psi_i(\mathrm{id} \wedge \psi_i) = \psi_i$, which yields immediately that $\psi(\mathrm{id} \wedge \psi) \geqslant \bigvee_{i \in I} \psi_i = \psi$. Therefore, ψ is an inf-overfilter. ■

In [5] one can find detailed results concerning the lattice structure of the class of filters and (inf-) overfilters; see also [17, 18, 23].

PROPOSITION 5–16. *If ψ is an overfilter (inf-overfilter, underfilter, sup-underfilter), then ψ^n is such as well, for every $n \geqslant 1$.*

PROOF. For overfilters, the result is obvious. Now assume that ψ is an inf-overfilter. Then

$$\psi^n(\mathrm{id} \wedge \psi^n) \geqslant \psi^n(\mathrm{id} \wedge \psi) = \psi^{n-1}\psi(\mathrm{id} \wedge \psi) \geqslant \psi^{n-1}\psi = \psi^n.$$

Here we have used that $\psi^n \geqslant \psi$, if ψ is an (inf-) overfilter. ■

The next result shows that inf-overfilters provide a useful tool for the construction of openings.

PROPOSITION 5–17. *If ψ is an inf-overfilter, then* $\mathrm{id} \wedge \psi$ *is an opening.*

PROOF. Let ψ be an inf-overfilter and $\alpha = \mathrm{id} \wedge \psi$. It is evident that α is increasing and anti-extensive. It is also idempotent, for

$$\alpha^2 = (\mathrm{id} \wedge \psi)(\mathrm{id} \wedge \psi) = \mathrm{id} \wedge \psi \wedge \psi(\mathrm{id} \wedge \psi) = \mathrm{id} \wedge \psi = \alpha.$$

This proves the result. ■

The dual statement says that $\mathrm{id} \vee \psi$ is a closing when ψ is a sup-underfilter. Combining the latter two propositions gives that $\mathrm{id} \wedge \psi^n$ is an opening for every $n \geqslant 1$ if ψ is an inf-overfilter. Since $\psi \leqslant \psi^2 \leqslant \psi^3 \leqslant \cdots$, we find that

$$\mathrm{id} \wedge \psi \leqslant \mathrm{id} \wedge \psi^2 \leqslant \mathrm{id} \wedge \psi^3 \leqslant \cdots .$$

The next result is obvious.

PROPOSITION 5–18. *Suppose that $\mathcal{L}$ possesses a negation. If ψ is an (inf-) overfilter, then ψ^* is a (sup-) underfilter.*

In the following proposition we sum up various ways to construct overfilters and inf-overfilters.

PROPOSITION 5–19.

(a) *Let (ε, δ) be an adjunction between $\mathcal{L}$ and $\mathcal{M}$ and let $\psi : \mathcal{M} \to \mathcal{L}$ be an increasing operator such that $\psi \geqslant \delta$, then $\psi\varepsilon$ is an inf-overfilter.*

(b) *Let (ε, δ) and (ε', δ') be adjunctions between $\mathcal{L}$ and $\mathcal{M}$ such that $\varepsilon' \leqslant \varepsilon$ and $\delta' \geqslant \delta$. If ψ is an (inf-) overfilter on $\mathcal{M}$, then $\delta'\psi\varepsilon$ is an (inf-) overfilter on $\mathcal{L}$.*

(c) *Let α be an opening and $\alpha \leqslant \psi$, then $\alpha\psi$ and $\psi\alpha\psi$ are overfilters, whereas $\psi\alpha$ and $\alpha\psi\alpha$ are inf-overfilters.*

(d) *Let ψ be an (inf-) overfilter and $\phi \geqslant \mathrm{id}$ then $\phi\psi$ is an (inf-) overfilter.*

(e) *If ψ is an overfilter and $\phi \geqslant \psi$, then $\phi\psi$ and $\psi\phi$ are overfilters.*

(f) *If ψ is an inf-overfilter and $\phi \geqslant \mathrm{id} \wedge \psi$, then $\phi\psi$ is an inf-overfilter.*

(g) *If ψ is an overfilter and β a closing, then $\beta\psi$, $\psi\beta\psi$, $\psi\beta$, $\beta\psi\beta$ are overfilters.*

(h) *If ψ is an inf-overfilter and β a closing, then $\beta\psi$, $\psi\beta\psi$ are inf-overfilters.*

PROOF. For a full proof we refer to [3]. Here we only prove (a) and (f).

(a) Let ε, δ, ψ be as stated. Then

$$\psi\varepsilon(\mathrm{id} \wedge \psi\varepsilon) \geqslant \psi\varepsilon(\mathrm{id} \wedge \delta\varepsilon) = \psi\varepsilon\delta\varepsilon = \psi\varepsilon,$$

where we have used that $\delta\varepsilon \leqslant \mathrm{id}$ and that $\varepsilon\delta\varepsilon = \varepsilon$.

(f) In this case,

$$\begin{aligned} \phi\psi(\mathrm{id} \wedge \phi\psi) &\geqslant \phi\psi(\mathrm{id} \wedge (\mathrm{id} \wedge \psi)\psi) \\ &= \phi\psi(\mathrm{id} \wedge \psi \wedge \psi^2) \\ &= \phi\psi(\mathrm{id} \wedge \psi) = \phi\psi, \end{aligned}$$

where we have used that $\psi^2 \geqslant \psi$, since ψ is an overfilter. From (a) [and also from (b)] we get that $\delta'\varepsilon$ is an inf-overfilter, if $\delta' \geqslant \delta$. ∎

5.6.2 Rank-max Openings

Ronse and Heijmans [19] (see also [5, Sect. 6–6]) have shown that every translation-invariant inf-overfilter on $\mathcal{P}(\mathbb{E}^d)$ is of the form

$$\psi(X) = \bigcap_{k\in K} \bigcup_{j\in J} (X \ominus B_j) \oplus A_{kj}, \tag{5–7}$$

where A_{kj}, B_j are structuring elements such that $B_j \subseteq A_{kj}$ for $k \in K$ and $j \in J$. As an illustration of this result we discuss the rank-max opening, first discussed by Ronse [15]; see also [5, 19].

Let A be a finite structuring element containing n points, and let $\mathcal{B}_k$ contain all subsets of A which contain k points (where $k \leqslant n$). It is evident that

$$\bigcup_{B\in\mathcal{B}_k} X \ominus B = \rho_{A,k}(X),$$

where $\rho_{A,k}$ is the kth rank operator which is defined by: $h \in \rho_{A,k}(X)$ if and only if $X \cap A_h$ contains at least k points [5]. The composition $\delta_A \varepsilon_B$ is an inf-overfilter, for $B \in \mathcal{B}_k$, hence

$$\bigvee_{B\in\mathcal{B}_k} \delta_A \varepsilon_B = \delta_A\Big(\bigvee_{B\in\mathcal{B}_k} \varepsilon_B\Big) = \delta_A \rho_{A,k}$$

is an inf-overfilter, too; here we have used Proposition 5–15. Note that the previous expression is a special case of Eq. 5–7.

The opening $\alpha_{A,k} = \mathrm{id} \wedge \delta_A \rho_{A,k}$ is called a *rank-max opening*. For $k = n$ it coincides with the structural opening $X \mapsto X \circ A$, whereas for $k = 1$ it yields the identity operator.

The rank-max openings are "more flexible" than the structural opening: the latter one preserves translates of A which fit entirely inside X. The rank-max opening $\rho_{A,k}$ preserves those portions $X \cap A_h$ which contain at least k points. It is evident that

$$\alpha_{A,n} \leqslant \alpha_{A,n-1} \leqslant \cdots \leqslant \alpha_{A,1} = \mathrm{id}.$$

Let us denote the dual closing, the *rank-min closing* by $\beta_{A,k}$:

$$\beta_{A,k} = \mathrm{id} \vee \varepsilon_{\check{A}} \rho_{A,n+1-k}.$$

It follows easily that $(\alpha_{A,k})^* = \beta_{A,k}$. We have

$$\beta_{A,n} \geqslant \beta_{A,n-1} \geqslant \cdots \geqslant \beta_{A,1} = \mathrm{id}.$$

We can use the rank-max openings and rank-min closings to construct AS-filters. Define e.g.,

$$(\beta_A \alpha_A)_m = \beta_{A,m}\alpha_{A,m}\beta_{A,m-1}\alpha_{A,m-1}\cdots\beta_{A,2}\alpha_{A,2},$$

where $m \leqslant n$. The AS-filter $(\alpha_A \beta_A)_m$ is defined analogously.

In Fig. 5–5 we give an illustration of these two filters for the case that A is the 3×3 square (hence $n = 9$).

Figure 5–5. $(\beta_A \alpha_A)_n(X)$ (left) and $(\alpha_A \beta_A)_n(X)$ (right); $n = 5$ in the first row, $n = 7$ in the second row, and $n = 9$ in the third row.

5.7 Generalized AS-Filters

In this section we extend the class of AS-filters obtained by composing openings and closings, as discussed in Section 5.5. The basic idea is to use overfilters instead of openings and underfilters instead of closings. The exposition in this section is extracted from [3]; see also [6, 7].

PROPOSITION 5–20. *Assume that ϕ is an overfilter, that ψ is an underfilter, and that $\phi \leqslant \psi$. The compositions $\phi\psi$, $\psi\phi$, $\phi\psi\phi$, $\psi\phi\psi$ are filters, and*

$$\phi \leqslant \phi\psi\phi \leqslant \left\{ \begin{matrix} \psi\phi \\ \phi\psi \end{matrix} \right\} \leqslant \psi\phi\psi \leqslant \psi.$$

PROOF. That, e.g., $\psi\phi$ is a filter follows from $\psi\phi\psi\phi \leqslant \psi^3\phi \leqslant \psi\phi$ and $\psi\phi\psi\phi \geqslant \psi\phi^3 \geqslant \psi\phi$. In the same fashion, one shows that the other compositions are filters, too. Furthermore,

$$\phi \leqslant \phi^3 \leqslant \phi\psi\phi \leqslant \phi\psi^2 \leqslant \phi\psi \leqslant \phi^2\psi \leqslant \psi\phi\psi \leqslant \psi^3 \leqslant \psi.$$

The inequalities with $\psi\phi$ instead of $\phi\psi$ follow analogously. ■

We consider translation-invariant operators on $\mathcal{P}(\mathbb{E}^d)$. Let $A \subseteq A'$; then $\phi(X) = (X \ominus A) \oplus A'$ defines an inf-overfilter. Dually, if $B \subseteq B'$, then $\psi(X) = (X \oplus B) \ominus B'$ defines a sup-underfilter. Now $\phi \leqslant \psi$ iff

$$(X \ominus A) \oplus A' \subseteq (X \oplus B) \ominus B',$$

for every $X \subseteq \mathbb{E}^d$. It is easy to see that this condition holds iff $A' \oplus B' \subseteq A \oplus B$. Since the reverse inclusion is trivially satisfied, we arrive at the following set of conditions:

$$A \subseteq A', \quad B \subseteq B', \quad A \oplus B = A' \oplus B'. \tag{5–8}$$

PROPOSITION 5–21. *Suppose that (A_i, A'_i), $i \in I$, and (B_j, B'_j), $j \in J$, are pairs of structuring elements such that*

$$A_i \subseteq A'_i, \quad B_j \subseteq B'_j, \quad A_i \oplus B_j = A'_i \oplus B'_j,$$

for every $i \in I$ and $j \in J$. Then

$$\phi(X) = \bigcup_{i \in I} (X \ominus A_i) \oplus A'_i$$

is an inf-overfilter,

$$\psi(X) = \bigcap_{j \in J}(X \oplus B_j) \ominus B'_j$$

is a sup-underfilter, and $\phi \leqslant \psi$.

We present two examples.

EXAMPLE 5–1. Let $A' = B'$ be the 3×3 square and

$$A = B = \begin{array}{|ccc|} \hline \bullet & \bullet & \bullet \\ \bullet & \cdot & \bullet \\ \bullet & \bullet & \bullet \\ \hline \end{array}$$

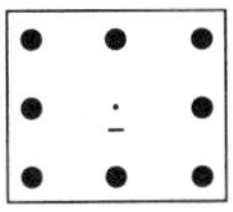

Let $\phi(X) = (X \ominus A) \oplus A'$ and $\psi(X) = (X \oplus B) \ominus B'$.

In the first row of Fig. 5–6 we depict $(\psi\phi)(X)$ and $(\phi\psi)(X)$. Compare these images with $(\beta\alpha)_1(X)$ and $(\alpha\beta)_1(X)$, respectively, in the first row of Fig. 5–4.

EXAMPLE 5–2. Let

$$A_1 = \begin{array}{|ccc|}\hline \bullet & \bullet & \bullet \\ \bullet & \cdot & \cdot \\ \bullet & \bullet & \bullet \\ \hline\end{array}, \quad A_2 = \begin{array}{|ccc|}\hline \bullet & \cdot & \bullet \\ \bullet & \cdot & \bullet \\ \bullet & \bullet & \bullet \\ \hline\end{array},$$

$$B_1 = \begin{array}{|ccc|}\hline \bullet & \bullet & \bullet \\ \cdot & \cdot & \bullet \\ \bullet & \bullet & \bullet \\ \hline\end{array}, \quad B_2 = \begin{array}{|ccc|}\hline \bullet & \bullet & \bullet \\ \bullet & \cdot & \bullet \\ \bullet & \cdot & \bullet \\ \hline\end{array},$$

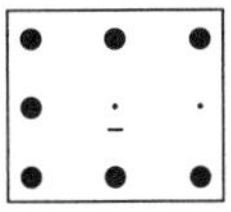

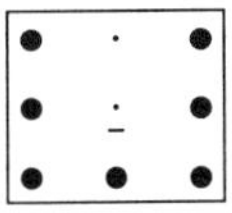

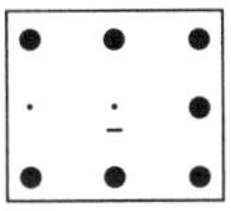

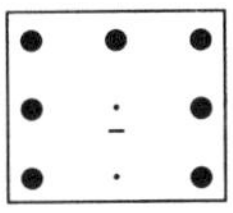

and let $A' = B'$ be the 3×3 square. Define $\phi(X) = ((X \ominus A_1) \cup (X \ominus A_2)) \oplus A'$ and $\psi(X) = ((X \oplus B_1) \cap (X \oplus B_2)) \ominus B'$. The images $(\psi\phi)(X)$ and $(\phi\psi)(X)$ are depicted in the second row of Fig. 5–6. In [7] we present a variant of this latter example where we have rotation invariance.

Now we consider generalized AS-filters that use more than one overfilter and one underfilter.

PROPOSITION 5–22. *Assume that $\phi_1, \phi_2, \ldots$ are overfilters and that $\psi_1, \psi_2, \ldots$ are underfilters and that the following conditions are satisfied:*

$$\begin{aligned} \phi_n &\leqslant \psi_n, \\ \phi_n\phi_{n-1} &\geqslant \phi_n, \\ \psi_n\psi_{n-1} &\leqslant \psi_n. \end{aligned} \tag{5–9}$$

Figure 5–6. $(\psi\phi)(X)$ and $(\phi\psi)(X)$ of Example 5–1 (top row) and Example 5–2 (bottom row).

Then $(\phi\psi)_n, (\psi\phi)_n, (\phi\psi\phi)_n, (\psi\phi\psi)_n$ *are absorbing sequences of filters and*

$$(\phi\psi\phi)_n \leqslant \left\{ \begin{matrix} (\phi\psi)_n \\ (\psi\phi)_n \end{matrix} \right\} \leqslant (\psi\phi\psi)_n. \qquad (5\text{–}10)$$

PROOF. To show that $(\phi\psi)_n$ is a filter, we must show that it is an underfilter and an overfilter at the same time. Note first that

$$(\psi)_n = \psi_n \cdots \psi_3\psi_2\psi_1 \leqslant \psi_n \cdots \psi_3\psi_2 \leqslant \psi_n \cdots \psi_3 \leqslant \cdots \leqslant \psi_n.$$

We find that

$$\begin{aligned} (\phi\psi)_n(\phi\psi)_n &= \phi_n\big(\psi_n(\phi\psi)_n\phi_n\psi_n\big)(\phi\psi)_{n-1} \\ &\leqslant \phi_n\big(\psi_n(\psi\psi)_n\psi_n^2\big)(\phi\psi)_{n-1} \\ &\leqslant \phi_n\big(\psi_n(\psi)_n\psi_n\big)(\phi\psi)_{n-1} \\ &\leqslant \phi_n(\psi_n^3)(\phi\psi)_{n-1} \\ &\leqslant \phi_n\psi_n(\phi\psi)_{n-1} \\ &= (\phi\psi)_n. \end{aligned}$$

This proves that the composition$(\phi\psi)_n$ is an underfilter. To show that it is an overfilter, we use that

$$(\phi)_n = \phi_n\phi_{n-1}\cdots\phi_1 \geqslant \phi_n.$$

Therefore,

$$\begin{aligned}(\phi\psi)_n(\phi\psi)_n &\geqslant (\phi\phi)_n(\phi\psi)_n \\ &\geqslant (\phi)_n(\phi\psi)_n \\ &\geqslant \phi_n(\phi\psi)_n \\ &= \phi_n^2\psi_n(\phi\psi)_{n-1} \\ &\geqslant \phi_n\psi_n(\phi\psi)_{n-1} \\ &= (\phi\psi)_n.\end{aligned}$$

Thus $(\phi\psi)_n$ is a filter.

To show that the other three compositions define filters, one can use similar arguments. Furthermore, e.g.,

$$(\phi\psi\phi)_n \leqslant (\phi\psi\psi)_n \leqslant (\phi\psi)_n,$$

since every ψ_n is an underfilter. ■

The conditions in Eq. 5–9 hold if we make the (stronger) assumption that

$$\cdots \leqslant \phi_3 \leqslant \phi_2 \leqslant \phi_1 \leqslant \psi_1 \leqslant \psi_2 \leqslant \psi_3 \leqslant \cdots. \tag{5–11}$$

We present some examples.

EXAMPLE 5–3. Let ϕ be an overfilter and ψ an underfilter such that $\phi \leqslant \psi$. Fix $N \geqslant 1$ and define, for $n = 1, 2, \ldots, N$:

$$\phi_n = \phi^{N+1-n}, \quad \psi_n = \psi^{N+1-n}.$$

Then Eq. 5–11 holds.

For the next example we need some preparation. Let ψ be an increasing translation-invariant operator on $\mathcal{P}(\mathbb{E}^d)$, and let (ε, δ) be a translation-invariant adjunction on $\mathcal{P}(\mathbb{E}^d)$. Thus δ is of the form $\delta(X) = X \oplus A$, for some structuring element $A \subseteq \mathbb{E}^d$. It follows that

$$\begin{aligned}\psi\delta(X) &= \psi\Big(\bigcup_{a\in A} X_a\Big) \supseteq \bigcup_{a\in A} \psi(X_a) \\ &= \bigcup_{a\in A} [\psi(X)]_a = \psi(X) \oplus A = (\delta\psi)(X).\end{aligned}$$

Figure 5–7. The images $(\psi\phi)_n(X)$ and $(\psi\phi)_n(X)$ of Example 5–5: $n = 1, 2, 3$ in the first, second, and third row, respectively.

Thus we find that $\psi\delta \geqslant \delta\psi$. Similarly, it follows that $\psi\varepsilon \leqslant \varepsilon\psi$. These relations yield that

$$\delta\psi\varepsilon \leqslant \psi \quad \text{and} \quad \varepsilon\psi\delta \geqslant \psi. \tag{5–12}$$

EXAMPLE 5–4. Consider the following translation-invariant operators on $\mathcal{P}(\mathbb{E}^d)$: an adjunction (ε, δ), an overfilter ϕ, and an underfilter ψ. Assume, moreover, that $\phi \leqslant \psi$. By Proposition 5–19(b) we know that $\phi_n = \delta^n\phi\varepsilon^n$ are overfilters, and that $\psi_n = \varepsilon^n\psi\delta^n$ are underfilters. Furthermore, by Eq. 5–12, $\phi_n = \delta\phi_{n-1}\varepsilon \leqslant \phi_{n-1}$, and dually, $\psi_n \geqslant \psi_{n-1}$. Therefore, the conditions in Eq. 5–11 hold.

EXAMPLE 5–5. Let α_n, β_n be openings and closings, respectively, and let ξ be an increasing operator such that

$$\cdots \leqslant \alpha_2 \leqslant \alpha_1 \leqslant \xi \leqslant \beta_1 \leqslant \beta_2 \leqslant \cdots. \tag{5–13}$$

Define $\phi_n = \alpha_n \xi$ and $\psi_n = \beta_n \xi$. From Proposition 5–19(c) we derive that ϕ_n are overfilters; dually, ψ_n are underfilters. It is obvious that Eq. 5–11 holds.

Suppose, for example, that ξ is the median operator on $\mathcal{P}(\mathbb{Z}^2)$ using the rhombus as structuring element (origin and four horizontal and vertical neighbors). Let α_n, β_n be the opening and closing, respectively, with the $(2n+1) \times (2n+1)$ square, and define $\phi_n = \alpha_n \xi$ and $\psi_n = \beta_n \xi$. It is easy to see that the conditions in Eq. 5–13 are satisfied. In Fig. 5–7 we depict the corresponding AS-filters $(\psi\phi)_n$ and $(\phi\psi)_n$ for $n = 1, 2, 3$. Comparing these images with those in Fig. 5–4, we see that the new AS-filters introduced here perform substantially better than the classical ones described in Section 5.5; see [6].

5.8 ITERATION

In this section we explain how to construct morphological filters by iteration of increasing operators which are not idempotent. Though we restrict attention to operators on $\mathcal{P}(E)$, most of the results can be extended to complete lattices; refer to [11] and [5, Chap. 13].

5.8.1 CONVERGENCE

Let $X_n \subseteq E$, $n \geqslant 1$, and $X \subseteq E$: we say that $X_n \to X$ (X_n converges to X) if $X_n(h) \to X(h)$ as $n \to \infty$, for every $h \in E$. Here $X(\cdot)$ is the characteristic function associated with the set X. It is easy to see that the following assertions are equivalent:

(i) $X_n \to X$,

(ii) $h \in X$ iff $h \in X_n$ for n large enough.

DEFINITION 5–5. An operator ψ on $\mathcal{P}(E)$ is said to be *continuous* if $X_n \to X$ implies that $\psi(X_n) \to \psi(X)$.

Let ψ, ψ_n be operators on $\mathcal{P}(E)$, $n \geqslant 1$: we say that $\psi_n \to \psi$ (ψ_n converges to ψ) if $\psi_n(X) \to \psi(X)$ for every $X \in \mathcal{P}(E)$. In this section we are concerned mostly with sequences ψ^n consisting of iterates of a given operator ψ.

5.8.2 FINITE WINDOW OPERATORS

DEFINITION 5–6. Let ψ be an increasing operator on $\mathcal{P}(E)$, and assume that $W(h) \subseteq E$ is a finite set for every $h \in E$. We say that ψ is a finite window operator with window W if

$$h \in \psi(X) \iff h \in \psi\big(X \cap W(h)\big),$$

for $h \in E$ and $X \subseteq E$.

Note that if $E = \mathbb{E}^d$ and ψ is translation invariant, we can take $W(h) = W_h$, where $W \subseteq \mathbb{E}^d$ is a finite set. If ψ is a finite window operator, then its dual ψ^* is such as well. Furthermore, compositions, finite suprema, and finite infima of finite window operators are finite window operators.

For our purposes, the main property of a finite window operator is given by the following result; a proof can be found in [11].

PROPOSITION 5–23. *Every finite window operator is continuous.*

5.8.3 ITERATION AND IDEMPOTENCE

Assume that ψ is a continuous operator on $\mathcal{P}(E)$ and that $\psi^n \to \psi^\infty$, where ψ^∞ is another operator on $\mathcal{P}(E)$. Note that we do not assume that ψ or ψ^∞ are increasing, nor that ψ^∞ is continuous. Then $\psi^n = \psi\psi^{n-1} \to \psi\psi^\infty$, as ψ is continuous. This yields that $\psi\psi^\infty = \psi^\infty$; hence $\psi^n\psi^\infty = \psi^\infty$ for every $n \geqslant 1$. Letting $n \to \infty$, we get $\psi^\infty\psi^\infty = \psi^\infty$. Thus we arrive at the following result:

PROPOSITION 5–24. *If ψ is a continuous operator on $\mathcal{P}(E)$ and $\psi^n \to \psi^\infty$, then ψ^∞ is idempotent. In particular, if ψ is also increasing, then ψ^∞ is a filter.*

PROOF. On the one hand $\psi^{n+1} \to \psi^\infty$ as $n \to \infty$, but on the other hand

$$\psi^{n+1} = \psi\psi^n \to \psi\psi^\infty \text{ as } n \to \infty,$$

by the continuity of ψ. This yields that $\psi\psi^\infty = \psi^\infty$, and more generally, $\psi^n\psi^\infty = \psi^\infty$, for every $n \geqslant 1$. Letting $n \to \infty$, this yields

$$\psi^\infty\psi^\infty = \psi^\infty,$$

meaning that ψ^∞ is idempotent. It is obvious that ψ^∞ is increasing, and we conclude that ψ^∞ is a filter. ■

For example, if $\psi \geqslant \text{id}$ (i.e., ψ is extensive), then $\psi^2 \geqslant \psi$, hence $\psi^3 \geqslant \psi^2$, etc. This yields immediately that $\psi^n \to \psi^\infty$, where ψ^∞ is given by $\psi^\infty(X) = \bigcup_{n\geqslant 1} \psi^n(X)$. If ψ is also increasing, then ψ^∞ is a closing.

We present a simple example of an increasing, extensive operator ψ for which ψ^∞ is not idempotent; it is easy to see that this operator is not continuous. Let $\mathbb{N} = \{0, 1, 2, \ldots\}$ and $\overline{\mathbb{N}} = \mathbb{N} \cup \{+\infty\}$. For $X \subseteq \overline{\mathbb{N}}$, we define $X + 1 = \{x + 1 \mid x \in X\}$, where $+\infty + 1 = +\infty$. Let the operator ψ on $\mathcal{P}(\overline{\mathbb{N}})$ be given by

$$\psi(X) = \begin{cases} X \cup (X+1), & \text{if } X \neq \mathbb{N} \\ \overline{\mathbb{N}}, & \text{if } X = \mathbb{N}. \end{cases}$$

Obviously, ψ is increasing and extensive. For $k = 1, 2, \ldots$ we have $\psi^k(\{0\}) = \{0, 1, \ldots, k\}$, hence $\psi^\infty(\{0\}) = \mathbb{N}$. However, $\psi\psi^\infty(\{0\}) = \psi(\mathbb{N}) = \overline{\mathbb{N}}$. This implies that $(\psi^\infty)^2(\{0\}) = \overline{\mathbb{N}} \neq \psi^\infty(\{0\})$.

Most of the previous results can be extended to gray-scale functions. We recall the following definition.

DEFINITION 5–7. A partially ordered set $\mathcal{T}$ is called a *chain* if for every two elements $s, t \in \mathcal{T}$ we have $s \leqslant t$ or $t \leqslant s$. It is called a *complete chain* if it is both a chain and a complete lattice.

For example, $\overline{\mathbb{Z}}$ and $\overline{\mathbb{R}}$ with the natural ordering are complete chains.

Consider the space $\mathrm{Fun}(E, \mathcal{T})$, where $\mathcal{T}$ is a subset of $\overline{\mathbb{R}}$ which is a complete chain. (In [5, Chap. 13] we consider also the case where $\mathcal{T}$ is an arbitrary complete lattice.) We say that the sequence of functions F_n converges to F, $F_n \to F$, if $F_n(x) \to F(x)$ as $n \to \infty$, for every $x \in E$.

Definition 5–5 generalizes easily to function operators: an operator ψ on $\mathrm{Fun}(E, \mathcal{T})$ is *continuous* if $F_n \to F$ implies that $\psi(F_n) \to \psi(F)$. Proposition 5–24 remains valid in this case.

PROPOSITION 5–25. *If ψ is a continuous operator on* $\mathrm{Fun}(E, \mathcal{T})$ *and $\psi^n \to \psi^\infty$, then ψ^∞ is idempotent. If, furthermore, ψ is increasing, then ψ^∞ is a filter.*

In Section 5.9.3 we present a class of operators, the so-called activity-extensive operators, for which the sequence ψ^n converges.

5.9 ACTIVITY ORDERING AND CENTER OPERATOR

5.9.1 ACTIVITY ORDERING

Activity ordering is a partial ordering on $\mathcal{O}(\mathcal{L})$, the complete lattice of operators on $\mathcal{L}$ (where $\mathcal{L}$ is a complete lattice), which provides a tool to compare the effect of two different operators. The notion of "activity ordering" is due to Serra [23]; see also [12]. A comprehensive discussion can also be found in [5].

DEFINITION 5–8. Given two operators ϕ, ψ on the complete lattice $\mathcal{L}$, we say that ψ is *more active than* ϕ, denoted by $\phi \preccurlyeq \psi$, if

$$\mathrm{id} \wedge \psi \leqslant \mathrm{id} \wedge \phi \quad \text{and} \quad \mathrm{id} \vee \psi \geqslant \mathrm{id} \vee \phi.$$

For example, if ϕ and ψ are both extensive, then $\phi \preccurlyeq \psi$ if and only if $\phi \leqslant \psi$. However, if both operators are anti-extensive, then $\phi \preccurlyeq \psi$ iff $\phi \geqslant \psi$. It is evident that every operator is more active than id, the identity operator. On the other hand, if $\mathcal{L} = \mathcal{P}(E)$, then the complement operator $X \mapsto X^c$ is more active than any other operator. However, this last observation cannot be generalized to arbitrary negations.

PROPOSITION 5–26. *If $\mathcal{L} = \mathcal{P}(E)$ or $\mathrm{Fun}(E, \mathcal{T})$, where $\mathcal{T}$ is a complete chain, then '$\preccurlyeq$' defines a partial ordering on $\mathcal{O}(\mathcal{L})$.*

PROOF. We consider the case $\mathcal{L} = \mathcal{P}(E)$. The proof for $\mathcal{L} = \mathrm{Fun}(E, \mathcal{T})$ is quite similar. It is obvious that $\preccurlyeq$ is reflexive and transitive. We show that it is anti-symmetric. Let ψ, ϕ be two operators such that $\phi \preccurlyeq \psi$ and $\psi \preccurlyeq \phi$, and take $X \subseteq E$. Then $X \cap \phi(X) = X \cap \psi(X)$ and $X \cup \phi(X) = X \cup \psi(X)$, hence $\phi(X) = \psi(X)$. ■

Suppose that $\mathcal{L}$ possesses a negation ν. If $\phi \preccurlyeq \psi$ then $\mathrm{id} \wedge \psi \leqslant \mathrm{id} \wedge \phi$, hence $(\mathrm{id} \wedge \psi)\nu \leqslant (\mathrm{id} \wedge \phi)\nu$. This gives us $\nu \wedge \psi\nu \leqslant \nu \wedge \phi\nu$. Applying ν at both sides yields $\nu^2 \vee \nu\psi\nu \geqslant \nu^2 \vee \nu\phi\nu$. Using that $\nu^2 = \mathrm{id}$ and $\nu\psi\nu = \psi^*$, we get that $\mathrm{id} \vee \psi^* \geqslant \mathrm{id} \vee \phi^*$. Similarly, we find that $\mathrm{id} \vee \psi \geqslant \mathrm{id} \vee \phi$ implies that $\mathrm{id} \wedge \psi^* \leqslant \mathrm{id} \wedge \phi^*$. Thus we arrive at the following result.

PROPOSITION 5–27. *Let $\mathcal{L}$ be a complete lattice which possesses a negation. Then $\phi \preccurlyeq \psi$ if and only if $\phi^* \preccurlyeq \psi^*$.*

5.9.2 CENTER OPERATOR

In Proposition 5–26 we have seen that the relation '$\preccurlyeq$' defines a partial ordering if $\mathcal{L} = \mathcal{P}(E)$ or $\mathrm{Fun}(E, \mathcal{T})$, with $\mathcal{T}$ a complete chain. In the first case, a much stronger result holds.

PROPOSITION 5–28. *The family of operators on $\mathcal{P}(E)$ endowed with the activity ordering '$\preccurlyeq$' is a complete lattice. Given a collection of operators ψ_i, $i \in I$, on $\mathcal{P}(E)$, the activity infimum and supremum are given, respectively, by*

$$\curlywedge_{i\in I} \psi_i = \left[\mathrm{id} \wedge \left(\bigvee_{i\in I} \psi_i\right)\right] \vee \left(\bigwedge_{i\in I} \psi_i\right), \tag{5–14}$$

$$\curlyvee_{i\in I} \psi_i = \left[\nu \wedge \left(\bigvee_{i\in I} \psi_i\right)\right] \vee \left(\bigwedge_{i\in I} \psi_i\right). \tag{5–15}$$

The operator $\curlywedge_{i\in I} \psi_i$ is called the *center* of the operators ψ_i; the operator $\curlyvee_{i\in I} \psi_i$ is called the *anti-center* [5, 12, 23]. It is obvious that the center is an increasing operator, given that every ψ_i is increasing. For the anti-center this is not true in general. In this paper, our interest only concerns the center. Observe that the annular filter ω discussed in Section 5.4 is the center operator of ε_B and δ_A.

It is straightforward to show that (cf. Proposition 5–27):

$$\curlywedge_{i\in I} \psi_i^* = \big(\curlywedge_{i\in I} \psi_i\big)^*, \qquad \curlyvee_{i\in I} \psi_i^* = \big(\curlyvee_{i\in I} \psi_i\big)^*.$$

The center of the family ψ_i has the form

$$\gamma = (\mathrm{id} \wedge \psi) \vee \phi = (\mathrm{id} \vee \phi) \wedge \psi, \tag{5–16}$$

where $\phi = \bigwedge_{i\in I} \psi_i$ and $\psi = \bigvee_{i\in I} \psi_i$. Note that $\phi \leqslant \psi$.

If $\mathcal{L} = \mathrm{Fun}(E, \mathcal{T})$, where $\mathcal{T}$ is a complete chain, then the center γ given by Eq. 5–16 also has the interpretation of the activity infimum. Because $\mathrm{Fun}(E, \mathcal{T})$ possesses no complement operator (though it may possess a negation), the activity supremum does not exist in general.

The center operator on $\mathrm{Fun}(E, \mathcal{T})$ has an interesting geometric interpretation that explains why this operator is called center. Let ψ_1, ψ_2 be operators on $\mathrm{Fun}(E, \mathcal{T})$; put $\phi = \psi_1 \wedge \psi_2$ and $\psi = \psi_1 \vee \psi_2$. Define γ by Eq. 5–16; thus γ is the center of ψ_1 and ψ_2. Define the mapping $m : \mathcal{T}^3 \to \mathcal{T}$ as follows: given t_1, t_2, t_3, let $m(t_1, t_2, t_3)$ be the value t_i which lies between the other two. In fact, m is given by the formula

$$m(t_1, t_2, t_3) = (t_1 \wedge t_2) \vee (t_1 \wedge t_3) \vee (t_2 \wedge t_3).$$

It is not difficult to show that

$$\gamma(F)(x) = m\big(F(x), \psi_1(F)(x), \psi_2(F)(x)\big);$$

see [5, Sect. 3–6]. Refer to Fig. 5–8 for an illustration.

DEFINITION 5–9. A lattice $\mathcal{L}$ is said to be *modular* if the condition

$$(X \vee Y) \wedge Y' = \big(X \wedge Y'\big) \vee Y \quad \text{if} \quad Y \leqslant Y'$$

holds, for $X, Y, Y' \in \mathcal{L}$.

It is obvious that on a modular lattice the identity $(\mathrm{id} \wedge \psi) \vee \phi = (\mathrm{id} \vee \phi) \wedge \psi$ holds for any two operators ϕ, ψ with $\phi \leqslant \psi$. Thus, we can extend the definition of the center γ given in Eq. 5–16 to arbitrary modular lattices. We point out that distributivity of a lattice implies modularity. In particular, $\mathcal{P}(E)$ and $\mathrm{Fun}(E, \mathcal{T})$, where $\mathcal{T}$ is a chain, are modular.

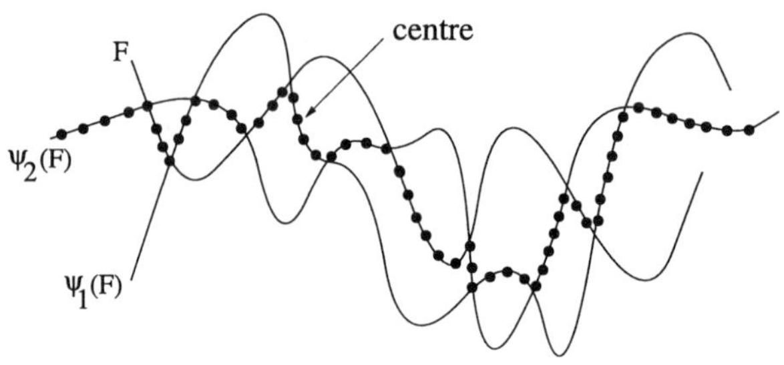

Figure 5–8. The center of the operators ψ_1 and ψ_2 is indicated by the black dots.

5.9.3 ACTIVITY-EXTENSIVE OPERATORS

We start with a definition.

DEFINITION 5–10. An operator ψ is called activity-extensive if $\psi^n \preccurlyeq \psi^{n+1}$, for every $n \geqslant 1$.

It is evident that every increasing operator which is extensive or anti-extensive is activity-extensive, but the converse is not true.

On $\mathrm{Fun}(E, \mathcal{T})$ there exists an interesting characterization of activity-extensive operators.

DEFINITION 5–11. Assume that $\mathcal{T}$ is a complete lattice. The sequence F_n, $n \geqslant 1$, in $\mathrm{Fun}(E, \mathcal{T})$ is called *pointwise monotone* if the sequence $F_n(x)$ is either increasing or decreasing for every $x \in E$.

If $\mathcal{T} = \{0, 1\}$, then $\mathrm{Fun}(E, \mathcal{T})$ is isomorphic to $\mathcal{P}(E)$. In this case, a sequence X_n is pointwise monotone if, for fixed $h \in E$, the sequence $X_n(h)$, where $X_n(\cdot)$ is the characteristic function of the set X_n, is of the form $0, 0, \ldots, 0, 1, 1, \ldots$ or $1, 1, \ldots, 1, 0, 0, \ldots$.

PROPOSITION 5–29. *The operator ψ on* $\mathrm{Fun}(E, \mathcal{T})$*, where $\mathcal{T}$ is a complete lattice, is activity-extensive if and only if the sequence $\psi^n(F)$ is pointwise monotone, for every function F.*

PROOF. We prove only the only if-statement. Let $F \in \mathrm{Fun}(E, \mathcal{T})$ and put $F_n = \psi^n(F)$ for $n \geqslant 0$. Given $x \in E$, assume that $n \geqslant 1$ is such that $F_0(x) = \cdots = F_{n-1}(x) < F_n(x)$. From $F_{n+1}(x) \vee F(x) \geqslant F_n(x) \vee F(x)$ it follows that $F_{n+1}(x) \geqslant F_n(x)$. Repeating this argument, we find that $F_n(x) \leqslant F_{n+1}(x) \leqslant F_{n+2}(x) \leqslant \cdots$. This proves the assertion. The case that $F_{n-1}(x) > F_n(x)$ is treated analogously. ■

An operator ψ on $\mathcal{P}(E)$ is activity-extensive if, for every X, the sequence $X \cap \psi^n(X)$ is decreasing, and the sequence $X^c \cap \psi^n(X)$ is increasing. It is well known [5, 16, 22] that an important class of operators on $\mathrm{Fun}(E)$ is the *flat operators*, i.e., operators generated by a set operator.

PROPOSITION 5–30. *An increasing operator on $\mathcal{P}(E)$ is activity-extensive if and only if its flat extension to* $\mathrm{Fun}(E)$ *is activity-extensive.*

The proof of this result can be found in [5, Prop. 13–44].

If ψ is an activity-extensive operator on $\mathrm{Fun}(E, \mathcal{T})$, where $\mathcal{T}$ is an arbitrary complete lattice, then $\psi^n(F)$ is pointwise monotone, for every function F. Fix $x \in E$; define $\psi^\infty(F)$ to be $\bigvee_{n \geqslant 1} \psi^n(F)(x)$ if the sequence $\psi^n(F)(x)$ is increasing, and $\bigwedge_{n \geqslant 1} \psi^n(F)(x)$ if it is decreasing. It follows immediately that $\psi^n \to \psi^\infty$ as $n \to \infty$.

PROPOSITION 5–31. *Let $\mathcal{T}$ be an arbitrary complete lattice and ψ an activity-extensive operator on* $\mathrm{Fun}(E, \mathcal{T})$; *then the sequence of iterates ψ^n converges (pointwise).*

We exploit this fact in Section 5.10.2. The next result, which is the main result of this section, and which is due to Serra [23], shows a simple, yet general, way to construct nontrivial activity-extensive operators.

PROPOSITION 5–32. *Let $\mathcal{L}$ be a complete modular lattice, ϕ an overfilter and ψ an underfilter such that $\phi \leqslant \psi$. The center operator $\gamma = (\mathrm{id} \wedge \psi) \vee \phi$ is activity-extensive.*

PROOF. To prove that $\gamma^{n-1} \preccurlyeq \gamma^n$, we show that

$$\gamma^n = \left(\mathrm{id} \wedge \psi\gamma^{n-1}\right) \vee \phi\gamma^{n-1} = \left(\mathrm{id} \vee \phi\gamma^{n-1}\right) \wedge \psi\gamma^{n-1}, \tag{5–17}$$

for every $n \geqslant 1$. Using the modularity of $\mathcal{L}$, we infer that

$$\begin{aligned} \mathrm{id} \wedge \gamma^n &= \mathrm{id} \wedge \psi\gamma^{n-1}, \\ \mathrm{id} \vee \gamma^n &= \mathrm{id} \vee \phi\gamma^{n-1}. \end{aligned}$$

In particular, using that $\gamma \leqslant \psi$ and that ψ is an underfilter, we get

$$\mathrm{id} \wedge \gamma^n \leqslant \mathrm{id} \wedge \psi^2\gamma^{n-2} \leqslant \mathrm{id} \wedge \psi\gamma^{n-2} \leqslant \mathrm{id} \wedge \gamma^{n-1}.$$

Analogously,

$$\mathrm{id} \vee \gamma^n \geqslant \mathrm{id} \vee \gamma^{n-1}.$$

Thus we have demonstrated that $\gamma^{n-1} \preccurlyeq \gamma^n$.

We prove Eq. 5–17 by induction. For $n = 1$ the result is obvious. Assume that it holds for n, then

$$\begin{aligned}\gamma^{n+1} &= \big((\mathrm{id}\wedge\psi)\vee\phi\big)\gamma^n\\ &= \big(\gamma^n\wedge\psi\gamma^n\big)\vee\phi\gamma^n\\ &= \big[\big((\mathrm{id}\vee\phi\gamma^{n-1})\wedge\psi\gamma^{n-1}\big)\wedge\psi\gamma^n\big]\vee\phi\gamma^n.\end{aligned}$$

Since $\psi\gamma^{n-1}\geqslant\psi^2\gamma^{n-1}\geqslant\psi\gamma^n$, we get

$$\begin{aligned}\gamma^{n+1} &= \big[(\mathrm{id}\vee\phi\gamma^{n-1})\wedge\psi\gamma^n\big]\vee\phi\gamma^n\\ &= \big(\mathrm{id}\vee\phi\gamma^{n-1}\vee\phi\gamma^n\big)\wedge\psi\gamma^n\\ &= \big(\mathrm{id}\vee\phi\gamma^n\big)\wedge\psi\gamma^n.\end{aligned}$$

Here we used the modularity of $\mathcal{L}$ and the fact that $\phi\gamma^{n-1}\leqslant\phi^2\gamma^{n-1}\leqslant\phi\gamma^n$. ■

Let ψ be an increasing operator, $\alpha\leqslant\psi$ an opening and $\beta\geqslant\psi$ a closing. From Proposition 5–19(c) and its dual we get that $\alpha\psi$ is an overfilter and that $\beta\psi$ is an underfilter. It is obvious that

$$\alpha\psi\leqslant\psi\leqslant\beta\psi.$$

Thus, Proposition 5–32 applies, and we arrive at the following result [8]:

PROPOSITION 5–33. *Let $\mathcal{L}$ be a modular lattice, ψ an increasing operator, α an opening, β a closing, and assume that $\alpha\leqslant\psi\leqslant\beta$. Then*

$$\pi = (\mathrm{id}\wedge\beta\psi)\vee\alpha\psi$$

is activity-extensive and $\pi\preccurlyeq\psi$. Furthermore, if α' is an opening and β' a closing such that $\alpha'\leqslant\alpha$ and $\beta'\geqslant\beta$, and if $\pi'=(\mathrm{id}\wedge\beta'\psi)\vee\alpha'\psi$, then $\pi'\preccurlyeq\pi$.

We point out that $(\mathrm{id}\wedge\psi\beta)\vee\psi\alpha$ is activity-extensive as well. However, this modification of ψ turns out to be less interesting than π; see [8]. Proposition 5–33 forms the basis for the construction of self-dual filters, as discussed in the following section.

5.10 SELF-DUAL FILTERS

This section discusses self-dual filters, that is, morphological filters which satisfy $\psi^*=\psi$. We have seen one instance of a self-dual filter in Section 5–4, namely, the annular filter for which the structuring elements governing foreground and background adjacency coincide. This particular filter is given by a simple explicit expression. As we observed earlier, it is the center of a dilation δ_A and its negative erosion $\varepsilon_A = \delta_A^*$, where A is a symmetric structuring element which does not contain the origin. The self-dual filters considered in this section are not as simple as the annular filter; they are all obtained by iteration of an increasing operator which is self-dual.

5.10.1 SELF-DUAL OPERATORS

Before we discuss the construction of self-dual filters by iteration, we explain how to build self-dual operators, as these form the main ingredient for this iteration procedure. The center operator is the most important tool for the construction of self-dual operators. This is the content of our first proposition.

PROPOSITION 5–34. *If ψ_i, $i \in I$, is a family of operators on $\mathcal{P}(E)$ such that with every ψ_i the negative operator ψ_i^* is also a member of the family, then the center $\gamma = \curlywedge_{i\in I} \psi_i$ is a self-dual operator.*

The proof is easy if one uses the explicit expressions in Eq. 5–14. Throughout the remainder of this section we restrict ourselves to translation-invariant operators on $\mathcal{P}(\mathbb{E}^d)$.

The median operator is the best-known example of a self-dual operator. More generally, if ψ_b is the morphological operator derived from a structuring element $A = \{a_1, a_2, \ldots, a_n\}$ and a Boolean function b of n variables, i.e.,

$$h \in \psi_b(X) \quad \text{if} \quad b\big(X(a_1+h), \ldots, X(a_n+h)\big) = 1,$$

then ψ_b is self-dual if and only if $b^* = b$.

EXAMPLE 5–6. Let A be the 3×3 square, and let $\rho_{A,s}$ be the corresponding rank operators (see Sect. 5.6.2). It is easy to see that

$$\rho_{A,s}^* = \rho_{A,10-s}, \qquad s = 1, 2, \ldots, 9.$$

Furthermore, $\rho_{A,10-s} \leqslant \rho_{A,s}$ if $s \leqslant 5$. The center of $\rho_{A,s}$ and $\rho_{A,10-s}$, written as η_s, is given by

$$\eta_s = (\mathrm{id} \wedge \rho_{A,s}) \vee \rho_{A,10-s}, \qquad s = 1, 2, 3, 4, 5.$$

Evidently, η_s is a self-dual operator. Furthermore, one can easily show that

$$\mathrm{id} = \eta_1 \precsim \eta_2 \precsim \eta_3 \precsim \eta_4 \precsim \eta_5.$$

Figure 5–9. From left to right and top to bottom: $\eta_2(X), \eta_3(X), \eta_4(X), \eta_5(X)$. This figure shows clearly that η_{s+1} is more active than η_s.

The operator η_5 is the median operator. In Fig. 5–9 we apply $\eta_2, \eta_3, \eta_4, \eta_5$ to our test image.

In [5, Chap. 13] and [8] we have presented a comprehensive treatment of the construction of self-dual operators based on the concept of *switch operator*. Here we only summarize some of the main results.

PROPOSITION 5–35. *An increasing, translation-invariant operator on $\mathcal{P}(\mathbb{E}^d)$ is self-dual if and only if it is of the form $\psi = \psi_{\mathcal{A}}$, where*

$$\psi_{\mathcal{A}}(X) = \left(X \cap \bigcap_{A \in \mathcal{A}} X \oplus \check{A}\right) \cup \bigcup_{A \in \mathcal{A}} X \ominus A, \tag{5–18}$$

where $\mathcal{A} \subseteq \mathcal{P}(\mathbb{E}^d)$ is a collection of structuring elements which satisfy

$$0 \notin A \quad and \quad A \cap B \neq \emptyset,$$

for $A, B \in \mathcal{A}$.

If $A \cap B \neq \emptyset$, then $X \ominus B \subseteq X \oplus \check{A}$. This yields that the operator given by Eq. (5–18) is the center of $X \mapsto \bigcup_{A \in \mathcal{A}} X \ominus A$ and its negative $X \mapsto \bigcap_{A \in \mathcal{A}} X \oplus \check{A}$.

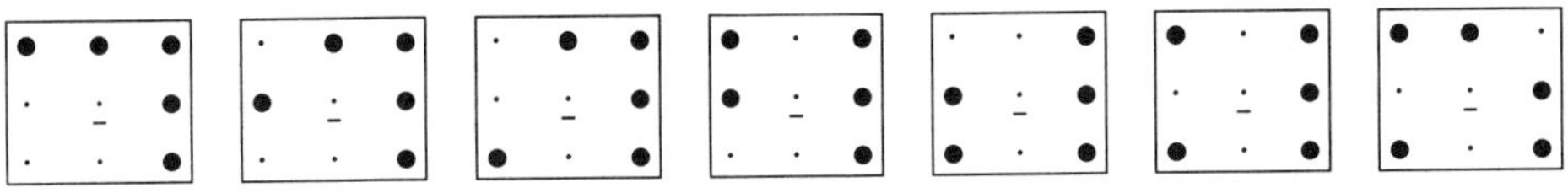

Figure 5–10. Collection $\mathcal{A}$ of structuring elements associated with the median operator.

There exists the following interpretation of Eq. 5–18. If a point h lies not in X, then h lies in the transformed image $\psi_{\mathcal{A}}(X)$ if and only if $A_h \subseteq X$ for some $A \in \mathcal{A}$. Dually, if $h \in X$ then $h \notin \psi_{\mathcal{A}}(X)$ if $A_h \subseteq X^c$ for some $A \in \mathcal{A}$.

For the median operator associated with the 3×3 square, the collection $\mathcal{A}$ contains the structuring elements depicted in Fig. 5–10.

The collection $\mathcal{A}$ in Eq. 5–18 is not uniquely determined by ψ. For example, if

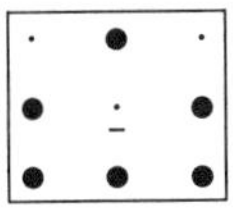

is added to the collection in Fig. 5–10, one still obtains the median operator. But we have the following result:

PROPOSITION 5–36. *Let $\mathcal{A}, \mathcal{B}$ be two collections of structuring elements in $\mathcal{P}(\mathbb{E}^d)$. The following two assertions are equivalent:*

(i) *for every $A \in \mathcal{A}$ there exists a $B \in \mathcal{B}$ such that $B \subseteq A$;*

(ii) $\psi_{\mathcal{A}} \preccurlyeq \psi_{\mathcal{B}}$.

In particular, this result implies that with any subcollection $\mathcal{B}$ of the structuring elements depicted in Fig. 5–10, there corresponds a self-dual operator $\psi_{\mathcal{B}}$ which is less active than the median operator. In the next subsection we discuss an alternative way to diminish the activity of a self-dual operator.

EXAMPLE 5–7. The self-dual operators η_k can be represented as in Eq. 5–18 using a collection of structuring elements $\mathcal{A}_k$ consisting of all subsets A of the 3×3 square with $0 \notin A$ and containing $10 - k$ points. For example, $\mathcal{A}_4$ consists of 28 structuring elements, namely:

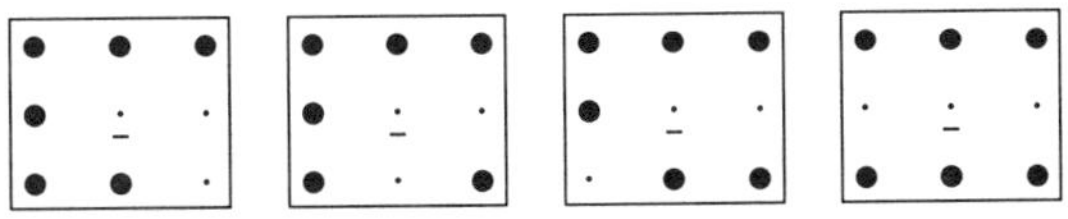

and rotations.

5.10.2 Construction of Self-dual Filters

Suppose we have an increasing, translation-invariant operator ψ which is self-dual. The next result gives an easy criterion for the activity-extensivity of ψ; see [8, Prop. 6–3].

PROPOSITION 5–37. *The operator $\psi_{\mathcal{A}}$ given by Eq. 5–18 is activity-extensive if and only if $0 \in \psi^n(A)$, for every $A \in \mathcal{A}$ and $n \geqslant 1$.*

PROOF. "only if": if $A \in \mathcal{A}$, then $0 \notin A$ and $0 \in \psi(A)$. If there exists an integer $n > 1$ such that $0 \notin \psi^n(A)$, then the sequence $\psi^k(A)$ is not pointwise monotone, hence ψ is not activity-extensive.

"if": suppose that $0 \in \psi^n(A)$ for $A \in \mathcal{A}$ and $n \geqslant 1$, and that ψ is not activity-extensive. Then there is a set X such that $\psi^n(X)$ is not pointwise monotone. Without loss of generality, we can assume that $0 \notin X$, $0 \in \psi(X)$ and $0 \notin \psi^m(X)$, for some $m > 1$. Using Eq. 5–18, we derive that $0 \in \bigcup_{A \in \mathcal{A}} X \ominus A$, which means that $A \subseteq X$ for some $A \in \mathcal{A}$. By assumption, $0 \in \psi^n(A) \subseteq \psi^n(X)$ for $n \geqslant 1$, a contradiction. This yields the result. ■

To obtain self-dual operators which are activity-extensive, we combine Proposition 5–33 and Proposition 5–34. (The continuity of these operators will be guaranteed by the fact that we restrict ourselves to finite window operators.)

Let ψ be a self-dual operator and let α be an opening with $\alpha \leqslant \psi$. The negative closing $\beta = \alpha^*$ satisfies $\beta \geqslant \psi$. Now Proposition 5–33 gives us that the center

$$\pi = (\mathrm{id} \wedge \beta\psi) \vee \alpha\psi \tag{5–19}$$

is activity-extensive, whereas Proposition 5–34 guarantees that π is self-dual (for $(\beta\psi)^* = \alpha\psi$).

REMARK 5–1. Alternatively, one can start with an increasing operator ψ such that $\psi \leqslant \psi^*$ and an opening $\alpha \leqslant \psi$. The negative closing $\beta = \alpha^*$ satisfies $\beta \geqslant \psi^*$, and the center of $\alpha\psi$ and $\beta\psi^*$ is self-dual and activity-extensive.

Before we present some examples on $\mathcal{P}(\mathbb{Z}^2)$, we give a criterion which guarantees that $\alpha \leqslant \psi$. The structural opening $\alpha_B(X) = X \circ B$ satisfies $\alpha_B \leqslant \psi$ if $B \subseteq \psi(B)$. We say that B is *persistent with respect to* ψ if $B \subseteq \psi(B)$. The following result is obvious.

PROPOSITION 5–38. *Let ψ be an increasing, translation-invariant operator on $\mathcal{P}(\mathbb{E}^d)$, and let B_i, $i \in I$, be persistent with respect to ψ. Then the opening $\alpha(X) = \bigcup_{i \in I} X \circ B_i$ satisfies $\alpha \leqslant \psi$.*

EXAMPLE 5–8. Let μ be the median operator with the 3×3 square as structuring element. The effect of μ on our test image can be seen in Fig. 5–9; recall that $\mu = \eta_5$. The set B given by

$B =$ 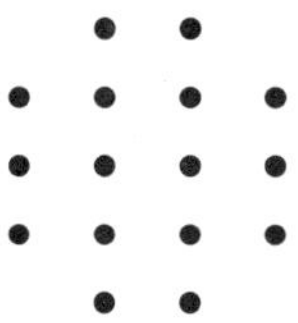

is persistent with respect to μ. Let $\alpha(X) = X \circ B$ and $\beta(X) = X \bullet B$. The modification π given by Eq. 5–19 is self-dual and activity-extensive. In the first row of Fig. 5–11 we depict $\pi(X)$ and $\pi^{\infty}(X)$.

EXAMPLE 5–9. The median operator discussed in the previous example is of the form $\mu = \psi_{\mathcal{A}}$, where $\mathcal{A}$ are the structuring elements depicted in Fig. 5–10. Consider the subcollection $\mathcal{B}$ of $\mathcal{A}$ which lacks the first structuring element and its three 90°-rotations. Now Proposition 5–36 yields that $\psi_{\mathcal{B}} \preccurlyeq \psi_{\mathcal{A}}$. The operator $\psi_{\mathcal{B}}$ is not activity-extensive; for example, the following pattern oscillates with period 2.

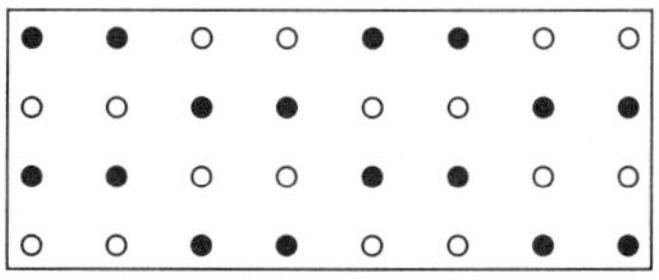

It is evident that the structuring element B is persistent with respect to $\psi_{\mathcal{B}}$, but we can find a smaller structuring element with this property, namely

$B' =$ 

Let α be the structural opening $\alpha(X) = X \circ B'$ and let β be the negative closing. Again, the modification π of $\psi_{\mathcal{B}}$ given by Eq. 5–19 is self-dual and activity-extensive. The images $\pi(X)$ and $\pi^{\infty}(X)$ are depicted in the second row of Fig. 5–11.

EXAMPLE 5–10. Consider the operator η_4 introduced in Example 5–6. This operator can be represented in the form Eq. 5–18 where $\mathcal{A}$ is given by

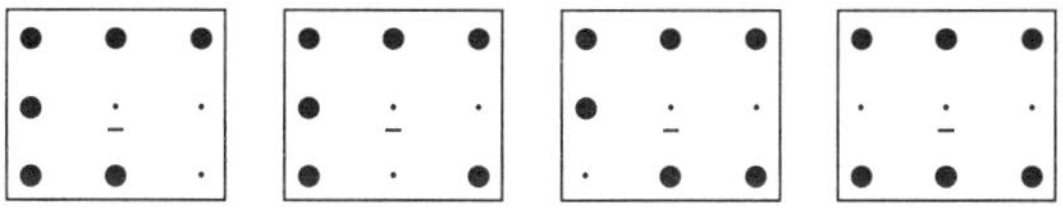

Figure 5–11. First three rows: $\pi(X)$ and $\pi^{\infty}(X)$ of Examples 5–8, 5–9, and 5–10. Bottom row: $\psi_{\mathcal{B}}(X)$ and $\psi_{\mathcal{B}}^{\infty}(X)$ of Example 5–11.

and their 45°-rotations. The operator η_4 is not activity-extensive since the pattern

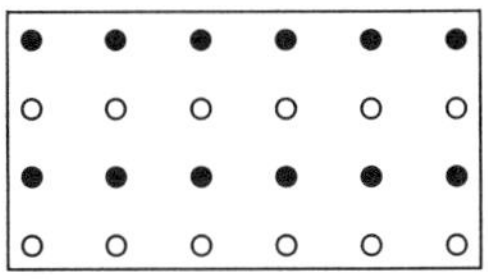

etc.

is 2-periodic. The structuring elements

 and 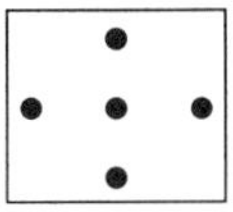

are persistent. Let α be the union of the structural openings associated with these two structuring elements; let $\beta = \alpha^*$ and let π be the modification of η_4 obtained from Eq. 5–19. The action of π and π^∞ can be seen from the third row in Fig. 5–11.

EXAMPLE 5–11. Consider the subcollection $\mathcal{B}$ of of the collection $\mathcal{A}$ of the previous example which is obtained by deleting the elements

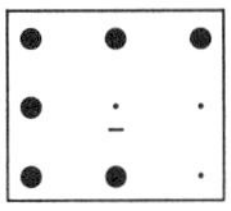 + 45°-rotations,

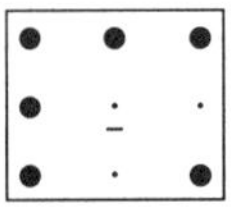 and 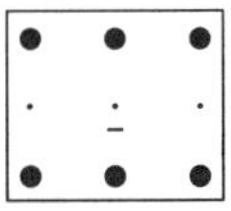 + 90°-rotations.

We use Proposition 5–37 to show that $\psi_{\mathcal{B}}$ is activity-extensive. Therefore we must consider the sequences $\psi^n_{\mathcal{B}}(B)$ for $B \in \mathcal{B}$. First we note that the triangles

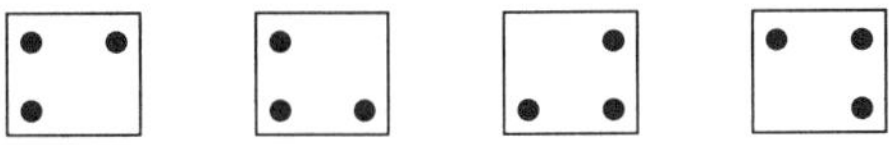

are invariant under $\psi_{\mathcal{B}}$. Now

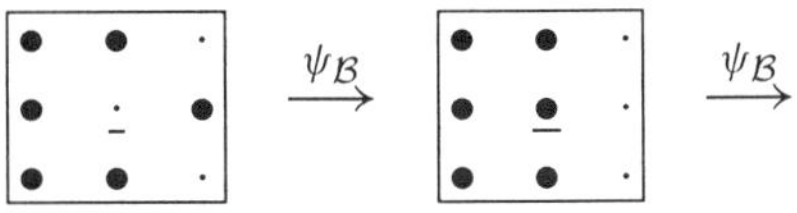

idem

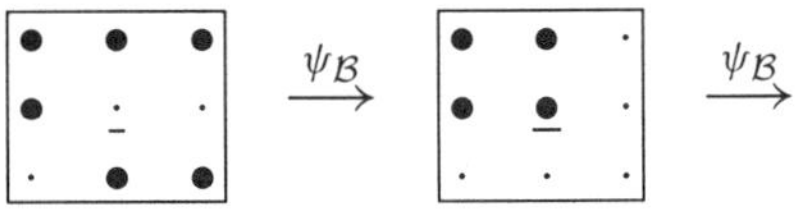

idem

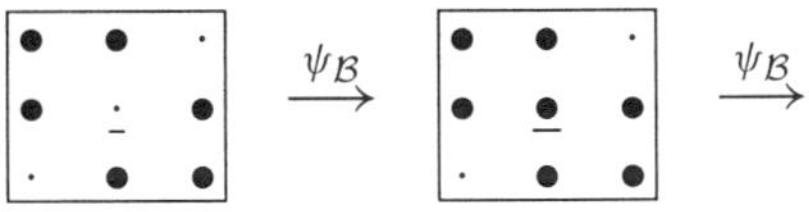

idem

This implies that $\psi_{\mathcal{B}}$ is activity-extensive. The bottom row in Fig. 5–11 depicts the transformed sets $\psi_{\mathcal{B}}(X)$ and $\psi_{\mathcal{B}}^{\infty}(X)$. These figures clearly show the invariance of the triangles depicted above.

REFERENCES

[1] Birkhoff, G., *Lattice Theory*, 3rd Ed., Vol. 25, American Mathematical Society, Providence, RI, 1967.

[2] Charif-Chefchaouni, M. and D. Schonfeld, "Morphological Representation of Nonlinear Filters," *J. Math. Imaging and Vision* **4**, 215–232, 1994.

[3] Heijmans, H. J. A. M., "Composing Morphological Filters," *IEEE Trans. Image Process.* **6** (5), 713–723, 1997.

[4] Heijmans, H. J. A. M., "Theoretical Aspects of Gray-Level Morphology," *IEEE Trans. Patt. Anal. Mach. Intell.* **13**, 568–582, 1991.

[5] Heijmans, H. J. A. M., *Morphological Image Operators*, Academic Press, Boston, 1994.

[6] Heijmans, H. J. A. M., "A New Class of Alternating Sequential Filters," in *1995 IEEE Workshop on Nonlinear Signal and Image Processing*, I. Pitas, ed., Vol. I, *Proc. IEEE*, 30–33, 1995.

[7] Heijmans, H. J. A. M., "Morphological Filters for Dummies," in *Mathematical Morphology and its Application to Image and Signal Processing*, P. Maragos, R. W. Shafer, and M. A. Butt, eds., Kluwer Academic Publishers, Boston, 127–137, 1996.

[8] Heijmans, H. J. A. M., "Self-Dual Morphological Operators and Filters," *J. Math. Imaging and Vision* **6**(1), 15–36, 1996.

[9] Heijmans, H. J. A. M., and C. Ronse, "The Algebraic Basis of Mathematical Morphology – Part I: Dilations and Erosions," *Comp. Vis. Graph. Image Process.* **50**, 245–295, 1990.

[10] Heijmans, H. J. A. M., and C. Ronse, "Annular Filters for Binary Images," *IEEE Trans. Image Process.*, to appear.

[11] Heijmans, H. J. A. M., and J. Serra, "Convergence, Continuity and Iteration in Mathematical Morphology," *J. Vis. Comm. Image Rep.* **3**, 84–102, 1992.

[12] Meyer, F., and J. Serra, "Contrasts and Activity Lattice," *Sign. Proc.* **16**, 303–317, 1989.

[13] Roerdink, J. B. T. M., "Mathematical Morphology with Non-Commutative Symmetry Groups," in *Mathematical Morphology in Image Processing*, E. R. Dougherty, ed., Chap. 7, pp. 205–254, Marcel Dekker, New York, 1993.

[14] Roerdink, J. B. T. M., "On the Construction of Translation and Rotation Invariant Morphological Operators," in *Mathematical Morphology: Theory and Hardware*, R. M. Haralick, ed., Oxford University Press, in press.

[15] Ronse, C., "Extraction of Narrow Peaks and Ridges in Images by Combination of Local Low Rank and Max Filters: Implementation and Applications to Clinical Angiography," Working Document WD47, Philips Research Laboratory, Brussels, Belgium, 1988.

[16] Ronse, C., "Why Mathematical Morphology Needs Complete Lattices," *Sign. Proc.* **21**, 129–154, 1990.

[17] Ronse, C., "Lattice-Theoretical Fixpoint Theorems in Morphological Image Filtering," *J. Math. Imaging and Vision* **4**, 19–41, 1994.

[18] Ronse, C., "Openings: Main Properties, and How to Construct Them," in *Mathematical Morphology: Theory and Hardware*, R. M. Haralick, ed., Oxford University Press, in press.

[19] Ronse, C., and H. J. A. M. Heijmans, "The Algebraic Basis of Mathematical Morphology – Part II: Openings and Closings," *CVGIP: Image Understanding* **54**, 74–97, 1991.

[20] Ronse, C., and H. J. A. M. Heijmans, "A Lattice-Theoretical Framework for Annular Filters in Morphological Image Processing," *Applicable Algebra in Engineering, Communication, and Computing* **9**, 45–89, 1998.

[21] Schonfeld, D., and J. Goutsias, "Optimal Morphological Pattern Restoration from Noisy Binary Images," *IEEE Trans. Patt. Anal. Mach. Intell.* **13**, 14–29, 1991.

[22] Serra, J., *Image Analysis and Mathematical Morphology*, Academic Press, London, 1982.

[23] Serra, J., ed., *Image Analysis and Mathematical Morphology*, Vol. II: *Theoretical Advances*, Academic Press, London, 1988.

[24] Serra, J., and L. Vincent, "On Overview of Morphological Filtering," *IEEE Trans. Circ. Sys. Sign. Proc.* **11**, 47–108, 1992.

CHAPTER 6

INTRODUCTION TO CONNECTED OPERATORS

Henk J. A. M. Heijmans
CWI
Amsterdam, The Netherlands

6.1 INTRODUCTION

Connectivity, in all its manifestations, has always been an important notion in the field of image processing [17]. This is even more true for methods from mathematical morphology because of their intrinsic topological and geometrical nature. A simple but extremely important instance of a morphological operation based on connectivity is the *reconstruction* of a point marker inside a set by successive dilations [21]. This reconstruction operator forms the basis for an approach in mathematical morphology which goes under the name *geodesic methods* [11, 12]. Such methods rely upon the connectivity of the underlying space; in the discrete case, one obtains a connectivity by imposing a graph structure. Another important step toward a systematic study of connected morphological operators was made by Matheron and Serra [22, Chap. 7] who discussed openings and closings and, more generally, strong filters that are connected (in a specific sense). Some years later, the first systematic studies on connected operators by Serra, Salembier, Crespo, Schafer, and others [5, 8, 13, 18, 20, 24] appeared in the literature. A major impetus to the current research on connected operators was given by the work of Vincent, providing for the first time efficient algorithms for gray-scale reconstruction [25, 26] and the area opening [27].

A connected operator is an operator that acts on the level of the flat zones of an image, rather than on the level of individual pixels. By flat zone we mean a maximal connected region where the gray-level is constant. In the binary case this means that such an operator cannot break connected components (grains) of the foreground or the background. Connected operators cannot introduce new discontinuities and as such they are eminently suited for applications where contour information is important. Image segmentation is such an application. In morphology, the main approach toward segmentation is provided by the watershed algorithm. However, this algorithm, when it is applied to an image, or rather its gradient, usually gives rise to a dramatic oversegmentation. To circumvent this problem, one might modify the image using an appropriate set of markers. Connected operators have proved to be useful for determining such markers automatically [5, 20]. Motion is another major field where connected operators have proved their usefulness. The reader is referred to [13, 19] for further details.

The goal of this chapter is to introduce the reader to the relatively new area of connected morphological operators. Apart from Section 6.6, where we treat gray-scale images, we are exclusively concerned with binary images on the two-dimensional square grid provided with 8-connectivity. In Section 6.2 we discuss the reconstruction operator and introduce the notion of *partition* associated with a binary image. This notion is used in Section 6.3 to give a formal definition of a connected operator. We present some elementary properties as well as some methods for their construction. In Section 6.3 we also introduce the concept of a *zonal graph*, also known in the literature as the *region adjacency graph*. The zonal graph concept enables a rather intuitive interpretation of connected operators on the one hand and efficient implementation of such operators on the other. An important class of connected operators is formed by the so-called *grain operators* introduced in Section 6.4. The effect of a grain operator at a given point depends exclusively on the foreground or background component to which this point belongs. In other words, grain operators use only local (or rather, regional) information. The basic examples are the area opening and closing. In Section 6.5 we show, among other things, that every grain operator is uniquely determined by two *criteria*, one for the foreground and one for the background components. As an illustration, we present a class of self-dual morphological filters which generalize the *annular filter* defined in Chapter 5. Up to this point, this chapter is exclusively concerned with binary images. In Section 6.6 we briefly discuss extensions to gray-scale images; particular attention will be given to the area opening and its implementation based on the zonal graph representation. We illustrate, by means of an example, the use of connected operators in image segmentation. We conclude with some final remarks in Section 6.7.

Some remarks about notation are in order. As we said earlier, to a large extent this paper will be concerned with binary two-dimensional images. In other words, our image space is $\mathcal{P}(\mathbb{Z}^2)$, the collection of all subsets of $\mathbb{Z}^2$. If $X \subseteq \mathbb{Z}^2$ and $h \in \mathbb{Z}^2$, then $X(h) = 1$ if $h \in X$ and $X(h) = 0$, otherwise. In other words, $X(\cdot)$ denotes the indicator function. If S is a statement then $[S]$ denotes the Boolean value (0 or 1) indicating whether S is true or false. Thus we can write $[h \in X]$ instead of $X(h)$. For other unknown notation and terminology, the reader may refer to Chapter 5.

6.2 Connectivity and Reconstruction

By $\mathcal{C}$ we denote the subcollection of all subsets of $\mathbb{Z}^2$ which are 8-connected. The family $\mathcal{C}$ is a typical example of a *connectivity class*; see also Section 6.7.

A set $X \in \mathcal{C}$ is called *connected*. Every set X is the disjoint union of the *connected components*, henceforth called *grains*, contained in X. We write $C \Subset X$ if C is a grain of X. We introduce the following notation: if $h \in \mathbb{Z}^2$, then

$$\gamma_h(X) = \begin{cases} \text{grain of } X \text{ which contains } h, & \text{if } h \in X, \\ \emptyset, & \text{if } h \notin X. \end{cases}$$

In Section 6.7 we will give a formal definition of γ_h for arbitrary connectivity classes. There, also the following result will be proved.

PROPOSITION 6–1. *γ_h is an opening, for every $h \in \mathbb{Z}^2$.*

We refer to γ_h as the *connectivity opening*. There is a simple algorithm to compute $\gamma_h(X)$ when h and X are given. In fact, it uses the notion of *reconstruction* which we describe now.

If $X, Y \subseteq \mathbb{Z}^2$, then $\rho(Y \mid X)$, the *reconstruction of Y with respect to X*, is the union of all grains of X that intersect with Y. Alternatively, we can write:

$$\rho(Y \mid X) = \bigcup_{h \in Y} \gamma_h(X). \qquad (6\text{–}1)$$

In most practical cases, Y is a subset of X. As a matter of fact, it is obvious that $\rho(Y \mid X) = \rho(Y \cap X \mid X)$. If $Y \cap X = \emptyset$, then $\rho(Y \mid X) = \emptyset$. From the expression in Eq. 6–1 one easily gets the following result.

PROPOSITION 6–2. *For every collection Y_i, $i \in I$, in $\mathcal{P}(\mathbb{Z}^2)$ one has*

$$\rho\Big(\bigcup_{i \in I} Y_i \mid X\Big) = \bigcup_{i \in I} \rho(Y_i \mid X),$$

in other words, the mapping $Y \mapsto \rho(Y \mid X)$ is a dilation, for a fixed set $X \subseteq \mathbb{Z}^2$.

The reconstruction $\rho(Y \mid X)$ can be computed easily by means of the following propagation algorithm:

$Q = \emptyset$; $R = Y \cap X$;
`while` $Q \neq R$ do {
    $Q = R$;
    $R = (Q \oplus B) \cap X$
}
$\rho(Y \mid X) = R$

Here B is the 3×3 square. The algorithm is illustrated in Fig. 6–1.

The sets X and Y in $\rho(Y \mid X)$ are called the *mask* (*image*) and *marker* (*image*), respectively. Obviously, $\gamma_h(X) = \rho(\{h\} \mid X)$, meaning that the opening γ_h can be computed with the aid of the algorithm given above.

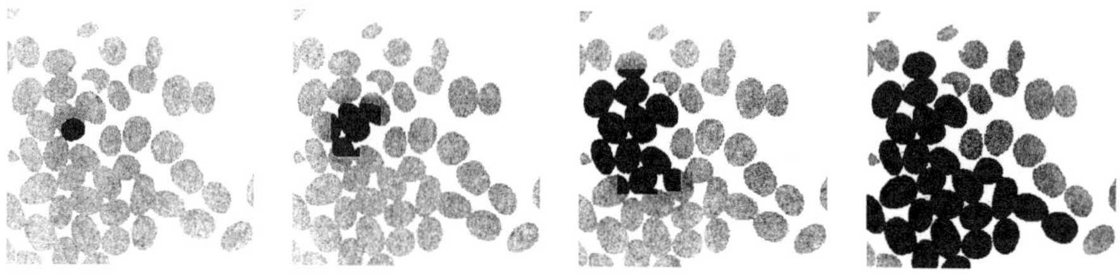

Figure 6–1. Reconstruction algorithm. From left to right: the mask image X (gray) and the marker image Y (black); 15 iterations; 50 iterations; final result $\rho(Y \mid X)$.

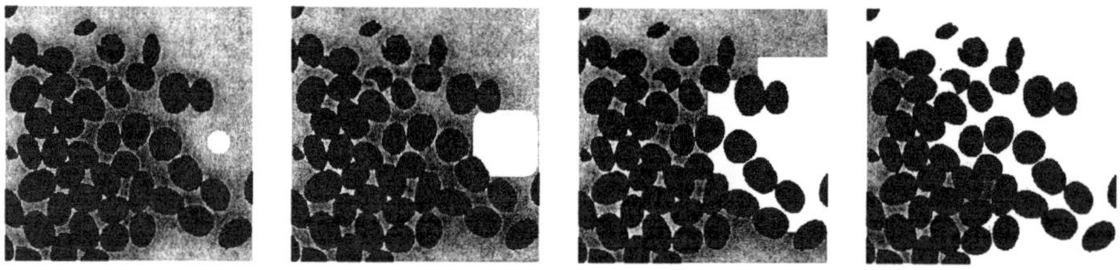

Figure 6–2. Dual reconstruction algorithm. From left to right: the mask image X (black) and the marker image Y (gray and black); 20 iterations; 75 iterations; final result $\rho^*(Y \mid X)$ (gray and black).

The reconstruction operator yields a reconstruction of the foreground. Instead, we can also perform a reconstruction of the background. We call the resulting operator the *background reconstruction* or *dual reconstruction*, and denote it by ρ^*:

$$\rho^*(Y \mid X) = \left[\rho\left(Y^c \mid X^c\right)\right]^c.$$

Now the converse of Proposition 6–2 holds, namely

$$\rho^*\left(\bigcap_{i \in I} Y_i \mid X\right) = \bigcap_{i \in I} \rho^*(Y_i \mid X),$$

that is, the mapping $Y \mapsto \rho^*(Y \mid X)$ is an erosion. The dual reconstruction is illustrated in Fig. 6–2.

Observe that

$$\rho(Y \mid X) \subseteq X \subseteq \rho^*(Y \mid X),$$

for any two sets $X, Y \subseteq \mathbb{Z}^2$.

By a *partition* of the space $\mathbb{Z}^2$ we mean a subdivision of this space into disjoint parts. A partition can be represented by a function $P : \mathbb{Z}^2 \to \mathcal{P}(\mathbb{Z}^2)$ which has the following properties:

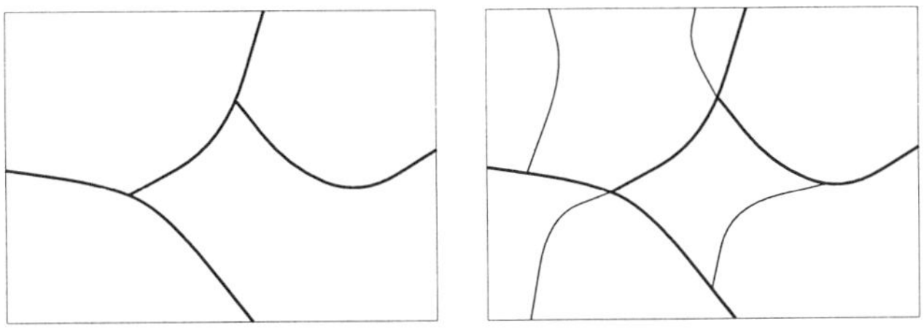

Figure 6–3. The partition at the left is coarser than the one at the right.

- $x \in P(x)$
- $P(x) = P(y)$ or $P(x) \cap P(y) = \emptyset$, for any two points $x, y \in \mathbb{Z}^2$.

Thus, $P(x)$ is the part of the partition that contains the point x. The family of all partitions of $\mathbb{Z}^2$ forms a complete lattice [22] under the partial ordering given by

$$P \sqsubseteq P' \quad \text{if } P'(h) \subseteq P(h), \text{ for every } h \in \mathbb{Z}^2.$$

We say that P is *coarser* than P', or that P' is *finer* than P. Fig. 6–3 shows an example.

A partition P is said to be *connected* if every part $P(h)$ is a connected set. Every binary image $X \subseteq \mathbb{Z}^2$ yields a unique connected partition $P(X)$, where the parts of $P(X)$ are the grains of X and X^c. Writing $P(X, h) = P(X)(h)$, we have

$$P(X, h) = \begin{cases} \gamma_h(X), & \text{if } h \in X \\ \gamma_h(X^c), & \text{if } h \notin X^c. \end{cases}$$

It is evident that $P(X) = P(X^c)$, for every set X. Note, however, that most partitions of $\mathbb{Z}^2$ are not of the form $P(X)$, with X a subset of $\mathbb{Z}^2$. This applies in particular for the partitions depicted in Fig. 6–3.

6.3 CONNECTED OPERATORS

By $X \setminus Y$ we denote the set difference of X and Y. Furthermore $X \triangle Y$ denotes the symmetric difference.

DEFINITION 6–1. An operator ψ on $\mathcal{P}(\mathbb{Z}^2)$ is *connected* if the partition $P(\psi(X))$ is coarser than $P(X)$, for every set $X \subseteq \mathbb{Z}^2$.

Below we will present an alternative formulation of this property. We start with some simple examples. Obviously, the identity operator id as well as the complement operator $X \to X^c$ are connected. Furthermore, every connectivity opening γ_h is connected.

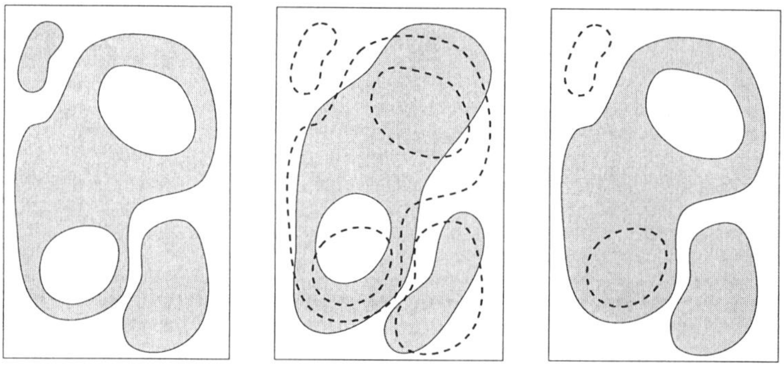

Figure 6–4. A connected operator applied to the left image can result in the image at the right but not in the one in the middle.

A connected operator acts on the grains of the foreground and background in an all-or-nothing way: either the grain is left untouched or deleted altogether. This means in particular that the borders in the image cannot be broken or changed, but only deleted. An illustration is given in Fig. 6–4: the middle image cannot be the output of a connected operator applied to the image at the left. However, the right image may result from a connected operator.

PROPOSITION 6–3. *An operator ψ is connected if and only if $X \Delta \psi(X)$ consists of grains of X and X^c, for every $X \subseteq \mathbb{Z}^2$.*

PROOF. "only if": assume that ψ is connected; then $P(\psi(X))$ is coarser than $P(X)$. We must prove that for every $h \in X\Delta\psi(X)$, the entire part $P(X, h)$ lies in $X\Delta\psi(X)$. We have to consider two cases: $h \in X$ and $h \notin X$.

$h \in X$: thus $h \notin \psi(X)$. Then $P(X,h) \subseteq P(\psi(X),h)$ leads to $\gamma_h(X) \subseteq \gamma_h(\psi(X)^c)$. But this means that $\gamma_h(X) \subseteq X\Delta\psi(X)$.

$h \notin X$: then $h \in \psi(X)$, and $P(X,h) \subseteq P(\psi(X),h)$ leads to $\gamma_h(X^c) \subseteq \gamma_h(\psi(X))$. That is, $\gamma_h(X^c) \subseteq X\Delta\psi(X)$.

"if": let $X \subseteq \mathbb{Z}^2$, we must show that $P(X,h) \subseteq P(\psi(X),h)$, for every $h \in \mathbb{Z}^2$. Again, we must distinguish between the cases $h \in X$ and $h \notin X$. We consider only the first case; the second is treated analogously. If $h \in X$, then $P(X,h) = \gamma_h(X)$. We must show that $\gamma_h(X) \subseteq P(\psi(X),h)$. Suppose $\gamma_h(X) \not\subseteq \psi(X)$; then there is a point k such that $k \in \gamma_h(X)$ and $k \notin \psi(X)$. Now $k \in X\Delta\psi(X)$, which yields that $\gamma_k(X) \subseteq X\Delta\psi(X)$. However, $\gamma_k(X) = \gamma_h(X)$, whence we conclude that $\gamma_h(X) \subseteq \psi(X)^c$, and thus $\gamma_h(X) \subseteq \gamma_h(\psi(X)^c) = P(\psi(X),h)$. ■

In fact, the condition that $X\Delta\psi(X)$ consists of grains of X and X^c, consists of two parts, namely that $X \setminus \psi(X)$ consists of grains of X, and that $\psi(X) \setminus X$ consists of grains of X^c.

PROPOSITION 6–4. *An operator ψ is connected if and only if its negative ψ^* is connected.*

PROOF. Assume that ψ is connected; then $P(\psi(X)) \sqsubseteq P(X)$, for every $X \subseteq \mathbb{Z}^2$. Substituting X^c yields that

$$P\big(\psi(X^c)\big) \sqsubseteq P\big(X^c\big).$$

Using that $P(\psi^*(X)) = P(\psi(X^c)^c) = P(\psi(X^c))$, and that $P(X^c) = P(X)$, we get that

$$P\big(\psi^*(X)\big) \sqsubseteq P(X).$$

This proves the result. ■

We present a number of results that show how to build new connected operators from known ones using composition, supremum, infimum, as well as other "Boolean combinations."

PROPOSITION 6–5. *If ψ_1, ψ_2 are connected, then $\psi_2\psi_1$ is connected, too.*

PROOF. This result is a simple consequence of Definition 6–1:

$$P\big(\psi_2\psi_1(X)\big) \sqsubseteq P\big(\psi_1(X)\big) \sqsubseteq P(X),$$

for every $X \subseteq \mathbb{Z}^2$, if ψ_1 and ψ_2 are connected. ■

PROPOSITION 6–6. *If ψ_i is a connected operator for every i in some index set I, then the infimum $\bigwedge_{i \in I} \psi_i$ and the supremum $\bigvee_{i \in I} \psi_i$ are connected, too.*

PROOF. We first prove the result for the infimum. Let $\psi = \bigwedge_{i \in I} \psi_i$, then $X \setminus \psi(X) = \bigcup_{i \in I}(X \setminus \psi_i(X))$ and $\psi(X) \setminus X = \bigcap_{i \in I}(\psi_i(X) \setminus X)$. As every set $X \setminus \psi_i(X)$ is a union of grains of X, $X \setminus \psi(X)$ is also a union of grains of X. Similarly, we have that $\psi(X) \setminus X$ is a union of grains of X^c, and we conclude that ψ is connected.

The result for the supremum can be obtained analogously, but it also follows from the observation that

$$\bigvee_{i \in I} \psi_i = \Big(\bigwedge_{i \in I} \psi_i^*\Big)^*,$$

in combination with Proposition 6–4. ■

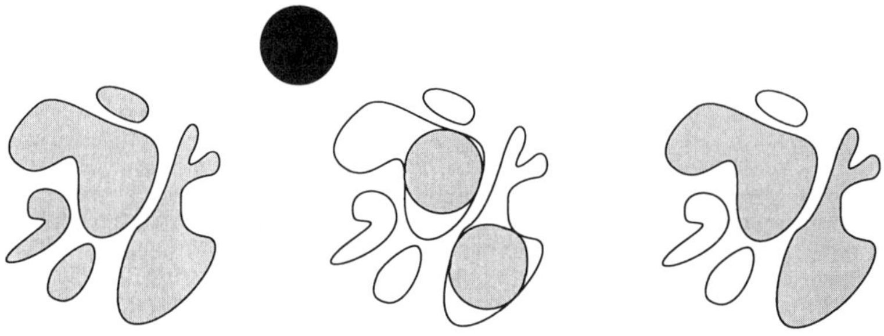

Figure 6–5. Opening by reconstruction: the original opening is an opening by a disk (in black). From left to right: X, $\alpha(X)$, and $\breve{\alpha}(X)$.

Recall that a *Boolean function* (of n variables) is a function $b : \{0, 1\}^n \to \{0, 1\}$. Given a Boolean function b and n operators $\psi_1, \ldots, \psi_n$ on $\mathcal{P}(\mathbb{Z}^2)$, we can define a new operator

$$\psi = b(\psi_1, \ldots, \psi_n)$$

as follows:

$$\psi(X)(h) = b\big(\psi_1(X)(h), \ldots, \psi_n(X)(h)\big);$$

here $X(h)$ equals 1 if $h \in X$ and 0 otherwise. For example, if $b(u_1, \ldots, u_n) = u_1 \cdot u_2 \cdot \cdots \cdot u_n$, then $b(\psi_1, \ldots, \psi_n) = \psi_1 \wedge \cdots \wedge \psi_n$.

PROPOSITION 6–7. *Given a Boolean function b of n variables and n connected operators $\psi_1, \psi_2, \ldots, \psi_n$, then the operator $\psi = b(\psi_1, \psi_2, \ldots, \psi_n)$ is connected as well.*

PROOF. The proof becomes obvious by the observation that the value of $\psi_i(X)(h)$ is constantly 0 or 1 on parts of the partition $P(X)$ (this value only depending on i). As a result, $\psi(X)(h)$ is constant on parts of $P(X)$, too, and therefore ψ is a connected operator. ■

An opening which is a connected operator is called a *connected opening* (same for closings). One can construct connected openings by starting with arbitrary openings and performing a reconstruction afterward: let α be an opening on $\mathcal{P}(\mathbb{Z}^2)$ and define

$$\breve{\alpha}(X) = \rho(\alpha(X) \mid X). \tag{6–2}$$

In Fig. 6–5 we show an example; there $\alpha(X) = X \circ B$, where B is a disk.

PROPOSITION 6–8. *If α is an opening, then $\breve{\alpha}$ is a connected opening. Moreover, α is a connected opening if and only if $\alpha = \breve{\alpha}$.*

Figure 6–6. Opening by reconstruction. From left to right: the original image X (gray); the opening $\alpha(X)$ by a horizontal line segment (length is 40 pixels); the reconstruction $\breve{\alpha}(X) = \rho(\alpha(X) \mid X)$.

PROOF. Assume α is an opening; we show that $\breve{\alpha}$ is an opening, too. It is obvious that $\breve{\alpha}$ is increasing. It is also obvious that $\alpha(X) \subseteq \breve{\alpha}(X) \subseteq X$. Therefore $\breve{\alpha}^2 \leqslant \breve{\alpha}$. We must show that $\breve{\alpha}^2 \geqslant \breve{\alpha}$:

$$\begin{aligned}\breve{\alpha}^2(X) &= \rho\big(\alpha\breve{\alpha}(X) \mid \breve{\alpha}(X)\big) \supseteq \rho\big(\alpha^2(X) \mid \breve{\alpha}(X)\big) \\ &= \rho\big(\alpha(X) \mid \breve{\alpha}(X)\big) = \bigcup_{h \in \alpha(X)} \gamma_h\big(\breve{\alpha}(X)\big).\end{aligned}$$

Using that $\gamma_h(\breve{\alpha}(X)) = \gamma_h(X)$ for $h \in \alpha(X)$, we get that

$$\breve{\alpha}^2(X) \supseteq \bigcup_{h \in \alpha(X)} \gamma_h(X) = \breve{\alpha}(X).$$

This proves that $\breve{\alpha}$ is an opening. By definition, $\breve{\alpha}(X)$ is a union of grains of X, hence $\breve{\alpha}$ is a connected operator.

Conversely, assume that α is a connected opening, hence $X \Delta \alpha(X) = X \setminus \alpha(X)$ is a union of grains of X. But this yields that $\alpha(X)$ is a union of grains of X, too. Therefore, $\alpha = \breve{\alpha}$. ■

For closings β we define $\breve{\beta}(X) = \rho^*(\beta(X) \mid X)$, and we can prove the dual statement of the proposition above. Note that the following duality relations hold:

$$\big(\alpha^*\big)^{\vee} = \big(\breve{\alpha}\big)^* \quad \text{and} \quad \big(\beta^*\big)^{\vee} = \big(\breve{\beta}\big)^*.$$

Fig. 6–6 shows an example where α is an opening with a horizontal line segment.

To get additional insight into the way connected operators behave, we introduce the concept of a *zonal graph*, sometimes called *region adjacency graph* in the literature [1]. At this point we restrict attention to the binary case. In Section 6.6 we will briefly discuss gray-scale images.

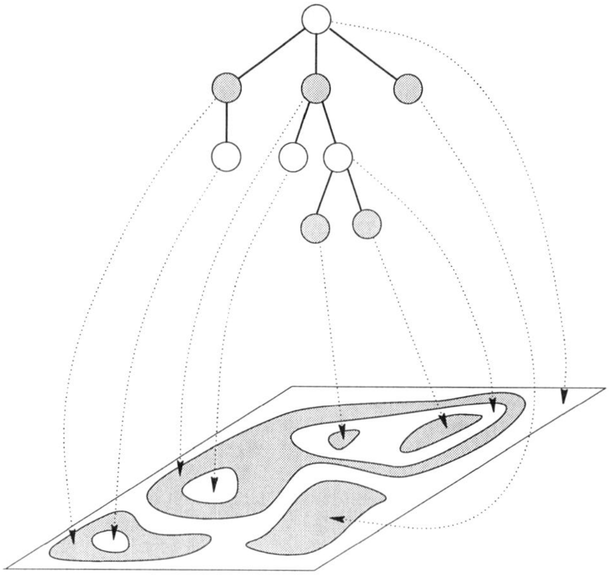

Figure 6–7. Zonal graph associated with a binary image.

As we have seen, a binary image $X \in \mathcal{P}(\mathbb{Z}^2)$ yields a colored partition (with colors 0 and 1) of the underlying space $\mathbb{Z}^2$. Now, consider the parts of $P(X)$ as the vertices of a graph. Two parts $C_1, C_2 \in P(X)$ are said to be *adjacent*, denoted by $C_1 \sim C_2$, if $C_1 \cup C_2$ is a connected set. To determine X from this graph, we have to specify for every vertex whether it is a subset of X or X^c. To this end we use the coloring $I_X : P(X) \to \{0, 1\}$:

$$I_X(C) = \begin{cases} 1, & \text{if } C \Subset X \\ 0, & \text{if } C \Subset X^c. \end{cases} \tag{6–3}$$

Note that because two adjacent vertices must have different colors, it suffices to specify the color of only one vertex; this, however, is no longer true in the gray-scale case.

The triple $(P(X), \sim, I_X)$ is called the *zonal graph* associated with X; see Fig. 6–7 for an illustration.

In the case considered here ($\mathbb{Z}^2$ with 8-connectivity), the following result holds (recall that a *tree* is a graph without cycles [2]):

PROPOSITION 6–9. *The graph $(P(X), \sim)$ is a tree, for every set $X \subseteq \mathbb{Z}^2$.*

A connected operator on $\mathcal{P}(\mathbb{Z}^2)$ amounts to a recoloring of the zonal graph associated with the set $X \subseteq \mathbb{Z}^2$. For example, the opening γ_h gives color 1 to the vertex

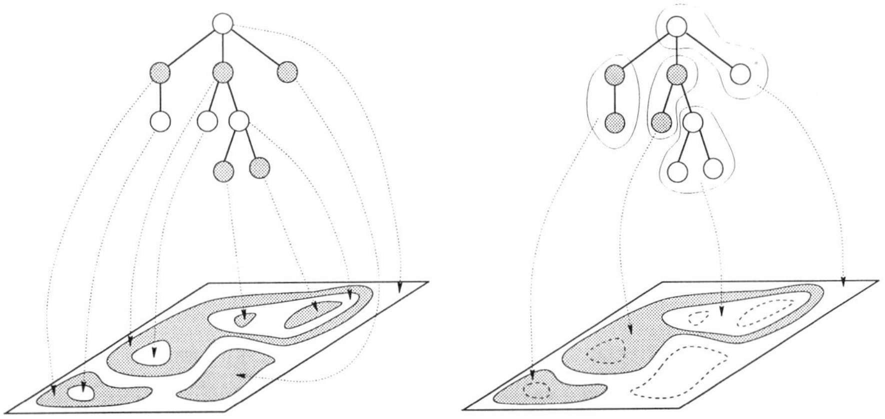

Figure 6–8. The leaves of the tree receive the color of their neighbor.

$C \in P(X)$ which satisfies $h \in C$ and $I_X(C) = 1$ (note that such a vertex exists if and only if $h \in X$), and 0 to all other vertices.

The converse is also true: every recoloring, followed by a merging of vertices with the same color, defines a connected operator. For precise statements we refer to future publications; here we only present an example.

Recall that a vertex in a tree is called a *leaf* if it possesses exactly one neighbor. For example, the tree in Fig. 6–7 contains 5 leaves. We define a recoloring as follows: the color at the leaves is flipped (from 0 to 1 and vice versa), but the colors of the other vertices are left unaltered. We apply this recoloring to the zonal graph depicted in Fig. 6–7, and merge adjacent vertices with the same color. The outcome is depicted in Fig. 6–8. The operator associated with this recoloring is connected (and self-dual).

6.4 GRAIN OPERATORS

This section discusses a particular class of connected operators, the so-called grain operators. We point out that Crespo and Schafer [7] call such operators *connected-component local operators*. Some of the basic results presented below [e.g., Proposition 6–10(a) and (c)] can also be found in their work.

DEFINITION 6–2. A connected operator $\psi : \mathcal{P}(\mathbb{Z}^2) \to \mathcal{P}(\mathbb{Z}^2)$ is called a *grain operator* if it has the following property: if $h \in \mathbb{Z}^2$ and $X, Y \subseteq \mathbb{Z}^2$ are such that $X(h) = Y(h)$ and $P(X, h) = P(Y, h)$, then $\psi(X)(h) = \psi(Y)(h)$.

Thus given a grain operator ψ, the value $\psi(X)(h)$ is completely determined by the information provided by $X(h)$ and $P(X, h)$, which is exactly the information

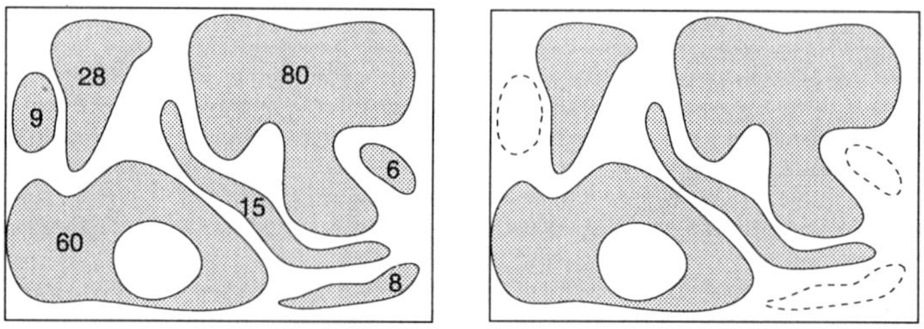

Figure 6–9. Area opening α_{10}. The numbers printed inside the grains represent their area.

"stored" at the vertex $P(X, h)$ of the associated zonal graph. In other words, the recoloring of the zonal graph corresponding with a grain operator is based entirely upon local information stored at individual vertices; all other information, e.g., about adjacent vertices, is irrelevant. For example, the recoloring considered in the last paragraph of the previous section, where the colors of the leaves are flipped, does not correspond to a grain operator. Indeed, in this case, the evaluation of a vertex being a leaf or not requires information about its neighbors: Is there one or more than one neighbor?

The identity operator, the complement operator, and the connectivity openings are grain operators. We discuss another very important example.

EXAMPLE 6–1 (Area opening). If C is a component of X, then area(C) denotes the area of C, i.e., the number of pixels. If C is unbounded, then area$(C) = +\infty$. The *area opening* α_S is the operator that deletes all grains from a set X with an area smaller than a given threshold S. Thus

$$\alpha_S(X) = \bigcup \{C \mid C \Subset X \text{ and } \text{area}(C) \geqslant S\}.$$

It is obvious that α_S is a grain operator, and that it is increasing, anti-extensive, and idempotent, hence an opening. The area opening (with $S = 10$) is illustrated in Fig. 6–9.

A grain operator which is also an opening (closing, filter) is called a *grain opening* (closing, filter). The next result states some basic properties of grain operators.

PROPOSITION 6–10.

(a) *If ψ_i is a grain operator for every $i \in I$, then $\bigvee_{i\in I} \psi_i$ and $\bigwedge_{i\in I} \psi_i$ are grain operators.*

(b) *If $\psi_1, \psi_2, \ldots, \psi_n$ are grain operators and b is a Boolean function of n variables, then $b(\psi_1, \psi_2, \ldots, \psi_n)$ is a grain operator.*

(c) *If ψ is a grain operator, then ψ^* is a grain operator.*

PROOF. (a) We show that $\psi = \bigvee_{i\in I} \psi_i$ is a grain operator. The proof for the infimum is analogous. Let h, X, Y be such that $X(h) = Y(h)$ and $P(X,h) = P(Y,h)$; we show that $\psi(X)(h) = \psi(Y)(h)$. Obviously,

$$\psi(X)(h) = \Big(\bigvee_{i\in I} \psi_i(X)\Big)(h) = \bigvee_{i\in I} \psi_i(X)(h).$$

Since every ψ_i is a grain operator, this last expression equals

$$\bigvee_{i\in I} \psi_i(Y)(h) = \Big(\bigvee_{i\in I} \psi_i(Y)\Big)(h) = \psi(Y)(h),$$

and this shows the result.

(b) Let $\psi = b(\psi_1, \psi_2, \ldots, \psi_n)$. By definition

$$\psi(X)(h) = b\big(\psi_1(X)(h), \psi_2(X)(h), \ldots, \psi_n(X)(h)\big),$$

and with this relation, the proof becomes similar to the proof of (a).

(c) This follows easily if one uses the relation $\psi^*(X)(h) = 1 - \psi(X^c)(h)$. ■

The composition of two grain operators, however, is not a grain operator, in general. In Fig. 6–10 we depict an example of a grain operator ψ for which ψ^2 is not a grain operator. The operator acts as follows: for every part of the partition $P(X)$ it switches the color from 1 to 0, or vice versa, if the area of this part is below a given threshold (20 in this specific example). The value $\psi^2(X)(h)$ is different in the upper and lower figure, although $X(h) = Y(h) = 1$ and $P(X,h) = P(Y,h)$. Therefore, ψ^2 is not a grain operator.

By the duality principle (see Chapter 5), every statement about openings has a dual version concerning closings. In what follows, we restrict ourselves to openings.

PROPOSITION 6–11. *Every anti-extensive grain operator is idempotent. In particular, an increasing grain operator is an opening if and only if it is anti-extensive.*

PROOF. Let ψ be an anti-extensive grain operator. We must show that ψ is idempotent. The nontrivial part of the proof consists of showing that $\psi^2 \geqslant \psi$. Assume that $h \in \psi(X)$; we must show that $h \in \psi^2(X)$. Put $Y = \psi(X)$, then $X(h) = Y(h) = 1$. This implies that

$$P(Y,h) = \gamma_h(Y) \subseteq \gamma_h(X) = P(X,h).$$

On the other hand,

$$P(X,h) \subseteq P(Y,h),$$

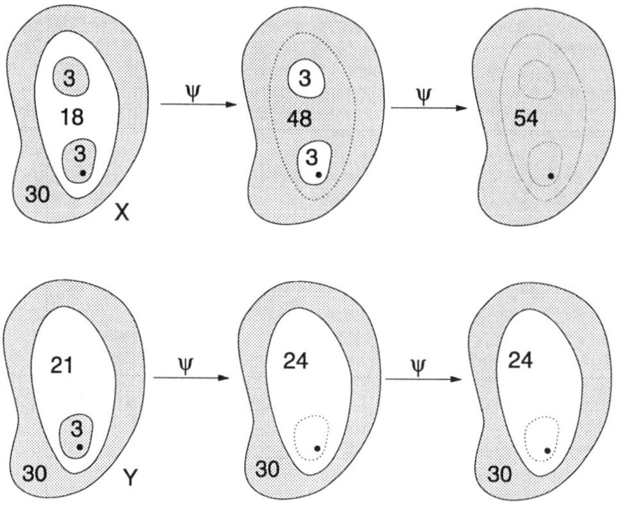

Figure 6–10. The operator ψ that changes the color of parts with areas less than 20 is a grain operator, but ψ^2 is not. The black point indicates the location of the point h.

since ψ is a connected operator, hence $P(Y)$ is coarser than $P(X)$. Thus we find that $P(X, h) = P(Y, h)$. Using that ψ is a grain operator, we get that

$$\psi(X)(h) = \psi(Y)(h) = 1.$$

This proves the result. ∎

In the previous section, we defined the opening by reconstruction; see Eq. 6–2. An interesting question is the following: For which openings α is the opening by reconstruction $\breve{\alpha}$ a grain opening? Under the extra assumption that α is a structural opening [10, 16], we can give a complete characterization.

PROPOSITION 6–12. *Let α be the structural opening given by $\alpha(X) = X \circ B$. Then $\breve{\alpha}$ is a grain opening if and only if B is connected.*

PROOF. Assume first that B is not connected and let Y be a grain of B. Furthermore, pick $h \in Y$ and put $X = B$. Then $X(h) = Y(h) = 1$ and $P(X, h) = P(Y, h) = Y$. Obviously, $\alpha(X) = X$ and $\alpha(Y) = \emptyset$, hence $\breve{\alpha}(X) = X$ and $\breve{\alpha}(Y) = \emptyset$. Thus $\breve{\alpha}$ is not a grain operator, as this would imply that $\breve{\alpha}(X)(h) = \breve{\alpha}(Y)(h)$.

Assume now that B is connected; we show that $\breve{\alpha}$ is a grain operator. It is easy to see that $\breve{\alpha}(X) = \bigcup\{C \mid C \Subset X \text{ and } C \circ B \neq \emptyset\}$. From this, we derive that

$$\breve{\alpha}(X)(h) = X(h) \wedge [P(X, h) \circ B \neq \emptyset],$$

whence it follows immediately that $\breve{\alpha}$ is a grain opening. ∎

PROPOSITION 6–13. *An opening* α *is a grain opening if and only if* $\alpha = \bigvee_{h\in\mathbb{Z}^2} \alpha\gamma_h$.

PROOF. Assume that α is a grain opening. It is trivial that $\alpha \geqslant \bigvee_{x\in\mathbb{Z}^2} \alpha\gamma_x$; therefore, we only have to show that $\alpha \leqslant \bigvee_{x\in\mathbb{Z}^2} \alpha\gamma_x$. Suppose that $h \in \alpha(X)$. Define $Y = \gamma_h(X)$, then $X(h) = \gamma_h(X)(h) = 1$. Furthermore, $P(X,h) = \gamma_h(X) = P(Y,h)$. Since α is a grain operator, we get that $\alpha(X)(h) = \alpha(Y)(h)$, and this implies that $h \in \alpha(Y) = \alpha\gamma_h(X) \subseteq \bigvee_{x\in\mathbb{Z}^2} \alpha\gamma_x(X)$.

To prove the converse, assume that $\alpha = \bigvee_{x\in\mathbb{Z}^2} \alpha\gamma_x$; we show that α is a grain operator. Suppose that $X(h) = Y(h)$ and $P(X,h) = P(Y,h)$. We must show that $\alpha(X)(h) = \alpha(Y)(h)$. If $X(h) = 0$, this is trivial; we consider the case that $X(h) = 1$. Thus $P(X,h) = \gamma_h(X) = \gamma_h(Y)$. Now

$$\begin{aligned}\alpha(X)(h) &= \Big(\bigvee_{x\in\mathbb{Z}^2} \alpha\gamma_x(X)\Big)(h) = \bigvee_{x\in\mathbb{Z}^2} \alpha\gamma_x(X)(h)\\ &= \alpha\gamma_h(X)(h) = \alpha\gamma_h(Y)(h)\\ &= \alpha(Y)(h).\end{aligned}$$

This concludes the proof. ■

Recall that a set B is *invariant* under an operator ψ if $\psi(B) = B$. We will show that any subset of grains of a set X invariant under a grain filter ψ is also invariant under ψ. Refer to [22, Chap. 7] for some related results. We start with a lemma.

LEMMA 6–1. *Let* ψ *be an increasing connected operator on* $\mathcal{P}(\mathbb{Z}^2)$*, and let* $X \subseteq \mathbb{Z}^2$ *satisfy* $\psi(X) \subseteq X$. *If* Y *is a union of grains of* X*, then* $\psi(Y) \subseteq Y$.

PROOF. Suppose that $\psi(Y) \nsubseteq Y$. Since ψ is connected, $\psi(Y) \setminus Y$ consists of grains of Y^c. Let D be a grain of Y^c contained in $\psi(Y) \setminus Y$. We show that $D \cap X^c \neq \emptyset$. Suppose that $D \subseteq X$. The grain D must be adjacent to a grain C of Y, meaning that $C \cup D$ is connected. However, $C \cup D \subseteq X$, and we conclude that C cannot be a grain of X. But this contradicts our assumption that Y consists of grains of X. Thus $D \cap X^c \neq \emptyset$.

Since $D \subseteq \psi(Y)$ and ψ is increasing, also $D \subseteq \psi(X)$. This yields that $\psi(X) \cap X^c \neq \emptyset$, i.e., $X \cap X^c \neq \emptyset$, a contradiction. We conclude that $\psi(Y) \subseteq Y$, as asserted. ■

PROPOSITION 6–14. *Let* ψ *be a grain filter and* $\psi(X) = X$.

(a) *If* Y *is a union of grains of* X*, then* $\psi(Y) = Y$.

(b) *If* Y *is a union of grains of* X^c*, then* $\psi(Y^c) = Y^c$.

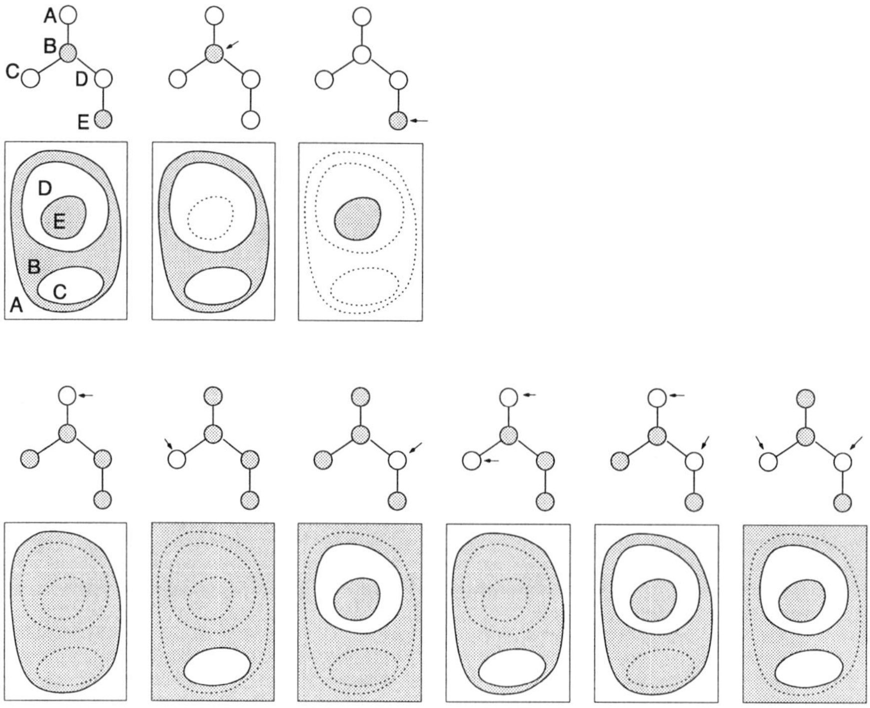

Figure 6–11. The left figure in the first row shows a set invariant with respect to some grain filter ψ. Proposition 6–14 states that the other sets (in gray) shown in this figure are invariant, too.

PROOF. (a) From the previous lemma we conclude that $\psi(Y) \subseteq Y$. We show that $Y \subseteq \psi(Y)$. Take $h \in Y$, then $X(h) = Y(h) = 1$. Furthermore, $P(X,h) = P(Y,h) = \gamma_h(X)$. From the fact that ψ is a grain operator we get that $\psi(X)(h) = \psi(Y)(h)$. But $\psi(X) = X$ and we conclude that $\psi(Y)(h) = X(h) = 1$, that is, $h \in \psi(Y)$. This shows that $Y \subseteq \psi(Y)$.

(b) If ψ is a grain filter, then ψ^* is a grain filter too. Furthermore, $\psi^*(X^c) = X^c$. If Y is a union of grains of X^c, then $\psi^*(Y) = Y$, by (a). But this means $\psi(Y^c) = Y^c$. ■

We illustrate this proposition by means of Fig. 6–11. The left figure at the first row shows a set X (along with its zonal graph) which is assumed to be invariant under a given grain filter. Our proposition gives us that the other sets depicted in this figure are invariant, too. The top row depicts the sets that are built of grains of X. The arrows in the zonal graph indicate which grains are used as building blocks. In the bottom row we consider sets that are built by using background grains, again indicated by arrows in the zonal graph.

6.5 GRAIN OPERATORS AND GRAIN CRITERIA

In this section we show that every grain operator is uniquely determined by two grain criteria. By a *grain criterion* we mean a mapping $u : \mathcal{C} \to \{0, 1\}$. Suppose we are given two (grain) criteria, u for the foreground and v for the background. Define an operator $\psi = \psi_{u,v}$ as follows: ψ preserves grains C of the foreground for which $u(C) = 1$ and grains C of the background for which $v(C) = 1$. In other words,

$$\psi(X) = \bigcup \big\{C \mid (C \Subset X \text{ and } u(C) = 1) \text{ or } (C \Subset X^c \text{ and } v(C) = 0)\big\}. \quad (6\text{–}4)$$

See Fig. 6–12 for an illustration.

Note that any operator so defined acts on the level of the grains, and it follows easily that ψ is a grain operator. But the converse can also be proved: every grain operator is of the form Eq. 6–4.

PROPOSITION 6–15. *An operator ψ is a grain operator if and only if there exist foreground and background criteria u, v such that $\psi = \psi_{u,v}$. The criteria u and v are given by*

$$u(C) = \big[C \subseteq \psi(C)\big]$$
$$v(C) = \big[C \subseteq \psi^*(C)\big].$$

PROOF. Let ψ be a grain operator and define u, v as above. We show that $\psi = \psi'$, where $\psi' := \psi_{u,v}$.

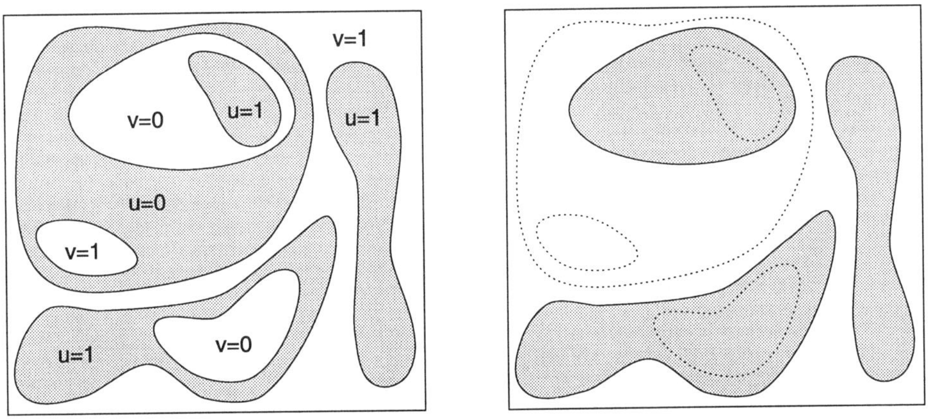

Figure 6–12. A binary image X (left) and its transform $\psi_{u,v}(X)$. In every foreground (resp. background) grain it is printed whether the grain criterion u (resp. v) equals 0 or 1.

First we show that $\psi(X) \subseteq \psi'(X)$ for every set X. Let C be a part of $P(X)$ and $C \subseteq \psi(X)$; we show that $C \subseteq \psi'(X)$. We distinguish two cases.

(1). $C \Subset X$. We show that $u(C) = 1$, for then $C \subseteq \psi'(X)$. We must prove that $C \subseteq \psi(C)$. Let $h \in C$, then $C(h) = X(h) = 1$ and $P(X, h) = P(C, h) = C$. Since ψ is a grain operator, we may conclude that $\psi(X)(h) = \psi(C)(h)$. Since $C \subseteq \psi(X)$, this expression equals 1, and we conclude that $h \in \psi(C)$. Therefore, $C \subseteq \psi(C)$.

(2). $C \Subset X^c$. We show that $C \subseteq \psi(C^c)$, which yields that $v(C) = 0$. Let $h \in C$, then $C^c(h) = X(h) = 0$. Furthermore, $P(C^c, h) = P(X, h) = C$, and using that ψ is a grain operator, we get that $\psi(X)(h) = \psi(C^c)(h)$. Since $C \subseteq \psi(X)$, this expression equals 1, and we conclude that $h \in \psi(C^c)$. Therefore, $C \subseteq \psi(C^c)$.

Next we show that $\psi'(X) \subseteq \psi(X)$. Assume that C is a part of $P(X)$ and that $C \subseteq \psi'(X)$. We show that $C \subseteq \psi(X)$. Again, we distinguish two cases.

(1). $C \Subset X$. Then $u(C) = 1$, hence $C \subseteq \psi(C)$. Let $h \in C$, then $C(h) = X(h) = 1$ and $P(X, h) = P(C, h) = C$, and, by the fact that ψ is a grain operator, we get that $\psi(X)(h) = \psi(C)(h) = 1$. This shows that $C \subseteq \psi(X)$.

(2). $C \Subset X^c$. Then $v(C) = 0$, which yields that $C \subseteq \psi(C^c)$. Let $h \in C$, then $C^c(h) = X(h) = 0$ and $P(X, h) = P(C^c, h) = C$, and we get that $\psi(X)(h) = \psi(C^c)(h)$. Since $C \subseteq \psi(C^c)$, this expression equals 1, and we conclude that $C \subseteq \psi(X)$.

This conclude our proof. ■

A criterion u is called *increasing* if $C, C' \in \mathcal{C}$ and $C \subseteq C'$ implies that $u(C) \leqslant u(C')$. One might expect that $\psi = \psi_{u,v}$ is increasing when both criteria u and v are increasing. The following example shows that this is not true in general. Let $X \subseteq Y$ be as in Fig. 6–13. Let C be a grain of X^c and $D \subseteq C$ a grain of Y. Suppose that u, v are increasing criteria (e.g., area criteria) such that $v(C) = u(D) = 0$. Then, by the increasingness of v, we have $v(C') = 0$ for the grain $C' = C \setminus D$ of Y^c, and it follows that $\psi(X) \not\subseteq \psi(Y)$.

PROPOSITION 6–16. *The grain operator $\psi_{u,v}$ is increasing if and only if both u and v are increasing criteria, and the following condition holds:*

$$u\big(\gamma_h(X \cup \{h\})\big) \vee v\big(\gamma_h(X^c \cup \{h\})\big) = 1, \tag{6–5}$$

if $X \subseteq \mathbb{Z}^2$ and $h \in \mathbb{Z}^2$.

PROOF. "if": assume that u, v satisfy the conditions above; we show that $\psi = \psi_{u,v}$ is increasing. Let $X \subseteq Y$; we must show that $\psi(X) \subseteq \psi(Y)$. Let $h \in \psi(X)$. We distinguish three cases.

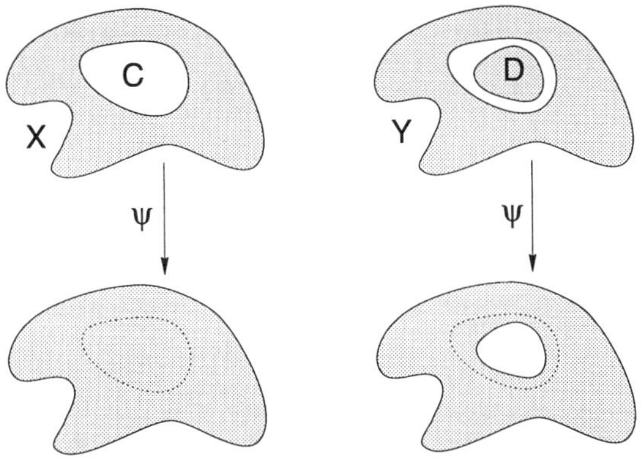

Figure 6–13. ψ is not increasing. Indeed, $X \subseteq Y$ but $\psi(X) \nsubseteq \psi(Y)$.

1. $h \in X$: put $C = \gamma_h(X)$, then $C \Subset X$ and $C \subseteq C' = \gamma_h(Y)$. As $h \in \psi(X)$, we have $u(C) = 1$, and since u is increasing $u(C') = 1$, giving that $h \in \psi(Y)$.

2. $h \notin Y$: put $C' = \gamma_h(Y^c)$ and $C = \gamma_h(X^c)$, then $C' \subseteq C$ since $Y^c \subseteq X^c$. From the fact that $h \in \psi(X)$ we conclude that $v(C) = 0$ and thus $v(C') = 0$, yielding that $h \in \psi(Y)$.

3. $h \in Y$ and $h \notin X$: suppose $h \notin \psi(Y)$, then $u(\gamma_h(Y)) = 0$. Now Eq. 6–5 implies that $v(\gamma_h(Y^c \cup \{h\})) = 1$. Obviously, $\gamma_h(Y^c \cup \{h\}) \subseteq \gamma_h(X^c)$, and since v is increasing, we get that $v(\gamma_h(X^c)) = 1$. However, this implies that the grain $\gamma_h(X^c)$ does not lie in $\psi(X)$, contradicting $h \in \psi(X)$. Thus we conclude that $h \in \psi(Y)$.

"only if": assume that $\psi = \psi_{u,v}$ is increasing. First we show that u is an increasing grain criterion. The proof that v is increasing is analogous. Let $C \subseteq C'$ be connected, then $\psi(C) \subseteq \psi(C')$. Suppose that $u(C) = 1$, then $C \subseteq \psi(C)$, hence $C \subseteq \psi(C')$. Thus we get that $C \subseteq C' \cap \psi(C')$, and we conclude that $u(C') = 1$ since otherwise $C' \cap \psi(C') = \emptyset$. Thus it remains to show Eq. 6–5. Let $X \subseteq \mathbb{Z}^2$ and $u(\gamma_h(X \cup \{h\})) = 0$; we must show that $v(\gamma_h(X^c \cup \{h\})) = 1$. Indeed, since $h \notin \psi(X \cup \{h\})$ and ψ is increasing, it follows that $h \notin \psi(X \setminus \{h\})$. This means that $v(P(X \setminus \{h\}, h)) = 1$. Now

$$P(X \setminus \{h\}, h) = \gamma_h\big((X \setminus \{h\})^c\big) = \gamma_h\big(X^c \cup \{h\}\big).$$

This yields the result. ■

We write $u \equiv 1$ if the criterion u is identically 1, i.e., $u(C) = 1$ for every grain C. When $v \equiv 1$, we write $\psi_{u,1}$ for $\psi_{u,v}$. Similarly, $\psi_{1,v}$ denotes the grain operator for which the foreground criterion u is identically 1.

EXAMPLE 6–2. We present some examples of grain criteria.

- $u(C) = C(h)$. Now $\psi_{u,1}$ equals the connectivity opening γ_h.
- $u(C) = [\text{area}(C) \geqslant S]$. The operator $\psi_{u,1}$ is the area opening considered in Example 6–1.
- $u(C) = [\text{perimeter}(C) \geqslant S]$, where perimeter$(C)$ equals the number of boundary pixels in C. This criterion is not increasing.
- $u(C) = [\text{area}(C)/(\text{perimeter}(C))^2 \geqslant k]$. Note that this criterion provides a measure for the circularity of C. This criterion is not increasing.
- $u(C) = [C \ominus B \neq \emptyset]$, which gives the outcome 1 if some translate of B fits inside C. If B is connected, then $\psi_{u,1} = \breve{\alpha}$, where $\alpha(X) = X \circ B$; cf. Proposition 6–12. However, if B is not connected, then $\psi_{u,1}$ is an opening that is smaller than $\breve{\alpha}$, i.e., $\psi_{u,1} \leqslant \breve{\alpha}$.

Breen and Jones [4] discuss various other increasing and nonincreasing criteria. We state some other useful properties.

PROPOSITION 6–17.

(a) $\psi^*_{u,v} = \psi_{v,u}$.

(b) *Given grain operators* ψ_{u_i,v_i}, *for* $i \in I$, *then*

$$\bigwedge_{i\in I} \psi_{u_i,v_i} = \psi_{\bigwedge_{i\in I} u_i, \bigvee_{i\in I} v_i} \quad \text{and} \quad \bigvee_{i\in I} \psi_{u_i,v_i} = \psi_{\bigvee_{i\in I} u_i, \bigwedge_{i\in I} v_i}.$$

(c) *Let* ψ_{u_i,v_i} *be grain operators for* $i = 1, 2, \ldots, n$, *and let* b *be a Boolean function of* n *variables, then*

$$b(\psi_{u_1,v_1}, \ldots, \psi_{u_n,v_n}) = \psi_{b(u_1,\ldots,u_n), b^*(v_1,\ldots,v_n)}.$$

Here b^* *denotes the negative of* b *given by* $b^*(u_1, \ldots, u_n) = 1 - b(1 - u_1, \ldots, 1 - u_n)$.

PROOF. We prove only (c). In Proposition 6–10(b) we have seen that $\psi = b(\psi_{u_1,v_1}, \ldots, \psi_{u_n,v_n})$ is a grain operator. Therefore, ψ is of the form $\psi_{u,v}$. From Proposition 6–15 we know that the foreground criterion u is given by $u(C) = [C \subseteq \psi(C)]$. Since ψ is a grain operator, $[C \subseteq \psi(C)] = [h \in \psi(C)]$, for every $h \in C$. But this last expression equals $b(\psi_{u_1,v_1}(C)(h), \ldots, \psi_{u_n,v_n}(C)(h))$. Using that $\psi_{u_i,v_i}(C)(h) = [C \subseteq \psi_{u_i,v_i}(C)] = u_i(C)$, we finally arrive at the identity $u(C) = b(u_1(C), \ldots, u_n(C))$. In a similar way we find that $v(C) = b^*(v_1(C), \ldots, v_n(C))$, and the result is proved. ■

The following result is obvious:

PROPOSITION 6–18. *The grain operator $\psi_{u,v}$ is extensive if and only if $u \equiv 1$. It is anti-extensive if and only if $v \equiv 1$.*

We have seen that a composition of grain operators is not a grain operator, in general. However, it is easy to see that a composition of (anti-) extensive grain operators is an (anti-) extensive grain operator. To be precise

$$\psi_{u_2,1}\psi_{u_1,1} = \psi_{u_1,1}\psi_{u_2,1} = \psi_{u_1 \wedge u_2,1} = \psi_{u_1,1} \wedge \psi_{u_2,1} \tag{6–6}$$

$$\psi_{1,v_2}\psi_{1,v_1} = \psi_{1,v_1}\psi_{1,v_2} = \psi_{1,v_1 \wedge v_2} = \psi_{1,v_1} \vee \psi_{1,v_2}. \tag{6–7}$$

Rather than presenting a formal proof of these relations (in fact, such a proof is rather straightforward, and we leave it as an exercise for the reader), we sketch only the underlying idea. The composition $\psi_{u_2,1}\psi_{u_1,1}$ cannot add background grains, but only delete foreground grains. A foreground grain C will be deleted in either of the two following situations: (i) $u_1(C) = 0$; (ii) $u_1(C) = 1$ but $u_2(C) = 0$. In other words, C is deleted if at least one of the criteria u_1 or u_2 is not satisfied. Therefore, the foreground grain criterion for the composition $\psi_{u_2,1}\psi_{u_1,1}$ is $u = u_1 \wedge u_2$.

Taking $u_1 = u_2 = u$ in Eq. 6–6, we find that $\psi^2_{u,1} = \psi_{u,1}$. Note in particular that this provides an alternative proof for Proposition 6–11.

As a special case of Eq. 6–6 we mention the identity

$$\alpha\gamma_h = \gamma_h\alpha,$$

for every grain opening α. Taking the supremum over $h \in \mathbb{Z}^2$ and using that $\bigvee_{h\in\mathbb{Z}^2} \gamma_h = \mathrm{id}$, we arrive at the identity in Proposition 6–13.

In a forthcoming paper we will examine the construction of morphological filters that are connected. Here we discuss one specific example, namely the generalization of the (self-dual) annular filter on $\mathcal{P}(\mathbb{Z}^2)$ (see Chap. 5). We start with a lemma.

LEMMA 6–2. *Given $X \subseteq \mathbb{Z}^2$ and $h \in \mathbb{Z}^2$, put $C = P(X, h)$. If area$(C) \leqslant 7$ then C is a leaf of the zonal graph (tree) of X and the unique neighbor C' of C satisfies area$(C') \geqslant 8$. Moreover, $C' \cup \{h\}$ is connected.*

Verification of the validity of this lemma is just a matter of checking all possibilities, and is left as an exercise for the reader. The estimate area $(C') \geqslant 8$ is sharp only if C comprises one pixel. It is easy to verify that we can replace the value 8 by 10, 12, 12, 14, 14, 16 for area$(C) = 2, 3, 4, 5, 6, 7$, respectively. But for our purposes, the estimate in the lemma is good enough.

Consider the increasing area criterion

$$u_S(C) = [\mathrm{area}(C) \geqslant S],$$

and define

$$\omega_S = \psi_{u_S,u_S}.$$

It is evident that ω_S is a self-dual grain operator. Note that $\omega_1 = \mathrm{id}$.

PROPOSITION 6–19. *If $S \leqslant 8$, then ω_S is a self-dual grain filter.*

PROOF. We must show that ω_S is increasing and idempotent. To show that ω_S is increasing, we apply Proposition 6–16. It is evident that u_S is increasing. We must show that condition 6–5 holds for $u = v = u_S$. Take $X \subseteq \mathbb{Z}^2$. Without loss of generality we may assume that $h \in X$. Put $C = \gamma_h(X)$, and suppose that $u_S(C) = 0$. Now, the previous lemma yields that C is a leaf of the zonal tree of X, which has a unique neighbor $C' \Subset X^c$. Furthermore, $C' \cup \{h\}$ is connected, thus $\gamma_h(X^c \cup \{h\}) = C' \cup \{h\}$, and the lemma says that the area of this component is greater than or equal to 9. Thus $u_S(\gamma_h(X^c \cup \{h\})) = 1$, which had to be shown. We conclude that ω_S is increasing.

Now we show that ω_S is idempotent. Since ω_S is self-dual, it is sufficient to show that $\omega_S \leqslant \omega_S^2$. Let C be a part of $P(X)$ and $C \subseteq \omega_S(X)$. We must show that $C \subseteq \omega_S^2(X)$. We distinguish two cases.

1. $C \Subset X$: since C is preserved by ω_S, it follows that $u_S(C) = 1$. Since u_S is increasing, the grain C' of $\omega_S(X)$ that contains C automatically obeys $u_S(C') = 1$, and therefore $C \subseteq C' \subseteq \omega_S^2(X)$.

2. $C \Subset X^c$: in view of the fact that $C \subseteq \omega_S(X)$, we have $u_S(C) = 0$. Thus $\mathrm{area}(C) < S$, in particular, $\mathrm{area}(C) \leqslant 7$. Now the previous lemma yields that C is a leaf of the zonal graph of X, and that its (unique) neighbor C' satisfies $\mathrm{area}(C') \geqslant 8$, hence $u_S(C') = 1$. This means that the connected set $C \cup C'$ lies in $\omega_S(X)$. Let C'' be the grain of $\omega_S(X)$ containing $C \cup C'$, then $u_S(C'') = 1$. Therefore $C'' \subseteq \omega_S^2(X)$, in particular $C \subseteq \omega_S^2(X)$. ■

The filter ω_2 is the annular filter discussed in Chapter 5. Note that ω_S, for $S \leqslant 8$, is the composition of the area opening $\alpha_S = \psi_{u_S,1}$ and the area closing $\beta_S = \psi_{1,u_S}$, that is,

$$\omega_S = \alpha_S \beta_S = \beta_S \alpha_S.$$

The compositions $\alpha_S \beta_S$ and $\beta_S \alpha_S$ are filters for every $S \geqslant 1$. In general, these two compositions are different. However, by our previous result, they coincide and define a self-dual grain operator if $S \leqslant 8$. An illustration is given in Fig. 6–14.

Figure 6–14. From left to right: a binary image X (see Chapter 5) and the results after filtering with ω_S for $S = 1, 4, 7$, respectively.

6.6 GRAY-SCALE IMAGES

Up to this point, we have been concerned exclusively with connected operators for binary images. In this section we describe briefly the extension to gray-scale images. Readers who want to know more details are referred to the literature [4–6, 20, 24].

In this section we restrict ourselves to images that can be modelled by numerical functions $f : \mathbb{Z}^2 \to \mathcal{T}$, where $\mathcal{T} = \{0, 1, 2, \ldots, T\}$. We denote by $\mathrm{Fun}(\mathbb{Z}^2)$ the set of all such functions. It is well known that $\mathrm{Fun}(\mathbb{Z}^2)$ defines a complete lattice (see also Chapter 5).

DEFINITION 6–3. Given a gray-scale function $f \in \mathrm{Fun}(\mathbb{Z}^2)$, a connected set $C \subseteq \mathbb{Z}^2$ is called a *flat zone* of f at level t if C is a grain of the level set $\{x \in \mathbb{Z}^2 \mid f(x) = t\}$.

In other words, a flat zone of a function is a maximal connected region where the function is constant. The flat zones of a function f yield a connected partition of the underlying space $\mathbb{Z}^2$; this partition will be denoted by $P(f)$. Observe that the partition of the indicator function of a set X coincides with the partition of the set defined in Section 6.2. We write $P(f, h) = P(f)(h)$ for $h \in \mathbb{Z}^2$. The following definition is a straightforward generalization of Definition 6–1.

DEFINITION 6–4. An operator Ψ on $\mathrm{Fun}(\mathbb{Z}^2)$ is *connected* if the partition $P(\Psi(f))$ is coarser than $P(f)$, for every function f.

Many of the results on connected operators for binary images can be extended to the gray-scale case (in particular Propositions from 6–4 to 6–7). However, we are not aiming at a comprehensive discussion of connected gray-scale operators in this section. Rather our goal is to give the reader a global impression of some aspects of such operators, e.g., their construction, their implementation, and their application in segmentation algorithms.

An important class of gray-scale operators is formed by the so-called *flat operators* [9, 10]. Given an increasing operator ψ on $\mathcal{P}(\mathbb{Z}^2)$, there exists a unique increasing operator Ψ on $\mathrm{Fun}(\mathbb{Z}^2)$ such that the following relation holds:

$$X(\Psi(f), t) = \psi(X(f, t)),$$

for $f \in \mathrm{Fun}(\mathbb{Z}^2)$ and $t \in \mathcal{T}$. Here $X(f, t) = \{x \in \mathbb{Z}^2 \mid f(x) \geqslant t\}$ is the threshold set associated with f. We say that Ψ is generated by ψ. The following result can be established (a formal proof will be given in a future publication).

PROPOSITION 6–20. *Let ψ be an increasing connected operator on $\mathcal{P}(\mathbb{Z}^2)$. Then the flat operator Ψ on* $\mathrm{Fun}(\mathbb{Z}^2)$ *generated by ψ is connected, too.*

In Section 6.3 we introduced the zonal graph for binary images. In fact, the same definition carries over to the gray-scale case. In this case, however, the coloring defined in Eq. 6–3 becomes a function $I_f : P(f) \to \mathcal{T}$. The zonal graph representation of a gray-scale function is tailor-made for the implementation of connected operators. This is best illustrated by means of an example. We consider the (flat extension of the) area opening with threshold S, and present an algorithm for its computation based upon the zonal graph representation of a gray-scale function f. Starting at the maximum gray-level (or color) $t = T$, we determine for each flat zone at level t if its area is greater than or equal to S. If so, then the output image receives the color t for every pixel in this zone, insofar as it hasn't been set at a previous step. If not, that is, if the area is less than S, then the color t is diminished by one. After this last step, a flat zone may have one or more neighbors with the same color. Such zones are then merged into one new flat zone. This procedure has to be repeated until the minimum color is attained. Thus we arrive at the following algorithm.

```
• initialization
  input image I;
  output image I'(x) ← 0, all x;

• find all flat zones corresponding with I;
  /* now I is defined at flat zones */

• t ← T;    /* T is maximum gray value */

• while t ≠ 0 do {
    for every flat zone C with I(C) = t do {
      if area(C) ⩾ S then
        for every x ∈ C: I'(x) ← max{I'(x), t};
      I(C) ← t − 1;
      merge C with neighbors C' with I(C') = t − 1;
      /* areas of flat zones that are merged
         can be added */
    }
    t ← t − 1;
  }
```

Our algorithm is illustrated in Fig. 6–15. A fast algorithm for the computation of the area opening for gray-scale images using the standard pixel representation has been given by Vincent [27].

A connected operator on Fun($\mathbb{Z}^2$) acts on the level of the zonal graph in the sense that it amounts to a recoloring of the vertices, followed by a merging of adjacent vertices that have obtained the same color [14]. As in the binary case, recoloring may be based on (grain) criteria. In the example given above we used a criterion

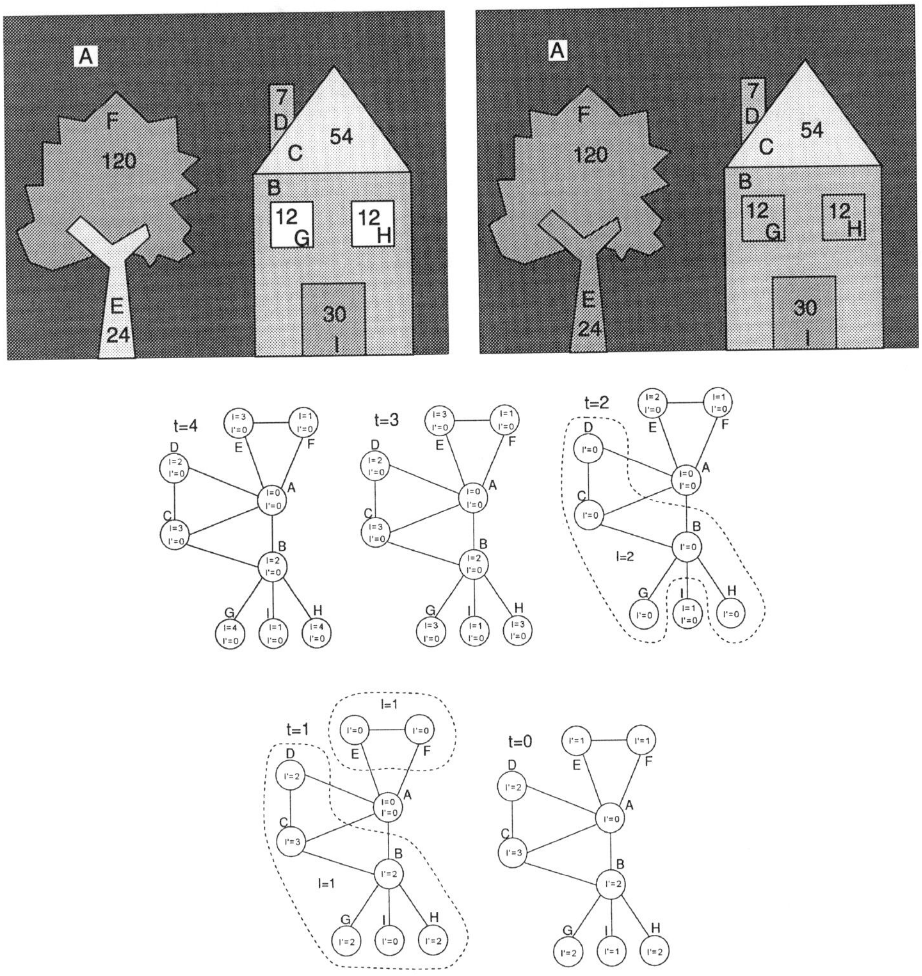

Figure 6–15. The area opening with $S = 25$ for the toy image at the top left results in the image at the right. The areas of the different flat zones are displayed. The five zonal graphs (second row, left to right) represent the successive steps in the algorithm. The input image I has gray levels between 0 and 4 (the higher the level, the brighter the image). The area opening darkens bright zones with small areas.

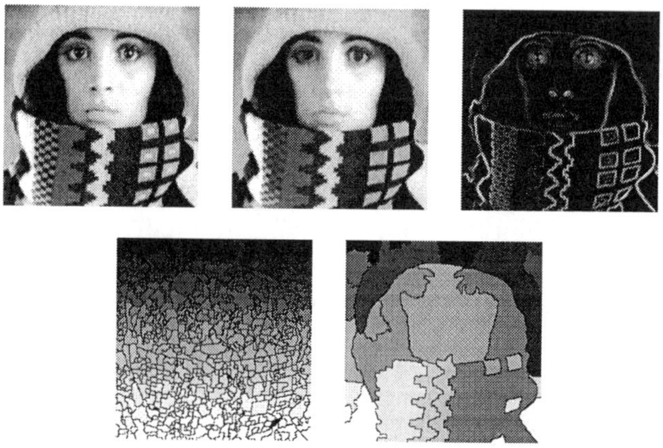

Figure 6–16. From left to right: original image, filtered image (area open-close with $S = 100$), gradient of original image, segmentation without markers, and segmentation with markers.

based on area. However, in the gray-scale case we can introduce another class of criteria, namely those based on (local) contrast; see e.g. [14, 18, 20]. In future publications we will discuss connected gray-scale operators based on contrast criteria in more detail.

We conclude this section with an application showing how to use connected operators for marker extraction in segmentation. In mathematical morphology, segmentation is often done by a watershed algorithm [3]. Usually this algorithm is not applied to the image itself, but rather to its gradient transform. Due to the noise present in the data, such a procedure often results in a huge oversegmentation; see Fig. 6–16.

This can be avoided by using markers that indicate the location of subsets of the different segments. Given a set of markers, one can then modify the gradient image, forcing the markers to become the local minima; since the watershed algorithm uses these local minima as a starting point, we have got rid of the oversegmentation in this way. Thus we are faced with the problem of marker detection. It is at this point that connected operators can be helpful. Rather than going into details, we refer again to Fig. 6–16. We have computed the area-based open-close filter with threshold $S = 100$. Then we computed the gradient of the filtered image, applied a thresholding (at level 10), and an area opening (to the binary image). The background parts can be considered as the homogeneous regions of the original image. This suggests that we can use the background grains as markers. Indeed, we use them to modify the original gradient image, and apply the watershed algorithm. As Fig. 6–16 clearly shows, we arrive at a much better segmentation.

6.7 CONCLUDING REMARKS

As the title of this chapter suggests, it contains an introduction to the theory of connected morphological operators. Our exposition is restricted to the case of binary images on a two-dimensional discrete 8-connected grid, with the exception of the previous section, which contains some results for the gray-scale case. We point out, however, that many of our results carry over to other image spaces and/or other notions of connectivity.

Serra [22] has introduced the notion of *connectivity class* for the complete Boolean lattice $\mathcal{P}(E)$, where E is an arbitrary set; see also [10, 15].

DEFINITION 6–5. A family $\mathcal{C} \subseteq \mathcal{P}(E)$ is a connectivity class if

- $\emptyset \in \mathcal{C}$ and $\{h\} \in \mathcal{C}$, for every $h \in E$;
- if $X_i \in \mathcal{C}$, $i \in I$, and $\bigcap_{i \in I} X_i \neq \emptyset$, then $\bigcup_{i \in I} X_i \in \mathcal{C}$.

This definition includes the class of 8-connected sets in $\mathbb{Z}^2$ (as well as the 4-connected sets), but one can find many other examples; see [15] for some interesting ones. Recently, Serra [23] has given an extension of the definition of connectivity class to other complete lattices than $\mathcal{P}(E)$.

Given a connectivity class $\mathcal{C}$, we can define the connectivity openings $\gamma_h : \mathcal{P}(E) \to \mathcal{P}(E)$ as follows:

$$\gamma_h(X) = \bigcup\{C \in \mathcal{C} \mid h \in C \text{ and } C \subseteq X\}.$$

To show that γ_h is indeed an opening, we observe first that γ_h is increasing and anti-extensive, hence that $\gamma_h^2 \leqslant \gamma_h$. On the other hand, $\gamma_h(X)$ is a union of sets $C \in \mathcal{C}$ with $h \in C \subseteq X$. Every C with this property satisfies $C \subseteq \gamma_h(X)$, and this yields that $\gamma_h(X) \subseteq \gamma_h^2(X)$. Now we can define the reconstruction operator ρ as in Eq. 6–1. As a matter of fact, many of the definitions and results stated in this paper carry over to this general framework. We will not elaborate on this theme here.

Another issue that we have not explored in this chapter is the theory of connected filters other than openings and closings. Readers interested in this topic are referred to [8, 13, 20, 22].

ACKNOWLEDGMENT

The author gratefully acknowledges interesting discussions with Jose Crespo, Fernand Meyer, Philippe Salembier, and Jean Serra about various aspects of connected operators at the seminar on connected operators, June 14–15, 1996, in Barcelona.

REFERENCES

[1] Ballard, D. H., and M. Brown, *Computer Vision*, Prentice-Hall, Englewood Cliffs, NJ, 1982.

[2] Berge, C., *Graphs*, 2nd ed., North-Holland, Amsterdam, 1985.

[3] Beucher, S., and F. Meyer, "The morphological approach to segmentation: the watershed transformation," in *Mathematical Morphology in Image Processing*, E. R. Dougherty, ed., Ch. 12, pp. 433–481, Marcel Deker, New York, 1993.

[4] Breen, E., and R. Jones, "An attribute-based approach to mathematical morphology," in *Mathematical Morphology and its Applications to Image and Signal Processing*, P. Maragos, R. W. Schafer and M. A. Butt, eds., pp. 41–48, Kluwer Academic Publishers, Boston, 1996.

[5] Crespo, J., Morphological connected filters and intra-region smoothing for image segmentation, PhD thesis, Georgia Institute of Technology, Atlanta, 1993.

[6] Crespo, J., and R. W. Schafer, "The flat zone approach and color images," in *Mathematical Morphology and its Applications to Image Processing*, J. Serra and P. Soille, eds., pp. 85–92, Kluwer Academic Publishers, 1994.

[7] Crespo, J., and R. W. Schafer, "Locality and adjacency stability constraints for morphological connected operators," *J. Math. Imaging and Vision* **7** (1), 1997, pp. 85–102.

[8] Crespo, J., J. Serra, and R. W. Schafer, "Theoretical aspects of morphological filters by reconstructions," *Sign. Proc.* **47** (2), 201–225, 1995.

[9] Heijmans, H. J. A. M., "Theoretical aspects of gray-level morphology," *IEEE Trans. Patt. Anal. Mach. Intell.* **13**, 568–582, 1991.

[10] Heijmans, H. J. A. M., *Morphological Image Operators*, Academic Press, Boston, 1994.

[11] Lantuéjoul, C., and S. Beucher, "On the use of the geodesis metric in image analysis," *J. Microscopy* **121**, 29–49, 1980.

[12] Lantuéjoul, C., and F. Maisonneuve, "Geodesic methods in quantitative image analysis," *Patt. Recogn.* **17**, 177–187, 1984.

[13] Pardàs, M., J. Serra, and L. Torres, "Connectivity filters for image sequences," *SPIE Proceedings*, Vol. 1769, pp. 318–329, 1992.

[14] Potjer, F. K., "Region adjacency graphs and connected morphological operators," in *Mathematical Morphology and its Applications to Image and Signal Processing*, P. Maragos, R. W. Schafer and M. A. Butt, eds., pp. 111–118, Kluwer Academic Publishers, Boston, 1996.

[15] Ronse, C., "Set-theoretical algebraic approaches to connectivity in continuous or digital spaces," *J. Math. Imaging and Vision* **8**, 1998, pp. 41–58.

[16] Ronse, C., and H. J. A. M. Heijmans, "The algebraic basis of mathematical morphology – Part II: Openings and closings," *CVGIP: Image Understanding* **54**, 74–97, 1991.

[17] Rosenfeld, A., "Connectivity in digital pictures," *J. Assoc. Comp. Mach.* **17**, 146–160, 1970.

[18] Salembier, P., and M. Kunt, "Size-sensitive multiresolution decomposition of images with rank order based filters," *Sign. Proc.* **27** (2), 205–241, 1992.

[19] Salembier, P., and A. Oliveras, "Practical extensions of connected operators," in *Mathematical Morphology and its Applications to Image and Signal Processing*, P. Maragos, R. W. Schafer and M. A. Butt, eds., pp. 97–110, Kluwer Academic Publishers, Boston, 1996.

[20] Salembier, P., and J. Serra, "Flat zones filtering, connected operators, and filters by reconstruction," *IEEE Trans. on Image Proc.* **4** (8), 1153–1160, 1995.

[21] Serra, J., *Image Analysis and Mathematical Morphology*, Academic Press, London, 1982.

[22] Serra, J., ed., *Image Analysis and Mathematical Morphology*, Vol. II: *Theoretical Advances*, Academic Press, London, 1988.

[23] Serra, J., "Connectivity on complete lattices," in *Mathematical Morphology and its Applications to Image and Signal Processing*, P. Maragos, R. W. Schafer and M. A. Butt, eds., pp. 81–96, Kluwer Academic Publishers, Boston, 1996.

[24] Serra, J., and P. Salembier, "Connected operators and pyramids," *SPIE Proceedings*, Vol. 2030, pp. 65–76, 1993.

[25] Vincent, L., "Morphological algorithms," in *Mathematical Morphology in Image Processing*, E. R. Dougherty, ed., Ch. 8, pp. 255–288, Marcel Dekker, New York, 993.

[26] Vincent, L., "Morphological grayscale reconstruction in image analysis: efficient algorithms and applications," *IEEE Trans. Image Proc.* **2**, 176–201, 1993.

[27] Vincent, L., "Morphological area openings and closings for gray-scale images," in "*Shape in Picture*," Y.-L. O, A. Toet, D. Foster, H. J. A. M. Heijmans and P. Meer, eds., pp. 197–208, Springer, Berlin, 1994.

CHAPTER 7

REPRESENTATION AND OPTIMIZATION OF STACK FILTERS

Jaakko T. Astola
Pauli Kuosmanen
Tampere Univ. of Technology
Tampere, Finland

7.1 REPRESENTATION OF STACK FILTERS

Stack filters are a class of sliding window nonlinear filters first introduced by Wendt *et al.* [46]. They perform well in many situations where linear filters fail. Thus, stack filters have been used in many applications; cf. [35].

7.1.1 DEFINITION OF STACK FILTERS

Let $X(t)$ denote a signal to be filtered. Now, the argument t in this paper is a time and/or spatial index. At time t the filter has available a fixed number, N, of samples of the signal $X(\cdot)$. For one-dimensional signals the available samples are typically obtained as follows: $\mathbf{x} = (X(t - N_1), \ldots, X(t - 1), X(t), X(t + 1), \ldots, X(t + N_2)) = (X_1(t), X_2(t), \ldots, X_N(t))$, where $N = N_1 + N_2 + 1$. Thus the filter can be understood as an operation whose window slides over the signal and at every time instant t operates on signal values inside the filter window. In image processing applications, we assume that the image samples are arranged inside the vector $\mathbf{x}$ in a known and fixed order.

The output of a stack filter at each window position is the result of a sum of a stack of binary operations operating on thresholded versions of the samples appearing in the filter's window. The key to the analysis of stack filters comes from their definition by threshold decomposition [46, 50]. By threshold decomposition we can divide the analysis of stack filters into smaller and simpler parts. In other words, most of the analysis can be done by studying binary signals.

Consider a vector $\mathbf{x} = (x_1, x_2, \ldots, x_N)$, where $x_i \in \{0, 1, \ldots, M - 1\}$. The *threshold decomposition* of $\mathbf{x}$ involves decomposing $\mathbf{x}$ into $M - 1$ binary vectors

$\mathbf{x}^1, \mathbf{x}^2, \ldots, \mathbf{x}^{M-1}$, according to the thresholding rule

$$x_n^m = T_m(x_n) = \begin{cases} 1, & \text{if } x_n \geqslant m, \\ 0, & \text{otherwise.} \end{cases} \tag{7–1}$$

Thus, the binary vector $\mathbf{x}^m$ is obtained by thresholding the input vector at the level m, for $1 \leqslant m \leqslant M-1$. An element x_n^k of the binary vector $\mathbf{x}^k$ takes on the value 1 whenever the element of the input vector x_n is greater than or equal to k.

It is important to note that this thresholding process can also be applied to all signals that are quantized to a finite number of arbitrary levels.

From Eq. 7–1 we see that the original multivalued (M-ary) vector can be reconstructed from its binary vectors

$$\mathbf{x} = \sum_{m=1}^{M-1} \mathbf{x}^m \quad \text{or equivalently} \quad x_n = \sum_{m=1}^{M-1} x_n^m.$$

Let $\mathbf{x}$ and $\mathbf{y}$ be binary vectors (signals) of fixed length. Define

$$\mathbf{x} \leqslant \mathbf{y} \quad \text{if and only if} \quad x_n \leqslant y_n \text{ for all } n. \tag{7–2}$$

Since the relation defined by Eq. 7–2 is reflexive, antisymmetric, and transitive, it defines a partial ordering in the set of binary vectors of fixed length. Now consider a signal $\mathbf{x}$ and its thresholded binary signals $\mathbf{x}^1, \mathbf{x}^2, \ldots, \mathbf{x}^{M-1}$. Clearly,

$$\mathbf{x}^i \leqslant \mathbf{x}^j \quad \text{if } i \geqslant j.$$

Thus the binary signals $\mathbf{x}^1, \mathbf{x}^2, \ldots, \mathbf{x}^{M-1}$ are nonincreasing in the sense of the partial ordering of Eq. 7–2.

It has turned out that by defining filtering operations based on those binary operators $f(\cdot)$ for which

$$f(\mathbf{x}^i) \leqslant f(\mathbf{x}^j) \quad \text{if } i \geqslant j \tag{7–3}$$

holds, we obtain a class of filters with many useful properties. This requirement led to the definition of stack filters based on positive Boolean functions [46]. Boolean functions are discussed in more detail in Section 7.1.3.

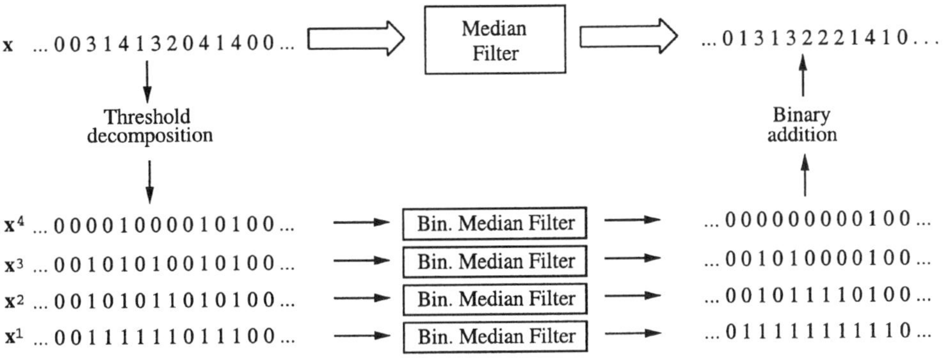

Figure 7–1. Illustration of stack filtering operation using threshold decomposition. The broad arrows show the overall filtering operation. The slender arrows show the same operation in the threshold decomposition architecture. The Boolean function used in the illustration is $f(\mathbf{x}) = x_{-1}x_0 + x_{-1}x_1 + x_0x_1$, which corresponds to the three-point median filter.

A Boolean function $f(\cdot)$ is called a *positive Boolean function* (PBF) if it can be written as a Boolean expression that contains only uncomplemented input variables. For a PBF $f(\cdot)$ it holds that

$$f(\mathbf{x}) \geqslant f(\mathbf{y}) \quad \text{if } \mathbf{x} \geqslant \mathbf{y}. \tag{7–4}$$

The property 7–4 is called the *stacking property*. See [33] for a complete proof of Eq. 7–4. Since 7–4 holds for PBFs, Eq. 7–3 also holds for them as required.

DEFINITION 7–1. A *stack filter* $S_f(\cdot)$ is defined by a positive Boolean function $f(\cdot)$ as follows

$$S_f(\mathbf{x}) = \sum_{m=1}^{M-1} f(\mathbf{x}^m). \tag{7–5}$$

Thus, filtering a vector $\mathbf{x}$ with a stack filter $S_f(\cdot)$ based on the PBF $f(\cdot)$ is equivalent to decomposing $\mathbf{x}$ to binary vectors $\mathbf{x}^m$, $1 \leqslant m \leqslant M-1$, by thresholding, filtering each threshold level with the binary filter $f(\cdot)$ and reconstructing the output vector as the sum 7–5. This procedure is depicted in Fig. 7–1, where we have used a three-point *median filter* as an example. In median filtering, the samples within the moving window are sorted by magnitude and the centermost value, the median of the samples within the window, is the filter output.

By Eq. 7–5 stack filters are completely characterized by their operation on binary vectors. The importance of this property arises from the fact that binary vectors are easier to analyze than multivalued vectors. Also, filtering each binary vector independently allows the operation to be done in parallel, and single binary filters are easy to implement.

7.1.2 CONTINUOUS STACK FILTERS

In this section the definition of continuous stack filter is reviewed [50]. Continuous stack filters operate on real signals. An attractive property of these filters is the possibility of deriving analytical results for their statistical properties.

DEFINITION 7–2. The output of the continuous stack filter defined by a positive Boolean function $f(x_1, x_2, \ldots, x_N)$ with input vector $\mathbf{x} = (X_1, X_2, \ldots, X_N)$ is given by

$$S_f(\mathbf{x}) = \max\{\beta \in \mathbf{R} : f(T_\beta(X_1), \ldots, T_\beta(X_N)) = 1\},$$

where the thresholding function is defined by Eq. 7–1.

The following proposition yields an important isomorphism between the continuous stack filter $S_f(\cdot)$ and the PBF $f(\cdot)$ it corresponds to [42].

PROPOSITION 7–1. *Let* $\mathbf{x} = (X_1, X_2, \ldots, X_N)$ *be an input vector to a stack filter* $S_f(\cdot)$ *defined by a positive Boolean function* $f(x_1, x_2, \ldots, x_N)$*. Then*

$$f(x_1, x_2, \ldots, x_N) = \sum_{i=1}^{K} \prod_{j \in P_i} x_j,$$

where P_i *are subsets of* $\{1, 2, \ldots, N\}$*, if and only if the stack filter* $S(\cdot)$ *corresponding to* $f(x_1, \ldots, x_N)$ *is*

$$S_f(\mathbf{x}) = \max\big\{\min\{X_j : j \in P_1\}, \min\{X_j : j \in P_2\}, \ldots, \min\{X_j : j \in P_K\}\big\}.$$

Thus, the real domain stack filter corresponding to a PBF can be expressed by replacing AND and OR with MIN and MAX, respectively. For example, the three-point median filter over real variables X_{-1}, X_0, and X_1 (see also Fig. 7–1) is a stack filter defined by the PBF $f(x_{-1}, x_0, x_1) = x_{-1}x_0 + x_{-1}x_1 + x_0x_1$, i.e.,

$$\mathrm{MED}\{X_{-1}, X_0, X_1\} = \mathrm{MAX}\big\{\mathrm{MIN}\{X_{-1}, X_0\}, \mathrm{MIN}\{X_{-1}, X_1\}, \mathrm{MIN}\{X_0, X_1\}\big\}.$$

The output distribution of a stack filter can be expressed using the following proposition [50].

PROPOSITION 7–2. *Let the input values X_b, in the window B of a stack filter $S_f(\cdot)$ defined by a positive Boolean function $f(\cdot)$ be independent random variables having the distribution functions $\Phi_b(t)$, respectively. Then the output distribution function $\Psi(t)$ of the stack filter $S_f(\cdot)$ is*

$$\Psi(t) = \sum_{\mathbf{x} \in f^{-1}(0)} \prod_{b \in B} \left(1 - \Phi_b(t)\right)^{x_b} \Phi_b(t)^{1-x_b},$$

where $f^{-1}(0)$ is the pre-image of 0, i.e., $f^{-1}(0) = \{\mathbf{x} : f(\mathbf{x}) = 0\}$ and binary values in the exponents are to be understood as real 0s and 1s.

In the case of independent and identically distributed (i.i.d) input values we get the following corollary.

COROLLARY 7–1. *Let the input values X_b, in the window B, $|B| = N$, of a stack filter $S_f(\cdot)$ defined by a positive Boolean function $f(\cdot)$ be independent, identically distributed random variables having a common distribution function $\Phi(t)$. Then the distribution function of the output $\Psi(t)$ of the stack filter $S_f(\cdot)$ is*

$$\Psi(t) = \sum_{i=0}^{N} A_i \left(1 - \Phi(t)\right)^i \Phi(t)^{N-i}, \qquad (7\text{–}6)$$

where the numbers A_i are defined by

$$A_i = \left|\{\mathbf{x} : f(\mathbf{x}) = 0, w_H(\mathbf{x}) = i\}\right| \qquad (7\text{–}7)$$

and $w_H(\mathbf{x})$ denotes the number of 1s in $\mathbf{x}$, i.e., its Hamming weight.

In order to guarantee that the stack filter is not defined by the trivial PBFs $f(\mathbf{x}) \equiv 0$ or $f(\mathbf{x}) \equiv 1$, we require

$$A_0 = 1 \quad \text{and} \quad A_N = 0. \qquad (7\text{–}8)$$

As $A_N = 0$, we can leave out $i = N$ from the sum 7–6.

PROPOSITION 7–3. *The numbers A_i satisfy*

$$0 \leqslant A_i \leqslant \binom{N}{i}, \quad i = 1, 2, \ldots, N.$$

7.1.3 Boolean Function Representations

In this section we present some well-known facts of the theory of Boolean functions.

A vector $\boldsymbol{\alpha} = (\alpha_1, \alpha_2, \ldots, \alpha_N)$, with components from $B = \{0, 1\}$, is called a binary of a Boolean vector; N is the length of the vector. The set of all Boolean vectors of length N is called N-dimensional cube B^N; $\boldsymbol{\alpha}$ is a vertex of B^N. Each Boolean vector $\boldsymbol{\alpha}$ has a unique representation as the number

$$\alpha = \nu(\boldsymbol{\alpha}) = \sum_{i=1}^{N} \alpha_i 2^{N-i}.$$

The function $f(x_1, x_2, \ldots, x_N) = f(\mathbf{x})$ of N variables $x_1, x_2, \ldots, x_N$ is called a completely defined Boolean function if $f(\mathbf{x}) : B^N \to B$, in other words, if it is a function, with arguments, as the function itself, taking values from $\{0, 1\}$. A function $f(x_1, x_2, \cdots, x_N) = f(\mathbf{x})$ of N variables $x_1, x_2, \cdots, x_N$ is called an incompletely (partially) defined Boolean function if f takes values from $B \cup \{-\}$, where "–" is interpreted so that the function is not defined for those vectors ("don't care" vectors). There are many ways of representing Boolean functions. Here we consider some of them.

The geometrical representation of a Boolean function arises from the representation of a Boolean space in the form of an N-dimensional cube (in Fig. 7–2 a 3-cube and 4-cube of Hasse are shown), with vertices corresponding to all possible collections of values of variables $x_1, x_2, \cdots, x_N$ of Boolean function $f(\mathbf{x})$. A vertex of the N-cube, which is indicated with a dark circle, defines a true (one) value of Boolean function and a vertex which is not indicated, a false (zero) value. An incompletely defined Boolean function is not defined on those vertices of an n-cube which are indicated with a light circle.

The geometrical representation of a Boolean function of N variables, which is illuminating for $N \leqslant 4$, becomes too complicated for larger values of N. The vector cubic representation of a Boolean function (completely or incompletely defined) is closely connected with the geometrical representation. This (vector cubic) representation consists of the set of ternary vectors $\mathbf{v}^{(j)}$, $j = 1, 2, \ldots, k$, of length N corresponding to the 1 of the completely defined Boolean function (corresponding to the 1 and "don't care" values of the incompletely defined Boolean function). Let the ternary vectors $\mathbf{v}^{(j)}$, $j = 1, 2, \ldots, k$, correspond to the 1s of the Boolean function $f(\mathbf{x})$. Then the Boolean function $f(\boldsymbol{\alpha}) = 1$ for binary vectors $\boldsymbol{\alpha} = (\alpha_1, \ldots, \alpha_N)$, obtained from $\mathbf{v}^{(j)}$ by any changes of entries of 2 in the $\mathbf{v}^{(j)}$ to 0 or 1.

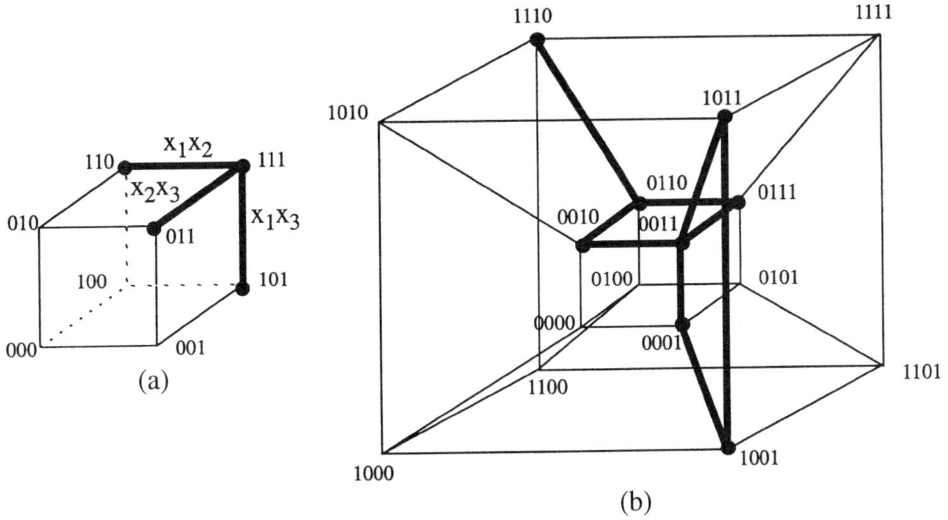

Figure 7–2. The geometrical representation (a) of 3-cube $f(\mathbf{x}) = x_1x_2 \vee x_1x_3 \vee x_2x_3$ (b) of 4-cube $f(\mathbf{x}) = \overline{x_1}x_3 \vee \overline{x_2}x_4 \vee x_2x_3\overline{x_4}$.

Table 7.1. Minimum Vector Cubic Representation of $f(\mathbf{x}) = x_1x_2 \vee x_1x_3 \vee x_2x_3$

x_1	x_2	x_3	y
2	1	1	1
1	2	1	1
1	1	2	1

Table 7.2. Minimum Vector Cubic Representation of $f(\mathbf{x}) = \overline{x_1}x_3 \vee \overline{x_2}x_4 \vee x_2x_3\overline{x_4}$

x_1	x_2	x_3	x_4	y
0	2	1	2	1
2	0	2	1	1
2	1	1	0	1

The vector cubic representation of a Boolean function with a minimum number of ternary vectors and maximum of the total number of entries equal to 2 in the collections is called the minimum vector cubic representation. In Tables 7.1 and 7.2 the minimum vector cubic representations of the Boolean functions shown in Fig. 7–2 are given. The vector cubic representations of the Boolean functions are convenient for description on a computer, especially for those Boolean functions that have the value 1 on a small subset of vectors.

Table 7.3. Tabular Representation of $f(\mathbf{x}) = x_1x_2 \vee x_1x_3 \vee x_2x_3$

$\nu(\mathbf{x})$	x_1	x_2	x_3	$f(\mathbf{x}) = f(x_1, x_2, x_3)$
0	0	0	0	0
1	0	0	1	0
2	0	1	0	0
3	0	1	1	1
4	1	0	0	0
5	1	0	1	1
6	1	1	0	1
7	1	1	1	1

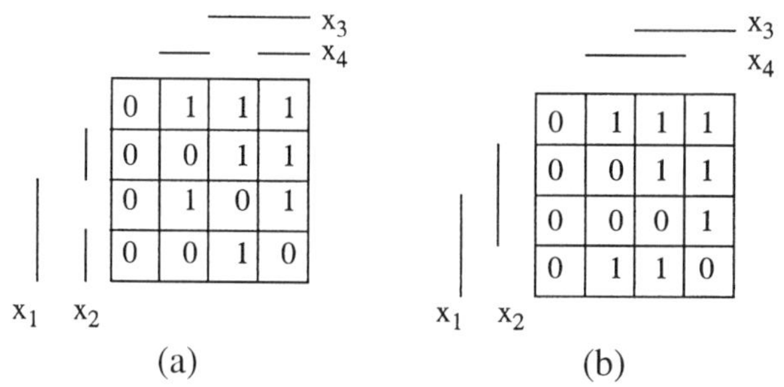

Figure 7–3. The Veich (a) and Karnaugh (b) diagrams of the Boolean function $f(\mathbf{x}) = \overline{x_1}x_3 \vee \overline{x_2}x_4 \vee x_2x_3\overline{x_4}$.

Another way of representing Boolean functions is the truth table or tabular form. In Table 7.3 the tabular representation of the Boolean function presented in Fig. 7–2 is shown in the form of a column vector.

The tabular representation of a Boolean function can be written in the form of a matrix. Historically two main matrix representations of Boolean functions were created: the Veich diagram and the Karnaugh map. The set of variables $X = \{x_1, x_2, \ldots, x_N\}$ of Boolean function $f(\mathbf{x})$ is divided into 2 subsets: the low-order $X_{low} = \{x_1, x_2, \ldots, x_{\lfloor \frac{N}{2} \rfloor}\}$ variables and the high-order $X_{high} = \{x_{\lceil \frac{N}{2} \rceil}, \ldots, x_N\}$ variables. The low-order variables specify the columns of the matrix, high-order ones, specify the rows. The following convention is used: the value 1 of the variable is represented by the line above the column (near the line), and zero by the absence of the line.

When the rows and columns are ordered according to the binary representation, we get the Veich diagram [Fig. 7–3 (a)], and when they are ordered according to the Grey-code we get the Karnaugh map [Fig. 7–3 (b)].

Table 7.4. Elementary Boolean Functions of One Variable

x	0	1	x	$\overline{x}$
0	0	1	0	1
1	0	1	1	0

Table 7.5. Elementary Boolean Functions of Two Variables

			$\wedge$	$\vee$	$\oplus$	$\equiv$	$\Rightarrow$	$\mid$	$\downarrow$
$\nu(\mathbf{x})$	x_1	x_2	f_1	f_2	f_3	f_4	f_5	f_6	f_7
0	0	0	0	0	0	1	1	1	1
1	0	1	0	1	1	0	1	1	0
2	1	0	0	1	1	0	0	1	0
3	1	1	1	1	0	1	1	0	0

In Table 7.4 the elementary Boolean functions are given: 0 and 1 are, respectively, identical to zero and one, x and $\bar{x}$ the function x and its negation $\overline{x}$, respectively. In Table 7.5 the following elementary functions are given in consecutive order: conjunction ($f_1 = x_1 \wedge x_2$ or x_1x_2), disjunction ($f_2 = x_1 \vee x_2$ or $x_1 + x_2$), modulo 2 sum ($f_3 = x_1 \oplus x_2$), equivalence ($f_4 = x_1 \equiv x_2$), implication ($f_5 = x_1 \Rightarrow x_2$), Sheffer's touch ($f_6 = x_1 \mid x_2$) and Pierce's arrow ($f_7 = x_1 \downarrow x_2$).

A Boolean function can also be expressed as an algebraic formula. For example, the same Boolean function given in Tables 7.1 and 7.2 can be realized using any of the formulas: $f(\mathbf{x}) = x_1x_2 \vee x_1x_3 \vee x_2x_3$, $f(\mathbf{x}) = (x_1 \vee x_2)(x_1 \vee x_3)(x_2 \vee x_3)$, $f(\mathbf{x}) = (1 \Rightarrow x_1)(1 \Rightarrow x_2) \vee (1 \Rightarrow x_1)(1 \Rightarrow x_3) \vee (1 \Rightarrow x_2)(1 \Rightarrow x_3)$.

The following equalities play an important role in the study of Boolean functions:

The rule of commutativity,

$$x \,\rho\, y = y \,\rho\, x; \quad \rho = \wedge, \vee,$$

the rule of associativity,

$$(x \,\rho\, y) \,\rho\, z = x \,\rho\, (y \,\rho\, z); \quad \rho = \wedge, \vee,$$

the rule of absorption,

$$x \,\rho_1\, (x \,\rho_2\, y) = x; \quad \rho_1 = \wedge, \rho_2 = \vee \text{ or } \rho_1 = \vee, \rho_2 = \wedge,$$

the rule of distributivity,

$$x\ \rho_1\ (y\ \rho_2\ z) = (x\ \rho_1\ y)\ \rho_2\ (x\ \rho_1\ z); \qquad \rho_1 = \wedge, \rho_2 = \vee \text{ or } \rho_1 = \vee, \rho_2 = \wedge,$$

the rules of negation,

$$x \wedge \overline{x} = 0,$$

$$x \vee \overline{x} = 1,$$

and

$$x \Rightarrow y = \overline{x} \vee y,$$

$$x \equiv y = (x \wedge y) \vee (\overline{x} \wedge \overline{y}).$$

Note, that the number of possible different Boolean functions of N independent variables is 2^{2^N}.

DEFINITION 7–3. Let f be written in the disjunctive normal form (minimum sum of products form)

$$f(x_0, x_1, \ldots, x_{N-1}) = \sum_{(i_0, i_1, \ldots, i_{N-1})} \phi(i_0, i_1, \ldots, i_{N-1}) x_0^{i_0} x_1^{i_1} \cdots x_{N-1}^{i_{N-1}},$$

where $\phi(\cdot)$ takes values 0 or 1. Then the vector representation of $\phi(\cdot)$ is called the characteristic vector of the disjunctive normal form of $f(\cdot)$.

There exists a simple algorithm for deciding the positivity of a Boolean function f, given as a truth table $\mathbf{f}$, requiring $N2^{N-1}$ additions and 2^N comparisions [2].

PROPOSITION 7–4. *For a Boolean function* $\mathbf{f}$ *of* N *variables to be positive, it is necessary and sufficient that*

$$\delta(\mathbf{K}_N \cdot \mathbf{f}) = \delta(\mathbf{f}),$$

where $\mathbf{K}_N$ *is the conjunctive matrix and* $\delta(\mathbf{g}) = (\delta(g_0), \delta(g_1), \ldots, \delta(g_{2^N-1}))$ *is the vector Kronecker function,*

$$\delta(g_i) = \begin{cases} 1, & \text{if } g_i = 0, \\ 0, & \text{otherwise}, \end{cases}$$

for $i = 0, 1, \ldots, 2^N - 1$.

ALGORITHM 7–1. Positivity of a Boolean function.
Step 1. Compute the conjunctive spectrum $\mathbf{K}_n \cdot \mathbf{f}$ of the Boolean function $\mathbf{f}$.
Step 2. If the set of zeroes of spectral coefficients $\mathbf{K}_n \cdot \mathbf{f}$ coincides with the set of zeroes of $\mathbf{f}$, then $\mathbf{f}$ is a positive Boolean function; otherwise $\mathbf{f}$ is not positive.

EXAMPLE 7–1. Let $\mathbf{f} = (0, 0, 0, 0, 1, 1, 1, 1)^T$. Now, $\mathbf{K}_n \cdot \mathbf{f} = (0, 0, 0, 0, 1, 1, 2, 4)^T$ and $\mathbf{f}$ is a positive Boolean function.

7.1.4 SOME PARTICULAR STACK FILTERS

During the past few years many different subclasses of stack filters have been developed for different signal and image processing approaches. In this section we briefly illustrate different approaches by considering a few classes of stack filters that are useful in signal and image processing. These subclasses serve as examples of stack filters which can be represented without any reference to Boolean functions and threshold decomposition architecture.

7.1.4.1 Median and Order Statistic Filters

Median filters form one of the most popular and simplest robust families of stack filters [44, 45]. They have been successfully used in several areas of digital signal processing, including image enhancement [20, 37, 39] and speech processing [19, 38]. In median filters, a filter window slides across the data and the median value of the samples inside the window is chosen to be the output of the filter. Let us denote the samples in the window by $X_1, X_2, \ldots, X_{2k+1}$ and the sorted (in increasing order) list of the samples by $X_{(1)}, X_{(2)}, \ldots, X_{(2k+1)}$. Then $X_{(t)}$ is called the tth *order statistic* and the median is the $(k + 1)$th order statistic. If instead of the median sample the tth order statistic of the values inside the window is chosen to be the output of the filter, the filter is called the tth *order statistic filter* or *ranked order filter*.

An important class of stack filters is defined by a generalization of the median filter. This class is called *weighted median filters* (WM). With WM filters, a different emphasis is assigned to different samples in the window by repeating (weighting) them suitably many times. The median filtering procedure can be stated as follows: sort the samples inside the filter window, duplicate each sample to the number of the corresponding weight and choose the median value from the new sequence. Weighted order statistic filters (WOS) are defined in an analogous manner. The output of a WOS filter is calculated by repeating each input sample up to the number of the corresponding weight, sorting the resulting multiset, and then choosing the Tth largest value from the sorted multiset, where T is the threshold. In this paper we use the notation of Yli-Harja *et al.* [50], where WOS filters are referred to by listing the weights separated by commas and the threshold separated by a semicolon in angle brackets as follows: $\langle w_1, w_2, \ldots, w_N ; T \rangle$. For example, the output

$Y(\mathbf{x})$ of the WOS filter $\langle 1, 2, 3, 2, 1; 3\rangle$ with input vector $\mathbf{x} = (X_1, \ldots, X_5)$ can be expressed as

$$Y(\mathbf{x}) = \text{the third largest value of the multiset } \{X_1, 2\Diamond X_2, 3\Diamond X_3, 2\Diamond X_4, X_5\},$$

where $\Diamond$ denotes the repetition (duplication) operation of any object, i.e.,

$$r\Diamond x = \overbrace{x, \ldots, x}^{r \text{ times}}.$$

Naturally, for 2-dimensional filters it is more convenient to represent the weights in matrix form.

WOS often balance between detail preservation and noise reduction.

Median and all rank-order operators of any dimension can be realized by the threshold decomposition architecture [11, 12]. This is also true for any composition or rank-order operators [12]. Therefore, rank-order operators and their compositions all are stack filters.

Figure 7–4 shows an image restored by a median and a weighted median filter. The window size is 5×5 in this illustration and the weights of the WM filters are given by

$$\begin{pmatrix} 1 & 1 & 1 & 1 & 1 \\ 1 & 2 & 2 & 2 & 1 \\ 1 & 2 & \mathbf{17} & 2 & 1 \\ 1 & 2 & 2 & 2 & 1 \\ 1 & 1 & 1 & 1 & 1 \end{pmatrix}.$$

From Fig. 7–4 we see that the 5×5 median filter removed plenty of important small details from the image. This is called the blurring effect of median-type filters. The weighted median filter in turn removed most of the noise and maintained details well at the same time. This was obtained by giving the largest weight to the sample which contains the most relevant information, that is, to the sample at the location of the estimate. This sample is called *midpoint* or *center* in this paper.

As already mentioned, WM filters belong to the class of stack filters. Thus, any WM filter can also be represented by a PBF in the binary domain. For example, the WM filter $\langle 1, 2, 3, 2, 1; 5\rangle$ corresponds to the following PBF

$$f(x_1, x_2, x_3, x_4, x_5) = x_2x_3 + x_3x_4 + x_1x_3x_5 + x_1x_2x_4 + x_2x_4x_5.$$

It is obvious that all PBFs do not define WM filters. In fact WM filters are stack filters defined by self-dual linearly separable positive Boolean functions [50]. In the following paragraphs we discuss the concepts of self-duality and linear separability.

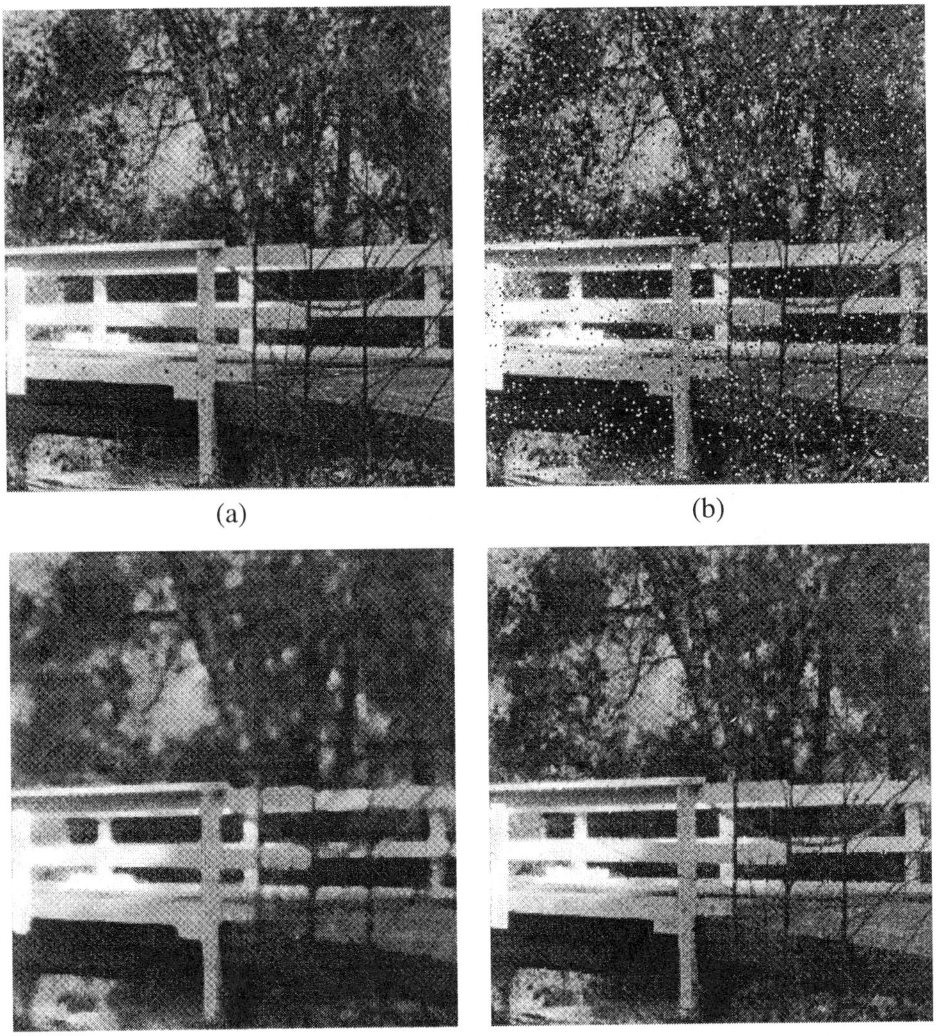

(c) (d)

Figure 7–4. Filtering a test image corrupted by additive impulsive noise (10%) with 5×5 windows. (a) Original image, (b) noisy image, (c) median filtered image, and (d) weighted median filtered image.

DEFINITION 7–4. The *dual* $f^D(\mathbf{x})$ of a Boolean function $f(\mathbf{x})$ is defined by the relation

$$f^D(\mathbf{x}) = \overline{f(\overline{\mathbf{x}})}.$$

A Boolean function $g(\mathbf{x})$ is *self-dual* if and only if

$$f(\mathbf{x}) = f^D(\mathbf{x}).$$

A stack filter which is defined by a self-dual positive Boolean function is called a *self-dual stack filter.*

The following proposition establishes an interesting and useful interrelation between the numbers A_i of self-dual stack filters.

PROPOSITION 7–5. *For a self-dual stack filter with window size N, we have*

$$A_i + A_{N-i} = \binom{N}{i}, \qquad i = 1, 2, \ldots, N. \tag{7–9}$$

PROOF. According to the properties of the duals of Boolean functions, it is true that if $\mathbf{x} \in \{\mathbf{u}: f(\mathbf{u}) = 1, w_H(\mathbf{u}) = i\}$, then $\overline{\mathbf{x}} \notin \{\mathbf{u}: f(\mathbf{u}) = 1, w_H(\mathbf{u}) = N - i\}$, and vice versa. This gives Eq. 7–9. ■

DEFINITION 7–5. A Boolean function $f(\mathbf{x})$ represented in the form

$$f(x_1, x_2, \ldots, x_N) = \begin{cases} 1 & \text{if } \sum_{i=1}^{N} w_i x_1 \geqslant T, \\ 0 & \text{otherwise,} \end{cases}$$

where x_i are binary variables and the weights w_i and the threshold T are constants, is called a *linearly separable Boolean function*. The values of the parameters $w_1, w_2, \ldots, w_N, T$ uniquely define a linearly separable Boolean function.

7.1.4.2 Morphological Filters

Morphological filters are nonlinear transformations that locally modify geometrical features of signals. They stem from the basic operations of a set-theoretical method for image processing and analysis, called *mathematical morphology*, which was introduced by Matheron [31] and Serra [41]. In morphological filtering, geometrical features of a signal are modified by convolving the signal "morphologically" with a *structuring set*, which is a finite set chosen for its geometrical properties. Several types of morphological filters have been defined and have been used in digital signal processing for a number of years. Discrete flat morphological filters are stack filters. Henceforth, we always mean by the term *morphological filter* a discrete flat morphological filter. The basic morphological filters are *dilation* and *erosion*. Four compound operations: *closing*, *opening*, *close-opening*, and *open-closing* are defined based on these operations. The definitions of these operations are given in the following paragraphs [41].

DEFINITION 7–6. Let $f : \mathbf{Z}^m \mapsto \mathbf{R}$ be a signal and $B \subset \mathbf{Z}^m$. Then the dilation of f by B is denoted by $f \oplus B$, and is defined by

$$f \oplus B(x) = \max_{y \in B_x}\{f(y)\}$$

for all $x \in \mathbf{Z}^m$.

DEFINITION 7–7. Let $f : \mathbf{Z}^m \mapsto \mathbf{R}$ be a signal and $B \subset \mathbf{Z}^m$. Then the erosion of f by B is denoted by $f \ominus B$, and is defined by

$$f \ominus B(x) = \min_{y \in B_x} \{f(y)\}$$

for all $x \in \mathbf{Z}^m$.

Thus the dilation (erosion) of f by B at any point x is obtained by shifting the set B to location x and taking the maximum (minimum) of f inside the shifted set. So the structuring set plays the same kind of role as a moving window. Dilation and erosion can also be called *extreme order statistic filters*.

DEFINITION 7–8. Let $f : \mathbf{Z}^m \mapsto \mathbf{R}$ be a signal $B \subset \mathbf{Z}^m$ and B^s the symmetric set of B. Then the closing of f by B is denoted by f^B, and is defined by

$$f^B(x) = \left[(f \oplus B) \ominus B^s\right](x)$$

for all $x \in \mathbf{Z}^m$.

DEFINITION 7–9. Let $f : \mathbf{Z}^m \mapsto \mathbf{R}$ be a signal $B \subset \mathbf{Z}^m$ and B^s the symmetric set of B. Then the opening of f by B is denoted by f_B, and is defined by

$$f_B(x) = \left[(f \ominus B) \oplus B^s\right](x)$$

for all $x \in \mathbf{Z}^m$.

In the same way that closing and opening were defined as dilation followed by erosion and erosion followed by dilation, the close-opening by the structuring set B is defined as the closing by B followed by the opening by B, and open-closing by B is defined as the opening by B followed by the closing by B.

Since morphological filters are based on order statistics, they are stack filters [46]. Explicit stack filter expressions for morphological filters are given in [24].

It is straightforward to show that dilation by B is the dual of erosion by B, closing by B is the dual of opening by B, and close-opening by B is the dual of open-closing by B.

7.2 OPTIMIZATION OF STACK FILTERS

The design of stack filters is naturally one of the most important issues in stack filtering and has been the subject of various papers and reports. The design problem

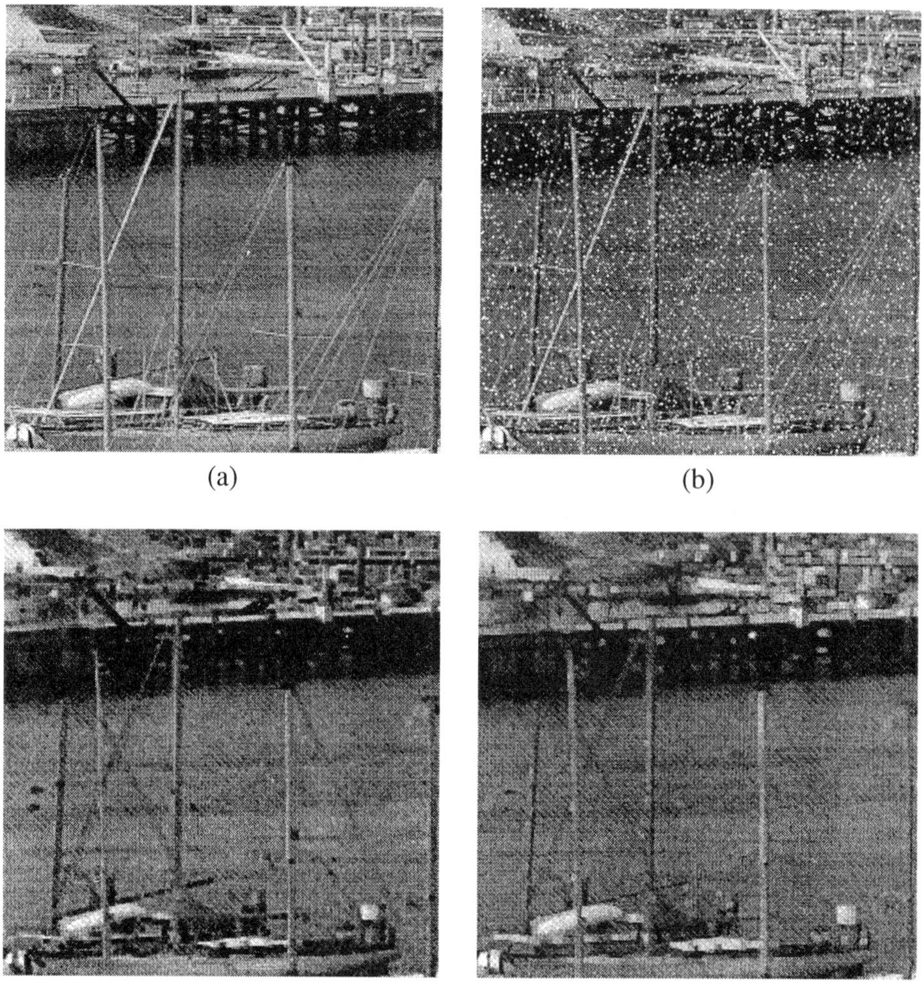

(a) (b) (c) (d)

Figure 7–5. Filtering a test image corrupted by additive impulsive noise (10%) with 3×3 windows. (a) Original image, (b) noisy image, (c) filtered image by morpohological erosion, and (d) filtered image by morphological opening.

can be formulated as follows: find the stack filter which restores the noisy signal close to the ideal document. Various closeness criteria have been developed; some are classic [e.g., the mean square error (MSE), the mean absolute error (MAE) and the minimax error], while some are relatively newer (e.g., associative memory [51] and shape preservation criteria [28]). Naturally, the resulting optimal filter depends on the optimality/closeness criterion (or cost function) used, which should be carefully chosen – a perfect optimization method with a wrong optimization criterion will give undesirable results.

Also, various optimal design techniques have been proposed for selecting the best filter in the class of stack filters. In almost all the classical papers, the approach taken was an analytical one, based on the availability of signal and noise models. This approach is often referred to as the *model-based approach*. Another approach, the *training approach*, is based on a representative training set containing the ideal signal and the corrupted signal. The efficiency of the first approach depends on the model's precision. Unfortunately, increased precision usually increases the model complexity at the same time. This is the modelling dilemma that should be solved in every case when stack filters are designed by using models for signals and noise. The efficiency of the second approach in turn depends on how sufficient the training set is and how well it generalizes to other similar problems. Our opinion is that this particular problem is almost completely unsolved at the moment.

In this section we give a couple of examples of these two approaches. First, we study optimization of stack filters with a simplified scenario: the ideal signal is constant and the noise distribution is known [26]. In this approach the aim is to seek an optimal stack filter, that produces the best noise attenuation and at the same time satisfies the given constraints. We show that the optimal stack filter, which achieves the best noise attenuation subject constraints, can usually be obtained in closed form for self-dual filters. We present also an algorithm for finding this closed form. Second, we study training-based optimization. During the past few decades there has been a growing interest in algorithms that are based on analogies with natural processes. Among these are many popular optimization and programming paradigms such as neural networks, simulated annealing, and genetic algorithms. They are easy to understand, because examples of them can be seen even in everyday life, but they are still efficient. We illustrate how simulated annealing and genetic algorithms can be used in stack filter design as training methods.

7.2.1 OPTIMIZATION WITH CONSTANT IDEAL SIGNAL AND KNOWN NOISE DISTRIBUTION

In the theory of linear systems, the power spectrum of the input signal together with the transfer function of the system determine the power spectrum of the output signal. It is not possible to get equally simple strong connections between the input process and the output process for stack filters. Explicit information on the statistical properties of the output can only be derived for the case of constant signal plus noise. Even then we need to assume the noise to be white. However, this result makes possible numerical optimization of noise attenuation of stack filters under different constraints, which quarantee that the filter satisfies prescribed specifications.

We first review a method to calculate the output moments of stack filters by using the coefficients A_i [26].

Let the input values $\mathbf{x} = (X_1, X_2, \ldots, X_N)$ of a stack filter $S_f(\cdot)$ be independent, identically distributed random variables having a common distribution function $\Phi(t)$ and density $\phi(t)$. Then the γ-order moment about the origin of the output of a stack filter can be expressed as

$$\alpha^\gamma = E\{Y_{out}^\gamma\} = \sum_{i=0}^{N-1} A_i M(\Phi, \gamma, N, i),$$

where

$$M(\Phi, \gamma, N, i) = \int_{-\infty}^{\infty} x^\gamma \frac{d}{dx}\left((1 - \Phi(x))^i \Phi(x)^{N-i}\right) dx, \quad i = 0, 1, \ldots, N-1.$$

By using the output moments about the origin we easily obtain output central moments, denoted by $\mu^\gamma = E\{(Y_{out} - E\{Y_{out}\})^\gamma\}$; for example, the second-order central output moment equals

$$\mu^2 = \sum_{i=0}^{N-1} A_i M(\Phi, 2, N, i) - \left(\sum_{i=0}^{N-1} A_i M(\Phi, 1, N, i)\right)^2. \qquad (7\text{–}10)$$

The second-order central output moment is quite often used as a measure of the noise attenuation capability of a filter. It quantifies the spread of the input samples with respect to their mean value. Equation 7–10 gives an expression for the second-order central output moment. $M(\cdot)$ is a function of the input distribution Φ, the window size N and index i.

In the following paragraphs the properties (without proofs) of the numbers $M(\Phi, \gamma, N, i)$ are reviewed [26]. Henceforward we assume that the input distribution $\Phi(t)$ is symmetric with respect to its mean μ_x, which is assumed to exist. We assume that the set $Q = \{t : \phi(t) > 0\}$ is a union of countable number of disjoint intervals of positive measure. This means that if $\phi(t) \neq 0$, then there exists an interval $I_t \subseteq (-\infty, \infty)$ of positive measure so that $t \in I_t$ and $\phi(u) > 0$ for all $u \in I_t$ and that there are countable number of intervals I_t. Without loss of generality, we assume

$$\mu_x = 0.$$

Therefore,

$$\Phi(t) = 1 - \Phi(-t),$$

which implies

$$\phi(t) = \phi(-t).$$

PROPOSITION 7–6. *The numbers $M(\Phi, \gamma, N, i)$ have the following recurrence formula*

$$M(\Phi, \gamma, N, i) = M(\Phi, \gamma, N-1, i-1) - M(\Phi, \gamma, N, i-1), \quad 1 \leqslant i \leqslant N$$

with initial values

$$M(\Phi, \gamma, N, 0) = \int_{-\infty}^{\infty} x^{\gamma} \frac{d}{dx}\left(\Phi(x)^N\right) dx, \quad i = 0, 1, \ldots, N-1, \ 0 \leqslant N.$$

PROPOSITION 7–7. *The numbers $M(\Phi, \gamma, N, i)$ satisfy*

$$M(\Phi, \gamma, N, i) = \sum_{j=0}^{i} \binom{i}{j} (-1)^{i-j} M(\Phi, \gamma, N-j, 0), \quad 0 \leqslant i \leqslant N.$$

PROPOSITION 7–8. *The numbers $M(\Phi, \gamma, N, i)$ satisfy the following symmetry property*

$$M(\Phi, \gamma, N, N-i) = (-1)^{\gamma+1} M(\Phi, \gamma, N, i).$$

PROPOSITION 7–9. *If γ is odd, then*

$$M(\Phi, \gamma, N, i) \begin{cases} > 0, & i = 0, N, \\ < 0, & \text{otherwise,} \end{cases}$$

and if γ is even, then

$$M(\Phi, \gamma, N, i) \begin{cases} = 0, & i = N/2, \\ > 0, & i = 0 \text{ or } N/2 < i < N, \\ < 0, & \text{otherwise.} \end{cases}$$

PROPOSITION 7–10. *If γ is odd, then*

$$\begin{aligned} M(\Phi, \gamma, N, i) &< M(\Phi, \gamma, N, i+1), \quad 0 < i < N/2 - 1, \\ M(\Phi, \gamma, N, i) &> M(\Phi, \gamma, N, i+1), \quad i = 0 \text{ or } N/2 \leqslant i < N-1, \end{aligned}$$

and if γ is even, then

$$\begin{aligned} M(\Phi, \gamma, N, i) &< M(\Phi, \gamma, N, i+1), \quad 0 < i < N-1, \\ M(\Phi, \gamma, N, i) &> M(\Phi, \gamma, N, i+1), \quad i = 0, N-1. \end{aligned}$$

The following proposition gives an estimate of the increase of $M(\Phi, 2, N, i+1) - M(\Phi, 2, N, i)$.

PROPOSITION 7–11. *Let $\Phi(t)$ be a distribution function such that its density function $\phi(t)$ satisfies the following conditions*:

(1) $\phi(t) = \phi(-t)$ *for all* $t \in \mathbf{R}$,

(2) $\phi(t)$ *is piecewise twice differentiable and the first derivative* $\phi'(t)$ *satisfies*

$$\phi'(t) \begin{cases} \geqslant 0, & t \leqslant 0, \\ \leqslant 0, & t > 0. \end{cases}$$

Then

$$M(\Phi, 2, N, i+1) > \frac{i+1}{N-i} M(\Phi, 2, N, i), \quad \text{for all } 0 < i < N-1. \tag{7–11}$$

When we write Eq. 7–11 in the form

$$\binom{N}{i+1} M(\Phi, 2, N, i+1) > \binom{N}{i} M(\Phi, 2, N, i)$$

and notice that $A_i \leqslant \binom{N}{i}$, we can observe that if the aim is to minimize the second-order moment about the origin, the smaller values of i have a more important role than the larger values of i.

Now we return to the calculation of the second-order central output moment of the output of a stack filter $S_f(\cdot)$. Let

$$\mathbf{a} = (A_0, A_1, \ldots, A_{N-1})$$

denote the $\mathbf{a}$-vector of a stack filter $S_f(\cdot)$ and

$$\mathbf{M}_\gamma = \big(M(\Phi, \gamma, N, 0), M(\Phi, \gamma, N, 1), \ldots, M(\Phi, \gamma, N, N-1)\big)$$

denote the $\mathbf{M}_\gamma$-vector. Then Eq. 7–10 can be rewritten as

$$\mu^2 = \mathbf{a}\mathbf{M}_2^T - \big(\mathbf{a}\mathbf{M}_1^T\big)^2, \tag{7–12}$$

where T denotes matrix transpose.

In the case of self-dual filters Eq. 7–12 is reduced to

$$\mu^2 = \mathbf{a}\mathbf{M}_2^T. \tag{7–13}$$

Note that while minimizing second-order central output moment we are in fact minimizing the mean square error in the following situation [26, 48]. Assume that the input X_i of a stack filter $S_f(\cdot)$ with window length N is a constant signal s plus additive white noise n_i, that is,

$$X_i = s + n_i, \tag{7–14}$$

where i stands for the ith sample. We denote the N samples inside the filter window $X_1, X_2, \ldots, X_N$. The output $\hat{s} = S_f(X_1, X_2, \ldots, X_N)$ of the stack filter $S_f(\cdot)$ is an estimate of s. The mean square error is defined as follows

$$E\{(s - \hat{s})^2\}.$$

Since s in Eq. 7–14 is a constant signal, we can recast $E\{(s - \hat{s})^2\}$ into the form 7–12. Thus, when the second-order central output moment is minimized, the mean square error $E\{(s - \hat{s})^2\}$ is also minimized. This explains why the second-order central output moment can be used as a measure of the noise attenuation capability of a filter.

The vectors $\mathbf{M}_i$ in Eqs. 7–12 and 7–13 are independent of the filter, while the vector $\mathbf{a}$ can be understood as a function of the filter. In the optimization process we aim at finding the vector $\mathbf{a} = (A_0, A_1, \ldots, A_{N-1})$, minimizing Eqs. 7–12 or 7–13.

In the following paragraphs, we provide an improved version of the optimization problem for self-dual filters. Let the size of the moving window, N, be odd. We write Eq. 7–13 as

$$\mu^2 = \mathbf{a}_{med}\mathbf{M}_2^T - \mathbf{a}_0\mathbf{M}_2^T + \mathbf{a}_1\mathbf{M}_2^T,$$

where $\mathbf{a}_{med}$ is the $\mathbf{a}$-vector of the standard N point median filter, that is,

$$\mathbf{a}_{med} = \left(1, \binom{N}{1}, \binom{N}{2}, \ldots, \binom{N}{\lfloor \frac{N-1}{2} \rfloor}, 0, 0, \ldots, 0\right), \tag{7–15}$$

and the vectors $\mathbf{a}_0 = (0, A_1^0, A_2^0, \ldots, A^0_{\lfloor \frac{N-1}{2} \rfloor}, 0, 0, \ldots, 0)$ and $\mathbf{a}_1 = (0, 0, \ldots, 0, A^1_{\lfloor \frac{N+1}{2} \rfloor}, A^1_{\lfloor \frac{N+3}{2} \rfloor}, \ldots, A^1_{N-1})$ satisfy

$$A_i^0 = A^1_{N-i}, \quad 1 \leqslant i \leqslant N-1. \tag{7–16}$$

Now, $\mathbf{a}_{med}\mathbf{M}_2^T$ is independent on the filter and therefore it is a constant which can be left out in optimization. Thus the optimization of self-dual filters is reduced to finding vectors $\mathbf{a}_0$ and $\mathbf{a}_1$, minimizing

$$-\mathbf{a}_0\mathbf{M}_2^T + \mathbf{a}_1\mathbf{M}_2^T. \tag{7–17}$$

Proposition 7–8, together with Eq. 7–16, implies that

$$-A_i M(\Phi, 2, N, i) = A_{N-i} M(\Phi, 2, N, N-i), \quad 1 \leqslant i \leqslant \left\lfloor \frac{N+1}{2} \right\rfloor.$$

Thus Eq. 7–17 is further reduced to

$$\mathbf{a}_0' \mathbf{M}_2'^T,$$

where $\mathbf{a}_0'$ and $\mathbf{M}_2'$ are truncated $\mathbf{a}_0$ and $\mathbf{M}_2$ vectors given by

$$\mathbf{a}_0' = \left(0, A_1^0, A_2^0, \ldots, A^0_{\lfloor \frac{N-1}{2} \rfloor}\right) \tag{7–18}$$

and

$$\mathbf{M}_2' = \left(M(\Phi, 2, N, 0), M(\Phi, 2, N, 1), \ldots, M\left(\Phi, 2, N, \left\lfloor \frac{N-1}{2} \right\rfloor\right)\right),$$

respectively.

The coefficient A_i^0 gives the number of binary vectors of Hamming weight i, $1 \leqslant i \leqslant \lfloor \frac{N-1}{2} \rfloor$, which the stack filter maps to 1, that is,

$$A_i^0 + A_i = \binom{N}{i}.$$

Yang *et al.* [48] used M_i to denote A_i^0.

The above considerations show that half of the numbers A_i completely define the second-order central output moment for self-dual stack filters.

7.2.1.1 Constraints for the Numbers A_i

Some basic conditions that A_i has to satisfy are considered in Kuosmanen *et al.* [26]. These conditions are listed below and are called *basic constraints* (BC).

(BC1):
$$A_i \in \mathbf{Z}_+,$$

(BC2):
$$A_0 = 1, \qquad A_N = 0,$$

(BC3):
$$0 \leqslant A_i \leqslant \binom{N}{i}, \quad i = 1, 2, \ldots, N,$$

(BC4):
$$A_{i+1} \leqslant \frac{N-i}{i+1} A_i, \quad i = 0, 1, \ldots, N-1.$$

The constraint (BC2) guarantees that the resulting optimal stack filter is not defined by the trivial positive Boolean functions $f(\mathbf{x}) = 0, \forall \mathbf{x} \in \{0, 1\}^N$ or $f(\mathbf{x}) = 1, \forall \mathbf{x} \in \{0, 1\}^N$.

We also have a special constraint for self-dual filters:

(SDC): If it is required that the filter be self-dual, then

$$A_i = \binom{N}{i} - A_{N-i}, \quad i = 0, 1, \ldots, N.$$

Different kinds of constraints have been studied, cf. [26, 27, 48]. In the following paragraphs we briefly review several such constraints. First we review so-called *rank selection constraints* (RSC) which determine some output characteristics that the optimal filter ought to have. In other words, by rank selection constraints we limit our search to a set of stack filters with some common statistical description.

Prasad *et al.* [36] defined so-called *rank selection probabilities* as follows:

DEFINITION 7–10. Let Y be the output of a stack filter defined by a positive Boolean function $f(\cdot)$. Then the ith *rank selection probability (RSP)* is denoted by $P[Y = X_{(i)}]$, $1 \leqslant i \leqslant N$, and is the probability that the output $Y = X_{(i)}$. The *rank selection probability vector* is the row vector $\mathbf{r} = (r_1, r_2, \ldots, r_N)$, where $r_i = P[Y = X_{(i)}]$, $1 \leqslant i \leqslant N$.

Kuosmanen *et al.* derived a simple relationship between coefficients A_i and rank selection probabilities r_i [25]. This connection is stated in the following proposition.

PROPOSITION 7–12. *The ith rank selection probability of a stack filter with window size N is given by*

$$r_j = \frac{A_{N-j}}{\binom{N}{j}} - \frac{A_{N-j+1}}{\binom{N}{j-1}}, \quad j = 1, 2, \ldots, N. \tag{7–19}$$

From Proposition 7–12 we obtain for instance the following corollaries; the proofs can be found in Kuosmanen *et al.* [26].

COROLLARY 7–2. *The coefficients A_i of a stack filter $S(\cdot)$ of window size N and rank selection vector $\mathbf{r} = (r_1, r_2, \ldots, r_N)$ satisfy*

$$A_{i+1} = \binom{N}{i+1}\left[\frac{A_i}{\binom{N}{i}} - r_{N-i}\right] = \frac{N-i}{i+1}A_i - \binom{N}{i+1} r_{N-i}, \quad i = 0, 1, \ldots, N-1,$$

$$A_{j+k} = \binom{N}{j+k}\left[\frac{A_j}{\binom{N}{j}} - \sum_{i=N-j-k+1}^{N-j} r_i\right], \quad 1 \leqslant k \leqslant N-j$$

and

$$\binom{N}{j} \sum_{i=N-j-k+1}^{N-j} r_i \leqslant A_j, \quad 1 \leqslant k \leqslant N-j.$$

COROLLARY 7–3. *The coefficients A_i of a stack filter $S(\cdot)$ of window size N satisfy*

$$A_{i+1} \leqslant \frac{N-i}{i+1} A_i, \quad i = 0, 1, \ldots, N-1$$

and

$$A_{i+1} \leqslant A_i, \quad \textit{when } i \geqslant \frac{N-1}{2}.$$

By using the above proposition and corollaries, the following constraints can be obtained.

(RSC1): If it is required that

$$r_1 = r_2 = \cdots = r_k = 0,$$

then

$$A_{N-1} = A_{N-2} = \cdots = A_{N-k} = 0.$$

(RSC2): If it is required that

$$r_k = r_{k+1} = \cdots = r_N = 0,$$

then

$$A_j = \binom{N}{j}, \quad j = 1, 2, \ldots, N-k+1.$$

Before presenting the third rank selection constraint, we need to fix some notations. Let the Greek letters denote arbitrary binary relations for a while. Then, for example, we use

$$x \begin{Bmatrix} \alpha^1 \\ \beta^2 \\ \gamma^3 \end{Bmatrix} y \quad \text{if and only if} \quad X \begin{Bmatrix} A^1 \\ B^2 \\ \Gamma^3 \end{Bmatrix} Y$$

to denote, that x is in relation α with y if and only if X is in relation A with Y; x is in relation β with y if and only if X is in relation B with Y; x is in relation γ with y if and only if X is in relation Γ with Y. Thus, the superscripts indicate what two relations are connected.

(RSC3): If it is required that for some i, $1 \leqslant i \leqslant N$, and $a \in [0, 1]$, it holds

$$r_i \left\{ \begin{matrix} >^1 \\ \geqslant^2 \\ =^3 \\ \leqslant^4 \\ <^5 \end{matrix} \right\} a,$$

then

$$A_{N-i+1} \left\{ \begin{matrix} <^1 \\ \leqslant^2 \\ =^3 \\ \geqslant^4 \\ >^5 \end{matrix} \right\} \frac{i}{N-i+1} A_{N-i} - \binom{N}{i-1} a.$$

(RSC4): If it is required that for some i, $1 \leqslant i \leqslant N$, and $a \in [0, 1]$, it holds

$$r_i \left\{ \begin{matrix} >^1 \\ \geqslant^2 \end{matrix} \right\} a,$$

then

$$A_{N-j} \left\{ \begin{matrix} >^1 \\ \geqslant^2 \end{matrix} \right\} \binom{N}{j} a, \quad \text{for all } i \leqslant j \leqslant N.$$

Rank selection constraints are not limited to those listed above and it is quite straightforward to find new ones for specific needs.

It is also possible to optimize stack filters under some prespecified set of so-called *structural constraints* [13, 26, 48, 49]. The goal of the structural constraints is to preserve some desired signal details (e.g., pulses in 1-D signals, lines in images) and to remove undesired signal patterns. The structural constraints consist of a list of different structures to be preserved, deleted or modified. Since stack filters obey the threshold decomposition, the structural constraints only need to be considered in the context of binary signals. That is, they can be specified by a set of binary vectors and their outputs. We divide these binary vectors into two subsets, *type 1* constraints and *type 0* constraints.

DEFINITION 7–11. A binary vector which is specified by the structural constraints is called a *type 1* constraint if the filter output is 1; otherwise it is called a *type 0* constraint.

We denote the set of all *type 1* constraints by $\Gamma_1 = \{\mathbf{x}_1, \mathbf{x}_2, \ldots, \mathbf{x}_p\}$ and the set of all *type 0* constraints by $\Gamma_0 = \{\mathbf{y}_1, \mathbf{y}_2, \ldots, \mathbf{y}_q\}$.

Structural constraints induce two new constraints for the coefficients A_i:

(SC1): Let the number of vectors $\mathbf{x} \in \Gamma_1$ with $w_H(\mathbf{x}) = i$ be $\gamma_i^{(1)}$ for all $1 \leqslant i \leqslant N-1$. Then

$$A_i \leqslant \binom{N}{i} - \gamma_i^{(1)}, \quad 1 \leqslant i \leqslant N-1.$$

(SC1): Let the number of vectors $\mathbf{x} \in \Gamma_0$ with $w_H(\mathbf{x}) = i$ be $\gamma_i^{(0)}$ for all $1 \leqslant i \leqslant N-1$. Then

$$A_i \geqslant \gamma_i^{(0)}, \quad 1 \leqslant i \leqslant N-1.$$

Breakdown point constraints (BPoC) form the third class of constraint used in this paper. The breakdown point is a simple quantitative global robustness measure widely used in statistics. Loosely speaking, it is the smallest fraction of free contamination which can carry the value of the estimator over all bounds [14, 15]. For example, the arithmetical mean has breakdown point $1/N$, where N is the amount of the samples. This means that even a single outlier can carry the mean to infinity. Often the breakdown point of the arithmetical mean is given to equal 0, which is obtained by letting $N \to \infty$. In general, the breakdown point lies between 0 and 1; a positive breakdown point is closely related (but not identical) to robustness of the estimator. It can be said that the breakdown point directly measures the global reliability of a statistic, which is one of the most important robustness aspects.

Since we are dealing with finite samples included in the input window, we adopt here a finite-sample definition of breakdown point. There are also other definitions for the finite-sample definition of breakdown point that we use, cf. [9, 15].

DEFINITION 7–12. The *finite-sample breakdown point* ε_N^* of the estimator $T_N(\cdot)$ at the sample $\{x_1, x_2, \ldots, x_N\}$, $x_i \neq \pm\infty$, is given by

$$\varepsilon_N^* = \frac{1}{N} \min\left\{m : \sup_{y_1,\ldots,y_m} |T_N(z_1, \ldots, z_N)| = \infty\right\}, \tag{7–20}$$

where the sample $\{z_1, \ldots, z_N\}$ is obtained by replacing any m sample points $x_{i_1}, \ldots, x_{i_m}$ by the arbitrary values $y_1, \ldots, y_m$.

Note that the finite-sample breakdown point Eq. 7–20 usually does not depend on the sample $\{x_1, x_2, \ldots, x_N\}$.

In the following paragraph a list of breakdown point constraints is given [27].

(BPoC1): If it is required that for $a \in [0, 1]$, it holds $\varepsilon_N^* > a$, then

$$A_{N-1} = A_{N-2} = \cdots = A_{N-\lfloor aN \rfloor} = 0 \quad \text{and} \quad A_i = \binom{N}{j}, \; j = 1, 2, \ldots, \lfloor aN \rfloor.$$

(BPoC2): If it is required that for $a \in [0, 1]$, it holds $\varepsilon_N^* \geqslant a$, then

$$A_{N-1} = A_{N-2} = \cdots = A_{N-\lceil aN \rceil + 1} = 0$$

and

$$A_i = \binom{N}{j}, \quad j = 1, 2, \ldots, \lceil aN \rceil - 1.$$

Naturally, constraints $\varepsilon_N^* < a$ and $\varepsilon_N^* \leqslant a$ are not very useful in real problems and they are skipped.

The problem of finding of the **a** vector of the stack filter which minimizes the second-order central output moment under constraints is stated as follows:

Minimize $\mathbf{aM}_2^T$
subject to
(BC1)-(BC4) basic constraints
(SDC) self-duality constraint
(possible) rank selection constraints
(possible) structural constraints
(possible) breakdown point constraints
(possible) other constraints

A solution for this optimization problem is given in Kuosmanen *et al.* [26]. In the following section we review the solution by representing the problem in a lattice-theoretic way which leads to the algorithm presented in Kuosmanen *et al.* [26].

7.2.1.2 Lattice-Theoretic Representation of the Optimization Problem

We need to recall some more basic lattice terminology [40]. Let L be a lattice. A *lower ideal* of L is a subset P of L such that if $\mathbf{x} \in P$ and $\mathbf{y} \leqslant \mathbf{x}$, then $\mathbf{y} \in P$. An *upper ideal* of L is a subset Q of L such that if $\mathbf{x} \in Q$ and $\mathbf{y} \geqslant \mathbf{x}$, then $\mathbf{y} \in Q$. Let $\mathbf{y}_1, \mathbf{y}_2, \ldots, \mathbf{y}_k \in L$; then $P = \{\mathbf{x} \in L : \exists \mathbf{y}_i, i = 1, 2, \ldots, k : \mathbf{x} \leqslant \mathbf{y}_i\}$ is called the *lower ideal generated by* $\mathbf{y}_1, \mathbf{y}_2, \ldots, \mathbf{y}_k$, and similarly, $Q = \{\mathbf{x} \in L : \exists \mathbf{y}_i, \; i = 1, 2, \ldots, k : \mathbf{x} \geqslant \mathbf{y}_i\}$ is called the *upper ideal generated by* $\mathbf{y}_1, \mathbf{y}_2, \ldots, \mathbf{y}_k$. Let $A \subseteq$

L; then $\mathbf{x} \in L$ is called a *maximal element* of A, if $\mathbf{y} \in A$, $\mathbf{y} \geqslant \mathbf{x}$ implies $\mathbf{x} = \mathbf{y}$; a *minimal element* of A is similarly defined. It is clear that an upper ideal is generated by its minimal elements and a lower ideal is generated by its maximal elements.

If in the optimization there are no constraints, then the optimal filter has $\mathbf{a}_0' = (0, 0, \ldots, 0)$ and therefore the $\mathbf{a}$ vector of the optimal filter equals the $\mathbf{a}$ vector of the median filter given in Eq. 7–15. In this case all vectors with their Hamming weight less than $\frac{N+1}{2}$ are mapped to 0 and other vectors are mapped to 1. When there are constraints, for some vectors $\mathbf{x}$, $w_H(\mathbf{x}) \leqslant N/2$, it holds that $f(\mathbf{x}) = 1$ and because of the self-duality, $f(\overline{\mathbf{x}}) = 0$, where $f(\cdot)$ is the positive Boolean function defining the optimal self-dual stack filter. Let Ω_1 be the upper ideal generated by the vectors $\mathbf{x}$, and Ω_0 be the lower ideal generated by the vectors $\overline{\mathbf{x}}$, that is,

$$\Omega_1 = \{\mathbf{x} \in B_N : w_H(\mathbf{x}) \leqslant N/2\} \cap f^{-1}(1)$$

and

$$\Omega_0 = \{\mathbf{x} \in B_N : w_H(\mathbf{x}) \geqslant N/2\} \cap f^{-1}(0).$$

Thus $\Omega_1 \subseteq f^{-1}(1)$ and $\Omega_0 \subseteq f^{-1}(0)$. In order that there exist a positive Boolean function $f(\cdot)$ (or a stack filter) with Ω_0 and Ω_1, it must hold that $\Omega_0 \cap \Omega_1 = \emptyset$. If the constraints are such that it is possible to form sets Ω_0 and Ω_1 in such a way that $\Omega_0 \cap \Omega_1 = \emptyset$, the constraints are feasible.

Furthermore, we denote $\Omega_{\text{med}} = \{\mathbf{x} \in B_N : w_H(\mathbf{x}) \geqslant N/2\}$. Then the filter $S_f(\cdot)$ can be described by

$$f^{-1}(1) = (\Omega_1 \cup \Omega_{\text{med}}) \cap (B_N \setminus \Omega_0). \tag{7–21}$$

Equation 7–21 expressed as a Boolean function yields

$$f(\mathbf{x}) = \left(\sum_{i=1}^{p} f_i(\mathbf{x}) + f_{\text{med}}(\mathbf{x}) \right) \prod_{i=1}^{p} f_i^D(\mathbf{x}), \tag{7–22}$$

where $f_{\text{med}}(\cdot)$ is the Boolean function of the N point standard median filter, $f_i(\cdot)$ is the elementary conjunction of an element of Ω_1, and p is the number of elements in the set Ω_1.

We can reduce Eq. 7–22 to a more simplified form, because we need only to sum over the minimal elements of Ω_1 instead of summing over all members of Ω_1. As proved in Yli-Harja *et al.* [50], the complements of the minimal elements of Ω_1 are the maximal elements of Ω_0.

The idea of our algorithm is to choose the minimal elements of Ω_1 in such a way that the constraints are satisfied and that the cardinalities of the sets Ω_1 and Ω_0 are kept as small as possible. This aim is achieved by adding new minimal elements only to the levels [by ith level we mean the vectors $\mathbf{x}$ with $w_H(\mathbf{x}) = i$] where they are necessary and adding as few of them as possible. By choosing as few elements as possible, we leave the maximum amount of freedom for the next levels, because we have maximized the number of vectors which may or may not belong to the sets Ω_1 and Ω_0 in the next levels. This freedom usually makes it easy to satisfy given constraints.

7.2.1.3 An Algorithm to Minimize Second-Order Central Output Moment over Self-Dual Filters under Constraints

Now we are ready to review an algorithm for the minimization of the second-order central output moment of self-dual filters under constraints [26]. A detailed discussion of the algorithm can be found in Kuosmanen *et al.* [26].

ALGORITHM 7–2. Find an optimal self-dual stack filter under given constraints. Let the Boolean variables be indexed as $x_1, x_2, \ldots, x_N$.

Step 1. Find the indices $1 \leqslant i \leqslant \left\lfloor \frac{N-1}{2} \right\rfloor$ of the coefficients A_i for which there are constraints other than basic constraints and the self-duality constraint. Let such indices be $i_1 < i_2 < \cdots < i_r$.

Step 2. Let $A_i = \binom{N}{i}$ for $1 \leqslant i < i_1$.

Step 3. For the index i_1 find the possible maximum value $A_{\max}$ under given constraints. Let $A_{i_1} = A_{\max}$ and $d = \binom{N}{i_1} - A_{\max}$. Now d gives the number of binary vectors of Hamming weight i_1 which should be minimal elements of the upper ideal Ω_1 at level i_1.

Step 4. Choose d binary vectors $\mathbf{x}_1, \mathbf{x}_2, \ldots, \mathbf{x}_d$ to be minimal elements of the upper ideal Ω_1. The vectors should satisfy $\mathbf{x}_i \not\leqslant \overline{\mathbf{x}_j}$, for all $i \neq j$, and not violate any structural constraint. If there are structural constraints that require specific vectors $\mathbf{x}$, $w_H(\mathbf{x}) = i_1$, to be in Ω_1, they are chosen. We discuss in more detail how the optimal choice of these vectors can be made below. Let $g(\cdot)$ equal the sum of the elementary conjunctions of the chosen vectors.

Steps (5)–(8) are repeated for each k, $2 \leqslant k \leqslant r$, in numerical order.

Step 5. For each $i_{k-1} < i \leqslant i_k$ find the maximum number of A_i given by the Boolean function $g(\cdot)$ by finding the cardinality of the set $\{\mathbf{y} : g(\mathbf{y}) = 0, w_H(\mathbf{y}) = i\}$. Denote that maximum value by A_i^g. In the calculation of A_i^g it is possible to use the inclusion-exclusion principle. We give also a direct spectral method for the calculation of the numbers A_i^g below. For $i_{k-1} < i < i_k$, let $A_i = A_i^g$; that is, in the levels where there are no constraints, minimal elements are not added.

Step 6. For the index i_k find the possible maximum value $A_{i_k}^c$ under given constraints. If the constraints do not limit A_{i_k} above and $A_{i_k}^g$ satisfies the constraints concerning A_{i_k}, then let $A_{i_k}^c = A_{i_k}^g$. Let $d = A_{i_k}^g - A_{i_k}^c$. Now d gives the number of binary vectors of Hamming weight i_k which should be minimal elements of the upper ideal Ω_1 at level i_k.

Step 7. If $d < 0$ or d is not defined, then there does not exist a self-dual filter satisfying the given constraints. Else let $A_{i_k} = A_{i_k}^c$.

Step 8. Choose d binary vectors $\mathbf{x}_1, \mathbf{x}_2, \ldots, \mathbf{x}_d$ to be minimal elements of the upper ideal Ω_1. A chosen vector $\mathbf{x}_i$ has to satisfy $g(\mathbf{x}_i) = 0$, $g^D(\mathbf{x}_i) = 1$ and $\mathbf{x}_i \not\leqslant \overline{\mathbf{x}_j}$, for all $1 \leqslant j \leqslant d$, $j \neq i$. Otherwise, either $\mathbf{x}_i$ would already belong to the set Ω_1 or $\Omega_0 \cap \Omega_1 \neq \emptyset$. Again, the chosen vectors must not be forbidden by some structural constraint and if there are structural constraints that require specific vectors $\mathbf{x}$, $w_H(\mathbf{x}) = i_1$, to be in Ω_1, they are chosen. If there are no d binary vectors satisfying the above requirements, then there does not exist a self-dual filter satisfying the given constraints. The optimal choice of these vectors is discussed below. Let $g'(\cdot)$ equal the sum of the elementary conjunctions of the chosen vectors and let $g(\cdot) = g(\cdot) + g'(\cdot)$.

When all the indices $1 \leqslant i \leqslant i_r$ have been run through, an optimal stack filter is defined by the Boolean function

$$f(\mathbf{x}) = \big(f_{med}(\mathbf{x}) + g(\mathbf{x})\big) g^D(\mathbf{x}). \qquad (7\text{–}23)$$

An optimal filter has the following interesting properties.

COROLLARY 7–4. *An optimal self-dual stack filter is optimal for any noise with symmetric density function.*

PROOF. This corollary follows since the choice of the vectors is independent of the underlying noise distribution, and the properties of the numbers $M(\Phi, \gamma, N, i)$ studied above hold for all symmetric density functions. ■

COROLLARY 7–5. *The optimal filter is not necessarily unique, because by choosing different vectors we would have obtained another optimal filter. In fact, every permutation of the variables $x_1, x_2, \ldots, x_N$ in the Boolean functions in Eq. 7–23 gives an optimal filter if there are no specific vectors which need to be in Ω_1 or Ω_0 by structuring constraints.*

EXAMPLE 7–2. Consider a self-dual stack filter ($N = 7$) with two constraints: $r_6 \geqslant 2/21$ and $r_5 \leqslant 2/10$. Let the Boolean variables be indexed by $x_1, x_2, \ldots, x_7$.

Step 1. The constraint $r_6 \geqslant 2/21$ gives by (RSC3) a constraint for A_2: $A_2 \leqslant 3A_1 - 2$ and the constraint $r_5 \leqslant 2/10$ gives a constraint for A_3: $A_3 \geqslant 5/3 \cdot A_2 - 7$. Thus, in this example $i_1 = 2$ and $i_2 = 3$.

Step 2. $A_1 = 7$.

Step 3. Constraint $A_2 \leqslant 3A_1 - 2$ gives $A_{\max} = 19$. Thus, $A_2 = 19$ and $d = 2$.

Step 4. We choose the vectors $\mathbf{x}_1 = (0, 0, 0, 1, 1, 0, 0)$ and $\mathbf{x}_2 = (0, 0, 1, 1, 0, 0, 0)$; then $g(\mathbf{x}) = x_4x_5 + x_3x_4$.

Step 5. There are 5 vectors $\mathbf{y}_i$ of weight 3, such that $\mathbf{y}_i \geqslant \mathbf{x}_1$, and 5 vectors $\mathbf{y}_j$, such that $\mathbf{y}_i \geqslant \mathbf{x}_2$, of which the vector $\mathbf{y} = (0, 0, 1, 1, 1, 0, 0)$ satisfies $\mathbf{y} \geqslant \mathbf{x}_1$ and $\mathbf{y} \geqslant \mathbf{x}_2$. Thus, $A_3^g = \binom{7}{3} - 2 \cdot 5 + 1 = 26$.

Step 6. Now the constraint $A_3 \geqslant 5/3 \cdot A_2 - 7$ gives $A_3 > 24$, which does not limit A_3 above and A_3^g satisfies $A_3^g > 24$. This gives $d = 0$.

Step 7. $A_3 = 26$.

Thus, an optimal self-dual stack filter ($N = 7$) satisfying $r_6 \geqslant 2/21$ and $r_5 \leqslant 2/10$ is defined by the Boolean function

$$f(\mathbf{x}) = (f_{\text{med}}(\mathbf{x}) + x_4x_5 + x_3x_4)(x_4 + x_5)(x_3 + x_4).$$

By Proposition 7–1 any stack filter can be implemented for multi-level signals by replacing AND and OR with MIN and MAX operations, respectively. This gives Corollary 7–6.

COROLLARY 7–6. *For real-valued signals, the optimal stack filter defined by the positive Boolean function given in Eq. 7–23 is a composition of the median filter and a set of maximum and minimum filters*:

$$S_f(\mathbf{X}) = \text{MIN}\big\{\text{MAX}\{\text{MED}\{\mathbf{X}\}, S_1(\mathbf{X})\}, S_2(\mathbf{X})\big\},$$

where

$$S_1(\mathbf{X}) = \text{MAX}\big\{S_{f_1}(\mathbf{X}), \ldots, S_{f_p}(\mathbf{X})\big\},$$
$$S_2(\mathbf{X}) = \text{MIN}\big\{S_{f_1}^D(\mathbf{X}), \ldots, S_{f_p}^D(\mathbf{X})\big\},$$

$S_{f_i}(\mathbf{X})$, $i = 1, 2, \ldots, p$, are stack filters for real-valued signals corresponding to the elementary conjuctions of $g(\mathbf{x})$. Likewise, $S^D_{f_i}(\mathbf{X})$, $i = 1, 2, \ldots, p$, correspond to the duals of the elementary conjuctions of $g(\mathbf{x})$.

Corollary 7–6 implies that the optimal self-dual stack filters defined in this paper are multistage rank-order-based filters. The idea of multistage filtering has been used to generate detail-preserving rank-order-based filters, cf. [3, 34]. It has been shown that these filters perform well when they are applied to image processing [3]. However, the design of these filters is not based on any optimization criterion. The optimal stack filters derived in this paper produce the best noise attenuation while satisfying some prespecified constraints.

7.2.1.4 The Optimal Choice of the Minimal Elements of Ω_1

In Steps (4) and (7) of Algorithm 7–2 the task was to choose d binary vectors $\mathbf{x}_1, \mathbf{x}_2, \ldots, \mathbf{x}_d$ to be minimal elements of the upper ideal Ω_1. If $\Omega_1 = \emptyset$, that is, we have not added any minimal elements to Ω_1, these chosen vectors should satisfy, $\mathbf{x}_i \not\leqslant \overline{\mathbf{x}_j}$, for all $i \neq j$. Otherwise $\Omega_0 \cap \Omega_1 \neq \emptyset$. If $\Omega_1 \neq \emptyset$, then in addition to the chosen vectors satisfying $\mathbf{x}_i \not\leqslant \overline{\mathbf{x}_j}$, for all $i \neq j$, a chosen vector $\mathbf{x}_i$ has to satisfy $g(\mathbf{x}_i) = 0$ and $g^D(\mathbf{x}_i) = 1$, where $g(\cdot)$ is the disjunction of the elementary conjunctions of the vectors that belong to Ω_1. Also, while choosing vectors, if there are structural constaints, it must hold for all chosen vectors $\mathbf{x}_i$ that $\mathbf{x}_i \not\leqslant \mathbf{y}$, where $\mathbf{y} \in \Gamma_0$. In addition, the vectors must be chosen in such a way that all vectors in Γ_1 are chosen to Ω_1.

We need the following properties of binary vectors.

PROPOSITION 7–13. *For two vectors $\mathbf{x} = (x_1, x_2, \ldots, x_N)$ and $\mathbf{y} = (y_1, y_2, \ldots, y_N)$ it holds that $\mathbf{x} \not\leqslant \overline{\mathbf{y}}$ and $\mathbf{y} \not\leqslant \overline{\mathbf{x}}$ if and only if there exists an entry i such that $x_i = y_i = 1$.*

PROOF. Assume first that there exist i such that $x_i = y_i = 1$. This means that for $\overline{\mathbf{x}} = (\overline{x_1}, \overline{x_2}, \ldots, \overline{x_N})$ and $\overline{\mathbf{y}} = (\overline{y_1}, \overline{y_2}, \ldots, \overline{y_N})$ it holds that $\overline{x_i} = \overline{y_i} = 0$ and thus $\mathbf{x} \not\leqslant \overline{\mathbf{y}}$ and $\mathbf{y} \not\leqslant \overline{\mathbf{x}}$. Assume then that $\mathbf{x} \not\leqslant \overline{\mathbf{y}}$ and $\mathbf{y} \not\leqslant \overline{\mathbf{x}}$. As $\mathbf{y} \not\leqslant \overline{\mathbf{x}}$, there must exist an index i, such that $y_i = 1$ and $\overline{x_i} = 0$. But from this we obtain that there exists an index i such that $x_i = y_i = 1$. ■

PROPOSITION 7–14. *The maximum cardinality of the set Θ of binary vectors of length N and Hamming weight i, $1 \leqslant i \leqslant \lfloor \frac{N-1}{2} \rfloor$, such that for every pair of vectors $\mathbf{x}, \mathbf{y} \in \Theta$ it holds that $\mathbf{x} \not\leqslant \overline{\mathbf{y}}$ and $\mathbf{y} \not\leqslant \overline{\mathbf{x}}$, is $\binom{N-1}{i-1}$. The set of maximal cardinality is formed by constructing all binary vectors of Hamming weight i with $x_j = 1$ for some fixed $1 \leqslant j \leqslant N$.*

Two proofs for Proposition 7–14 can be found in [6]. The original statement was in a paper by Erdös, Ko and Rado [10], which has turned out to be a milestone in the theory of extremal set systems according to many authorities. Proposition 7–14 implies that in order to guarantee that we can choose as many vectors as necessary for Ω_1, we should choose an index x_j, and let $x_j = 1$ in all chosen vectors. By doing this, Proposition 7–13 implies that $\Omega_1 \cap \Omega_0 = \emptyset$, as it should. Usually, if there are no structural constraints, it is sensible to choose the index corresponding to the value X_0 (midpoint) that we are estimating. By this choice we increase the selection probability of this sample, which often leads to an improved preservation of details.

Another problem is how to choose the vectors in level k in such a way that in levels l, $l > k$ there are as few vectors as possible. It is clear that for every vector $\mathbf{x}_i$ in level k, there exist $N-k$ vectors $\mathbf{y}_j$ in level $k+1$, such that $\mathbf{x}_i \leqslant \mathbf{y}_j$. If we choose a vector $\mathbf{x}_i$ to be a member of Ω_1, then by the stacking property Eq. 7–4 all these $N-k$ vectors in level $k+1$ are members of Ω_1 as well. This means that we should choose the vectors in level k in such a way that we obtain the same vectors as often as possible when we form the $N-k$ vectors corresponding to each of the chosen vectors. Otherwise, the coefficient A_{k+1} is not minimized.

By the distance between the vectors $\mathbf{x} = (x_1, x_2, \ldots, x_N)$ and $\mathbf{y} = (y_1, y_2, \ldots, y_N)$ is meant the number of the indices i, where $x_i \neq y_i$. It is clear that the distance of two vectors in the same level is at least 2. If we choose two vectors $\mathbf{x}$ and $\mathbf{y}$ from level k with distance 2, there is exactly one vector $\mathbf{z}$ in level $k+1$, such that $\mathbf{x} \leqslant \mathbf{z}$ and $\mathbf{y} \leqslant \mathbf{z}$, namely, the vector $\mathbf{z} = (x_1 + y_1, x_2 + y_2, \ldots, x_N + y_N)$. If the distance is greater than 2, then there are no vectors $\mathbf{z}$ in level $k+1$, such that $\mathbf{x} \leqslant \mathbf{z}$ and $\mathbf{y} \leqslant \mathbf{z}$. It is naturally appropriate to choose the vector $\mathbf{x}$, which has the maximum number of the vectors $\mathbf{y}$, in the same level and already in Ω_1, with distance 2 between $\mathbf{x}$ and $\mathbf{y}$. If there is more than one of these vectors, some tie-breaking rule must be applied. In this case, if there is an index such that the vectors in Ω_1 seldom have 1 in this index, then we break the tie in favor of a vector that has 1 in this index, if such a vector exists. Often it is recommended that the vectors be chosen, if possible, to be symmetric around the midpoint.

In Step 5 of Algorithm 7–2 we have to find the numbers A_i when we have added some vectors to Ω_1. This can be done directly as follows. First we need some definitions.

DEFINITION 7–13. Let $\mathbf{A} = (A_{ij})_{m \times m}$ be an $m \times m$ matrix and $\mathbf{B}$ an $n \times n$ matrix. The Kronecker product $\mathbf{A} \otimes \mathbf{B}$ of $\mathbf{A}$ and $\mathbf{B}$ is the $mn \times mn$ matrix given by

$$\mathbf{A} \otimes \mathbf{B} = \begin{pmatrix} a_{11}\mathbf{B} & a_{12}\mathbf{B} & \ldots & a_{1m}\mathbf{B} \\ a_{21}\mathbf{B} & a_{22}\mathbf{B} & \ldots & a_{2m}\mathbf{B} \\ \vdots & \vdots & \ddots & \vdots \\ a_{m1}\mathbf{B} & a_{m2}\mathbf{B} & \ldots & a_{mm}\mathbf{B} \end{pmatrix}.$$

DEFINITION 7–14. The conjunctive matrix of order N is the $2^N \times 2^N$ matrix $\mathbf{K}_N$ defined as

$$\mathbf{K}_N = \begin{pmatrix} 1 & 0 \\ 1 & 1 \end{pmatrix}^{\otimes N} = \underbrace{\begin{pmatrix} 1 & 0 \\ 1 & 1 \end{pmatrix} \otimes \cdots \otimes \begin{pmatrix} 1 & 0 \\ 1 & 1 \end{pmatrix}}_{N \text{ times}}.$$

EXAMPLE 7–3.

$$\mathbf{K}_2 = \begin{pmatrix} 1 & 0 & 0 & 0 \\ 1 & 1 & 0 & 0 \\ 1 & 0 & 1 & 0 \\ 1 & 1 & 1 & 1 \end{pmatrix} \quad \text{and} \quad \mathbf{K}_3 = \begin{pmatrix} 1 & 0 & 0 & 0 & 0 & 0 & 0 & 0 \\ 1 & 1 & 0 & 0 & 0 & 0 & 0 & 0 \\ 1 & 0 & 1 & 0 & 0 & 0 & 0 & 0 \\ 1 & 1 & 1 & 1 & 0 & 0 & 0 & 0 \\ 1 & 0 & 0 & 0 & 1 & 0 & 0 & 0 \\ 1 & 1 & 0 & 0 & 1 & 1 & 0 & 0 \\ 1 & 0 & 1 & 0 & 1 & 0 & 1 & 0 \\ 1 & 1 & 1 & 1 & 1 & 1 & 1 & 1 \end{pmatrix}.$$

DEFINITION 7–15. Let $\mathbf{f}$ be a column vector of length 2^N. Then the vector $\mathbf{K}_N \cdot \mathbf{f}$ is called the conjunctive transform of the vector $\mathbf{f}$.

The name of the *conjunctive* matrix $\mathbf{K}_N$ is connected with the fact that the jth column of this matrix, $j = 0, 1, \ldots, 2^N - 1$, is the truth table of the following *elementary conjunction*: $x_0^{j_0} x_1^{j_1} \ldots x_{N-1}^{j_{N-1}}$, where $\hat{j} = (j_0, j_1, \ldots, j_{N-1})$.

DEFINITION 7–16. Let $\mathbf{U}_N$ denote the *elementary Boolean matrix*, a rectangular $N + 1 \times 2^N$ matrix, defined by the following recurrent relation:

$$\mathbf{U}_0 = (1), \quad \mathbf{U}_1 = \begin{pmatrix} 1 & 0 \\ 0 & 1 \end{pmatrix} \quad \text{and} \quad \mathbf{U}_N = \begin{pmatrix} \mathbf{U}_{N-1} & \mathbf{0}_{N-1} \\ \mathbf{0}_{N-1} & \mathbf{U}_{N-1} \end{pmatrix},$$

where $\mathbf{0}_{N-1}$ is the row vector of zeroes of the length 2^{N-1}.

EXAMPLE 7–4.

$$\mathbf{U}_2 = \left(\begin{array}{cc|cc} 1 & 0 & 0 & 0 \\ \cline{3-4} 0 & 1 & 1 & 0 \\ \cline{1-2} 0 & 0 & 0 & 1 \end{array}\right) \quad \text{and} \quad \mathbf{U}_3 = \left(\begin{array}{cccc|cccc} 1 & 0 & 0 & 0 & 0 & 0 & 0 & 0 \\ \cline{5-8} 0 & 1 & 1 & 0 & 1 & 0 & 0 & 0 \\ 0 & 0 & 0 & 1 & 0 & 1 & 1 & 0 \\ \cline{1-4} 0 & 0 & 0 & 0 & 0 & 0 & 0 & 1 \end{array}\right).$$

The name of the *elementary Boolean* matrix $\mathbf{U}_N$ is connected with the fact that every $(i + 1)$th row of this matrix is the truth table of the ith elementary Boolean function, $i = 0, 1, \ldots, N$, i.e., the Boolean function $f^{(i)}(\mathbf{x})$, such that

$$f^{(i)}(\mathbf{x}) = \begin{cases} 1 & \text{if } w_H(\mathbf{x}) = i, \\ 0 & \text{otherwise}, \end{cases} \qquad i = 0, 1, \ldots, N.$$

PROPOSITION 7–15. *Let* $\mathbf{c}_\Omega$ *be the characteristic vector of the positive Boolean function* $f(\mathbf{x})$. *Then, the* $\mathbf{a}$ *vector can be obtained by*

$$\mathbf{a}^T = \left(\binom{N}{0}, \binom{N}{1}, \ldots, \binom{N}{N}\right)^T - \mathbf{U}_N \cdot T_1(\mathbf{K}_N \cdot \mathbf{c}_\Omega),$$

where the elementary Boolean matrix is defined in Definition 7–16, *the thresholding operation is given by Eq.* 7–1 *and the conjuctive transform is defined by Definition* 7–14.

EXAMPLE 7–5. Assume that $N = 4$ and we have added the vectors $\mathbf{x}_1 = (0, 1, 1, 0)$ and $\mathbf{x}_2 = (0, 1, 0, 1)$ to Ω_1. Therefore, $\Omega = \{(0, 1, 1, 0), (0, 1, 0, 1)\}$ and the characteristic vector equals $\mathbf{c}_\Omega = (0, 0, 0, 0, 0, 1, 1, 0, 0, 0, 0, 0, 0, 0, 0, 0)^T$. We easily obtain

$$\mathbf{K}_4 \cdot \mathbf{c}_\Omega = (0, 0, 0, 0, 0, 1, 1, 2, 0, 0, 0, 0, 0, 1, 1, 2)^T,$$
$$T_1(\mathbf{K}_4 \cdot \mathbf{c}_\Omega) = (0, 0, 0, 0, 0, 1, 1, 1, 0, 0, 0, 0, 0, 1, 1, 1)^T,$$
$$\mathbf{U}_4 \cdot T_1(\mathbf{K}_4 \cdot \mathbf{c}_\Omega) = (0, 0, 2, 3, 1)^T,$$

and finally

$$\mathbf{a}^T = (1, 4, 6, 4, 1)^T - (0, 0, 2, 3, 1)^T = (1, 4, 4, 1, 0)^T.$$

7.2.2 OPTIMIZATION BY SIMULATED ANNEALING

Simulated annealing is a statistical optimization method having its origins in condensed matter physics, where it was developed to find the ground state of a solid [21]. During the past decade it has been widely used to solve global combinatorial optimization problems [1, 29], including problems in signal processing, cf. [4, 7, 22, 28, 30]. It has shown efficiency especially in problems where a near-optimal solution (i.e., a solution close to global optimum) is also acceptable.

In the physical process denoted by annealing, a solid is heated up in a heat bath by increasing the temperature of the bath to a value at which the solid melts. Then the temperature is decreased until the particles arrange themselves in the low-energy ground state. The ground state can be achieved if the maximum temperature is sufficiently high and the cooling is carried out sufficiently slowly so that the solid can reach thermal equilibrium.

When simulated annealing is used in combinatorial optimization problems, the temperature is usually replaced with a *control parameter*, which has some *initial value*. Naturally there must also be some rule to decrease the value of the control parameter and some *stopping criterion*, for example, an *end value* for the control parameter. To reach the thermal equilibrium, we use the *Metropolis* [32] algorithm. In the Metropolis algorithm, a sequence of configurations (here representing stack filters) is generated until the equilibrium is reached. Each new configuration is selected from the neighborhood of the previous configuration, i.e., it is in some sense close to the previous configuration. The selected configuration is then accepted with some non-zero probability which depends on the value of the control parameter and on the difference of the values of the costs of the previous and the selected configurations. The *cost function* c is some error function whose value is calculated from the original image and the filtered image. In filter optimization, the cost function is typically MAE or MSE.

It can be shown [29] that simulated annealing reaches the global optimum if the cooling is made sufficiently slow. Unfortunately it may happen that we need to generate an infinite number of configurations. In practical situations this is of course not possible and so we have to restrict the number of the configurations somehow. In practice we can use, for example, the following simplified algorithm for simulated annealing.

ALGORITHM 7–3. Simulated annealing.

Step 1. Initialize the control parameter t and select a starting configuration s.

Step 2. Select a new configuration s' at random from the neighborhood of s. If $c(s') \leqslant c(s)$, set $s \leftarrow s'$, otherwise set $s \leftarrow s'$ with the probability $e^{-(c(s')-c(s))/t}$. Repeat this step k times.

Step 3. Decrease the value of the control parameter t. If the new value is greater than the end value, go to step 2, otherwise stop.

When the algorithm stops, s is the desired configuration and $c(s)$ the corresponding optimum value. From the algorithm we can see that in order to apply simulated annealing we need an initial value for the control parameter, some rule to decrease the value of the control parameter, some stopping criterion, and an upper bound for the number of iterations in step 2. These form the *cooling schedule* for simulated annealing. The starting configuration is usually selected at random.

The initial value for the control parameter is usually chosen in such a way that in the beginning almost all configurations would be accepted in the Metropolis algorithm, i.e., the acceptance ratio would be 0.95. The new value t_{n+1} of the control parameter is obtained by the rule

$$t_{n+1} = \alpha \cdot t_n,$$

where t_n is the previous value of the control parameter and α $(0 < \alpha < 1)$ is some constant, usually close to 1, e.g., $\alpha = 0.99$. The value of the control parameter is decreased if either $\eta \cdot \lambda$ $(0 < \eta < 1)$ configurations have been accepted or λ configurations have been generated, whichever occurs first. The purpose of the constant η is to make the cooling faster in the beginning, when the value of the control parameter is large. The parameter λ may be chosen to approximate the size of the largest neighborhood of all configurations, while the parameter η is usually chosen to be relatively small, e.g., $\eta = 0.1$. The algorithm terminates if there have not been any new accepted configurations for some fixed number of last values of the control parameter.

The definition of the neighborhood of a filter $S(\cdot)$ naturally depends on the representation of the positive Boolean function. Consider first the truth table representation of Boolean functions. Let $\alpha = (\alpha_1, \alpha_2, \ldots, \alpha_{2^N})$ be a PBF defining a stack filter $S_\alpha(\cdot)$. Then the simplest way of defining possible neighboring filters of $S_\alpha(\cdot)$ is as follows:

Let $\beta = (\beta_1, \beta_2, \ldots, \beta_{2^N})$ be a binary vector obtained from α by complementing some arguments of α. If the β obtained is positive, it is considered to define a stack filter in the neighborhood of $S_\alpha(\cdot)$. This definition is quite natural and the behavior of the filter $S_\beta(\cdot)$ obtained resembles the behavior of the filter $S_\alpha(\cdot)$ in many ways. However, by this definition only a few of the changes will lead to another positive Boolean function and another group of complementation operations must be tried. This problem can be solved by finding the characteristic vector form of α and complementing its elements. Methods for this are described in [2]. This yields to an increased complexity.

The drawbacks of representation by truth tables in optimization by simulated annealing are clearly the necessity of checking the positivity of every Boolean function and the amount of the storage needed for these functions. The second problem makes it practically impossible to use this representation for large window sizes.

Consider next the representation of positive Boolean functions by storing only minimal terms. Let

$$f(x_1, x_2, \ldots, x_N) = \sum_{i=1}^{K} \prod_{j \in P_i} x_j$$

define a stack filter $S_f(\cdot)$. Then the Boolean function

$$g(x_1, x_2, \ldots, x_N) = \sum_{i=1}^{L} \prod_{j \in Q_i} x_j$$

can be considered a neighbor of $f(x_1, x_2, \ldots, x_N)$ if it can be formed from $f(x_1, x_2, \ldots, x_N)$ with one of the following rules:

1. Add a new subset of $\{1, 2, \ldots, N\}$ to the set $\mathcal{P} = \{P_1, P_2, \ldots, P_K\}$.
2. Remove one set P_i from $\mathcal{P} = \{P_1, P_2, \ldots, P_K\}$.
3. Modify one set $P_i = \{P_{i_1}, P_{i_2}, \ldots, P_{i_z}\}$ from $\mathcal{P} = \{P_1, P_2, \ldots, P_K\}$ either by removing one index P_{i_k}, $1 \leqslant k \leqslant z$, or by adding a new index $P_{i_{z+1}}$.

It is easy to verify that these modifications retain the positivity of the new Boolean function. This representation is more suitable for the optimization of stack filters by simulated annealing, especially with large window sizes.

7.2.3 Optimization by Genetic Algorithms

The programming paradigm of genetic algorithms is based on the Darwinian laws of natural selection. In an optimization problem, a genetic algorithm generally maintains a population of objects in the search space. The population then changes under the laws of natural selection or some variants of them. Objects in the population behave just as some species in nature; they recombine to create new members of the population, and when there are too many individuals, the weakest do not survive. In addition, there has to be some method to prevent the population from degeneration, i.e., to get new chromosomes into the population members. In nature this is done by mutation, which changes the chromosomes of the individuals. After every step the population is evaluated and only the fittest ones survive to recombine and only their chromosomes stay in the population.

The type of data structure plays an important role in genetic algorithms. The first genetic algorithms, which are called *simple genetic algorithms* [16] represented objects as binary vectors. Since then, however, *real coding* using multivalued strings has been successfully used, because it is often easy to use and more natural. The basic structure of a genetic algorithm is as follows.

ALGORITHM 7–4. Genetic algorithm.

Step 1. Create and evaluate the initial population.

Step 2. Apply the operation of recombination to randomly selected individuals of the current population.

Step 3. Apply the operation of mutation to the current population.

Step 4. Evaluate the current population.

```
Parent 1:        0 0 1 1 1 0 0 0 0
Parent 2:        1 1 1 0 0 0 1 0 1
---------------------------------
Random String    0 1 0 0 1 1 1 0 0   probability of 1 is 0.5
---------------------------------
Offspring 1:     0 1 1 1 0 0 1 0 0
Offspring 2:     1 0 1 0 1 0 0 0 1
```

Figure 7–6. The uniform crossover.

Step 5. Select the individuals that are good enough to survive to the next generation.

Step 6. If the termination condition is not satisfied, then go back to step 2.

The two basic transforms of individuals are chosen here to be as follows. The *recombination* or *crossover* operation transforms two individuals of the population into new individuals in the following way. The parents are first selected randomly from the population in such a way that the probability of being selected is higher for fitter individuals. Since the first selection mechanisms, several variations and improvements of the fitness function have been proposed [8]. One suitable possibility is to use a method in which the fitness value is proportional to the rank, rather than actual evaluation values [5, 47] (computed directly from the error criterion used). Thus, the algorithm will not converge to a local minimum in case of a *super individual*. If the algorithm generates a very good individual in an otherwise poor population, the super individual will get a very good evaluation value compared with the others. Unless a rank-based selection method is used, the super individual will have a great number of offspring, and the population is likely to converge to that local optimum.

Consider first the truth table representation of Boolean functions. There are various ways to recombine binary vectors reported in the literature. Among the simplest are the following ones: let $\alpha = (\alpha_1, \alpha_2, \ldots, \alpha_{2^N})$ and $\beta = (\beta_1, \beta_2, \ldots, \beta_{2^N})$ be binary vectors. Uniform crossover [43] recombines the vectors according to a random zero-one sequence. At every location, a bit exchange between the parent vectors is performed if the random sequence contains 1 at that location, as shown in Fig. 7–6. It is obvious that the resulting offspring need not be positive and that they have to be checked. This is a serious drawback of this method.

A second way to recombine binary vectors is to randomly select a crossover point c, $1 \leqslant c \leqslant 2^N$, from the vectors. The vectors are then divided into two parts according to the crossover point. For the first offspring, the first c genes are copied from the first parent and the rest of the genes from the second parent. For the second offspring, the first c genes are copied from the second parent and the rest of the genes from the first parent. This procedure is illustrated in Fig. 7–7 In other words, let $\alpha = (\alpha_1, \alpha_2, \ldots, \alpha_{2^N})$ and $\beta = (\beta_1, \beta_2, \ldots, \beta_{2^N})$ be

```
Parent 1:        0 0 1|1 1 0 0 0 0
Parent 2:        1 1 1|0 0 0 1 0 1
---------------------+-----------
Offspring 1:     0 0 1|0 0 0 1 0 1
Offspring 2:     1 1 1|1 1 0 0 0 0
```

Figure 7–7. Crossover by one crossover point.

the selected parents. Then their offspring are $(\alpha_1, \alpha_2, \ldots, \alpha_c, \beta_{c+1}, \ldots, \beta_{2^N})$ and $(\beta_1, \beta_2, \ldots, \beta_c, \alpha_{c+1}, \ldots, \alpha_{2^N})$.

A generalization of this method would allow more crossover points, and finally we could use the uniform crossover discussed above. Again, nothing guarantees the positivity of the offspring.

The mutation can be done by going through all the elements of every vector (except some of the best in a so-called elitist strategy) and randomly selecting the elements to be mutated, i.e., complemented. This way one vector can have more than one mutation. One may allow only one mutation, but this may lead the optimization to get stuck in a local minimum more easily.

Consider now the representation of positive Boolean functions by storing only minimal terms. There are many ways to define the recombination and mutation. A straightforward way is as follows:

Let

$$\sum_{i=1}^{K} \prod_{j \in P_i} x_j \quad \text{and} \quad \sum_{i=1}^{L} \prod_{j \in Q_i} x_j$$

be the selected parents and $\mathcal{P} = \{P_1, P_2, \ldots, P_K\}$, $\mathcal{Q} = \{Q_1, Q_2, \ldots, Q_K\}$. Further, let $\mathcal{P}' \subseteq \mathcal{P}$ and $\mathcal{Q}' \subseteq \mathcal{Q}$ be randomly chosen. Let us denote $\mathcal{R} = \mathcal{P}' \cup \mathcal{Q}'$, $\mathcal{S} = \mathcal{P}' \cup (\mathcal{Q} \setminus \mathcal{Q}')$, $\mathcal{U} = \mathcal{Q}' \cup (\mathcal{P} \setminus \mathcal{P}')$, and $\mathcal{U} = (\mathcal{P} \setminus \mathcal{P}') \cup (\mathcal{Q} \setminus \mathcal{Q}')$. Then these collections of sets define the sets of minimal terms of offspring, which are easy to show to be positive Boolean functions. Thus, the positivity does not need to be checked.

The mutation operations can be realized by allowing the mutation to change one individual into a new individual belonging to the neighborhood of the PBF discussed in Section 7.2.2. Thus in the case of a truth table representation, after every mutation the mutated individual must be checked to verify its positivity, whereas in the case of the minimal term representation, this checking is not necessary.

From this discussion one can conclude that the minimal term representation seems to be a more appropriate representation of PBFs in the optimization of stack filters than the truth table representation. If the optimization parameters, such as population size and mutation rate, are properly selected, then the population will get better and better generation by generation and finally converge to the global optimum.

REFERENCES

[1] Aarts E. and Korst J., *Simulated Annealing and Bolzmann Machines*, Wiley, New York, 1989.

[2] Agaian S., Astola J., and Egiazarian K., *Binary Polynomial Transforms and Nonlinear Digital Filters*, Marcel Dekker, New York, 1995.

[3] Arce G. R. and Foster R. E., "Detail preserving ranked-order based filters for image processing," *IEEE Trans. Acoust. Speech Sig. Process.* **ASSP-37**, pp. 83–98, January 1989.

[4] Asano A., Matsumura K., Shudoh T., Itoh K., Ichioka Y., and Yokozeki S., "Design of morphological filters by learning," in *Proc. IEEE Winter Workshop on Nonlinear Digital Signal Processing*, Tampere, Finland, January 1993.

[5] Baker J. E., "Adaptive selection methods for genetic algorithms," in *Proc. of the First International Conference on Gen. Algorithms*, Lawrence Erlbaum Associates, Hillsdale, NJ, 1985.

[6] Bollobas B., Combinatorics: Set Systems, Hypergraphs, Families of Vectors and Combinatorial Probability, Cambridge University Press, Cambridge, 1986.

[7] Davidson J., "Simulated annealing and morphology neural networks," in *Proc. of SPIE Symp. on Image Algebra and Morphological Image Processing II*, Proc. SPIE **1568**, pp. 119–127, July 1992.

[8] DeJong K. A., An Analysis of the Behavior of a Class of Genetic Adaptive Systems, doctoral dissertation, Univ. of Michigan, 1975.

[9] Donoho D. L. and Huber P. J., "The notion of breakdown point," A festschrift for Erich L. Lehmann, P. J. Bickel, K. A. Doksum, J. L. Hodges, Jr., eds., pp. 157–184, Wadsworth, Belmont, CA, 1983.

[10] Erdös P., Ko C., and Rado R., "Intersection theorems for systems of finite sets," *Quart. J. Math. Oxford*, **12**, pp. 313–320, 1961.

[11] Fitch J. P., Coyle E. J., and Gallagher N. C. Jr., "Median filtering by threshold decomposition," *IEEE Trans. Acoust. Speech Sig. Process.* **ASSP-32**, pp. 1183–1188, December 1984.

[12] Fitch J. P., Coyle E. J., and Gallagher N. C. Jr., "Threshold decomposition of multidimensional rank order operators," *IEEE Trans. Circ. Syst.* **CAS-32**, pp. 445–450, May 1985.

[13] Gabbouj M. and Coyle E. J., "Minimum mean absolute error stack filtering with structuring constrains and goals," *IEEE Trans. Acoust. Speech Sig. Process.* **38** (6), pp. 955–968, June 1990.

[14] Hampel F. R., "The influence curve and its role in robust estimation," *J. Am. Stat. Assoc. Math. Stat.* **69**, pp. 383–393, 1974.

[15] Hampel F. R., Rousseeuw P. J., Ronchetti E. M, and Stahel W. A., *Robust Statistics: The Approach Based on Influence Functions*, Wiley, New York, 1986.

[16] Holland J., *Adaptation in Natural and Artificial Systems*, University of Michigan Press, Ann Arbor, 1975.

[17] Huber P. J., *Robust Statistics*, Wiley, New York, 1981.

[18] Huttunen H., Kuosmanen P., Koskinen L., and Astola J., "Optimization of soft morphological filters by genetic algorithms," in *Proc. of Image Algebra and Morphological Image Processing V*, Proc. SPIE **2300**, pp. 13–24, 1994.

[19] Jayant N. S., "Average and median-based smoothing techniques for improving digital speech quality in the presence of transmission errors," *IEEE Trans. Commun.* **24** (9), pp. 1043–1045, September 1976.

[20] Justusson B. I., "Median filtering: statistical properties," in T. S. Huang, ed., *Topics in Applied Physics. Two-Dimensional Digital Signal Processing II: Transforms and Median Filters*, Springer-Verlag, New York, 1981.

[21] Kirkpatrick S., Gelatt C. D. Jr., and Vecchi M. P., "Optimization by simulated annealing," in *Science* **220**, pp. 671–680, 1983.

[22] Koivisto P. and Kuosmanen P., "Design of soft morphological filters by learning," in *Proc. of Nonlinear Image Processing V*, Proc. SPIE **2180**, pp. 11–123, 1994.

[23] Koivisto P., Huttunen H., and Kuosmanen P., "Optimal Compositions of Soft Morphological Filters," in *Proc. of Nonlinear Image Processing VI*, Proc. SPIE **2424**, pp. 16–27, 1995.

[24] Koskinen L., Astola J., and Neuvo Y., "Statistical properties of discrete morphological filters," in *Proc. IEEE International Symposium on Circuits and Systems*, pp. 1219–1222, 1990.

[25] Kuosmanen P., Astola J., and Agaian S., "On rank selection probabilities," *IEEE Trans. Sig. Process.* **42** (11), pp. 3255–3258, November 1994.

[26] Kuosmanen P. and Astola J., "Optimal stack filters under rank selection and structural constraints," *Sig. Process.* **41** (3), pp. 309–338, February 1995.

[27] Kuosmanen P. and Astola J., "Breakdown points, breakdown probabilities, midpoint sensitivity curves and optimization of stack filters," in *Circ. Syst. Sig. Process.* **15** (2), pp. 165–211, February 1996.

[28] Kuosmanen P., Koivisto P., Huttunen H., and Astola J., "Shape preservation criteria and optimal soft morphological filtering", *J. Math. Imag. Vis.* **5** (4), pp. 319–336, December 1995.

[29] van Laarhoven P. J. M. and Aarts E. H. L., *Simulated Annealing: Theory and Applications*, D. Reidel, Dordrecht, 1987.

[30] Lee C., "Simulated annealing applied to acoustic signal tracking," in *Proc. of Nonlinear Image Processing III*, Proc. SPIE **1658**, pp. 344–355, 1992.

[31] Matheron P., *Random Sets and Integral Geometry*, Wiley, New York, 1975.

[32] Metropolis N., Rosenbluth A., Rosenbluth M., Teller A., and Teller E., "Equation of state calculations by fast computing machines," *J. Chem. Phys.* **21**, pp. 1087–1092, 1953.

[33] Muroga S., *Threshold Logic and Its Applications*, Wiley, New York, 1971.
[34] Nieminen A., Heinonen P., and Neuvo Y., "A new class of detail-preserving filters for image processing," *IEEE Trans. Patt. Anal. Mach. Intell.* **PAMI-9**, pp. 74–90, January 1987.
[35] Pitas I. and Venetsanopoulos A., *Nonlinear Digital Filters*, Kluwer Academic Publishers, New York, 1990.
[36] Prasad M. K. and Lee Y. H., "Stack filters and selection probabilities," in *Proc. of the IEEE International Symposium on Circuits and Systems*, pp. 1747–1750, May 1990.
[37] Pratt W. K., *Digital Image Processing*, Wiley, New York, 1978.
[38] Rabiner L. R., Sambur M. R., and Schmidt C. E., "Application of nonlinear smoothing algorithm to speech processing," *IEEE Trans. Acoust. Speech Sig. Process.* **23** (6), pp. 552–557, December 1975.
[39] Rosenfeld A. and Kak A. C., *Digital Picture Processing*, vol. 1, Academic Press, New York, 1982.
[40] Rutherford D. E., *Introduction to Lattice Theory*, Oliver and Boyd Ltd., Britain, 1965.
[41] Serra S., *Image Analysis and Mathematical Morphology*, Academic Press, New York, 1988.
[42] Sternberg S. R., "Grayscale morphology," in S. R. Sternberg and J. Serra, eds., Special Section on Mathematical Morphology, *Comp. Vis. Graph. Imag. Process.* **35**, pp. 333–355, 1986.
[43] Syswerda G., "Uniform Crossover in Genetic Algorithms," in *Proc. 3rd Int. Conf. Genetic Algorithms*, pp. 2–9, Morgan Kaufmann Publishers, Los Altos, CA, 1989.
[44] Tukey J. W., *Exploratory Data Analysis*, Addison-Wesley, Menlo Park, CA, 1971.
[45] Tukey J. W., "Nonlinear (Nonsuperposable) methods for smoothing data," in *Congr. Rec. EASCON*, p. 673, 1974.
[46] Wendt P., Coyle E., and Gallagher N., "Stack filters," *IEEE Trans. Acoust. Speech Sig. Process.* **34**, pp. 898–911, August 1986.
[47] Whitley D., "The GENITOR algorithm and selection pressure: Why rank-based allocation of reproductive trials is best," in *Proc. 3rd Int. Conf. Genetic Algorithms*, Morgan Kaufmann Publishers, Los Altos, CA, 1989.
[48] Yang R., Yin L., Gabbouj M., Astola J., and Neuvo Y., "Optimal weighted median filters under the structural constraints," *IEEE Trans. Sig. Process.* **43** (3), pp. 591–604, 1995.
[49] Yin L., "Optimal stack filter design: A structural approach," in *IEEE Trans. Sig. Process.* **43** (4), pp. 831–840, April 1995.
[50] Yli-Harja O., Astola J., and Neuvo Y., "Analysis of the properties of median and weighted median filters using threshold logic and stack filter representation," *IEEE Trans. Sig. Process.* **39**, pp. 395–410, February 1991.
[51] Yu P. T. and Coyle E., "On the existence and design of the best stack filter based on associative memory," *IEEE Trans. Circ. Syst.* **39** (3), pp. 171–184, March 1992.

Chapter 8

Invariant Signals of Median and Stack Filters

Jaakko T. Astola
Pauli Kuosmanen
Tampere Univ. of Technology
Tampere, Finland

Invariants of nonlinear filters play an important role, especially for those filters which can be represented by finite window logical operations. For median and stack filters, invariant signals (also called root signals) are used to describe the filtering properties of a particular filter. The features of a median-type filter which best describe its actual performance are its output distribution and its set of invariant signals. The main use of median-type filters is in smoothing out random disturbances, and the output distribution for arbitrary stochastic input would, in principle, give all information about the filter's noise attenuation power. Unfortunately the general output distribution can seldom be put into a manageable form. The invariants, for their part, reveal the deterministic behavior of the filter and in this respect the set of invariants resembles the passband of a linear frequency selective filter. In classical morphological image processing, the process in a sense starts with choosing the structuring elements and thus also the invariants. In median-type filtering, one usually pays attention first to the noise removal properties and second to the deterministic properties, which are mainly characterized by the invariants.

8.1 Invariants of 1-D Median and Ranked-Order Filters

One of the simplest and most widely used nonlinear filters is the median filter. By definition, a real discrete time signal is a function $X : \mathbf{Z} \to \mathbf{R}$. We say that X is a finite-length signal if there is n_0 such that $X(n) = 0$ for $|n| \geqslant n_0$. Let k be a nonnegative integer. A window width $2k + 1$ median filter is the operator $\mathcal{M}$ mapping the set of real into itself and defined by

$$\mathcal{M}[X](n) = Med[X(n-k), X(n-k+1), \ldots, X(n+k)] \qquad (8\text{–}1)$$

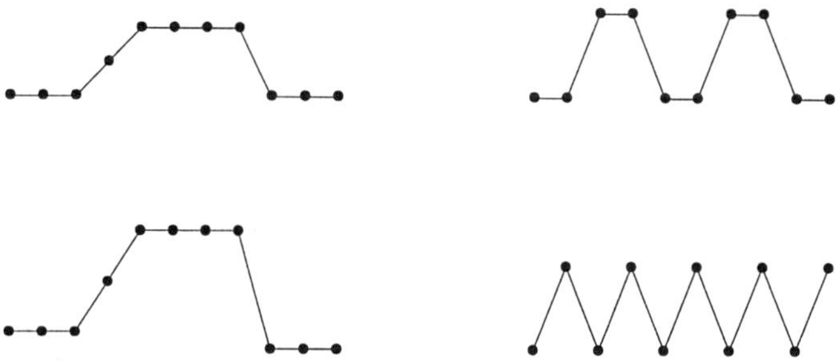

Figure 8–1. Invariants of median filters of lengths 3 and 5.

The invariants of the median filter of window width $2k + 1$ are, of course, the signals satisfying

$$\mathcal{M}[X](n) = X(n) \quad \text{for all } n. \tag{8–2}$$

In Figure 8–1 are depicted invariants of median filters of lengths 3 and 5.

We shall first discuss the invariants of median filters in historical order and after that we present a general theorem by [9] which completely characterizes the invariants of 1-D median filters and ranked-order filters. Similar results were later derived independently in [23] using different methods. In the latter part of the chapter we discuss the mainly unresolved question of the invariants of two- and higher-dimensional median-type filters. The problem of characterizing the invariants of median-type filters in two- and higher-dimensional spaces is essentially unresolved. The results for one-dimensional signals in particular cases can be directly extended to higher dimensional spaces. This chapter is not intended to be a complete survey of existing results. We have tried to choose results that give the main ideas of broad significance. Owing to the nature of the problem, there are many ingenious results on invariants of particular filtering methods which cannot be considered here.

Threshold Decomposition

Consider a real signal X and let Y be the signal obtained by a median filter of length $2k + 1$, i.e.

$$Y(n) = \mathcal{M}[X](n) \quad \text{for all } n.$$

It is obvious that for any $a \in \mathbf{R}$

$$Y(n) \geqslant a \Leftrightarrow \text{at least } k + 1 \text{ of } X(n - k), \ldots, X(n + k) \geqslant a. \tag{8–3}$$

This means that the Boolean signal indicating whether $Y(n) \geqslant a$ is obtained by median filtering the Boolean signal X_B:

$$X_B(n) = \begin{cases} 1 & \text{if } X(n) \geqslant a \\ 0 & \text{otherwise} \end{cases}. \tag{8–4}$$

In median filtering literature, the consequence of this fact is usually expressed in the following way. Let $X(n)$ be an M-level signal, i.e. $X : \mathbf{Z} \to \{0, 1, \ldots, M-1\}$, $m = 0, 1, \ldots, M-1$. Form $(M-1)$ binary signals $X_m : \mathbf{Z} \to \{0, 1\}$, $m = 0, 1, \ldots, M-1$ by

$$X_m(n) = \begin{cases} 1 & \text{if } X(n) \geqslant m \\ 0 & \text{otherwise} \end{cases}. \tag{8–5}$$

Obviously $X = X_1 + X_2 + \cdots + X_{M-1}$ and by what was said above, we have

$$\begin{aligned} \mathcal{M}_{2k+1}(X) &= \mathcal{M}_{2k+1}(X_1 + \cdots + X_{M-1}) \\ &= \mathcal{M}_{2k+1}(X_1) + \cdots + \mathcal{M}_{2k+1}(X_{M-1}). \end{aligned} \tag{8–6}$$

Thus median filtering satisfies the weak superposition 8–6. This observation has been the starting point of the theory of stack filters [34].

The threshold decomposition method is very useful in analyzing median-type filters. In many cases it means that it is enough to study the behavior of the filter only in the binary case. The general case then follows immediately.

Convergence to Invariant

We consider the following basic question: If a finite-length signal is repeatedly filtered, does it eventually turn into a invariant signal and if so how many filterings are needed? It is quite easy to see that any finite-length signal will turn into an invariant after sufficiently many filterings. Finding a good upper bound for the number of filterings requires a more careful analysis [33].

Let us first show the convergence. Assume that X is a binary finite-length signal and that $X(n) = 0$ for $n \leqslant 0$ or $n > K$. Let us run length code X as a sequence $\infty r_1 r_2 \ldots r_s \infty$, where $\infty, r_2, \ldots$ are the lengths of runs of 0's and $r_1, r_3, \ldots, r_s$ are the lengths of runs of 1's. We can further assume that the first run of 1's starts at index 1 and thus the second run of 0's starts at $r_1 + 1$ and so on.

PROPOSITION 8–1. *If X is an invariant, then $r_i \geqslant k+1$ for $i = 1, 2, \ldots, s$.*

PROOF. Assume that j is the smallest index such that $r_j \leqslant k$ and suppose that the median filter is centered at time index $r_1 + \cdots + r_{j-1} + 1$. Then there are in the window k bits from run r_{j-1} and at least one bit from run r_{j+1} implying that the first bit of run r_j (the bit in the center of the window) must change. This contradiction proves the proposition. ■

Notice that this argument also immediately shows that any finite-length signal will eventually turn into an invariant. This is because the first bit in the first run of length less than $k+1$ will always change and obviously can never change back. Thus we have:

THEOREM 8–1. *Any finite-length signal will turn into an invariant after a finite numbe of filterings by a median filter of length* $2k+1$.

The following theorem will give an upper bound on the number of filterings that are needed to turn any signal into an invariant signal. It is the best possible general bound because one can construct signals meeting the bound.

THEOREM 8–2. *Let* X *be a binary finite-length signal and* $X(n) = 0$ *for* $n \leqslant 0$ *or* $n > K$. *Then at most*

$$3\left\lceil \frac{L-2}{2k+4} \right\rceil \tag{8–7}$$

filterings by a median filter of length $2k+1$ *are needed to convert* X *into an invariant signal.*

In the following paragraphs we present the results by [9] completely characterizing the invariants of a median filter of length $2k+1$ (and also other ranked-order filters).

Assume that $X : \mathbf{Z} \to \{0, 1\}$ is an invariant signal. Let $\{r_i\}$ be the sequence of runlengths of X. We define two functions f and b in the following way:

$f(i)$ is the smallest nonnegative integer for which

$$\sum_{j=i}^{i+f(i)} r_j > k \tag{8–8}$$

and $b(i)$ is the smallest nonnegative integer for which

$$\sum_{j=i-b(i)}^{i} r_j > k. \tag{8–9}$$

The following proposition is the basis of the analysis [9].

PROPOSITION 8–2. *For an invariant signal X there is a nonnegative integer s called the window frequency of the invariant such that for any i,* $f(i) = b(i) = 2s$.

PROOF. Suppose that the center of the window is on the first bit of run r_i of 1's. The number of 1's must go from k to $k+1$ compared to the previous position. Therefore 1 has entered the window from the right and this bit belongs $r_{i+f(i)}$, which thus has to be a run of 1's, implying that $f(i)$ is even. If r_i is a run of 0's, we similarily see that $f(i)$ is even. Thus $f(i)$ is an even function.

Next we show that $b(i+f(i)) = f(i)$. From the definition of $f(i)$ it follows that $b(i+f(i)) \leqslant f(i)$. Suppose that strict inequality holds. Then

$$\sum_{j=i+1}^{i+f(i)} r_j > k, \tag{8–10}$$

and it follows that $f(i+1) < f(i)$. Since both numbers are even, we must have $f(i+1) \leqslant f(i) - 2$. But then

$$\sum_{j=i+1}^{i-1+f(i)} r_j > k \tag{8–11}$$

and adding r_i contradicts the minimality of $f(i)$. In the same way one sees that $f(i-b(i)) = b(i)$. Now, from $b(i+f(i)) = f(i)$ it follows that

$$\sum_{j=i+1}^{i+f(i)} r_j \leqslant k, \tag{8–12}$$

and so $f(i) \leqslant f(i+1)$. Similarily we get that $b(i) \geqslant b(i+1)$. Then we have

$$f(i) \geqslant \cdots \geqslant f(i-b(i)) = b(i) \geqslant \cdots \geqslant b(i+f(i)) = f(i), \tag{8–13}$$

implying $f(i) = b(i)$. From $f(i) \leqslant f(i+1) = b(i+1) = f(i)$ we see that f and b are constant functions proving the proposition. ■

The full characterization of invariants follows from the next theorem, which shows that the run lengths of an invariant satisfy a linear recurrence.

THEOREM 8–3. *For all i*

$$\sum_{j=1}^{s} r_{i+1-2j} = \sum_{j=1}^{s} r_{i+2j}. \tag{8–14}$$

PROOF. Without limiting generality, we may assume that r_i is a run of 1's. When the center of the window is on the last bit of r_i, the number of 1's in the window is $k+1$. Now by the definition of $f(i) = b(i) = 2s$ we have

$$\begin{array}{ll} \text{(i)} & \displaystyle\sum_{j=i+1}^{i+1+2s} r_j > k \\ \text{(ii)} & \displaystyle\sum_{j=i+1}^{i+2s} r_j \leqslant k \\ \text{(iii)} & \displaystyle\sum_{j=i-2s}^{i} r_j > k \\ \text{(iv)} & \displaystyle\sum_{j=i+1-2s}^{i} r_j \leqslant k \end{array} \quad . \tag{8–15}$$

The rightmost k places in the window contain $\sum_{j=1}^{s} r_{i+2j}$ 1's by (i) and (ii). The number of 0's in the leftmost $k+1$ places in the window is $\sum_{j=1}^{s} r_{i+1-2j}$. Thus the number of 1's in the window is

$$\sum_{j=1}^{s} r_{i+2j} + \left(k + 1 - \sum_{j=1}^{s} r_{i+1-2j}\right). \tag{8–16}$$

Equating 8–16 with $k+1$ gives 8–14. ■

Because 8–14 is a recurrence of the order $4k-1$, any $4k-1$ consecutive r_i determines all the r_i. The following result is a certain kind of converse of Theorem 8–3.

THEOREM 8–4. *Any sequence of natural numbers satisfying the recurrence* 8–14 *corresponds to a binary invariant having a window frequency s for some window size* $2k+1$.

PROOF. Let $\{r_i\}$ be a sequence of natural numbers satisfying the recurrence 8–14. Write

$$B_{i,h} = r_i + \cdots + r_{i+h-1}.$$

To get hold of a candidate window we define

$$k = \max_i B_{i,2s}. \tag{8–17}$$

With this window we can show that s behaves like a window frequency. The sum of any $2s$ consecutive terms is always at most k, so we need only to show that the sum of any $2s+1$ consecutive terms is greater than k. Adding two instances of recurrence 8–14 gives

$$B_{i-2s,2s} = B_{i+1,2s} \tag{8–18}$$

and so also

$$B_{i-2s,2s+1} = B_{i+1,2s+1}. \tag{8–19}$$

Now, suppose that $B_{i,2s+1} \leqslant k$ for some i. By definition of k, there is a j such that $B_{j,2s} = k$. Because the periods of $B_{i,2s+1}$ and $B_{i,2s}$ are relatively prime means that we can find a shifted version with the same sum in a larger block, which is a contradiction. Now that we have the window size and window frequency, we can use the arguments of Theorem 8–3 the other way round and count the number of 1's when the center of the window is placed in the end of a run of 1's. It is straightforward to check that the binary sequence corresponding to runlengths is an invariant. ■

From the fact that the sequence of runlengths corresponding to an invariant with positive window frequency s satisfies the nontrivial linear recurrence 8–14 we see that these invariants are in fact periodic signals. The characteristic polynomial $g_s(x)$ of recurrence 8–14 can be written as

$$g_s(x) = \frac{(x^{2s+1}-1)(x^{2s}-1)}{x^2-1}, \tag{8–20}$$

which means that the zeroes of g_s are all the $(2s+1)$th roots of unity $\{\alpha_i\}$ together with all the $(2s)$th roots of unity except -1 and 1 $\{\beta_i\}$. From the theory of linear difference equations it follows that

$$r_n = \sum_{i=1}^{2s+1} \alpha_i^n + \sum_{i=1}^{2s-2} \beta_i^n. \tag{8–21}$$

Because each sequence $\{\alpha_i^n\}$ is periodic, with a period dividing $2s+1$, and each sequence $\{\beta_i^n\}$ is periodic with a period dividing $2s$, the sequence $\{r_i\}$ must be periodic, with a period dividing $2s(2s+1)$.

The Nonbinary Case

Let X be an M-valued signal, i.e. $X : \mathbf{Z} \to \{0, 1, \ldots, M-1\}$. It is obvious that X is an invariant of the median filter with window size $2k+1$; then also the thresholded signals $X_1, X_2, \ldots, X_{M-1}$ are binary invariant signals. We call a run r_i *long* if it consists of at least $k+1$ digits. The following theorems characterize nonbinary invariants.

THEOREM 8–5. *If X_1 or X_{M-1} has a window a frequency zero, then all runs of* 0 *and all runs of* $(M-1)$ *are long. In particular, if X has a long run of* 0 *or a long run of* $(M-1)$, *then all* 0 *and* $(M-1)$ *runs are long.*

PROOF. Suppose that X_1 has a window frequency $s = 0$. This means that all runs in X_1 are long and the 0 runs of X_1 are exactly the 0 runs of X. Similarily the 1 runs of X_{M-1} are exactly the $(M-1)$ runs of X. The second part follows directly from this. ■

THEOREM 8–6. *Consider a signal $X : \mathbf{Z} \to \{0, 1, \ldots, M-1\}$. If X is invariant, then it is either bivalued or all runs of* $\min_n X(n)$ *and* $\max_n X(n)$ *are long.*

PROOF. We can assume that $\min_n X(n) = 0$ and $\max_n X(n) = M-1$. Consider a subsequence of the form (reversing X if necessary).

$$\ldots 0ab \ldots c(M-1)(M-1) \ldots \tag{8–22}$$

Here $ab \ldots c$ contains no 0's or $(M-1)$'s. If $\alpha \geqslant k$, we get the window frequency 0 by thresholding and so we can assume that $\alpha < k$. When the center of the window is on the last 0, the window contains $k+1$ 0's. At the first $(M-1)$ it contains $k+1$ $(M-1)$'s. Since these widows overlap, their $2k+2+\alpha$ places contain $k+1$ 0's, $k+1$ $(M-1)$'s and at least α other symbols. Since this adds up to the whole interval, it follows that the k places before A consist of 0's only and the k places after B consist of $(M-1)$'s only. This means that either there is a long run of 1's or there is $(1)(M-1)$ or $(M-1)(1)$ as a subsequence somewhere. In the first case we are done, so assume the last. Then the above gives a window with only 0's and $(M-1)$'s. This implies that X is a two-valued signal. ■

The following simple theorem characterizes invariants which have long runs.

THEOREM 8–7. *If X is not a two-valued signal with a positive window frequency, then all local extrema are long runs. Conversely, any such sequence is invariant.*

PROOF. We already know that all runs of 0 or $(M-1)$ are long. Let e be a long local minimum (say) and let a part of X be

$$e = \cdots = e < a_1 \leqslant a_2 \leqslant \cdots \leqslant a_{l-1} < a_l = \cdots = a_m > a_{m+1}. \tag{8–23}$$

When the center of the window is at time index k we see that $m - l \geqslant k$ because

$$e, \ldots, e, a_1, \ldots, a_{l-1}, a_{m+1} < a_l.$$

It follows from this that the next local extremum a_l is also long and so all local extrema are long runs. The last part of the proposition is easy to verify. ■

Other Ranked-Order Filters

Consider $X_1, \ldots, X_N$. Let $\pi = ((1), (2), \ldots, (N))$ be a permutation of $(1, 2, \ldots, N)$ for which

$$X_{(1)} \leqslant X_{(2)} \leqslant \cdots \leqslant X_{(N)}. \tag{8–24}$$

Define the pth rank-ordered element $R_p[x_1, \ldots, x_N]$ of $x_1, \ldots, x_N$ by

$$R_p[x_1, \ldots, x_N] = x_{(p)}. \tag{8–25}$$

Now, the ranked-order filters are defined as follows. Let $N = 2k + 1$ be an odd integer $\geqslant 3$ and $X : \mathbf{Z} \to \mathcal{R}$ a real signal. The output of the pth ranked-order filter $\mathcal{R}_p$ with window size N and center at time index n is

$$\mathcal{R}_p[X](n) = R_p[X(n-k), X(n-k+1), \ldots, X(n+k)]. \tag{8–26}$$

It is obvious that for all ranked-order filters (of which median filter is a special case) the thresholded signal of an invariant is also invariant. Suppose that X is an M-valued signal, i.e. $X : \mathbf{Z} \to \{0, 1, \ldots, M-1\}$. We see immediately that if 0 is the output, then the window contains at least p 0's and if $(M-1)$ is the output, then the window contains at least $N + 1 - p$ $(M-1)$'s. We define window frequency and long runs as before. The same proof as in the median case shows that a binary invariant has a window frequency. The following proposition holds.

PROPOSITION 8–3. *Assume that $\mathcal{R}_p$ is not the median filter and that X is a nonconstant invariant. Then no nonconstant thresholded version of X has a long run.*

PROOF. Assume that a thresholded binary version Z of X has a long run. Then it is a binary invariant with window frequency zero and so all runs are long. Without limiting generality, we may assume that $p > k + 1$. Let the center of the window be on the last bit of a run of 0's

$$\underbrace{0\ldots0}_{>k}\underbrace{1\ldots1}_{>k}.$$

Now the output is 1, which contradicts the invariance of Z.

The following theorem shows that nonmedian ranked-order filters do not have slowly varying invariants. ■

THEOREM 8–8. *Assume that $\mathcal{R}_p$ is not the median filter. Then all invariants are binary.*

PROOF. We can assume that $\min_n X(n) = 0$ and $\max_n X(n) = M - 1$. Consider a subsequence of the form (reversing X if necessary).

$$\ldots 0 \ldots \ldots 0 \underbrace{ab \ldots c}_{\alpha}(M-1) \ldots \ldots (M-1).$$

Here $ab \ldots c$ contains no 0's or $(M-1)$'s. If $\alpha \geqslant k$, we get window frequency 0 by thresholding and so we can assume that $\alpha < k$. When the center of the window is on the last 0, the window contains p 0's. At the first $(M-1)$ it contains $2k+2-p$ $(M-1)$'s. Since these windows overlap, their $2k+2+\alpha$ places contain p 0's, $2k+2-p$ $(M-1)$'s and at least α other symbols. Since this adds up to the whole interval, it follows that the k places before the last 0 consist of 0's only and the k places after the first $(M-1)$ consist of $(M-1)$'s only. This means that either there is a long run of 1's or there is $(1)(M-1)$ or $(M-1)(1)$ as a subsequence somewhere. In the first case we are done, so assume the last. Then the above gives a window with only 0's and $(M-1)$'s. This implies that X is a two-valued signal. ∎

The following theorem shows that the run lengths of binary invariants of ranked-order filters also satisfy linear difference equations, which is otherwise the same as for the median case but nonhomogeneous.

THEOREM 8–9. *Let X be a binary invariant of $\mathcal{R}_p$. Then for all i the run lengths r_i satisfy*

$$\sum_{j=1}^{s} r_{i+1-2j} = \sum_{j=1}^{s} r_{i+2j} = \begin{cases} k+1-p & \text{for } r_i \text{ a run of } 1\text{'s} \\ p-k-1 & \text{for } r_i \text{ a run of } 0\text{'s} \end{cases}. \tag{8–27}$$

PROOF. Assume that the center of the window is on the last bit of a run of 1's. The number of 1's in the window is $2k+2-p$ because the next output is 0. Let us count the number of 1's in another way using the same idea as in the median case. The rightmost k places of the window contain $\sum_{j=1}^{s} r_{i+2j}$ bits equal to 1. The number of 0's in the left $k+1$ places in the window is $\sum_{j=1}^{s} r_{i+1-2j}$. Subtracting this from $k+1$ and adding we get the equation

$$k+1-r = \sum_{j=1}^{s} r_{i+1-2j} - \sum_{j=1}^{s} r_{i+2j}.$$

In the same way, if r_i is a run of 0's, we get

$$r-k-1 = \sum_{j=1}^{s} r_{i+1-2j} - \sum_{j=1}^{s} r_{i+2j},$$

implying the result. ∎

EXAMPLE 8–1. Consider $\mathcal{R}_4$ for window size 9. The binary signal X:

$$\ldots 1110110011011101100110111 \ldots$$

is an invariant with window frequency $s = 1$ and the corresponding sequence of run lengths is

$$\ldots 3122213122213 \ldots$$

The difference equation 8–27 for the run lengths of an invariant for the ranked-order filters other than the median are otherwise like for the median but contain the nonhomogenity term on the right hand side. The theory of difference equations tells us that the solutions of general ranked-order filters are obtained by finding a solution of 8–27 and adding any solution of 8–27. As a final result, on one-dimensional invariants of ranked-order filters, we show that if p is too large or too small, the only invariants are the constant sequences.

THEOREM 8–10. *Nonconstant invariants with a positive window frequency* s *of* $\mathcal{R}_p$ *with window size* $2k+1$ *exist if and only if* $2s+1 \leqslant p \leqslant 2k+1-2s$.

PROOF. Suppose that $p \leqslant k$. We know that $r_0 + r_1 + \cdots + r_{2s-1} \leqslant k$ and $r_0 + r_{-1} + \cdots + r_{-2s+1} \leqslant k$. Suppose that r_0 is a block of 1's. Then at least one of the sums $r_0 + r_1 + \cdots + r_{2s-1} \leqslant k$ and $r_0 + r_{-1} + \cdots + r_{-2s+1} \leqslant k$ is less than $\lfloor p/2 \rfloor$. This implies $2s+1 \leqslant p$. The other inequality can be proved similarly. ■

Number of Invariant Signals

One interesting question is the number of invariant signals relative to all signals. This number gives an indication of how strongly the filter will affect random signals.

EXAMPLE 8–2. Consider the operation of a median filter of length 3 on binary signals. Let $A(N)$ be the set of all binary signals of length N, i.e.

$$A(N) = \{X : \mathbf{Z} \to \{0, 1\} \mid X(n) = 0 \text{ for } n < 0 \text{ or } n \geqslant N\},$$

and let $B(N) = \{X \in A(N) \mid X(N-1) = 0\}$ and $C(N) = \{X \in A(N) \mid X(N-1) = 1\}$. Let $\widehat{A}(N) = \{X \in A(N) \mid X \text{ is invariant}\}$ and define $\widehat{B}(N)$ and $\widehat{C}(N)$ in the same way. Now,

$$\begin{aligned} \widehat{B}(N) &= \widehat{B}(N-1) \cup \big(\widehat{C}(N-1) + \delta(N-1)\big) \\ \widehat{C}(N) &= \widehat{C}(N-1) \cup \big(\widehat{B}(N-2) + \delta(N-2) + \delta(N-1)\big). \end{aligned}$$

Because $\widehat{A}(N) \cap \widehat{B}(N) = \emptyset$, $\widehat{A}(N-1) \cap (\widehat{B}(N-1) + \delta(N-1)) = \emptyset$, and $\widehat{B}(N-1) \cap (\widehat{B}(N-1) + \delta(N-2) + \delta(N-1)) = \emptyset$, it follows, denoting $|\widehat{A}(N)| + |\widehat{B}(N)| = F(N)$, that

$$F(N) = F(N-1) + F(N-2).$$

By checking the initial conditions, we see that the number of binary invariants of length N for the median filter of length 3 in fact obeys the Fibonacci sequence $1, 1, 2, 3, 5, 8, 13, \ldots$ which can also be expressed in closed form as

$$F(N) = \frac{1}{\sqrt{5}}\left(\left(\frac{1+\sqrt{5}}{2}\right)^{N-1} - \left(\frac{1-\sqrt{5}}{2}\right)^{N-1}\right).$$

A related way of describing invariant signals is to generate them by finite-state machines or trellis diagrams. The invariants of the median filter of length 3 can be generated by a 4-state machine where the states are identified with the last two bits in the window.

A possible use of a median filter is as a prefilter before compression [5]. The signal is repeatedly filtered until it becomes invariant. Because the invariant is "less random" than the original, one can expect higher compression rates. The state model can be used to estimate the gain if the invariant is compressed instead of the original.

The above approach generalizes to median filters with longer windows. Denote as before

$$\begin{aligned} A(N) &= \{X : \mathbf{Z} \to \{0,1\} \mid X(N) = 0 \text{ for } n < 0 \text{ or } n \geqslant N\} \\ B(N) &= \{X \in A(N) \mid X(N-1) = 0\} \\ C(N) &= \{X \in A(N) \mid X(N-1) = 1\} \end{aligned}$$

and by $\widehat{A}(N)$, $\widehat{B}(N)$ and $\widehat{C}(N)$ the corresponding sets of invariants. It is straightforward to see that

$$\begin{aligned} \left|\widehat{B}(N)\right| &= \left|\widehat{B}(N-1)\right| + \left|\widehat{C}(N-1)\right| \\ \left|\widehat{C}(N)\right| &= \left|\widehat{C}(N-1)\right| + \left|\widehat{B}(N-k-1)\right| \end{aligned}$$

with appropriate initial conditions. These recursions can be solved with standard methods [27, 25], e.g., using generating functions.

EXAMPLE 8–3. For median filters of length 5 we obtain, denoting $b_N = |\widehat{B}(N)|$ and $c_N = \widehat{C}(N)$, the recursion

$$\begin{aligned} b_N &= b_{N-1} + c_{N-1} \\ c_N &= c_{N-1} + b_{N-3} \\ b_0 &= b_1 = b_2 = 1 \\ c_0 &= c_1 = c_2 = 0. \end{aligned} \tag{8–28}$$

If we write $b(z) = \sum_{i=0}^{\infty} b_i z^i$ and $c(z) = \sum_{i=0}^{\infty} b_i z^i$, Eq. 8–28 implies that

$$\begin{aligned} b(z) &= z\big(b(z) + c(z)\big) + 1 \\ c(z) &= zc(z) + z^3 b(z), \end{aligned}$$

from which we get

$$\begin{aligned} b(z) &= \frac{1-z}{1-2z+z^2-z^4} = 1 + z + z^2 + z^3 + 2z^4 + 4z^5 + \cdots \\ c(z) &= \frac{z^3}{1-2z+z^2-z^4} = z^3 + 2z^4 + 3z^5 + 4z^6 + \cdots \end{aligned}$$

The smallest zero in absolute value of $z^4 - z^2 + 2z - 1$ is $(\sqrt{5}-1)/2 \sim 0.618$ and thus asymptotically also the number of invariants of length at most N is $\sim((\sqrt{5}-1)/2)^N$.

The state description leads immediately to interesting connections with automata theory and formal languages. We will not go deeper in this topic here.

8.1.1 INVARIANTS OF TWO-DIMENSIONAL MEDIAN FILTERS

In this section we extend some results of one-dimensional invariants to two-dimensional signals or images. This situation is much more difficult. This follows from the much greater freedom there is in the operation of two-dimensional median-type filters. In the one-dimensional case, when the window moves one step ahead, one sample is discarded and a new one enters the window. For two-dimensional windows, everything depends on the shape of the window and the direction the window moves. Also, the useful concept of window frequency seems impossible to generalize in a truly meaningful manner. So many simple rules that were useful in the one-dimensional case are now useless. In this section we describe a few general facts of two-dimensional invariants. Again we are mostly concerned with the binary case because most multilevel properties can be reduced to the binary case.

Let $X : \mathbf{Z}^2 \to \{0, 1\}$ be a binary image and let W be a finite subset of $\mathbf{Z}^2$ such that

(i) $(0, 0) \in W$ and

(ii) $(a, b) \in W \Rightarrow W(-a, -b) \in W$.

From (i) and (ii) it follows that the cardinality of W $|W| = 2k + 1$ for an integer k. The operation of the median filter $\mathcal{M}_W$ with window W on the image X is defined by

$$\mathcal{M}_W[X](u, v) = \text{Median}\{X(u + i, v + j) \mid (i, j) \in W\}. \tag{8–29}$$

In the one-dimensional case we saw that repeated filtering turned any finite-length signal into an invariant in finitely many passes. A similar but weaker result holds for two-dimensional signals. First we need to specify what is meant by an image of finite size. A binary image X is of finite size if there is M such that $X(u, v) = 0$ for $|u| + |v| > M$. The following counterexample [21] shows that an image of finite size need not converge to an invariant.

EXAMPLE 8–4. Consider the two-dimensional signal

$$\begin{matrix} 11010111 \\ 11101011 \end{matrix}\text{'}$$

where all sites not otherwise marked are assumed to have zero value. After application of the median filter with $W = \{(0, 0), (0, 1), (1, 0), (0, -1), (-1, 0)\}$ we get

$$\begin{matrix} 11101011 \\ 11010111 \end{matrix}$$

and the result will oscillate between these two images in repeated filtering.

Goles and Olivos [21] have proved a general result on threshold functions which was used in [18] to deduce the following

THEOREM 8–11. *Let $X : \mathbf{Z}^2 \to \{0, 1\}$ be a binary image of finite size. Then there is s such that*

$$\mathcal{M}_W{}^{s+2} = \mathcal{M}_W{}^{s}$$

for any W satisfying (i) *and* (ii).

If a one-dimensional binary signal had window frequency zero, i.e. all runs had at least length $k + 1$ for a window of length $2k + 1$, then this signal was invariant. Such signals were called locally monotonic in [30]. Following [30] we extend one-dimensional local monotonicity to images and use this property to derive some sufficient conditions for invariant images.

PROPOSITION 8–4. *Consider an image X and let W be a window satisfying* (i) *and* (ii) *and let L be an arbitrary line in $\mathcal{R}^2$. If*

$$\text{Median}\{X(u, v) \mid (u, v) \in L \cap W\} = X(0, 0)$$

for all lines L passing through the origin, then $\mathcal{M}_W[X](0, 0) = X(0, 0)$.

PROOF. Obvious. In the sequel we assume that all windows W satisfy (i) and (ii). We say that an image X is locally monotonic with respect to W if for all (l, m) the one-dimensional signal $\{X(i, j) \mid (i, j) - (l, m) \in L \cap W\}$ is monotonic for all lines L passing through the origin. The ordering of the elements of $L \cap W$ is the one induced by the natural ordering of L. ■

THEOREM 8–12. *If an image of finite size is locally monotonic with respect to a window W, then it is an invariant of the median filter with window W.*

PROOF. Follows immediately from Proposition 8–4.

There are only sporadic results on the structure of two-dimensional (binary) invariants. For rectangular windows we have the following simple result. Let $X_1 : \mathbf{Z} \to \{0, 1\}$ be an invariant of the median filter with window $W_1 = \{-k_1, -k_1 + 1, \ldots, k_1\}$ and let $X_2 : \mathbf{Z} \to \{0, 1\}$ be an invariant of the median filter with window $W_2 = \{-k_2, -k_2 + 1, \ldots, k_2\}$. Then $X : \mathbf{Z}^2 \to \{0, 1\}$ defined by

$$X(i, j) = \big(1 - X_1(i) \text{ XOR } X_2(j)\big)$$

is an invariant of the two-dimensional median filter with window $W = W_1 \times W_2$.

This can be seen as follows. Let $X(i, j) = 1$. Let there be $u \geqslant k_1 + 1$ ones in W_1 when the center is on $X_1(i)$ and $v \geqslant k_2 + 1$ ones in W_2 when the center is on $X_2(j)$. The number of ones in W when the center is on $X(i, j)$ is $uv + (2k_1 + 1 - u)(2k_2 + 1 - v)$ and the number of zeroes is $u(2k_2 + 1 - v) + (v(2k_1 + 1 - u)$. Thus the difference of ones and zeroes is $(2u - 2k_1 - 1)(2v - 2k_2 - 1) > 0$, implying that the output is also equal to one. If $X(i, j) = 0$, the same argument shows that the output is zero. Figure 8–2 shows this type of invariant for window sizes 3×3 and 5×5. ■

Invariants of Other Filter Classes

Many different filter classes have been defined for various purposes. Each of them has its own invariant signal classes but usually very few general results are known. The invariants of FIR-median hybrid filters have been studied in [7]. The invariants of the filter class called WMMR filters (weighted majority of minimum range) were studied in [23]. This filter class has the interesting and useful property that its invariants are piecewise constant signals. The class of stack filters is quite general

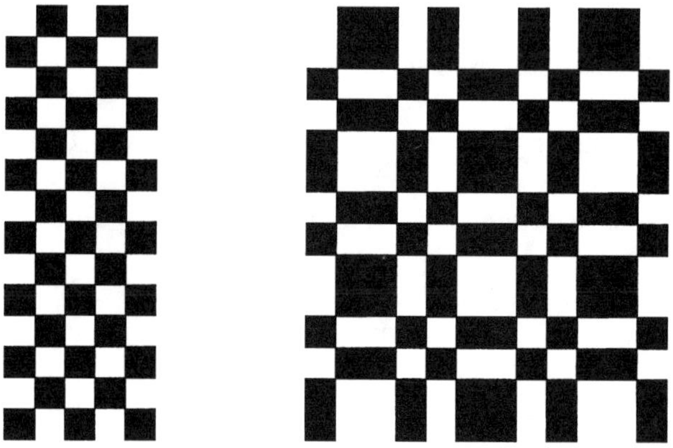

Figure 8–2. Invariants of two-dimensional median filters with window sizes 3×3 and 5×5.

because it includes all morphological filters with flat structuring elements as well as all ranked-order filters. Thus it is natural that nothing very deep can be said about their root signals except for special cases as median filters. The state description gives a natural way of generating invariant signals for stack filters. This approach has been used in [4, 2, 3].

REFERENCES

[1] Abramowitz M., and Stegun I., Handbook of Mathematical Functions, Dover, New York, 1968.

[2] Arce G., and Gallagher N., Jr., "Root-signal set analysis for median filters," in *Proc. 18th Annual Allerton Conf. on Communication, Control, and Computing*, Monticello, IL, Oct. 1980.

[3] Arce G., and Gallagher N., Jr., "Stochastic analysis for the multilevel root signal set of median filters," in *Proc. Conf. Inform. Sci. Syst.*, pp. 395–400, 1982.

[4] Arce G., and Gallagher N., Jr., "State description for the root-signal set of median filters," *IEEE Trans. Acoust Speech Sign. Process., ASSP-30*, pp. 894–902, 1982.

[5] Arce G., and Gallagher N., "BTC image coding using median filter roots," *IEEE Trans. Commun. COM-31*, pp. 784–793, 1983.

[6] Arce G., Gallagher N., Jr. and Nodes T., "Median filters: Theory for one- and two-dimensional filters," *Adv. Comput. Vis. Imag. Process.*, vol. 2, pp. 89–166, 1986.

[7] Astola J., Heinonen P., and Neuvo Y., "On the oscillatory roots of one and two dimensional median type filters," in *Proc. 1986 IEEE Int. Conf. Acoust. Speech, Signal Processing*, Tokyo, Japan, April, 1986.

[8] Astola J., Heinonen P., and Neuvo Y., "On root structures of median and median-type filters," *IEEE Trans. Acoust. Speech Sign. Process.*, **ASSP-35**, pp. 1199–1201, 1987.
[9] Brandt J., "Invariant Signals for Median Filters," *Utilitas Mathematica*, **31**, pp. 93–105, 1987.
[10] Brownrigg D., "The weighted median filter," *Commun. Assoc. Comput. Machinery*, **27**, pp. 807–818, 1984.
[11] Butz A., "A class of rank order smoothers," *IEEE Trans. Acoust Speech Sign. Process.*, **ASSP-34**, pp. 157–165, 1986.
[12] Butz A., "Some properties of a class of rank order smoothers," *IEEE Trans. Acoust Speech Sign. Process.*, **ASSP-34**, pp. 614–615, 1986.
[13] David H., *Order Statistics*, John Wiley Sons, New York, 1970.
[14] Yom D. H., and Ann S., "Directed graph representation for root-signal set of median filters," *Proc. IEEE*, **75**, pp. 1542–1544, 1987.
[15] Döhler H.-U., "Generation of roots of two dimensional median filters," Sign. Process. **18**, pp. 269–276, 1989.
[16] Fitch P., Coyle E., and Gallagher N., Jr., "Median filtering by threshold decomposition," *IEEE Trans. Acoust. Speech Sign. Process.*, **ASSP-32**, pp. 1183–1188, 1984.
[17] Fitch J., Coyle E., and Gallagher N., Jr., "Root properties and convergence rates of median filters," *IEEE Trans. Acoust Speech Sign. Process.*, **ASSP-33**, pp. 230–240, 1985.
[18] Fogelman F., Goles E., and Weisbuch G., "Transient length in sequential iteration of threshold functions," *Discrete Appl. Math.* **6**, pp. 95–98, 1983.
[19] Gabbouj M., Yu P. T., and Coyle E., "Convergence behavior and root signal sets of stack filters," *Circ. Syst. Sign. Process.*, Special Issue on Median and Morphological Filtering, 1991.
[20] Gallagher N., Jr., and Wise G., "A theoretical analysis of the properties of median filters," *IEEE Trans. Acoust. Speech Sign. Process.*, **ASSP-29**, pp. 1136–1141, 1981.
[21] Goles E., and Olivos J., "Comportement periodique des fonctions a seuil binaires & applicationes," *Discrete Appl. Math.* **3**, pp. 93–105, 1981.
[22] Haavisto P., Gabbouj M., and Neuvo Y., "Median based idenpotent filters," *J. Circ. Syst. Comput.*, **1**, pp. 125–148, 1991.
[23] Longbotham H., and Eberly D., "Statistical properties, fixed points and decomposition with WMMR," in *Nonlinear Image Processing*, E. Dougherty and J. Astola, eds., pp. 3–20, Kluwer Academic Publishers, Norwell, MA, 1993.
[24] Mallows C., "Some theory of nonlinear smoothers," *Ann. Stat.*, **8**, pp. 695–715, 1980.
[25] Mickens R., *Difference Equations*, Van Nostrand, New York, 1987.
[26] Nodes T., and Gallagher N., Jr., "Two-dimensional root structures and convergence properties of the separable median filter," *IEEE Trans. Acoust. Speech Sign. Process.*, **ASSP-31**, pp. 1350–1365, 1983.

[27] Stanley R., *Enumerative Combinatorics*, Volume I, Wadsworth Brooks/Cole, Belmont, CA, 1986.

[28] Tukey J., *Exploratory Data Analysis.*, Addison–Wesley, Menlo Park, CA, 1971/1977.

[29] Tukey J., "Nonlinear (nonsuperposable) methods for smoothing data," in *Congr. Rec. EASCON*, p. 673, 1974.

[30] Tyan S., "Median filtering: Deterministic properties," in *Topics in Applied Physics, Two-Dimensional Digital Signal Processing II*, T. Huang, ed., vol. 43, pp. 197–217, Springer-Verlag, Berlin, Germany, 1981.

[31] Wang Q., Gabbouj M., and Neuvo Y., "Root-signal sets of morphological filters and their use in variable-length BTC image coding," in *Nonlinear Image Processing*, pp. 59–76, E. Dougherty and J. Astola eds., Kluwer Academic Publishers, Norwell, MA, 1993.

[32] Wendt P., Coyle E., and Gallagher N., Jr., "Stack filters: Their definition and some initial properties," in *Proc. 19th Annual Conf. on Information Sciences and Systems*, Baltimore, MD, March 1985.

[33] Wendt P., Coyle E., and Gallagher N., Jr., "Some convergence properties of median filters," *IEEE Trans. Circuits Systems*, **CAS-33**, pp. 276–286, 1986.

[34] Wendt P., Coyle E., and Gallagher N., "Stack filters," *IEEE Trans. Acoust. Speech, Sign. Process.*, **ASSP-34**, pp. 898–911, 1986.

[35] Yli-Harja O., Astola J., and Neuvo Y., "Analysis of the properties of median and weighted median filters using threshold logic and stack decomposition," *IEEE Transactions on Signal Processing*, **SP-39**, pp. 395–410, 1991.

[36] Yu P. T., and Coyle E., "Convergence behavior and N-roots of stack filters," *IEEE Trans. Acoust. Speech and Sign. Process.*, **ASSP-38**, pp. 1529–1544, 1990.

CHAPTER 9

BINARY POLYNOMIAL TRANSFORMS AND LOGICAL CORRELATION

Karen O. Egiazarian
Jaakko T. Astola
Tampere University of Technology
Tampere, Finland

Sos S. Agaian
University of Texas/Austin
San Antonio, Texas

9.1 INTRODUCTION

Spectral analysis is a powerful tool in communication, signal/image processing and applied mathematics and there are many reasons behind its usefulness. First, in many applications it is convenient to transform a problem into another, easier problem; the spectral domain is essentially a steady-state viewpoint; this not only provides insight for analysis but also for design. Second, and more important; the spectral approach allows us to treat entire classes of signals which have similar properties in the spectral domain; it has excellent properties such as fast algorithms and high energy compaction of the transform data.

However, there are some problems in signal/image processing (especially if the input data is binary) where this tool cannot be applied directly. We may properly ask what is a basic transform for binary signal/image processing which plays the same role as spectral transform for linear processing. Similarly we may ask what is a basic transform for nonlinear signal/image processing which plays the same role as spectral transform for linear processing.

We also note that arithmetic and logic operations between pixels are used extensively in most branches of image processing, and that logical operations apply to binary images, while the arithmetic operations apply to multivalued pixels. We may ask what is a basic transform that is suitable for different kinds of data (binary and not binary) and different kinds of systems (linear and nonlinear).

In this chapter we define and develop a compact representation of signals/images for later use in digital logic and nonlinear signal/image analysis. Our approach is based on the introduction of binary parametric Rademacher functions and on generating a set of transforms using Rademacher functions and different kinds of arithmetic and logical operations. Emphasis is placed on the three classes of transforms introduced in this chapter:

– transforms generated by binary Rademacher functions using only one operation (arithmetic or logical);

– transforms generated by binary Rademacher functions using two operations (arithmetic or logical);

– transforms generated by three-valued Rademacher functions using two operations (arithmetic or logical).

We present the basic properties of these transforms. This development is motivated by potential applications in areas such as nonlinear signal/image processing and construction of new architectures of signal/image processing computers or systems suitable for simultaneous arithmetic or logical operations.

9.2 BINARY POLYNOMIAL FUNCTIONS AND MATRICES

9.2.1 RADEMACHER FUNCTIONS AND MATRICES

Let a and b be arbitrary integers. We form the classes $r_n(t,a,b)$ and $s_n(t,a,b)$ of real periodic functions in the following way. Let

$$r_0(t,a,b) \equiv s_0(t,a,b) \equiv 1, \tag{9–1}$$

$$r_{n+1}(t,a,b) = \begin{cases} a, & \text{for } t \in \bigcup_{m=0}^{2^n-1} [\frac{m}{2^n}, \frac{m}{2^n} + \frac{1}{2^{n+1}}), \\ b, & \text{for } t \in \bigcup_{m=0}^{2^n-1} [\frac{m}{2^n} + \frac{1}{2^{n+1}}, \frac{m+1}{2^n}), \end{cases} \tag{9–2}$$

and

$$s_{n+1}(t,a,b) = \begin{cases} a, & \text{for } t \in [0, \frac{1}{2^{n+1}}), \\ b, & \text{for } t \in [\frac{1}{2^{n+1}}, \frac{1}{2^n}), \\ 0, & \text{otherwise for } t \in [0,1), \end{cases} \qquad n = 0, 1, 2, \ldots. \tag{9–3}$$

Note that

$$r_1(t,a,b) = s_1(t,a,b) = \begin{cases} a, & \text{for } t \in [0, \frac{1}{2}), \\ b, & \text{for } t \in [\frac{1}{2}, 1). \end{cases}$$

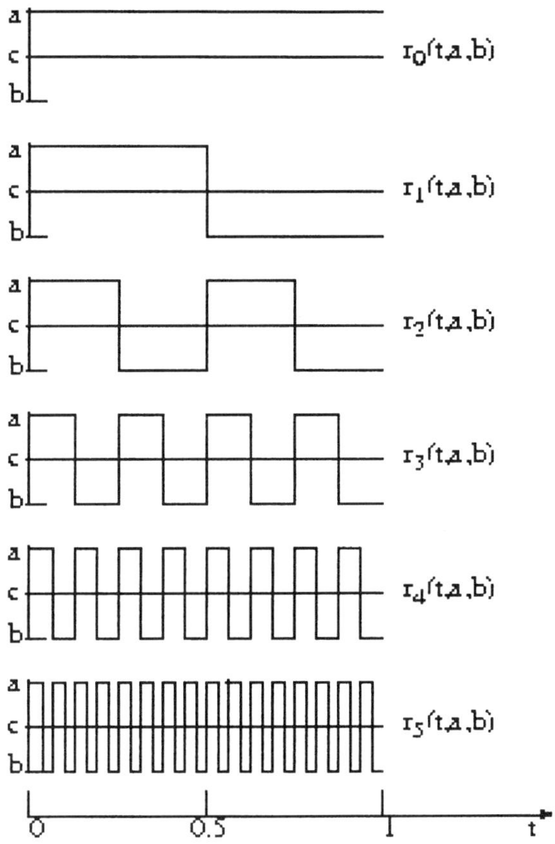

Figure 9–1. The first six Rademacher (a, b)-functions of I-type.

DEFINITION 9.2.1. The functions $r_n(t, a, b)$ are called *Rademacher (a, b)-functions of I-type* and the system of these functions is the *system of Rademacher (a, b)-functions of I-type*.

DEFINITION 9.2.2. The functions $s_n(t, a, b)$ are called *Rademacher (a, b)-functions of II-type* and the system of these functions is the *system of Rademacher (a, b)-functions of II-type*.

Note that when $a = 1$ and $b = -1$, the Rademacher (a, b)-functions of I-type coincide with the classical Rademacher functions [75]. In Figs. 9–1 and 9–2 the first six Rademacher (a, b)-functions of I-type and II-type are shown. Let us consider some properties of the generalized Rademacher functions defined above.

PROPERTY 9.2.1. *There are the following representations of the Rademacher (a, b)-functions of the I-type:*

$$1)\quad r_{n+1}(t, a, b) = a + (b - a)c_{n+1}(t), \quad t \in [0, 1), \tag{9–4}$$

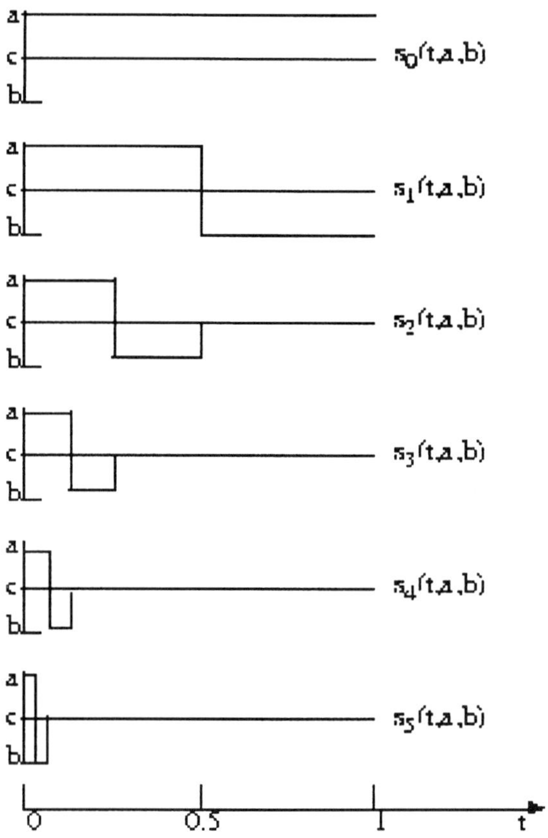

Figure 9–2. The first six Rademacher (a, b)-functions of II-type.

where c_{n+1} are the coefficients from

$$t = \sum_{n=1}^{\infty} c_n(t)2^{-n} = \sum_{n=1}^{\infty} c_n 2^{-n}. \tag{9–5}$$

PROOF. It is easy to verify that

$$c_{n+1}(t) = \begin{cases} 0, & \text{for } t \in \bigcup_{m=0}^{2^n-1} [\frac{m}{2^n}, \frac{m}{2^n} + \frac{1}{2^{n+1}}), \\ 1, & \text{for } t \in \bigcup_{m=0}^{2^n-1} [\frac{m}{2^n} + \frac{1}{2^{n+1}}, \frac{m+1}{2^n}), \end{cases}$$

bringing Eq. 9–4. ■

2) $r_n(t, a, b)$

$$= \frac{1}{2}\Big[(a-b)\text{sgn}[\sin(2^n \pi t)] + a + b\Big], \quad t \neq \frac{k}{2^n}, \; k = 0, \ldots, 2^{n-1}, \tag{9–6}$$

where $\operatorname{sgn}(y) = \begin{cases} 1, & \text{if } y > 0, \\ -1, & \text{if } y < 0. \end{cases}$

PROOF. Formula 9–6 follows from the representation of the $(1, -1)$-Rademacher functions of I-type via $r_n(t, 1, -1) = \operatorname{sgn}[\sin(2^n \pi t)]$. ■

3) *There is the following representation of the Rademacher functions of II-type*:

$$s_{n+1}(t, a, b) = \left[a + (b - a)c_{n+1}(t)\right] \prod_{i=1}^{n} c_i(t). \tag{9–7}$$

PROOF. It follows from the definition of Rademacher (a, b)-functions of II-type and is similar to proof of 1). ■

4) *There is the following relation between the Rademacher (a, b)-functions of I-type and II-type*:

$$r_{n+1}(t, a, b) = \sum_{m=0}^{2^{n-1}-1} s_{n+1}\left(t - \frac{m}{2^{n-1}}, a, b\right). \tag{9–8}$$

PROOF. Formula 9–8 follows from the representation of the Rademacher (a, b)-functions of II-type:

$$s_{n+1}\left(t - \frac{m}{2^{n-1}}, a, b\right) = \begin{cases} a, & \text{for } t \in [\frac{m}{2^{n-1}}, \frac{2m+1}{2^n}), \\ b, & \text{for } t \in [\frac{2m+1}{2^n}, \frac{m+1}{2^{n-1}}), \end{cases} \tag{9–9}$$

and formula 9–2. ■

For a natural number n and integers a and b we define the following discrete functions:

$$R_k^{(n)}(x, a, b) = r_k\left(\frac{2x+1}{2^{n+1}}, a, b\right),$$
$$k = 0, 1, \ldots, n-1;\ x = 0, 1, \ldots, 2^n - 1;\ n \geqslant 0. \tag{9–10}$$

$$S_k^{(n)}(x, a, b) = s_k\left(\frac{2x+1}{2^{n+1}}, a, b\right),$$
$$k = 0, 1, \ldots, n-1;\ x = 0, 1, \ldots, 2^n - 1;\ n \geqslant 0. \tag{9–11}$$

DEFINITION 9.2.3. The rectangular $(2^n \times (n+1))$ matrix $\mathbf{R}(n, a, b)$, whose (x, k)-th element is $R_k^{(n)}(x, a, b)$, $x = 0, 1, \ldots, 2^n - 1$, $k = 0, \ldots, n$, is called *Rademacher* (a, b) *matrix of I-type* (of order n).

DEFINITION 9.2.4. The rectangular $(2^n \times (n+1))$ matrix $\mathbf{S}(n,a,b)$, whose (x,k)-th element is $S_k^{(n)}(x,a,b)$, $x = 0, 1, \ldots, 2^n - 1$, $k = 0, \ldots, n$, is called *the Rademacher* (a,b) *matrix of II-type* (of order n).

REMARK 9.2.1. For each pair of integers (n,k), $n > 0$, $1 \leqslant k \leqslant n$, the kth column of $\mathbf{R}(n,a,b)$ defines a function $\{0, 1, \ldots, 2^n - 1\} \to \{a, b\}$. This set of functions is called the system of *discrete Rademacher* (a,b) *functions of I-type*. Analogously, for each pair of integers (n,k), $n > 0$, $1 \leqslant k \leqslant n$, the kth column of $\mathbf{S}(n,a,b)$ defines a function $\{0, 1, \ldots, 2^n - 1\} \to \{a, b\}$, the system of which is called the system of *discrete Rademacher* (a,b) *functions of II-type*.

EXAMPLE 9.2.1. When $n = 4$, the Rademacher (a,b)-matrices of I-type and II-type have the following forms, respectively:

$$\mathbf{R}^T(4,a,b) = \begin{pmatrix} 1 & 1 & 1 & 1 & 1 & 1 & 1 & 1 & 1 & 1 & 1 & 1 & 1 & 1 & 1 & 1 \\ a & a & a & a & a & a & a & a & b & b & b & b & b & b & b & b \\ a & a & a & a & b & b & b & b & a & a & a & a & b & b & b & b \\ a & a & b & b & a & a & b & b & a & a & b & b & a & a & b & b \\ a & b & a & b & a & b & a & b & a & b & a & b & a & b & a & b \end{pmatrix},$$

and the Rademacher (a,b)-matrices of II-type have the following forms:

$$\mathbf{S}^T(4,a,b) = \begin{pmatrix} 1 & 1 & 1 & 1 & 1 & 1 & 1 & 1 & 1 & 1 & 1 & 1 & 1 & 1 & 1 & 1 \\ a & a & a & a & a & a & a & a & b & b & b & b & b & b & b & b \\ a & a & a & a & b & b & b & b & 0 & 0 & 0 & 0 & 0 & 0 & 0 & 0 \\ a & a & b & b & 0 & 0 & 0 & 0 & 0 & 0 & 0 & 0 & 0 & 0 & 0 & 0 \\ a & b & 0 & 0 & 0 & 0 & 0 & 0 & 0 & 0 & 0 & 0 & 0 & 0 & 0 & 0 \end{pmatrix},$$

where T denotes matrix transposition.

The discrete Rademacher (a,b)-matrices of I-type and II-type have the following properties:

PROPERTY 9.2.2.

$$1)\quad R_k^{(n)}(x,a,b) = a + (b-a) \cdot x_k(x); \qquad k = 1, \ldots, n,$$

where n is a natural number, and x, $0 \leqslant x \leqslant 2^n - 1$, is written in the form

$$x = \sum_{j=0}^{n-1} x_j(x) \cdot 2^{n-j-1}, \qquad j = 0, 1, \ldots, n-1. \tag{9–12}$$

PROOF. Clearly

$$\frac{2x+1}{2^{n+1}} = \frac{1}{2^{n+1}}\left[\left(\sum_{j=0}^{n-1} x_j(x)\cdot 2^{n-j}\right)+1\right] = \sum_{k=1}^{n}\frac{x_{k-1}(x)}{2^{-k}} + \frac{1}{2^{n+1}}.$$

Therefore, from Eqs. 9–4 and 9–10 we get

$$R_k^{(n)}(x,a,b) = a + (b-a)\cdot x_k(x)$$

as required. ■

$$2)\quad \mathbf{R}^T(n,a,b)\cdot\mathbf{R}(n,a,b) = 2^{n-2}\big[(a-b)^2\cdot\mathbf{I}_{(n)} + (a+b)^2\cdot\mathbf{J}_{(n)}\big],$$

where $\mathbf{I}_{(n)}$ *is the identity matrix of order* n *and* $\mathbf{J}_{(n)}$ *is the* $(n\times n)$ *matrix all of whose entries are equal to* 1.

In particular, for the classic Rademacher matrices

$$\mathbf{R}^T(n,1,-1)\cdot\mathbf{R}(n,1,-1) = 2^n\cdot\mathbf{I}_{(n)}.$$

PROOF. Clearly, the (k,j)th element of the matrix $\mathbf{R}^T(n,a,b)\cdot\mathbf{R}(n,a,b)$ has the form:

$$\sum_{x=0}^{2^n-1} R_k^{(n)}(x,a,b)R_j^{(n)}(x,a,b) = \sum_{x=0}^{2^n-1}[a+(b-a)x_k(x)][a+(b-a)x_j(x)]$$

$$= 2^n a^2 + a(b-a)\sum_{x=0}^{2^n-1}[x_j(x)+x_k(x)] + (b-a)^2\sum_{x=0}^{2^n-1} x_k(x)x_j(x)$$

$$= 2^n\big[a^2+a(b-a)\big] + (b-a)^2\cdot\begin{cases}2^{n-2}, & \text{if } k\neq j\\ 2^{n-1}, & \text{if } k=j\end{cases}$$

$$= \begin{cases}2^{n-2}(a+b)^2, & \text{if } k\neq j\\ 2^{n-1}(a^2+b^2), & \text{if } k=j\end{cases}$$

and $2^{n-2}[(a-b)^2+(a+b)^2] = 2^{n-1}(a^2+b^2)$. ■

3) *Let* n *be a natural number,* $1\leqslant k\leqslant n$ *and suppose that* x, $0\leqslant x\leqslant 2^n-1$, *is written in the form* 9–12. *Then*

$$S_k^{(n)}(x,a,b) = \big[a+(b-a)\cdot x_k(x)\big]\prod_{i=0}^{k-1}[1-x_i(x)];\quad k=1,\ldots,n. \qquad (9\text{–}13)$$

PROOF. It is similar to proof of 1), and follows from Eq. 9–7. ■

$$4)\ \mathbf{S}^T(n,a,b)\cdot\mathbf{S}(n,a,b) = a(a+b)\cdot\left[\mathbf{J}_{(n)} + \sum_{i=1}^{n-2} 2^i \mathbf{P}_{(n)}^{(i)}\right] + b(b-a)\cdot\mathbf{D}_{(n)},$$

where $\mathbf{J}_{(n)}$ *is the* $(n\times n)$ *matrix all of whose entries are equal to* 1, $\mathbf{P}_{(n)}^{(i)}$ *is the* $(n\times n)$ *matrix whose leftmost* $(n-i)\times(n-i)$ *submatrix is* $\mathbf{J}_{(n-i)}$, $i=1,\ldots,n-2$, *and*

$$\mathbf{D}_{(n)} = \operatorname{diag}\left(2^{n-1}, 2^{n-2}, \ldots, 2, 1\right).$$

In particular, for the Rademacher $(1,-1)$*-matrices of II-type*:

$$\mathbf{S}^T(n,1,-1)\cdot\mathbf{S}(n,1,-1) = \operatorname{diag}\left(2^n, 2^{n-1}, \ldots, 2\right)$$

and for the Rademacher $(0,1)$*-matrices of II-type*:

$$\mathbf{S}^T(n,0,1)\cdot\mathbf{S}(n,0,1) = \operatorname{diag}\left(2^{n-1}, 2^{n-2}, \ldots, 2, 1\right).$$

PROOF. From Eq. 9–13, the (k,j)th (let $k\leqslant j$) element of the matrix $\mathbf{S}^T(n,a,b)\cdot\mathbf{S}(n,a,b)$ has the form:

$$\begin{aligned}
\Sigma &= \sum_{x=0}^{2^n-1} S_k^{(n)}(x,a,b)S_j^{(n)}(x,a,b) = \sum_{x=0}^{2^n-1}[a+(b-a)x_k(x)]\\
&\quad\times\prod_{i=1}^{k-1}[1-x_i(x)][a+(b-a)x_j(x)]\prod_{l=1}^{j-1}[1-x_l(x)]\\
&= \sum_{x=0}^{2^n-1}\prod_{i=0}^{j-1}[1-x_i(x)]\Bigg\{a^2+a(b-a)[x_j(x)+x_k(x)]\\
&\qquad\qquad + (b-a)^2\sum_{x=0}^{2^n-1}x_k(x)x_j(x)\Bigg\}.
\end{aligned}$$

Let $k=j$. Then

$$\begin{aligned}
\Sigma &= \sum_{x=0}^{2^n-1}\prod_{i=0}^{j-1}[1-x_i(x)]\left\{a^2+2a(b-a)x_j(x)+(b-a)^2x_j(x)\right\}\\
&= \sum_{x=0}^{2^n-1}\prod_{i=0}^{j-1}[1-x_i(x)]\left\{a^2+x_j(x)(b^2-a^2)\right\}\\
&= a^2\cdot 2^{n-j+1}+\left(b^2-a^2\right)\cdot 2^{n-j}.
\end{aligned}$$

Now, let $k < j$. Then

$$\begin{aligned}\Sigma &= \sum_{x=0}^{2^n-1}\prod_{i=0}^{j-1}[1-x_i(x)]\big[a^2+a(b-a)x_j(x)\big]\\ &= a^2\cdot 2^{n-j}+ab\cdot 2^{n-j}=a(a+b)2^{n-j}.\end{aligned}$$

■

9.2.2 $(a,b;\tau)$-POLYNOMIAL FUNCTIONS OF I-TYPE

Let τ be an arbitrary associative arithmetic or logical operation and $(m_0, m_1, \ldots, m_{n-1})$ be the binary representation of an integer $m = \sum_{i=0}^{n-1} m_i 2^{n-1-i}$, $0 \leqslant m \leqslant 2^n - 1$. Let also $\mathbf{H} \otimes \mathbf{G}$ be the Kronecker product of the matrix $\mathbf{H}$ and the matrix $\mathbf{G}$, and $\mathbf{H}^{\otimes n}$ be the Kronecker nth power of the matrix $\mathbf{H}$, i.e., $\mathbf{H}^{\otimes n} = \underbrace{\mathbf{H}\otimes\mathbf{H}\otimes\cdots\otimes\mathbf{H}}_{n\text{ times}}$.

We form a class of functions $\phi_m^{(n)}(x,a,b;\tau)$ from the system of discrete Rademacher (a,b)-functions of I-type (i.e., from $\{R_m^{(n)}(x,a,b)\}_{m=0}^{n-1}$) in the following way. Let

$$\begin{aligned}\phi_m^{(n)}(x,a,b;\tau) &= \phi_{m_0,m_1,\ldots,m_{n-1}}^{(n)}(x,a,b;\tau)\\ &= \big[R_0^{(n)}(x,a,b)\big]^{m_0}\tau\cdots\tau\big[R_{n-1}^{(n)}(x,a,b)\big]^{m_{n-1}},\end{aligned}\qquad (9\text{–}14)$$

where τ is an arithmetic or logical operation. If τ is a logical operation, then we naturally assume $a, b \in \{0, 1\}$.

DEFINITION 9.2.5. The functions $\phi_m^{(n)}(x,a,b;\tau)$, $m = 0,\ldots,n-1$, are called *$(a,b;\tau)$-polynomial functions*, and they form *the system of $(a,b;\tau)$-polynomial functions*.

DEFINITION 9.2.6. If τ is an arithmetic (logical) operation, the $(a,b;\tau)$-polynomial functions are called arithmetic (logical).

DEFINITION 9.2.7. The square matrix $\mathbf{\Phi}(n,a,b;\tau) = [\phi_m^{(n)}(x,a,b;\tau)]$, $x, m = 0, 1, \ldots, 2^n-1$, which has as columns the $(a,b;\tau)$-polynomial functions, is called *the $(a,b;\tau)$-polynomial matrix*.

In the sequel, when $a = 0$ and $b = 1$ the symbols a and b are omitted in the notations of $(a,b;\tau)$-polynomial functions and matrices.

DEFINITION 9.2.8. The matrix $\mathbf{V}_n(\tau) \overset{def}{=} \mathbf{\Phi}(n,0,1;\tau)$ is called *the τ-polynomial logical matrix*.

DEFINITION 9.2.9. The $(2^n \times 2^n)$ matrix $\widetilde{\mathbf{\Phi}}(n, a, b; \tau) = \mathbf{J}_n - \mathbf{\Phi}(n, a, b; \tau)$ of order 2^n is called *the complementary* $(a, b; \tau)$*-polynomial matrix.* $\mathbf{J}_n$ is the $(2^n \times 2^n)$ matrix whose entries are all equal to 1.

DEFINITION 9.2.10. The matrix $\widetilde{\mathbf{V}}_n(\tau) \stackrel{def}{=} \widetilde{\mathbf{\Phi}}(n, 0, 1; \tau)$ is called *the complementary* τ*-polynomial logical matrix.*

Now we will investigate particular cases of $(a, b; \tau)$-polynomial functions and matrices. Let τ be multiplication operation. The $(a, b, \times)$-polynomial functions $\phi_m^{(n)}$ are formed via Rademacher (a, b)-functions $R_m^{(n)}(x, a, b; \times)$ by the following formula:

$$\phi_m^{(n)}(x, a, b; \times) = \prod_{i=0}^{n-1} \left[R_i^{(n)}(x, a, b)\right]^{m_i},$$

where m_i is the ith bit in the binary representation of m via $m = \sum_{i=0}^{n-1} m_i 2^{n-1-i}$.

When $a = 1, b = -1$, the (a, b)-polynomial functions coincide with the Walsh functions, and when $a = 0, b = 1$, they coincide with the Reed–Muller (conjunctive) functions [8].

Based on Property 9.2.2, 1) we have the following expressions:

$$w_m(x) = \phi_m^{(n)}(x, 1, -1) = \prod_{i=0}^{n-1} (1 - 2x_i)^{m_i}, \tag{9–15}$$

for the Walsh functions, and

$$c_m(x) = \phi_m^{(n)}(x, 0, 1) = \prod_{i=0}^{n-1} x_i^{m_i}, \tag{9–16}$$

for the conjunctive functions.

The Walsh and conjunctive matrices are defined from the corresponding functions:

$$\begin{aligned} \mathbf{W}_n &= \left[W_{ij}^{(n)}\right] = \left[w_i^{(n)}(j)\right], \\ \mathbf{K}_n &= \left[\phi_{ij}^{(n)}\right] = \left[\phi_i^{(n)}(j)\right]. \end{aligned}$$

The set of Walsh functions is generally classified into three groups, namely, Hadamard, Paley, and Walsh. These differ from each other by the different order in which individual functions appear. The same holds also for the corresponding set of matrices (i.e., they differ by the ordering of their rows and columns). For ordering of the Walsh functions and corresponding matrices, see [8].

PROPERTY 9.2.3. *Let n be a natural number and let the integers m and x, $0 \leqslant m, x \leqslant N-1 = 2^n - 1$, have binary representations $m = (m_0, \ldots, m_{n-1})$, $x = (x_0, \ldots, x_{n-1})$. Then*

1) *Walsh–Hadamard functions can be generated using the relation*:

$$w_m^{(n)}(x) = \phi_m^{(n)}(x, 1, -1; \times) = (-1)^{\sum_{i=0}^{n-1} m_i x_i}. \tag{9–17}$$

2) *The system of Walsh functions $w_m^{(n)}(x)$ is orthogonal*:

$$\sum_{x=0}^{N-1} w_m^{(n)}(x) w_k^{(n)}(x) = N \cdot \delta_{m,k}, \quad \text{where } \delta_{m,k} = \begin{cases} 1, & \text{for } m = k, \\ 0, & \text{otherwise.} \end{cases}$$

3) *The functions $w_m^{(n)}(x)$ are symmetric with respect to m and x*:

$$w_m^{(n)}(x) = w_x^{(n)}(m).$$

4) *The Walsh functions of order n form a group with respect to multiplication. This group is isomorphic to the group of $\langle B_2^n, \oplus \rangle$ of all binary vectors of length n with respect to component-wise addition modulo 2. The isomorphism is given by*

$$w_m^{(n)}(x) \cdot w_k^{(n)}(x) = w_{m \oplus k}^{(n)}(x).$$

5) *Walsh matrices $\mathbf{W}_n$ of order $N = 2^n$ are orthogonal and symmetric matrices, i.e.,*

$$\mathbf{W}_n \cdot \mathbf{W}_n^T = \mathbf{W}_n^T \cdot \mathbf{W}_n = N \cdot \mathbf{I}_n; \qquad \mathbf{W}_n = \mathbf{W}_n^T,$$

where $\mathbf{I}_n$ is the identity matrix of order 2^n, and they can be determined by:

$$\mathbf{W}_n = \begin{pmatrix} 1 & 1 \\ 1 & -1 \end{pmatrix}^{\otimes n} = \mathbf{W}_1^{\otimes n}, \tag{9–18}$$

where $\mathbf{W}_1 = \left(\begin{smallmatrix} 1 & 1 \\ 1 & -1 \end{smallmatrix}\right)$ is the Walsh matrix of order 2, $\mathbf{W}_1^{\otimes n}$ is the nth Kronecker power of $\mathbf{W}_1$.

EXAMPLE 9.2.2. The Walsh $\mathbf{W}_3$ matrix in Hadamard ordering has the following form:

$$\mathbf{W}_3 = \begin{pmatrix} 1 & 1 & 1 & 1 & 1 & 1 & 1 & 1 \\ 1 & -1 & 1 & -1 & 1 & -1 & 1 & -1 \\ 1 & 1 & -1 & -1 & 1 & 1 & -1 & -1 \\ 1 & -1 & -1 & 1 & 1 & -1 & -1 & 1 \\ 1 & 1 & 1 & 1 & -1 & -1 & -1 & -1 \\ 1 & -1 & 1 & -1 & -1 & 1 & -1 & 1 \\ 1 & 1 & -1 & -1 & -1 & -1 & 1 & 1 \\ 1 & -1 & -1 & 1 & -1 & 1 & 1 & -1 \end{pmatrix}.$$

Now let τ be an arbitrary associative binary logical operation, $a = 0$ and $b = 1$. Let m and x have binary representations $(m_0, \ldots, m_{n-1})$ and $(x_0, \ldots, x_{n-1})$, respectively. Then, by Property 9.2.2, formula 9–14 becomes

$$\phi_m^{(n)}(x, 0, 1; \tau) = \phi_{m_0,\ldots,m_{n-1}}^{(n)}(x, 0, 1; \tau) = x_0^{m_0} \ \tau \ \cdots \ \tau \ x_{n-1}^{m_{n-1}}. \tag{9–19}$$

Examples of the τ-polynomial logical functions are:

a) conjunctive, if τ is the operation $\wedge$ (AND); denoted by $\phi_m^{(n)}(x, \wedge)$;

b) disjunctive, if τ is the operation $\vee$ (OR); denoted by $\phi_m^{(n)}(x, \vee)$;

c) antivalent, if τ is the operation $\oplus$ (XOR); denoted by $\phi_m^{(n)}(x, \oplus)$.

DEFINITION 9.2.11. The set of functions $\overline{\phi_m^{(n)}(x; \tau)}$ obtained by the complementation of τ-polynomial logical functions $\phi_m^{(n)}(x; \tau)$ is called *the system of complementary τ-polynomial logical functions.*

The systems of τ-polynomial logical functions and complementary τ-polynomial logical functions can be unified in the following way. Consider the threshold operator $\sigma_{c,d}$:

$$\sigma_{c,d}(x) = \begin{cases} c, & \text{if } x \geqslant 1, \\ d, & \text{if } x < 1. \end{cases} \tag{9–20}$$

Then the system of functions

$$\psi_m^{(n)}(x; \tau) = \sigma_{c,d}\big[\phi_m^{(n)}(x; \tau)\big], \tag{9–21}$$

where $\phi_m^{(n)}(x; \tau)$ is an arbitrary τ-polynomial logical function, includes:

Table 9.1. The discrete $(a, b; \tau)$-polynomial functions $\phi_m^{(3)}(x, a, b; \tau)$, with $r_j \stackrel{def}{=} R_j^{(3)}(x, a, b)$, $j = 1, 2, 3$.

No.	**Ordering by**		
	Paley	**Grey–Walsh**	**Hadamard**
0	$p_0(x) = 1\tau 1\tau 1$	$w_0 = 1\tau 1\tau 1$	$h_0 = 1\tau 1\tau 1$
1	$p_1(x) = r_1\tau 1\tau 1$	$w_1 = r_1\tau 1\tau 1$	$h_1 = r_3\tau 1\tau 1$
2	$p_2(x) = r_2\tau 1\tau 1$	$w_2 = r_1\tau r_2\tau 1$	$h_2 = r_2\tau 1\tau 1$
3	$p_3(x) = r_1\tau r_2\tau 1$	$w_3 = r_2\tau 1\tau 1$	$h_3 = r_2\tau r_3\tau 1$
4	$p_4(x) = r_3\tau 1\tau 1$	$w_4 = r_2\tau r_3\tau 1$	$h_4 = r_1\tau 1\tau 1$
5	$p_5(x) = r_1\tau r_3\tau 1$	$w_5 = r_1\tau r_2\tau r_3$	$h_5 = r_1\tau r_3\tau 1$
6	$p_6(x) = r_2\tau r_3\tau 1$	$w_6 = r_1\tau r_3\tau 1$	$h_6 = r_1\tau r_2\tau 1$
7	$p_7(x) = r_1\tau r_2\tau r_3$	$w_7 = r_3\tau 1\tau 1$	$h_7 = r_1\tau r_2\tau r_3$

1) the system of τ-polynomial logical functions when $c = 1,\ d = 0$;

2) the system of complementary τ-polynomial logical functions when $d = 1$, $c = 0$.

Since the indexing (ordering) of τ-polynomial logical functions can be made in different ways (Hadamard, Paley, or Walsh ordering), different systems of τ-polynomial logical functions may appear. Their general form can be seen from Example 9.2.3.

EXAMPLE 9.2.3. The systems $\{\phi_m^{(n)}(x, a, b; \tau)\}_{m=0}^{2^n-1}$ of discrete $(a, b; \tau)$-polynomial functions for the case $n = 3$ are shown in Table 9.1.

In the following, unless otherwise stated, the polynomial systems of functions and matrices will be considered in Hadamard ordering. We now briefly give the main properties of τ-polynomial logical matrices.

PROPERTY 9.2.4.

1) *The conjunctive matrix* $\mathbf{K}_n = \mathbf{V}_n(\wedge)$ *(in Hadamard ordering) of order* 2^n *is defined by*

$$\mathbf{K}_n = \begin{pmatrix} \mathbf{K}_{n-1} & 0 \\ \mathbf{K}_{n-1} & \mathbf{K}_{n-1} \end{pmatrix} = \begin{pmatrix} 1 & 0 \\ 1 & 1 \end{pmatrix}^{\otimes n} = \mathbf{K}_1^{\otimes n}, \qquad (9\text{–}22)$$

where $\mathbf{K}_1$ *is the conjunctive matrix of order* 2.

2) *The* (i, j)*th element of the conjunctive matrix* $\mathbf{K}_n$, $K_n(i, j) = 1$ *if and only if* $i_k \geqslant j_k$ *for all* $k = 0, \ldots, n-1$, *where* $(i_0, \ldots, i_{n-1})$ *and* $(j_0, \ldots, j_{n-1})$ *are the binary representations of the numbers* i *and* j, *respectively,* $i, j = 0, \ldots, 2^n - 1$.

3) *The matrix* $\mathbf{K}_n$ *is nonsingular. The inverse matrix* $\mathbf{K}_n^{-1}$ *has the form*:

$$\mathbf{K}_n^{-1} = \begin{pmatrix} 1 & 0 \\ -1 & 1 \end{pmatrix}^{\otimes n} = (\mathbf{K}_1^{-1})^{\otimes n}. \tag{9–23}$$

4) *The following relations hold*

$$\mathbf{K}_n \cdot \mathbf{K}_n^T = \bar{\mathbf{I}}_n \cdot \mathbf{K}_n^T \cdot \mathbf{K}_n \bar{\mathbf{I}}_n = \begin{pmatrix} 1 & 1 \\ 1 & 2 \end{pmatrix}^{\otimes n}, \tag{9–24}$$

$$\bar{\mathbf{I}}_n = \begin{pmatrix} 0 & 0 & \ldots & 0 & 1 \\ 0 & 0 & \ldots & 1 & 0 \\ & \ldots & \ldots & \ldots & \\ 1 & 0 & \ldots & 0 & 0 \end{pmatrix}$$

is the opposite diagonal identity matrix of order 2^n, *and*

$$\mathbf{W}_n = \mathbf{K}_n^{-1} \cdot \begin{pmatrix} 0 & 2 \\ 1 & 0 \end{pmatrix}^{\otimes n} \mathbf{K}_n. \tag{9–25}$$

5) *Let* $\mathbf{V}_n(\oplus)$ *be the* $\oplus$*-polynomial logic matrix of order* 2^n. *Then*

$$\mathbf{W}_n = \begin{cases} \bar{\mathbf{I}}_n[\mathbf{J}_n - 2\mathbf{V}_n(\oplus)], & \textit{if } n \text{ is even} \\ \bar{\mathbf{I}}_n[\mathbf{J}_n - 2\widetilde{\mathbf{V}}_n(\oplus)], & \textit{if } n \text{ is odd,} \end{cases} \tag{9–26}$$

where $\bar{\mathbf{I}}_n$ *is the opposite diagonal identity matrix of order* 2^n, $\mathbf{J}_n$ *is the square matrix of ones of order* 2^n *and* $\widetilde{\mathbf{V}}_n(\oplus) = \mathbf{J}_n - \mathbf{V}_n(\oplus)$.

6) *The following recurrent constructions for* τ*-polynomial logical matrices, in Hadamard ordering, hold*

$$\mathbf{D}_k = \mathbf{V}_k(\vee) = \begin{pmatrix} \mathbf{J}_{k-1} & \mathbf{V}_{k-1}(\vee) \\ \mathbf{J}_{k-1} & \mathbf{J}_{k-1} \end{pmatrix}, \tag{9–27}$$

$$\mathbf{V}_k(\oplus) = \begin{pmatrix} \widetilde{\mathbf{V}}_{k-1}(\oplus) & \mathbf{V}_{k-1}(\oplus) \\ \widetilde{\mathbf{V}}_{k-1}(\oplus) & \widetilde{\mathbf{V}}_{k-1}(\oplus) \end{pmatrix}, \tag{9–28}$$

and

$$\mathbf{V}_k(\equiv) = \begin{pmatrix} \mathbf{V}_{k-1}(\equiv) & \widetilde{\mathbf{V}}_{k-1}(\equiv) \\ \mathbf{V}_{k-1}(\equiv) & \mathbf{V}_{k-1}(\equiv) \end{pmatrix}, \tag{9–29}$$

where $\mathbf{V}_1(\tau) = \left(\begin{smallmatrix} 1 & 0 \\ 1 & 1 \end{smallmatrix}\right)$, $\widetilde{\mathbf{V}}_k(\tau) = \mathbf{J}_k - \mathbf{V}_k(\tau)$ *is complementary to* $\mathbf{V}_k(\tau)$, *for all considered logical operations* τ *that are disjunction* ($\vee$), *antivalence* ($\oplus$) *and equivalence* ($\equiv$).

EXAMPLE 9.2.4. The conjunctive (Reed–Muller) $\mathbf{K}_3$ matrix in Hadamard ordering has the following form:

$$\mathbf{K}_3 = \begin{pmatrix} 1 & 0 & 0 & 0 & 0 & 0 & 0 & 0 \\ 1 & 1 & 0 & 0 & 0 & 0 & 0 & 0 \\ 1 & 0 & 1 & 0 & 0 & 0 & 0 & 0 \\ 1 & 1 & 1 & 1 & 0 & 0 & 0 & 0 \\ 1 & 0 & 0 & 0 & 1 & 0 & 0 & 0 \\ 1 & 1 & 0 & 0 & 1 & 1 & 0 & 0 \\ 1 & 0 & 1 & 0 & 1 & 0 & 1 & 0 \\ 1 & 1 & 1 & 1 & 1 & 1 & 1 & 1 \end{pmatrix}.$$

There follows a simple extension of the conjunctive matrix.

DEFINITION 9.2.12. The matrix

$$\mathbf{K}_M^{(w)} = (\mathbf{K}_1^{(w)})^{\otimes M} = \begin{pmatrix} 1 & 0 \\ 2 & 1 \end{pmatrix}^{\otimes M}, \tag{9–30}$$

where $\otimes M$ denotes the Mth Kronecker power of the matrix, is called the *Hamming-weighted conjunctive matrix.*

EXAMPLE 9.2.5. The Hamming-weighted matrices of orders 4 and 8 are

$$\mathbf{K}_2^{(w)} = \begin{pmatrix} 1 & 0 & 0 & 0 \\ 2 & 1 & 0 & 0 \\ 2 & 0 & 1 & 0 \\ 4 & 2 & 2 & 1 \end{pmatrix}, \qquad \mathbf{K}_3^{(w)} = \begin{pmatrix} 1 & 0 & 0 & 0 & 0 & 0 & 0 & 0 \\ 2 & 1 & 0 & 0 & 0 & 0 & 0 & 0 \\ 2 & 0 & 1 & 0 & 0 & 0 & 0 & 0 \\ 4 & 2 & 2 & 1 & 0 & 0 & 0 & 0 \\ 2 & 0 & 0 & 0 & 1 & 0 & 0 & 0 \\ 4 & 2 & 0 & 0 & 2 & 1 & 0 & 0 \\ 4 & 0 & 2 & 0 & 2 & 0 & 1 & 0 \\ 8 & 4 & 4 & 2 & 4 & 2 & 2 & 1 \end{pmatrix}. \tag{9–31}$$

DEFINITION 9.2.13. Let an M-variable Boolean function $f(\cdot)$ be given by its truth table $\mathbf{f}$. The Hamming-weighted conjunctive transform (spectrum) of $\mathbf{f}$ is

$$\mathbf{h} = K_M^{(w)} \cdot \mathbf{f}. \tag{9–32}$$

Note the following properties of the Hamming-weighted conjunctive matrix and the Hamming-weighted conjunctive transform.

PROPOSITION 9.2.1.

1) $\mathbf{K}_M^{(w)} = \mathbf{K}_M^2$, *where* $\mathbf{K}_M$ *is the conjunctive or Reed–Muller matrix of order* 2^M *defined as Eq. 9–22.*

2) *Let* $K_{\alpha,\beta}^{(w)}$, $\alpha, \beta = 0, 1, \ldots, 2^M - 1$, *be the entries of the matrix* $\mathbf{K}_M^{(w)}$ *and let* $\boldsymbol{\alpha} = (\alpha_0, \ldots, \alpha_{M-1})$ *and* $\boldsymbol{\beta} = (\beta_0, \ldots, \beta_{M-1})$ *be the vectors corresponding to the binary codes of the numbers* α *and* (β). *Then*

$$\left(K_M^{(w)}\right)_{\alpha,\beta} = \begin{cases} 2^{w_H(\boldsymbol{\alpha},\boldsymbol{\beta})} & \text{if } \boldsymbol{\alpha} \geqslant \boldsymbol{\beta}, \\ 0 & \text{otherwise,} \end{cases} \qquad (9\text{–}33)$$

where $w_H(\boldsymbol{\alpha}, \boldsymbol{\beta})$ *is the Hamming distance.*

PROOF. 1) obviously follows from the definitions of matrices $\mathbf{K}_M^{(w)}$ and $\mathbf{K}_M$.

We prove 2) by induction on M. Let $M = 1$. The entries of the matrix

$$\mathbf{K}_1^{(w)} = \begin{pmatrix} 1 & 0 \\ 2 & 1 \end{pmatrix}^{\otimes M}$$

are

$$\left(K_1^{(w)}\right)_{\alpha,\beta} = \begin{cases} 1 & \text{if } \alpha = \beta, \\ 2 & \text{if } \alpha > \beta, \\ 0 & \text{otherwise,} \end{cases}$$

and Eq. 9–33 is true.

Let us suppose that Eq. 9–33 is true for $M = L$. By the definition of the matrix $\mathbf{K}_M^{(w)}$ there exists the following recurrent relation

$$\mathbf{K}_{L+1}^{(w)} = \begin{pmatrix} \mathbf{K}_L^{(w)} & 0 \\ 2\mathbf{K}_L^{(w)} & \mathbf{K}_L^{(w)} \end{pmatrix},$$

from which we find

$$\left(K_{L+1}^{(w)}\right)_{\alpha,\beta} = \begin{cases} (K_L^{(w)})_{\alpha',\beta'} & \text{if } \alpha_L = \beta_L, \\ 2(K_L^{(w)})_{\alpha',\beta'} & \text{if } \alpha_L > \beta_L, \\ 0 & \text{otherwise,} \end{cases} \qquad (9\text{–}34)$$

where $\boldsymbol{\alpha} = \alpha_0, \alpha_1, \ldots, \alpha_L$, $\boldsymbol{\alpha}' = \alpha_0, \ldots, \alpha_{L-1}$, and $\boldsymbol{\beta} = \beta_0, \beta_1, \ldots, \beta_L$, $\boldsymbol{\beta}' = \beta_0, \ldots, \beta_{L-1}$. From Eq. 9–34 and induction hypothesis we get Eq. 9–33. ∎

9.2.3 (a, b)-POLYNOMIAL FUNCTIONS OF II-TYPE

First define the shifted Rademacher (a, b)-functions of II-type, namely

$$s_m^{(k)}(x, a, b) = s^{(k)}\Big(x - \frac{m}{2^k}, a, b\Big), \tag{9–35}$$

where $m \in \{0, 1, \ldots, 2^k - 1\}$.

DEFINITION 9.2.14. The functions $h_0^{(n)}(x, a, b) = s^{(0)}(x, a, b)$, $h_1^{(n)}(x, a, b) = s^{(1)}(x, a, b)$, and $h_{2^k+m}^{(n)}(x, a, b) = s_m^{(k+1)}(x, a, b)$, $m = 0, 1, \ldots, 2^k - 1$, $k = 1, \ldots, n-1$, are called *(a, b)-polynomial functions of II-type*, and form *the system of (a, b)-polynomial functions of II-type*.

DEFINITION 9.2.15. The square matrix $\mathbf{H}(n, a, b) = [h_r^{(n)}(x, a, b)]$, $x, r = 0, 1, \ldots, 2^n - 1$, which has as columns the (a, b)-polynomial functions of II-type, is called *the (a, b)-polynomial matrix of II-type*.

Note that the classical Haar functions (matrices) are the particular case of the (a, b)-polynomial functions (matrices) of II-type where $a = 1, b = -1$. The Haar-conjunctive functions (matrices) will be (0, 1)-polynomial functions (matrices) of II type.

EXAMPLE 9.2.6. The Haar (unnormalized) matrix $\mathbf{H}_3$ and Haar-conjunctive $\mathbf{HK}_3$ matrix are:

$$\mathbf{H}_3 = \begin{pmatrix} 1 & 1 & 1 & 0 & 1 & 0 & 0 & 0 \\ 1 & 1 & 1 & 0 & -1 & 0 & 0 & 0 \\ 1 & 1 & -1 & 0 & 0 & 1 & 0 & 0 \\ 1 & 1 & -1 & 0 & 0 & -1 & 0 & 0 \\ 1 & -1 & 0 & 1 & 0 & 0 & 1 & 0 \\ 1 & -1 & 0 & 1 & 0 & 0 & -1 & 0 \\ 1 & -1 & 0 & -1 & 0 & 0 & 0 & 1 \\ 1 & -1 & 0 & -1 & 0 & 0 & 0 & -1 \end{pmatrix}.$$

$$\mathbf{HK}_3 = \begin{pmatrix} 1 & 0 & 0 & 0 & 0 & 0 & 0 & 0 \\ 1 & 0 & 0 & 0 & 1 & 0 & 0 & 0 \\ 1 & 0 & 1 & 0 & 0 & 0 & 0 & 0 \\ 1 & 0 & 1 & 0 & 0 & 1 & 0 & 0 \\ 1 & 1 & 0 & 0 & 0 & 0 & 0 & 0 \\ 1 & 1 & 0 & 0 & 0 & 0 & 1 & 0 \\ 1 & 1 & 0 & 1 & 0 & 0 & 0 & 0 \\ 1 & 1 & 0 & 1 & 0 & 0 & 0 & 1 \end{pmatrix}.$$

Note that the normalized orthogonal Haar matrix of order 8 is given by

$$\mathbf{H}_3 \cdot \operatorname{diag}\big(1, 1, \sqrt{2}, \sqrt{2}, 2, 2, 2, 2\big).$$

9.2.4 Binary Polynomial Logical Functions and Matrices. Constructions Using Two Operations

Let τ, σ be associative logical operations, s, n and p be natural numbers such that $p = n - s + 1$; and $1 \leqslant p, s \leqslant n$. As before, we write $\mathbf{m} = (m_0, m_1, \ldots, m_{n-1})$ as the binary representation of the number m, $0 \leqslant m \leqslant 2^n - 1$. Let us use the notations $x_k = R_k^{(n)}(x) = R_k^{(n)}(x, 0, 1)$ for Rademacher logical functions, i.e., the case $a = 0, b = 1$ of Rademacher (a, b)-functions of I-type.

DEFINITION 9.2.16. The set of functions $\rho_{m,s}^{(n)}(x; \tau, \sigma) : \{0, 1, \ldots, 2^n - 1\} \to \{0, 1\}$, $m = 0, 1, \ldots, 2^n - 1$, defined by

$$\rho_{m,s}^{(n)}(x; \tau, \sigma) = \mathop{\tau}_{0\leqslant k_1<k_2<\cdots<k_p\leqslant n-1} \Big(\big(R_{k_1}^{(n)}\big)^{m_{k_1}} \sigma \, \big(R_{k_2}^{(n)}\big)^{m_{k_2}} \sigma \cdots \sigma \, \big(R_{k_p}^{(n)}\big)^{m_{k_p}}\Big), \quad (9\text{–}36)$$

is called *the system of discrete* $(s, n; \tau, \sigma)$*-polynomial logical functions.*

REMARK 9.2.2. The system of discrete $(s, n; \tau, \sigma)$-polynomial logical functions contains as special cases, for $s = 1$ and $s = n$, the system of σ- and τ-polynomial logical functions, respectively.

DEFINITION 9.2.17. The system of discrete $(s, n; \vee, \wedge)$-polynomial logic functions:

$$\rho_{m,s}^{(n)}(x) = \bigvee_{0\leqslant k_1<k_2<\cdots<k_p\leqslant n-1} \Big(x_{k_1}^{m_{k_1}} \wedge \cdots \wedge x_{k_p}^{m_{k_p}}\Big), \quad (9\text{–}37)$$

$$x_j^{m_j} = \begin{cases} 1, & \text{if } m_j = 0, \\ x_j, & \text{if } m_j = 1, \end{cases} \quad (9\text{–}38)$$

where the disjunctions are over all conjunctions of p variables, $p = n - s + 1$, is called *the system of discrete* (s, n)*-polynomial disjunctive-conjunctive functions.*

The square matrix $\mathbf{L}_n^{(s)} = [\rho_{m,s}^{(n)}(x)]$, $x, m = 0, 1, \ldots, 2^n - 1$, of order 2^n with the columns representing the (s, n)-polynomial disjunctive-conjunctive functions is called *the* (s, n)*-polynomial disjunctive-conjunctive matrix.*

EXAMPLE 9.2.7. The $(2, 4)$-polynomial disjunctive-conjunctive system of functions has the form given in Table 9.2.

Table 9.2. The system of $(2,4)$-polynomial disjunctive-conjunctive functions.

m	$m_0m_1m_2m_3$	**Polynomial disjunctive-conjunctive functions** $\rho_{m,2}^{(4)}(x)$
0	0 0 0 0	1
1	0 0 0 1	1
2	0 0 1 0	1
3	0 0 1 1	$x_2 \vee x_3$
4	0 1 0 0	1
5	0 1 0 1	$x_1 \vee x_3$
6	0 1 1 0	$x_1 \vee x_2$
7	0 1 1 1	$x_1x_2 \vee x_1x_3 \vee x_2x_3$
8	1 0 0 0	1
9	1 0 0 1	$x_0 \vee x_3$
10	1 0 1 0	$x_0 \vee x_2$
11	1 0 1 1	$x_0x_2 \vee x_0x_3 \vee x_2x_3$
12	1 1 0 0	$x_0 \vee x_1$
13	1 1 0 1	$x_0x_1 \vee x_0x_3 \vee x_1x_3$
14	1 1 1 0	$x_0x_1 \vee x_0x_2 \vee x_1x_2$
15	1 1 1 1	$x_0x_1x_2 \vee x_0x_1x_3 \vee x_0x_2x_3 \vee x_1x_2x_3$

Let $\{\rho_{m,s}^{(n)}(x)\}_{m=0}^{2^n-1}$ be the system of discrete (s,n)-polynomial disjunctive-conjunctive functions.

DEFINITION 9.2.18. The system of functions $\{\mu_{m,s}^{(n)}(x)\} = \overline{\{\rho_{m,s}^{(n)}(x)\}}_{m=0}^{2^n-1}$, obtained by taking the complements of the functions $\rho_{m,s}^{(n)}(x)$, is called *the system of complementary* (s,n)*-polynomial disjunctive-conjunctive functions.*

The square matrix $\mathbf{M}_n^{(s)} \stackrel{def}{=} [\mu_{m,s}^{(n)}(x)]$, $x, m = 0, 1, \ldots, 2^n - 1$, of order 2^n with the columns representing the complementary (s,n)-polynomial disjunctive-conjunctive functions, is called *the complementary* (s,n)*-polynomial disjunctive-conjunctive matrix.*

Using De Morgan's rule (i.e., $\overline{(x \vee y)} = \bar{x} \wedge \bar{y}$), the complementary (s,n)-polynomial disjunctive-conjunctive functions $\mu_{m,s}^{(n)}(x)$ can be written in the fol-

Table 9.3. The complementary system of $(2, 4)$-polynomial disjunctive-conjunctive functions.

$m_0m_1m_2m_3$	**Complementary polynomial disjunctive-conjunctive functions** $\mu_{m,2}^{(4)}(t)$
0 0 0 0	0
0 0 0 1	0
0 0 1 0	0
0 0 1 1	$\bar{x}_2\bar{x}_3$
0 1 0 0	0
0 1 0 1	$\bar{x}_1\bar{x}_3$
0 1 1 0	$\bar{x}_1\bar{x}_2$
0 1 1 1	$(\bar{x}_1 \vee \bar{x}_2)(\bar{x}_1 \vee \bar{x}_3)(\bar{x}_2 \vee \bar{x}_3)$
1 0 0 0	0
1 0 0 1	$\bar{x}_0\bar{x}_3$
1 0 1 0	$\bar{x}_0\bar{x}_2$
1 0 1 1	$(\bar{x}_0 \vee \bar{x}_2)(\bar{x}_0 \vee \bar{x}_3)(\bar{x}_2 \vee \bar{x}_3)$
1 1 0 0	$\bar{x}_0\bar{x}_1$
1 1 0 1	$(\bar{x}_0 \vee \bar{x}_1)(\bar{x}_0 \vee \bar{x}_3)(\bar{x}_1 \vee \bar{x}_3)$
1 1 1 0	$(\bar{x}_0 \vee x_1)(\bar{x}_0 \vee \bar{x}_2)(\bar{x}_1 \vee \bar{x}_2)$
1 1 1 1	$(\bar{x}_0 \vee \bar{x}_1 \vee \bar{x}_2)(\bar{x}_0 \vee \bar{x}_1 \vee \bar{x}_3)(\bar{x}_0 \vee \bar{x}_2 \vee \bar{x}_3)(\bar{x}_1 \vee \bar{x}_2 \vee \bar{x}_3)$

lowing form:

$$\mu_{m,s}^{(n)}(x) = \bar{\rho}_{m,s}^{(n)}(x) = \overline{\bigvee_k (x_{k_1}^{m_{k_1}} \wedge x_{k_2}^{m_{k_2}} \wedge \cdots \wedge x_{k_p}^{m_{k_p}})}$$
$$= \bigwedge_k (\overline{x_{k_1}^{m_{k_1}}} \vee \overline{x_{k_2}^{m_{k_2}}} \vee \cdots \vee \overline{x_{k_p}^{m_{k_p}}}), \tag{9–39}$$

where

$$\overline{x_{k_i}^{m_{k_i}}} = \begin{cases} 0, & \text{if } m_i = 0, \\ \bar{x}_i, & \text{if } m_i = 1, \end{cases} \tag{9–40}$$

$i = 0, 1, \ldots, n-1, \quad p = n - s + 1.$

EXAMPLE 9.2.8. The complementary $(2, 4)$-polynomial disjunctive-conjunctive system of functions has the form given in Table 9.3.

Similar to the system of Walsh functions [21], the ordering of (s, n)-polynomial (and complementary polynomial) disjunctive-conjunctive functions can be made in different ways. In the sequel we will use the Hadamard ordering. Now we give the basic properties of discrete (s, n)-polynomial disjunctive-conjunctive functions.

PROPERTY 9.2.5. *Let* $\rho_{m,s}^{(n)}(x)$ *and* $\mu_{m,s}^{(n)}(x) = \overline{\rho_{m,s}^{(n)}}(x)$ *be the direct and complementary* (s, n)*-polynomial disjunctive-conjunctive functions, respectively. The following relations hold*:

1) $\rho_{m,s}^{(n)}(x)$ *can be obtained recursively as*

$$\begin{aligned}\rho_{m,s}^{(n)}(x) &= \rho_{m,s}^{(n)}(x_0, x_1, \ldots, x_{n-1}) \\ &= x_{n-1}^{m_{n-1}} \rho_{m',s}^{(n-1)}(x_0, x_1, \ldots, x_{n-2}) \vee \rho_{m',s-1}^{(n-1)}(x_0, x_1, \ldots, x_{n-2}),\end{aligned} \quad (9\text{–}41)$$

$$\begin{aligned}\mu_{m,s}^{(n)}(x) &= \mu_{m,s}^{(n)}(x_0, x_1, \ldots, x_{n-1}) \\ &= \overline{x_{n-1}^{m_{n-1}}} \mu_{m',s-1}^{(n-1)}(x_0, x_1, \ldots, x_{n-2}) \vee \mu_{m',s}^{(n-1)}(x_0, x_1, \ldots, x_{n-2}),\end{aligned} \quad (9\text{–}42)$$

where $x_{n-1}^{m_{n-1}}$ *and* $\overline{x_{n-1}^{m_{n-1}}}$ *are determined by Eqs. 9–38 and 9–40, respectively, and* $(m_0, m_1, \ldots, m_{n-2})$ *and* $(m_0, m_1, \ldots, m_{n-1})$ *are the binary representations of the numbers* m' *and* m*, respectively.*

PROOF. We prove Eq. 9–41. (The proof of Eq. 9–42 is similar.) From formula 9–37 we have

$$\begin{aligned}\rho_{m,s}^{(n)}(x) = x_{n-1}^{m_{n-1}}\Bigg(&\bigvee_{k,\, k_p=n-1} \left(x_{k_1}^{m_{k_1}} \wedge \cdots \wedge x_{k_{p-1}}^{m_{k_{p-1}}}\right)\Bigg) \\ \vee \Bigg(&\bigvee_{k,\, k_p\neq n-1} \left(x_{k_1}^{m_{k_1}} \wedge \cdots \wedge x_{k_p}^{m_{k_p}}\right)\Bigg),\end{aligned}$$

from which we obtain Eq. 9–41. ■

2) *The inclusions*

$$\varepsilon_{s-1}^{(n)} \subseteq \varepsilon_s^{(n)} \quad \textit{and} \quad \overline{\varepsilon_s^{(n)}} \subseteq \overline{\varepsilon_{s-1}^{(n)}} \quad (9\text{–}43)$$

hold where $\varepsilon_s^{(n)}$ *and* $\overline{\varepsilon_s^{(n)}}$ *are the sets of ones of the direct and complementary* (s, n)*-polynomial disjunctive-conjunctive functions, respectively, represented as Boolean functions of* $2n$ *variables* $x_0, x_1, \ldots, x_{n-1}, m_0, m_1, \ldots, m_{n-1}$.

PROOF. We show this using induction on n.

Let $n=2$. Since $\rho_{m,1}^{(2)}(x)=x_0^{m_0}x_1^{m_1}$ and $\rho_{m,2}^{(2)}(x)=x_0^{m_0}\vee x_1^{m_1}$, it is obvious that $\varepsilon_1^{(2)}\subseteq\varepsilon_2^{(2)}$. Let $n=3$. Then $\rho_{m,1}^{(3)}(x)=x_0^{m_0}x_1^{m_1}x_2^{m_1}$, $\rho_{m,2}^{(3)}(x)=x_0^{m_0}x_1^{m_1}\vee x_0^{m_0}x_2^{m_1}\vee x_1^{m_0}x_2^{m_1}$, and $\rho_{m,3}^{(3)}(x)=x_0^{m_0}\vee x_1^{m_1}\vee x_2^{m_2}$. Hence, $\varepsilon_1^{(3)}\subseteq\varepsilon_2^{(3)}\subseteq\varepsilon_3^{(3)}$ for all $1<s<n$. Suppose that $\varepsilon_{s-1}^{(n-1)}\subseteq\varepsilon_s^{(n-1)}$ for all $1<s<n$.

Then from Eq. 9–41 we have

$$\begin{aligned}\rho_{m,s-1}^{(n)}(x) &= x_{n-1}^{m_{n-1}}\cdot\rho_{m',s-1}^{(n-1)}(x_0,x_1,\ldots,x_{n-2})\\ &\quad\vee\rho_{m',s-2}^{(n-1)}(x_0,x_1,\ldots,x_{n-2}),\end{aligned}\tag{9–44}$$

from which we obtain $\varepsilon_{s-1}^{(n)}\subseteq\varepsilon_s^{(n)}$ for all $1<s\leqslant n$, using the induction hypothesis and comparing Eqs. 9–44 with 9–41.

Similarly, one can show that $\overline{\varepsilon_s^{(n)}}\subseteq\overline{\varepsilon_{s-1}^{(n)}}$. ■

3) *The functions* $\rho_{m,s}^{(n)}(x)$ *and* $\mu_{m,s}^{(n)}(x)$ *can be constructed recursively by*:

$$\rho_{m,s}^{(n)}(x)=\begin{cases}\rho_{m',s-1}^{(n-1)}(x_0,x_1,\ldots,x_{n-2}), & \textit{if } \bar{x}_{n-1}m_{n-1}=1,\\ \rho_{m',s}^{(n-1)}(x_0,x_1,\ldots,x_{n-2}), & \textit{otherwise,}\end{cases}\tag{9–45}$$

and

$$\mu_{m,s}^{(n)}(x)=\begin{cases}\mu_{m',s-1}^{(n-1)}(x_0,x_1,\ldots,x_{n-2}), & \textit{if } \bar{x}_{n-1}m_{n-1}=1,\\ \mu_{m',s}^{(n-1)}(x_0,x_1,\ldots,x_{n-2}), & \textit{otherwise,}\end{cases}\tag{9–46}$$

where m' *is as above.*

In the following discussion we consider the basic properties of (s,n)-polynomial (direct and complementary) disjunctive-conjunctive matrices.

PROPERTY 9.2.6. *Similarly to the second part of the Property* 9.2.5, *let the systems of the direct and complementary* (s,n)*-polynomial disjunctive-conjunctive functions be represented as a Boolean function of* $2n$ *variables* $x_0,x_1,\ldots,x_{n-1}$, $m_0,m_1,\ldots,m_{n-1}$. *Then*

1) *The matrices* $\mathbf{L}_n^{(s)}$ *and* $\mathbf{M}_n^{(s)}$ *will be represented as the truth tables of these Boolean functions. This is illustrated in Example* 9.2.9.

EXAMPLE 9.2.9. If $n=3$ and $s=2$, the matrices $\mathbf{L}_n^{(s)}$ and $\mathbf{M}_n^{(s)}$ are the truth tables of the Boolean functions $\rho_{2,j}^{(3)}(x)=x_0^{j_0}x_1^{j_1}\vee x_0^{j_0}x_2^{j_2}\vee x_1^{j_1}x_2^{j_2}$ and $\mu_{2,j}^{(3)}(t)=\overline{\rho_{2,j}^{(3)}(x)}$, respectively.

2) *The matrices* $\mathbf{L}_n^{(1)}$ *and* $\mathbf{L}_n^{(n)}$ *are, respectively, the conjunctive matrix* $\mathbf{K}_n$ *(which is also known as the Reed–Muller matrix) and disjunctive matrix* $\mathbf{D}_n$. *They are determined by Eqs. 9–22 and 9–27.*

3) *The square matrices* $\mathbf{L}_n^{(s)}$ *and* $\mathbf{M}_n^{(s)}$ *of order* 2^n *can be recurrently represented by*

$$\mathbf{L}_n^{(s)} = \begin{pmatrix} \mathbf{L}_{n-1}^{(s)} & \mathbf{L}_{n-1}^{(s-1)} \\ \mathbf{L}_{n-1}^{(s)} & \mathbf{L}_{n-1}^{(s)} \end{pmatrix} \tag{9–47}$$

and

$$\mathbf{M}_n^{(s)} = \begin{pmatrix} \mathbf{M}_{n-1}^{(s)} & \mathbf{M}_{n-1}^{(s-1)} \\ \mathbf{M}_{n-1}^{(s)} & \mathbf{M}_{n-1}^{(s)} \end{pmatrix}, \tag{9–48}$$

where $1 < s \leqslant n$, *and the initial matrices* $\mathbf{K}_n = \mathbf{L}_n^{(1)}$ *and* $\mathbf{D}_n = \mathbf{L}_n^{(n)}$ *are given by Eqs. 9–22 and 9–27, respectively,* $\mathbf{M}_n^{(1)} = \mathbf{J}_n - \mathbf{K}_n$ *and* $\mathbf{M}_n^{(n)} = \mathbf{J}_n - \mathbf{D}_n$, $\mathbf{J}_n$ *is the matrix of ones of order* 2^n.

Expressions 9–47 and 9–48 follow directly from the recurrent constructions of the corresponding functions given in Eqs. 9–45 and 9–46.

4) *The element* l_{ij} *of the matrix* $\mathbf{L}_n^{(s)}$ *[respectively, the element* r_{ij} *of the matrix* $\mathbf{M}_n^{(s)}$*], is equal to* 0 *(respective to* 1*) if and only if in the binary representation* $i = (i_{n-1}, i_{i-2}, \ldots, i_0)$ *and* $j = (j_{n-1}, j_{i-2}, \ldots, j_0)$ *there are no more than s components such that* $i_k = 0$ *and* $j_k \neq 0$, *for* $0 \leqslant k \leqslant n-1$.

5) *Let* $\nu(\mathbf{L}_n^{(s)})$ *be the number of zeroes in the matrix* $\mathbf{L}_n^{(s)}$. *Then,* $\nu(\mathbf{D}_n) = \nu(\mathbf{L}_n^{(n)}) = 1$, $\nu(\mathbf{L}_n^{(n-1)}) = 3n + 1, \ \ldots, \ \nu(\mathbf{L}_n^{(1)}) = \nu(\mathbf{K}_n) = 2^{2n} - 3^n = 2^{2n}(1 - (3/4)^n)$.

9.2.5 BINARY POLYNOMIAL LOGICAL FUNCTIONS AND MATRICES. EXTENSIONS OF DIMENSION

In the following section we give two modifications of the systems of conjunctive functions and matrices, obtained by enlarging the number of functions and the values of variables. These are called interval splicing and absorbing functions and matrices. We briefly describe the basic properties of interval matrices.

Interval Splicing Matrices

Consider the system $\phi_m^{(n)}(x)$ of conjunctive functions (or $\wedge$-polynomial logical functions), which is obtained from the logical (0, 1)-Rademacher functions according to Eq. 9–19.

Let $(\hat{m}_0, \hat{m}_1, \ldots, \hat{m}_{n-1})$ be the ternary representation of a number m, and x_j be the binary logical variables, i.e., $m = \sum_{j=0}^{n-1} \hat{m}_j 3^{n-j}$, $\hat{m}_j \in \{0, 1, 2\}$, $x_j \in \{0, 1\}$, $j = 0, \ldots, n-1$. Consider the following elementary polynomials defined by

$$\hat{x}_j^{\hat{m}_j} = \begin{cases} 1, & \text{if } \hat{m}_j = 2, \\ x_j, & \text{if } \hat{m}_j = 1, \\ \bar{x}_j, & \text{if } \hat{m}_j = 0. \end{cases} \tag{9–49}$$

DEFINITION 9.2.19. The set of functions $\{\theta_m^{(n)}(x)\} = \{\theta_m^{(n)}(x_0, \ldots, x_{n-1})\}$, $m = 0, 1, \ldots, 3^n - 1$, obtained by

$$\{\theta_m^{(n)}(x)\} = x_0^{\hat{m}_0} \wedge \cdots \wedge x_{n-1}^{\hat{m}_{n-1}}, \tag{9–50}$$

is called *the system of interval splicing functions*. The functions $\theta_m^{(n)}(x)$ themselves are called *the interval splicing functions*.

We call the $(2^n \times 3^n)$ matrix $\mathbf{A}_n = [\theta_m^{(n)}(x)]$, $x = 0, \ldots, 2^n - 1$; $m = 0, \ldots, 3^n - 1$, whose columns are the interval splicing functions, the *interval splicing matrix*.

REMARK 9.2.3. The name of the interval splicing matrix comes from its connection with coverings of ones of a Boolean function with intervals of the n-dimensional binary cube. These matrices are used when we construct the contracted disjunctive normal form from the truth table of a given Boolean function (BF). Using the matrix $\mathbf{A}_n$ we find all the intervals that cover the ones of the BF. This is the matrix analogue to the operation of generalized splicing of the intervals $xC_1 \vee \bar{x}C_2 = xC_1 \vee \bar{x}C_2 \vee C_1C_2$, where C_1 and C_2 are the elementary conjunctions in a disjunctive normal form of the BF.

In the following discussion we give some basic properties of interval splicing matrices. To this end, we need Definition 9.2.20.

DEFINITION 9.2.20. We say that the range $\boldsymbol{\alpha} = (\alpha_0, \ldots, \alpha_{n-1})$ is contained in the range (or precedes the range) $\boldsymbol{\beta} = (\beta_0, \ldots, \beta_{n-1})$, if $(\alpha_i, \beta_i) \in \{(0,0), (0,2), (1,1), (1,2), (2,2)\}$ for $i = 0, 1, \ldots, n-1$. The precedence relation is denoted by $\boldsymbol{\alpha} \preceq \boldsymbol{\beta}$.

Thus, for example, the range $\boldsymbol{\alpha} = (0, 1, 0, 0)$ is contained in the range $\boldsymbol{\beta} = (2, 2, 0, 2)$, i.e., $\boldsymbol{\alpha} \preceq \boldsymbol{\beta}$, and the range $\boldsymbol{\alpha}' = (1, 0, 1)$ is not contained in the range $\boldsymbol{\beta}' = (2, 1, 1)$, because $(\alpha_2', \beta_2') = (0, 1)$.

Now let us consider the properties of the interval splicing matrices.

PROPERTY 9.2.7.

1) *The interval splicing matrix* $\mathbf{A}_n$ *can be expressed as*

$$\mathbf{A}_n = \mathbf{A}_1^{\otimes n} = \begin{pmatrix} 1 & 0 & 1 \\ 0 & 1 & 1 \end{pmatrix}^{\otimes n}. \tag{9–51}$$

2) *The elements* $[A_n]_{ij}$ *of the matrix* $\mathbf{A}_n$, $i = 0, \ldots, 2^n - 1,\ j = 0, \ldots, 3^n - 1$, *can be calculated by*

$$[A_n]_{ij} = i_0^{\hat{j}_0} \cdots i_{n-1}^{\hat{j}_{n-1}}, \tag{9–52}$$

where i_k *is the kth component of the binary representation* $\mathbf{i} = (i_0, \ldots, i_{n-1})$ *and* $\hat{j}_k$ *is the kth component of the ternary representation* $\mathbf{j} = (\hat{j}_0, \ldots, \hat{j}_{n-1})$. *According to Eq.* 9–49 *we have*

$$0^1 = 1^0 = 0 \quad \textit{and} \quad 0^0 = 1^1 = 1^2 = 0^2 = 1 \tag{9–53}$$

in formula 9–52.

3) *The element* $[A_n]_{ij}$ *of the matrix* $\mathbf{A}_n$ *is equal to* 1 *if and only if* $\mathbf{i} \preceq \mathbf{j}$, *i.e., when the range* $\mathbf{i}$ *is contained in the range* $\mathbf{j}$, *where* $\mathbf{j} = (\hat{j}_0, \ldots, \hat{j}_{n-1})$ *is the ternary representation of* j, *and* $\mathbf{i} = (\hat{i}_0, \ldots, \hat{i}_{n-1})$ *is the ternary representation of* i.

Indeed, if $[A_n]_{ij} = 1$, then according to Eqs. 9–52 and 9–53, there is no such pair of components t in the binary representation of i and in the ternary representation of j, that $(i_t, j_t) = (0, 1)$ or $(i_t, j_t) = (1, 0)$, $0 \leqslant t \leqslant n - 1$. This implies $\mathbf{i} \preceq \mathbf{j}$. And vice versa, if $\mathbf{i} \preceq \mathbf{j}$, then by formulas 9–52 and 9–53 we obtain $[A_n]_{ij} = 1$.

Interval Absorbing Matrices

As above, let $\phi_m^{(n)}(t)$ be a system of conjunctive functions. Let $(\hat{m}_0, \hat{m}_1, \ldots, \hat{m}_{n-1})$ be the ternary representation of a number m; $\hat{y}_j$ be the ternary (three-digit) logical variables, i.e., $\hat{m}_j,\ \hat{y}_j \in \{0, 1, 2\}$, $j = 0, \ldots, n - 1$; and the following elementary polynomials be defined by:

$$\hat{y}_j^{\hat{m}_j} = \begin{cases} 1, & \text{if } \hat{m}_j = 2, \\ 0, & \text{if } \hat{m}_j \neq 2, \\ y_j, & \text{if } \hat{m}_j = 1;\ \hat{y}_j \neq 2, \\ \overline{\hat{y}}_j, & \text{if } \hat{m}_j = 0;\ \hat{y}_j \neq 2. \end{cases} \tag{9–54}$$

DEFINITION 9.2.21. The set of functions $\{\xi_{m,n}(y)\} = \{\xi_m^{(n)}(\hat{y}_0, \ldots, \hat{y}_{n-1}),\ m = 0, 1, \ldots, 3^n - 1\}$, where

$$\xi_m^{(n)}(y) = \hat{y}_0^{\hat{m}_0} \wedge \cdots \wedge \hat{y}_{n-1}^{\hat{m}_{n-1}}, \tag{9–55}$$

is called *the system of interval absorbing functions*. The functions $\xi_m^{(n)}(y)$ themselves are called *the interval absorbing functions*.

The *interval absorbing matrix* is the square matrix $\mathbf{B}_n = [\xi_m^{(n)}(x)],\ x, m = 0, 1, \ldots, 3^n - 1$ of order 3^n whose columns are the interval absorbing functions.

REMARK 9.2.4. The names of the interval absorbing (as well as the interval splicing) functions and matrices are connected with the coverings of ones of a Boolean function with intervals of the n-dimensional binary cube. These matrices are used when we construct the contracted disjunctive normal form from the truth table given Boolean function (BF). Using the matrix $\mathbf{B}_n$, we find among all intervals the maximal intervals that correspond to the operation of absorbing the intervals $C_1 \vee C_1C_2 = C_1$, where C_1 and C_2 are elementary conjunctions.

Now let us consider the basic properties of the interval absorbing matrices.

PROPERTY 9.2.8.

1) *The interval absorbing matrix* $\mathbf{B}_n$ *is defined by*:

$$\mathbf{B}_n = \mathbf{B}_1^{\otimes n} = \begin{pmatrix} 1 & 0 & 1 \\ 0 & 1 & 1 \\ 0 & 0 & 1 \end{pmatrix}^{\otimes n}. \tag{9–56}$$

2) *The elements* $[B_n]_{ij}$ $(i, j = 0, \ldots, 3^n - 1)$ *of the matrix* $\mathbf{B}_n$ *can be calculated by*

$$[B_n]_{ij} = \hat{i}_0^{\hat{j}_0} \cdots \hat{i}_{n-1}^{\hat{j}_{n-1}}, \tag{9–57}$$

where $\hat{i}_k$ *and* $\hat{j}_k$ *are, respectively, the kth component of the ternary representations* $\mathbf{i} = (\hat{i}_0, \ldots, \hat{i}_{n-1})$ *and* $\mathbf{j} = (\hat{j}_0, \ldots, \hat{j}_{n-1})$. *According to Eqs.* 9–49 *and* 9–54 *we have Eq.* 9–53 *in formula* 9–57.

3) *The element* $[B_n]_{ij}$ *of the matrix* $\mathbf{B}_n$ *is equal to* 1 *if and only if* $\mathbf{i} \preceq \mathbf{j}$, *where* $\mathbf{i} = (\hat{i}_0, \ldots, \hat{i}_{n-1})$ *and* $\mathbf{j} = (\hat{j}_0, \ldots, \hat{j}_{n-1})$ *are the ternary representations of* i *and* j, *respectively.*

Indeed, if $[B_n]_{ij} = 1$, then according to Eqs. 9–57 and 9–53, there is no such component k in the ternary representations of i and j, that either $(\hat{i}_k, \hat{j}_k) = (0, 1)$, or $(\hat{i}_k, \hat{j}_k) = (1, 0)$ or $(\hat{i}_k = 2, \hat{j}_k \neq 2)$ holds. Otherwise, $[B_n]_{ij}$ should be equal to zero. These conditions imply $\mathbf{i} \preceq \mathbf{j}$. Vice versa, we obtain $[B_n]_{ij} = 1$ by substituting the ternary representation components of the values of i and j, to Eq. 9–57 taking into account Eq. 9–53.

9.3 BINARY POLYNOMIAL TRANSFORMS

9.3.1 (a, b)-POLYNOMIAL TRANSFORMS OF II-TYPE AS BINARY WAVELET TRANSFORMS

Binary wavelet transform emerged from the application of wavelet theory to finite fields with two elements **GF**(2) [52, 81]. It is highly advantageous from a computational point of view, since the computations are performed by "exclusive OR" and "AND" operations. Although smoothness and vanishing moments properties of real discrete wavelet transforms [84] do not make sense in **GF**(2), BWT is useful for localizing data in time and frequency and separately encoding rapid and slow changes across the data. In that sense, an example for a one-stage BWT is defined as

$$BWT = \begin{pmatrix} \mathbf{I} & \mathbf{0} \\ \mathbf{0} & \mathbf{I} \end{pmatrix} \begin{pmatrix} \mathbf{I} & \mathbf{0} \\ \mathbf{I} & \mathbf{I} \end{pmatrix} \mathbf{L} \tag{9–58}$$

in [52], where $\mathbf{L}$ is a permutation matrix called "lazy wavelet transform" such that $L(x_1, x_2, \ldots) = [x_2, x_4, \ldots, x_1, x_3, \ldots]$. Wavelet vectors resulting from Eq. 9–58 are exactly the same as the columns of Haar-conjunctive matrix ((0, 1)-polynomial matrix of II-type) introduced in Section 9.2.3. The multiresolution property of a (0, 1)-polynomial transform of II-type can be seen from Example 9.2.6 where

$$\begin{aligned}
V_{-3} &= span(\{1, 1, 1, 1, 1, 1, 1, 1\}),\\
V_{-2} &= span(\{1, 1, 1, 1, 0, 0, 0, 0\}, \{0, 0, 0, 0, 1, 1, 1, 1\}),\\
V_{-1} &= span(\{1, 1, 0, 0, 0, 0, 0, 0\}, \{0, 0, 1, 1, 0, 0, 0, 0\}, \{0, 0, 0, 0, 1, 1, 0, 0\},\\
&\qquad \{0, 0, 0, 0, 0, 0, 1, 1\}),\\
V_0 &= span(\{1, 0, 0, 0, 0, 0, 0, 0\}, \{0, 1, 0, 0, 0, 0, 0, 0\}, \ldots,\\
&\qquad \{0, 0, 0, 0, 0, 0, 0, 1\}),
\end{aligned}$$

$$V_{-3} \subset V_{-2} \subset V_{-1} \subset V_0.$$

The first four columns act as a high-pass filter capturing rapid changes and the last four columns in Example 9.2.6 act as a low-pass filter capturing slow changes [81].

9.3.2 (a, b)-POLYNOMIAL FUNCTIONS OF I-TYPE AS DISCRETE WAVELET PACKET TRANSFORMS

Wavelet packet transform is a generalization of wavelet transform which offers a wider range of analysis possibilities for a signal [84]. It is associated with a best basis selection algorithm which selects a subdecomposition structure among all possible decomposition structures presented by the packet transform, subject to a criterion. Wavelet packet transform can be better visualized by a full binary tree, where left and right branchings represent low-pass and high-pass filterings, respectively, and the best basis selection corresponds to extracting a subtree of the binary tree (see Fig. 9–3).

Consider the Walsh matrix $\mathbf{W}_n$ of order n in Hadamard ordering (see Example 9.2.2). It corresponds to a wavelet packet transform with [1 1] and [1 $-$ 1] being the low-pass and high-pass filters, respectively. The conjunctive matrix $\mathbf{K}_n$ of order n (see Example 9.2.4) is also a wavelet packet transform in **GF**(2) with [1 0] and [1 1] acting as the low-pass and high-pass filters, respectively.

9.3.3 EFFICIENT COMPUTATION ALGORITHMS

Suppose that $\mu(\mathbf{H})$ arithmetic operations are needed to compute the discrete transform $\mathbf{g} = \mathbf{H} \cdot \mathbf{f}$ when the usual straightforward algorithm of multiplication of a matrix a by vector (the direct method) is used. The algorithms which require less than $\mu(\mathbf{H})$ arithmetic operations to compute the same transform are called fast algorithms.

Cooley and Tukey [27] introduced in signal processing the algorithm of fast Fourier transform (FFT), which requires $O(n2^n)$ arithmetic operations instead of 2^{2n} arithmetic operations of 2^n-point discrete transforms by orthogonal Fourier basis in the direct method of computation.

Later fast algorithms were developed also for other transforms. Mostly they are based on Good's theorem [41]. Let a square matrix $\mathbf{H}$ of order $N = N_1 N_2 \cdots N_k$ be represented as the Kronecker product of k matrices $\mathbf{V}^{(j)}$ of order N_j, $j =$

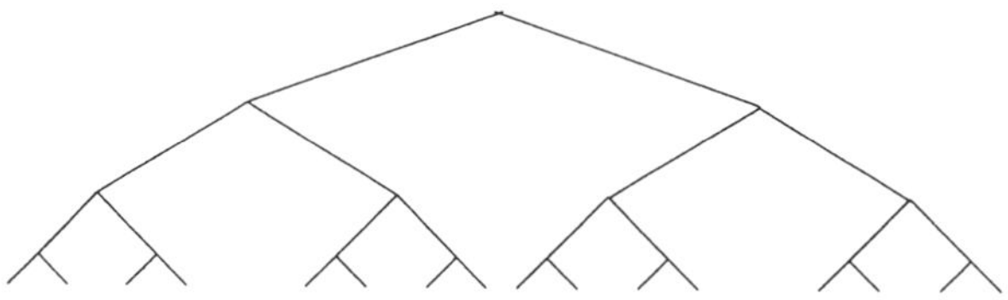

Figure 9–3. Wavelet packet transform structure of depth 4.

$1, 2, \ldots, k$, i.e., $\mathbf{H} = \mathbf{V}^{(1)} \otimes \mathbf{V}^{(2)} \otimes \cdots \otimes \mathbf{V}^{(k)}$. Then, by Good's theorem there exists a way of representing $\mathbf{H}$ as a product of k sparse matrices:

$$\mathbf{H} = \prod_{j=1}^{k} \mathbf{H}^{(j)},$$

where

$$\mathbf{H}^{(j)} = \mathbf{I}_{(M(j))} \otimes \mathbf{V}^{(j)} \otimes \mathbf{I}_{(L(j))}, \quad j = 1, 2, \ldots, k,$$

$\mathbf{I}_{(n)}$ is the identity matrix of order n, $M(j) = N_{j+1}N_{j+2}\cdots N_k$ and $L(j) = N_1 \times N_2 \cdots N_{j-1}$.

Let the transform matrix $\mathbf{H}$ be represented in the form of a product of k sparse matrices

$$\mathbf{H} = \mathbf{H}^{(1)} \cdot \mathbf{H}^{(2)} \cdots \mathbf{H}^{(k)}. \tag{9–59}$$

Then the discrete transform $\mathbf{g} = \mathbf{H} \cdot \mathbf{f}$ can be computed in k stages by the consecutive computation of k transforms by the matrices $\mathbf{H}^{(j)}$, $j = k, k-1, \ldots, 1$:

$$\mathbf{g} = \left(\mathbf{H}^{(1)} \cdots \left(\mathbf{H}^{(k-1)}(\mathbf{H}^{(k)} \cdot \mathbf{f})\right) \cdots\right).$$

Thus the following algorithm can be applied to the computation of discrete transforms.

ALGORITHM 9.3.1. The fast binary polynomial transform by matrix $\mathbf{H}$ represented in the form Eq. 9–59 consists of the following steps:

Step 1. Initialize a counter, $j = 0$;

Step 2. Set $\mathbf{f}^{(0)} = \mathbf{f}$, where $\mathbf{f}$ is the input vector;

Step 3. (Computation of the transform by the matrix $\mathbf{H}^{(k-j)}$.)

Compute the matrix product $\mathbf{f}^{(j+1)} = \mathbf{H}^{(k-j)} \cdot \mathbf{f}^{(j)}$;

Step 4. Set $j = j + 1$;

Step 5. If $j = k$ then STOP, else go to *Step 3*.

At the end of procedure the output $\mathbf{g} = \mathbf{f}^{(k)} = \mathbf{H} \cdot \mathbf{f}$ is obtained.

Based on this algorithm we obtain:

PROPOSITION 9.3.1. *For the computation of the binary polynomial transform* $\mathbf{g} = \mathbf{H} \cdot \mathbf{f}$ *by the matrix* $\mathbf{H}$ *represented in the form Eq. 9–59,*

$$\mu^+(\mathbf{H}) \leqslant \sum_{j=1}^{k} \left[Q(\mathbf{H}^{(j)}) - N^{(j)} \right] \tag{9–60}$$

additions are required, where $Q(\mathbf{H}^{(j)})$ *and* $N^{(j)}$ *are the number of nonzero elements and the number of nonzero rows in the matrix* $\mathbf{H}^{(j)}$*, respectively. If the matrix* $\mathbf{H}$ *is represented as the Kronecker product* $\mathbf{H} = \mathbf{V}^{(1)} \otimes \mathbf{V}^{(2)} \otimes \cdots \otimes \mathbf{V}^{(k)}$ *of the* $(n_1 \times n_2)$ *matrices* $\mathbf{V}^{(j)}$, $j = 1, 2, \ldots, k$*, then*

$$\mu^+(\mathbf{H}) \leqslant \sum_{j=1}^{k} (n_1)^{k-j} (n_2)^{j-1} \left(Q(\mathbf{V}^{(j)}) - n_1 \right). \tag{9–61}$$

PROOF. Only addition operations are needed for computation of binary polynomial transforms. In Algorithm 9.3.1 the transform by the matrix $\mathbf{H}$ is reduced to consecutive transforms by the matrices $\mathbf{H}^{(j)}$, $j = 1, 2, \ldots, k$. So the total number of operations, $\mu^+(\mathbf{H})$, which are executed in Algorithm 9.3.1 is equal to

$$\mu^+(\mathbf{H}) = \sum_{j=1}^{k} \mu^+(\mathbf{H}^{(j)}),$$

where $\mu^+(\mathbf{H}^{(j)})$ is the number of operations needed for computation of the transform by the matrix $\mathbf{H}^{(j)}$, $j = 1, 2, \ldots, k$. If the direct algorithm of multiplication of the matrix by a vector is used, then $\mu^+(\mathbf{H}^{(j)}) = Q(\mathbf{H}^{(j)}) - N^{(j)}$, where $Q(\mathbf{H}^{(j)})$ and $N^{(j)}$ are as in the proposition. So we have proved the equality in Eq. 9–60, which gives the complexity of the algorithm considered. The possibility of applying other algorithms gives the estimate 9–60.

By Good's theorem, when $\mathbf{H} = \mathbf{V}^{(1)} \otimes \mathbf{V}^{(2)} \otimes \cdots \otimes \mathbf{V}^{(k)}$, where $\mathbf{V}^{(j)}$ are $(n_1 \times n_2)$ matrices, $j = 1, 2, \ldots, k$, (and, therefore, $\mathbf{H}$ is the $(n_1^k \times n_2^k)$ matrix), then $\mathbf{H}$ can be represented in the form Eq. 9–59, where

$$\mathbf{H}^{(j)} = \mathbf{I}_{(n_2^{j-1})} \otimes \mathbf{V}^{(j)} \otimes \mathbf{I}_{(n_1^{k-j})}.$$

In the matrix $\mathbf{H}^{(j)}$ each element of the matrix $\mathbf{V}^{(j)}$ is repeated $n_1^{k-j} n_2^{j-1}$ times, so $Q(\mathbf{H}^{(j)}) = n_1^{k-j} n_2^{j-1} Q(\mathbf{V}^{(j)})$. After substitution of this equality into Eq. 9–60, we obtain Eq. 9–61. Here we exclude the trivial case when $\mathbf{V}^{(j)}$ contains a row with only zero elements. ■

REMARK 9.3.1. When $\mathbf{H}$ can be represented as the Kronecker product of $(n \times n)$ matrices, then Eq. 9–61 becomes the inequality

$$\mu^{+}(\mathbf{H}) \leqslant \sum_{j=1}^{k} n^{k-1}\big(Q(\mathbf{V}^{(j)}) - n\big). \tag{9–62}$$

As we have from Proposition 9.3.1, Algorithm 9.3.1 will be a fast algorithm when $\sum_{j=1}^{k}(Q(\mathbf{H}^{(j)}) - N^{(j)}) \leqslant Q(\mathbf{H})$. In the sequel we will consider fast algorithms of the concrete binary polynomial transforms.

Fast (0,1)-Rademacher Transforms of I Type

LEMMA 9.3.1. *The complete* (0, 1)*-Rademacher matrix* $\widetilde{\mathbf{R}}_n = \widetilde{\mathbf{R}}(n, 0, 1)$ *can be represented as*

$$\widetilde{\mathbf{R}}_n = \prod_{j=0}^{n-1} \mathbf{R}_n^{(j)}, \tag{9–63}$$

where

$$\mathbf{R}_n^{(0)} = \begin{pmatrix} \mathbf{I}_{n-1} & \mathbf{0}_{n-1} \\ \mathbf{I}_{n-1} & \mathbf{1}_{n-1} \end{pmatrix}, \tag{9–64}$$

$$\mathbf{R}_n^{(j)} = \begin{pmatrix} \mathbf{I}_{n-1-j} & \mathbf{0}_{n-1} & 0 \\ \mathbf{I}_{n-1-j} & \mathbf{1}_{n-1} & 0 \\ 0 & 0 & \mathbf{I}_{(j)} \end{pmatrix}, \quad j = 1, 2, \ldots, n-2, \tag{9–65}$$

$$\mathbf{R}_n^{(n-1)} = \begin{pmatrix} 1 & 0 & \cdots & 0 \\ 1 & & & \\ 0 & & & \\ \vdots & & \bar{\mathbf{I}}_{(n)} & \\ 0 & & & \end{pmatrix}, \tag{9–66}$$

$\mathbf{I}_j$ *(respectively,* $\mathbf{I}_{(j)}$*) is the identity matrix of order* 2^j *(respectively, of order* j*),* $\bar{\mathbf{I}}_{(n)}$ *is an opposite identity matrix of order* n *and* $\mathbf{0}_j$ *(respectively,* $\mathbf{1}_j$*) are column vectors of length* 2^j*, consisting of zeros (respectively, of ones).*

It follows from Lemma 9.3.1 that Algorithm 9.3.1 can be used for computation of a complete (0, 1)-Rademacher transform. In this case at Step 3 of Algorithm 9.3.1, matrices $\mathbf{R}_n^{(j)}$, $j = 0, 1, \ldots, n-1$, are used. The complexity of the resulting algorithm is estimated as in the following:

PROPOSITION 9.3.2. *Let* $\mathbf{f}$ *be a column vector of length* n. *Then the complete* $(0, 1)$*-Rademacher transform* $\mathbf{R}_n \cdot \mathbf{f}$ *can be computed with* $2^n - 1$ *addition operations.*

PROOF. The proposition follows from the fact that each matrix $\mathbf{R}_n^{(j)}$, $j = 0, 1, \ldots, n-1$, contains $Q(\mathbf{R}_n^{(j)}) = 3 \cdot 2^{n-1-j} + j$ nonzero elements, and all its $N^{(j)} = 2^{n-j} + j$ rows are nonzero rows. So for computation of the transform $\mathbf{f}^{(j+1)} = \mathbf{R}_n^{(n-1-j)} \cdot \mathbf{f}^{(j)}$, $\mu^+(\mathbf{R}_n^{(n-1-j)}) = 2^j$ additions are needed. Hence, in total, by using Algorithm 9.3.1 and Lemma 9.3.1,

$$\mu^+(\widetilde{\mathbf{R}}_n) = \sum_{j=0}^{n-1} 2^{n-1-j} = 2^n - 1$$

additions are needed for computation of the transform $\mathbf{g} = \widetilde{\mathbf{R}}_n \cdot \mathbf{f}$. ■

Note that $n2^n$ additions are required in the straightforward algorithm of the complete $(0, 1)$-Rademacher transform. Thus, Algorithm 9.3.1 based on the representation of the matrix $\widetilde{\mathbf{R}}_n$, by Lemma 9.3.1, is a fast complete $(0, 1)$-Rademacher transform algorithm.

EXAMPLE 9.3.1. Let us consider the fast algorithm of the complete $(0, 1)$-Rademacher transform $\mathbf{g} = \widetilde{\mathbf{R}}_3 \cdot \mathbf{f}$ of the column vector $\mathbf{f} = (2, 1, 3, 1)^T$. By Lemma 9.3.1, matrix $\widetilde{\mathbf{R}}_3$ can be represented as

$$\widetilde{\mathbf{R}}_3 = \widetilde{\mathbf{R}}_3^{(0)} \widetilde{\mathbf{R}}_3^{(1)} \widetilde{\mathbf{R}}_3^{(2)} = \begin{pmatrix} 1 & 0 & 0 & 0 & 0 \\ 0 & 1 & 0 & 0 & 0 \\ 0 & 0 & 1 & 0 & 0 \\ 0 & 0 & 0 & 1 & 0 \\ 1 & 0 & 0 & 0 & 1 \\ 0 & 1 & 0 & 0 & 1 \\ 0 & 0 & 1 & 0 & 1 \\ 0 & 0 & 0 & 1 & 1 \end{pmatrix} \begin{pmatrix} 1 & 0 & 0 & 0 \\ 0 & 1 & 0 & 0 \\ 1 & 0 & 1 & 0 \\ 0 & 1 & 1 & 0 \\ 0 & 0 & 0 & 1 \end{pmatrix} \begin{pmatrix} 1 & 0 & 0 & 0 \\ 1 & 0 & 0 & 1 \\ 0 & 0 & 1 & 0 \\ 0 & 1 & 0 & 0 \end{pmatrix}.$$

By Algorithm 9.3.1 we compute:

1. $j = 0$;

2. $\mathbf{f}^{(0)} = \mathbf{f} = (2, 1, 3, 1)^T$;

3. Compute the matrix product, $\mathbf{f}^{(1)} = \mathbf{R}_3^{(2)} \cdot \mathbf{f}^{(0)} = (2, 3, 3, 1)^T$;

4. $j = 1$;

5. Compute the matrix product, $\mathbf{f}^{(2)} = \mathbf{R}_3^{(1)} \cdot \mathbf{f}^{(1)} = (2, 3, 5, 6, 1)^T$;

6. $j = 2$;

7. Compute the matrix product, $\mathbf{f}^{(3)} = \mathbf{R}_3^{(0)} \cdot \mathbf{f}^{(2)} = (2, 3, 5, 6, 3, 4, 6, 7)^T$;

8. $j = 3$. STOP. The output column vector is

$$\mathbf{g} = \mathbf{f}^{(3)} = (2, 3, 5, 6, 3, 4, 6, 7)^T.$$

Note that 7 additions are executed in this fast algorithm instead of 12 additions as in the straightforward algorithm.

Fast $(a, b, \times)$-Polynomial Transforms of I-Type

Let us consider the transforms by the $(a, b, \times)$-polynomial matrices $\mathbf{\Phi}(n, a, b; \times)$.

LEMMA 9.3.2. *The $(a, b, \times)$-polynomial matrices $\mathbf{\Phi}_n = \mathbf{\Phi}(n, a, b; \times)$ can be expressed by the following formulae*:

$$\mathbf{\Phi}_n = (\mathbf{G}_n)^n, \tag{9–67}$$

where

$$\mathbf{G}_n = \begin{pmatrix} 1\ a & 0\ 0 & \dots & 0 & 0 \\ 0\ 0 & 1\ a & \dots & 0 & 0 \\ \dots & \dots & \dots & \dots \\ 0\ 0 & 0\ 0 & \dots & 1 & a \\ 1\ b & 0\ 0 & \dots & 0 & 0 \\ 0\ 0 & 1\ b & \dots & 0 & 0 \\ \dots & \dots & \dots & \dots \\ 0\ 0 & 0\ 0 & \dots & 1 & b \end{pmatrix} \tag{9–68}$$

or

$$\mathbf{\Phi}_n = \prod_{j=1}^{n} (\mathbf{I}_{j-1} \otimes \mathbf{\Phi}_1 \otimes \mathbf{I}_{n-j}), \tag{9–69}$$

where

$$\mathbf{\Phi}_1 = \mathbf{G}_1 = \begin{pmatrix} 1 & a \\ 1 & b \end{pmatrix}$$

and $\mathbf{I}_r$ is the identity matrix of order 2^r.

PROOF. Formula 9–69 follows directly from Good's theorem [41] and can be easily derived using the properties of Kronecker's product:

$$\begin{aligned} \mathbf{\Phi}_n &= (\mathbf{\Phi}_{n-1} \otimes \mathbf{I}_1)(\mathbf{I}_{n-1} \otimes \mathbf{\Phi}_1) = (\mathbf{\Phi}_{n-2} \otimes \mathbf{I}_2)(\mathbf{I}_{n-2} \otimes \mathbf{\Phi}_1 \otimes \mathbf{I}_1)(\mathbf{I}_{n-1} \otimes \mathbf{\Phi}_1) \\ &= \dots = \prod_{i=1}^{n} (\mathbf{I}_{i-1} \otimes \mathbf{\Phi}_1 \otimes \mathbf{I}_{n-1}). \end{aligned}$$

Formula 9–67 can be obtained after corresponding permutations of rows and columns of the matrices in the right side of Eq. 9–69. ■

Both representations 9–67 and 9–69 can be used for the construction of fast $(a, b, \times)$-polynomial I type transform algorithms like Algorithm 9.3.1. In these cases at Step 3 of Algorithm 9.3.1 the matrix $\mathbf{G}_n$ from Eq. 9–68, or matrices $\mathbf{G}_n^{(j)} = \mathbf{I}_{j-1} \otimes \mathbf{\Phi}_1 \otimes \mathbf{I}_{n-j}$ from Eq. 9–69 are used, respectively. The complexities of both algorithms coincide and are estimated as in the following:

PROPOSITION 9.3.3. *Let* $\mathbf{f}$ *be a column vector of length* $N = 2^n$. *Then the* $(a, b, \times)$*-polynomial I-type transform of* $\mathbf{f}$ *can be computed with* $O(N\log N)$ *addition operations.*

PROOF. The proposition follows from the fact that the number of factors on the right side of Eq. 9–67, as well as on the right side of Eq. 9–69, is equal to n, and at the same time requires 2^n additions to be transformed by each of these matrices. ■

Since the discrete Walsh transform and the discrete Reed–Muller transforms are from the class of the $(a, b, \times)$-polynomial I-type transforms (cases of $a = 1,\ b = -1$ and $a = 0,\ b = 1$, respectively), the fast Walsh and Reed–Muller transform algorithm comes directly from the representations 9–67 or 9–69.

EXAMPLE 9.3.2. Consider the fast Walsh transform algorithms of order 8 based on representations 9–67 and 9–69. For $n = 3$ $(N = 8)$ representation 9–67 takes the form

$$\mathbf{W}_3 = \mathbf{G}_3^3 = \begin{pmatrix} 1 & 1 & 0 & 0 & 0 & 0 & 0 & 0 \\ 0 & 0 & 1 & 1 & 0 & 0 & 0 & 0 \\ 0 & 0 & 0 & 0 & 1 & 1 & 0 & 0 \\ 0 & 0 & 0 & 0 & 0 & 0 & 1 & 1 \\ 1 & -1 & 0 & 0 & 0 & 0 & 0 & 0 \\ 0 & 0 & 1 & -1 & 0 & 0 & 0 & 0 \\ 0 & 0 & 0 & 0 & 1 & -1 & 0 & 0 \\ 0 & 0 & 0 & 0 & 0 & 0 & 1 & -1 \end{pmatrix}^3 . \qquad (9\text{–}70)$$

Let $\mathbf{f} = (1, 3, 2, 1, 0, 2, 1, 3)^T$ be the input vector. By Algorithm 9.3.1, where representation 9–70 is used, we compute

1. $j = 0$;

2. $\mathbf{f}^{(0)} = \mathbf{f} = (1, 3, 2, 1, 0, 2, 1, 3)^T$;

3. Compute the matrix product, $\mathbf{f}^{(1)} = \mathbf{G}_3 \cdot \mathbf{f}^{(0)} = (4, 3, 2, 4, -2, 1, -2, -2)^T$;

4. $j = 1$;

5. Compute the matrix product, $\mathbf{f}^{(2)} = \mathbf{G}_3 \cdot \mathbf{f}^{(1)} = (7, 6, -1, -4, 1, -2, -3, 0)^T$;

6. $j = 2$;

7. Compute the matrix product, $\mathbf{f}^{(3)} = \mathbf{G}_3 \cdot \mathbf{f}^{(2)} = (13, -5, -1, -3, 1, 3, 3, -3)^T$;

8. $j = 3$. STOP. The output column vector is

$$\mathbf{g} = \mathbf{f}^{(3)} = (13, -5, -1, -3, 1, 3, 3, -3)^T.$$

Now consider the fast Walsh transform algorithm based on representation 9–69. For $n = 3$, 9–69 becomes

$$\mathbf{W}_n = \mathbf{G}_n^{(1)}\mathbf{G}_n^{(2)}\mathbf{G}_n^{(3)} = \begin{pmatrix} 1 & 0 & 0 & 0 & 1 & 0 & 0 & 0 \\ 0 & 1 & 0 & 0 & 0 & 1 & 0 & 0 \\ 0 & 0 & 1 & 0 & 0 & 0 & 1 & 0 \\ 0 & 0 & 0 & 1 & 0 & 0 & 0 & 1 \\ 1 & 0 & 0 & 0 & -1 & 0 & 0 & 0 \\ 0 & 1 & 0 & 0 & 0 & -1 & 0 & 0 \\ 0 & 0 & 1 & 0 & 0 & 0 & -1 & 0 \\ 0 & 0 & 0 & 1 & 0 & 0 & 0 & -1 \end{pmatrix}$$

$$\times \begin{pmatrix} 1 & 0 & 1 & 0 & 0 & 0 & 0 & 0 \\ 0 & 1 & 0 & 1 & 0 & 0 & 0 & 0 \\ 1 & 0 & -1 & 0 & 0 & 0 & 0 & 0 \\ 0 & 1 & 0 & -1 & 0 & 0 & 0 & 0 \\ 0 & 0 & 0 & 0 & 1 & 0 & 1 & 0 \\ 0 & 0 & 0 & 0 & 0 & 1 & 0 & 1 \\ 0 & 0 & 0 & 0 & 1 & 0 & -1 & 0 \\ 0 & 0 & 0 & 0 & 0 & 1 & 0 & -1 \end{pmatrix} \begin{pmatrix} 1 & 1 & 0 & 0 & 0 & 0 & 0 & 0 \\ 1 & -1 & 0 & 0 & 0 & 0 & 0 & 0 \\ 0 & 0 & 1 & 1 & 0 & 0 & 0 & 0 \\ 0 & 0 & 1 & -1 & 0 & 0 & 0 & 0 \\ 0 & 0 & 0 & 0 & 1 & 1 & 0 & 0 \\ 0 & 0 & 0 & 0 & 1 & -1 & 0 & 0 \\ 0 & 0 & 0 & 0 & 0 & 0 & 1 & 1 \\ 0 & 0 & 0 & 0 & 0 & 0 & 1 & -1 \end{pmatrix}.$$

Using this representation in Algorithm 9.3.1 we compute

1. $j = 0$;

2. $\mathbf{f}^{(0)} = \mathbf{f} = (1, 3, 2, 1, 0, 2, 1, 3)^T$;

3. Compute the matrix product, $\mathbf{f}^{(1)} = \mathbf{G}_3^{(3)} \cdot \mathbf{f}^{(0)} = (4, -2, 3, 1, 2, -2, 4, -2)^T$;

4. $j = 1$;

5. Compute the matrix product, $\mathbf{f}^{(2)} = \mathbf{G}_3^{(2)} \cdot \mathbf{f}^{(1)} = (7, -1, 1, -3, 6, -4, -2, 0)^T$;

6. $j = 2$;

7. Compute the matrix product, $\mathbf{f}^{(3)} = \mathbf{G}_3^{(1)} \cdot \mathbf{f}^{(2)} = (13, -5, -1, -3, 1, 3, 3, -3)^T$;

8. $j = 3$. STOP. The output column vector is

$$\mathbf{g} = \mathbf{f}^{(3)} = (13, -5, -1, -3, 1, 3, 3, -3)^T.$$

Note that 24 additions are executed in both fast algorithms instead of 56 additions as in the straightforward algorithm.

EXAMPLE 9.3.3. Consider the fast conjunctive transform algorithms of order 8 of the binary column vector $\mathbf{f} = (1, 0, 0, 1, 1, 1, 0, 1)^T$ by representations 9–67 and 9–69. For $n = 3$ $(N = 8)$ representation 9–67 takes the form

$$\mathbf{K}_3 = \mathbf{Q}_3^3 = \begin{pmatrix} 1 & 0 & 0 & 0 & 0 & 0 & 0 & 0 \\ 0 & 0 & 1 & 0 & 0 & 0 & 0 & 0 \\ 0 & 0 & 0 & 0 & 1 & 0 & 0 & 0 \\ 0 & 0 & 0 & 0 & 0 & 0 & 1 & 0 \\ 1 & 1 & 0 & 0 & 0 & 0 & 0 & 0 \\ 0 & 0 & 1 & 1 & 0 & 0 & 0 & 0 \\ 0 & 0 & 0 & 0 & 1 & 1 & 0 & 0 \\ 0 & 0 & 0 & 0 & 0 & 0 & 1 & 1 \end{pmatrix}^3 . \tag{9–71}$$

By Algorithm 9.3.1, where representation 9–71 is used, we compute

1. $j = 0$;

2. $\mathbf{f}^{(0)} = \mathbf{f} = (1, 0, 0, 1, 1, 1, 0, 1)^T$;

3. Compute the matrix product, $\mathbf{f}^{(1)} = \mathbf{Q}_3 \cdot \mathbf{f}^{(0)} = (1, 0, 1, 0, 1, 1, 2, 1)^T$;

4. $j = 1$;

5. Compute the matrix product, $\mathbf{f}^{(2)} = \mathbf{Q}_3 \cdot \mathbf{f}^{(1)} = (1, 1, 1, 2, 1, 1, 2, 3)^T$;

6. $j = 2$;

7. Compute the matrix product, $\mathbf{f}^{(3)} = \mathbf{Q}_3 \cdot \mathbf{f}^{(2)} = (1, 1, 1, 2, 2, 3, 2, 5)^T$;

8. $j = 3$. STOP. The output column vector is

$$\mathbf{g} = \mathbf{f}^{(3)} = (1, 1, 1, 2, 2, 3, 2, 5)^T.$$

Now consider the fast conjunctive transform algorithm based on representation 9–69. For $n = 3$, 9–69 becomes

$$\mathbf{K}_n = \mathbf{K}_n^{(1)}\mathbf{K}_n^{(2)}\mathbf{K}_n^{(3)} = \begin{pmatrix} 1 & 0 & 0 & 0 & 0 & 0 & 0 & 0 \\ 0 & 1 & 0 & 0 & 0 & 0 & 0 & 0 \\ 0 & 0 & 1 & 0 & 0 & 0 & 0 & 0 \\ 0 & 0 & 0 & 1 & 0 & 0 & 0 & 0 \\ 1 & 0 & 0 & 0 & 1 & 0 & 0 & 0 \\ 0 & 1 & 0 & 0 & 0 & 1 & 0 & 0 \\ 0 & 0 & 1 & 0 & 0 & 0 & 1 & 0 \\ 0 & 0 & 0 & 1 & 0 & 0 & 0 & 1 \end{pmatrix}$$

$$\times \begin{pmatrix} 1 & 0 & 0 & 0 & 0 & 0 & 0 & 0 \\ 0 & 1 & 0 & 0 & 0 & 0 & 0 & 0 \\ 1 & 0 & 1 & 0 & 0 & 0 & 0 & 0 \\ 0 & 1 & 0 & 1 & 0 & 0 & 0 & 0 \\ 0 & 0 & 0 & 0 & 1 & 0 & 0 & 0 \\ 0 & 0 & 0 & 0 & 0 & 1 & 0 & 0 \\ 0 & 0 & 0 & 0 & 1 & 0 & 1 & 0 \\ 0 & 0 & 0 & 0 & 0 & 1 & 0 & 1 \end{pmatrix} \begin{pmatrix} 1 & 0 & 0 & 0 & 0 & 0 & 0 & 0 \\ 1 & 1 & 0 & 0 & 0 & 0 & 0 & 0 \\ 0 & 0 & 1 & 0 & 0 & 0 & 0 & 0 \\ 0 & 0 & 1 & 1 & 0 & 0 & 0 & 0 \\ 0 & 0 & 0 & 0 & 1 & 0 & 0 & 0 \\ 0 & 0 & 0 & 0 & 1 & 1 & 0 & 0 \\ 0 & 0 & 0 & 0 & 0 & 0 & 1 & 0 \\ 0 & 0 & 0 & 0 & 0 & 0 & 1 & 1 \end{pmatrix}.$$

Using this representation in Algorithm 9.3.1 we compute

1. $j = 0$;

2. $\mathbf{f}^{(0)} = \mathbf{f} = (1, 0, 0, 1, 1, 1, 0, 1)^T$;

3. Compute the matrix product, $\mathbf{f}^{(1)} = \mathbf{K}_3^{(3)} \cdot \mathbf{f}^{(0)} = (1, 1, 0, 1, 1, 2, 0, 1)^T$;

4. $j = 1$;

5. Compute the matrix product, $\mathbf{f}^{(2)} = \mathbf{K}_3^{(2)} \cdot \mathbf{f}^{(1)} = (1, 1, 1, 2, 1, 2, 1, 3)^T$;

6. $j = 2$;

7. Compute the matrix product, $\mathbf{f}^{(3)} = \mathbf{K}_3^{(1)} \cdot \mathbf{f}^{(2)} = (1, 1, 1, 2, 2, 3, 2, 5)^T$;

8. $j = 3$. STOP. The output column vector is

$$\mathbf{g} = \mathbf{f}^{(3)} = (1, 1, 1, 2, 2, 3, 2, 5)^T.$$

As one can see, 12 additions are executed in both fast algorithms instead of 19 additions as in the straightforward algorithm.

Now let us consider an efficient algorithm for the computation of a Hamming-weighted conjunctive transform.

LEMMA 9.3.3. *There is a representation of the Hamming-weighted conjunctive matrix in the form*:

$$\mathbf{K}_M^{(w)} = \prod_{j=1}^{M} \left(\widetilde{\mathbf{K}}_M^{(w)} \mathbf{P}_{2^M}\right) = \prod_{j=1}^{M} \left(\left(\bigoplus_{p=1}^{2^{M-n-1}} \bigoplus_{q=1}^{2^n} \mathbf{K}_1^{(w)} \right) \mathbf{P}_{2^M} \right), \tag{9–72}$$

where $\bigoplus$ *denotes the direct sum of matrices, and* $\mathbf{P}_{2^M}$ *is the matrix of the perfect shuffle operator* [80].

PROOF. There is the following property of the matrix of the perfect shuffle operator [80]

$$\mathbf{P}_{2^M}(\mathbf{U} \otimes \mathbf{V})\mathbf{P}_{2^M}^{-1} = \mathbf{V} \otimes \mathbf{U},$$

where $\mathbf{U}$ and $\mathbf{V}$ are square matrices of orders 2 and 2^{M-1}, respectively. In particular we have

$$\mathbf{I}_{2^{M-m}} \otimes \mathbf{K}_1^{(w)} \otimes \mathbf{I}_{2^{m-1}} = \mathbf{P}_{2^M}^{m-1} \cdot \left(\mathbf{I}_{2^{M-1}} \otimes \mathbf{K}_1^{(w)}\right) \cdot \mathbf{P}_{2^M}^{-m+1} = \mathbf{P}_{2^M}^{m-1} \cdot \widetilde{\mathbf{K}}_M^{(w)} \cdot \mathbf{P}_{2^M}^{-m+1}.$$

Also taking into account that $\mathbf{P}_{2^M}^{M} = \mathbf{I}_{2^M}$, now we can represent $\mathbf{K}_M^{(w)}$ in the form

$$\begin{aligned}\mathbf{K}_M^{(w)} &= \prod_{m=1}^{M} [\mathbf{P}_{2^M}^{m-1} \widetilde{\mathbf{K}}_1^{(w)} \mathbf{P}_{2^M}^{-m+1}] = \mathbf{P}_{2^M}^{0} \cdot \prod_{m=1}^{M-1} [\widetilde{\mathbf{K}}_1^{(w)} \mathbf{P}_{2^M}^{-m+1} \mathbf{P}_{2^M}^{m}] \cdot \widetilde{\mathbf{K}}_M^{(w)} \mathbf{P}_{2^M}^{-M+1} \\ &= \prod_{m=1}^{M} [\widetilde{\mathbf{K}}_M^{(w)} \mathbf{P}_{2^M}].\end{aligned}$$

The matrix $\widetilde{\mathbf{K}}_M^{(w)} \mathbf{I}_{2^{M-1}} \otimes \mathbf{K}_1^{(w)}$ is a block diagonal matrix with equal blocks $\mathbf{K}_1^{(w)}$ on the diagonal. So it can be presented in the form of a direct sum as in Eq. 9–72. The lemma is proved. ■

Fast (a, b)-Polynomial Transforms of II-Type

Let us consider transforms by the (a, b)-polynomial matrices of II-type $\mathbf{H}(n, a, b)$.

LEMMA 9.3.4. *The* (a, b)*-polynomial matrices of II-type* $\mathbf{H}_n = \mathbf{H}(n, a, b)$ *can be expressed by the following formula*:

$$\mathbf{H}_n = \mathbf{G}_n \prod_{j=1}^{n-1} \mathrm{diag}(\mathbf{G}_{n-j}, \mathbf{I}_j, \ldots, \mathbf{I}_{n-1}), \tag{9–73}$$

where

$$\mathbf{G}_k = \left[\mathbf{I}_{k-1} \otimes \begin{pmatrix} 1 \\ 1 \end{pmatrix} \;\; \mathbf{I}_{k-1} \otimes \begin{pmatrix} a \\ b \end{pmatrix} \right], \tag{9–74}$$

where

$$\mathbf{H}_1 = \mathbf{G}_1 = \begin{pmatrix} 1 & a \\ 1 & b \end{pmatrix},$$

and $\mathbf{I}_r$ *is the identity matrix of order* 2^r.

PROOF. It follows from the following recurrent construction of the (a, b)-polynomial matrices of II-type $\mathbf{H}_n$ is given by

$$\mathbf{H}_n = \left[\mathbf{H}_{n-1} \otimes \begin{pmatrix} 1 \\ 1 \end{pmatrix} \;\; \mathbf{I}_{n-1} \otimes \begin{pmatrix} a \\ b \end{pmatrix} \right]. \qquad \blacksquare \tag{9–75}$$

PROPOSITION 9.3.4. *Let* $\mathbf{f}$ *be a column vector of length* $N = 2^n$. *Then the* (a, b)-*polynomial II-type transform of* $\mathbf{f}$ *can be computed with* $O(N)$ *addition operations.*

PROOF. The transform by each matrix $\mathrm{diag}(\mathbf{G}_{n-j}, \mathbf{I}_j, \ldots, \mathbf{I}_{n-1})$ requires 2^{n-j} operations ($j = 1, \ldots, n-1$) plus 2^n operations for transform by $\mathbf{G}_n$. $\blacksquare$

EXAMPLE 9.3.4. Consider the fast Haar transform algorithm of order 8 of the binary column vector $\mathbf{f} = (1, 0, 0, 1, 1, 1, 0, 1)^T$ by representation 9–73. For $a = 1$, $b = -1$, $n = 3$ ($N = 8$), representation 9–73 takes the form

$$\mathbf{H}_3 = \mathbf{H}_3^{(1)}\mathbf{H}_3^{(2)}\mathbf{H}_3^{(3)} = \begin{pmatrix} 1 & 0 & 0 & 0 & 1 & 0 & 0 & 0 \\ 1 & 0 & 0 & 0 & -1 & 0 & 0 & 0 \\ 0 & 1 & 0 & 0 & 0 & 1 & 0 & 0 \\ 0 & 1 & 0 & 0 & 0 & -1 & 0 & 0 \\ 0 & 0 & 1 & 0 & 0 & 0 & 1 & 0 \\ 0 & 0 & 1 & 0 & 0 & 0 & -1 & 0 \\ 0 & 0 & 0 & 1 & 0 & 0 & 0 & 1 \\ 0 & 0 & 0 & 1 & 0 & 0 & 0 & -1 \end{pmatrix}$$

$$\times \begin{pmatrix} 1 & 0 & 1 & 0 & 0 & 0 & 0 & 0 \\ 1 & 0 & -1 & 0 & 0 & 0 & 0 & 0 \\ 0 & 1 & 0 & 1 & 0 & 0 & 0 & 0 \\ 0 & 1 & 0 & -1 & 0 & 0 & 0 & 0 \\ 0 & 0 & 0 & 0 & 1 & 0 & 0 & 0 \\ 0 & 0 & 0 & 0 & 0 & 1 & 0 & 0 \\ 0 & 0 & 0 & 0 & 0 & 0 & 1 & 0 \\ 0 & 0 & 0 & 0 & 0 & 0 & 0 & 1 \end{pmatrix} \begin{pmatrix} 1 & 1 & 0 & 0 & 0 & 0 & 0 & 0 \\ 1 & -1 & 0 & 0 & 0 & 0 & 0 & 0 \\ 0 & 0 & 1 & 0 & 0 & 0 & 0 & 0 \\ 0 & 0 & 0 & 1 & 0 & 0 & 0 & 0 \\ 0 & 0 & 0 & 0 & 1 & 0 & 0 & 0 \\ 0 & 0 & 0 & 0 & 0 & 1 & 0 & 0 \\ 0 & 0 & 0 & 0 & 0 & 0 & 1 & 0 \\ 0 & 0 & 0 & 0 & 0 & 0 & 0 & 1 \end{pmatrix}.$$

Using this representation in Algorithm 9.3.1 we compute

1. $j = 0$;

2. $\mathbf{f}^{(0)} = \mathbf{f} = (1, 0, 0, 1, 1, 1, 0, 1)^T$;

3. Compute the matrix product, $\mathbf{f}^{(1)} = \mathbf{H}_3^{(3)} \cdot \mathbf{f}^{(0)} = (1, 1, 0, 1, 1, 1, 0, 1)^T$;

4. $j = 1$;

5. Compute the matrix product, $\mathbf{f}^{(2)} = \mathbf{K}_3^{(2)} \cdot \mathbf{f}^{(1)} = (1, 1, 2, 0, 1, 1, 0, 1)^T$;

6. $j = 2$;

7. Compute the matrix product, $\mathbf{f}^{(3)} = \mathbf{K}_3^{(1)} \cdot \mathbf{f}^{(2)} = (2, 0, 2, 0, 2, 2, 1, -1)^T$;

8. $j = 3$. STOP. The output column vector is

$$\mathbf{g} = \mathbf{f}^{(3)} = (2, 0, 2, 0, 2, 2, 1, -1)^T.$$

EXAMPLE 9.3.5. Consider the fast Haar-conjunctive transform algorithm of order 8 of the binary column vector $\mathbf{f} = (1, 0, 0, 1, 1, 1, 0, 1)^T$ by representation 9–73. For $a = 0,\ b = 1,\ n = 3\ (N = 8)$, representation 9–73 takes the form

$$\mathbf{H}_3 = \mathbf{H}_3^{(1)}\mathbf{H}_3^{(2)}\mathbf{H}_3^{(3)} = \begin{pmatrix} 1 & 0 & 0 & 0 & 0 & 0 & 0 & 0 \\ 1 & 0 & 0 & 0 & 1 & 0 & 0 & 0 \\ 0 & 1 & 0 & 0 & 0 & 0 & 0 & 0 \\ 0 & 1 & 0 & 0 & 0 & 1 & 0 & 0 \\ 0 & 0 & 1 & 0 & 0 & 0 & 0 & 0 \\ 0 & 0 & 1 & 0 & 0 & 0 & 1 & 0 \\ 0 & 0 & 0 & 1 & 0 & 0 & 0 & 0 \\ 0 & 0 & 0 & 1 & 0 & 0 & 0 & 1 \end{pmatrix}$$

$$\times \begin{pmatrix} 1 & 0 & 0 & 0 & 0 & 0 & 0 & 0 \\ 1 & 0 & 1 & 0 & 0 & 0 & 0 & 0 \\ 0 & 1 & 0 & 0 & 0 & 0 & 0 & 0 \\ 0 & 1 & 0 & 1 & 0 & 0 & 0 & 0 \\ 0 & 0 & 0 & 0 & 1 & 0 & 0 & 0 \\ 0 & 0 & 0 & 0 & 0 & 1 & 0 & 0 \\ 0 & 0 & 0 & 0 & 0 & 0 & 1 & 0 \\ 0 & 0 & 0 & 0 & 0 & 0 & 0 & 1 \end{pmatrix} \begin{pmatrix} 1 & 0 & 0 & 0 & 0 & 0 & 0 & 0 \\ 1 & 1 & 0 & 0 & 0 & 0 & 0 & 0 \\ 0 & 0 & 1 & 0 & 0 & 0 & 0 & 0 \\ 0 & 0 & 0 & 1 & 0 & 0 & 0 & 0 \\ 0 & 0 & 0 & 0 & 1 & 0 & 0 & 0 \\ 0 & 0 & 0 & 0 & 0 & 1 & 0 & 0 \\ 0 & 0 & 0 & 0 & 0 & 0 & 1 & 0 \\ 0 & 0 & 0 & 0 & 0 & 0 & 0 & 1 \end{pmatrix}.$$

Using this representation in Algorithm 9.3.1 we compute

1. $j = 0$;

2. $\mathbf{f}^{(0)} = \mathbf{f} = (1, 0, 0, 1, 1, 1, 0, 1)^T$;

3. Compute the matrix product, $\mathbf{f}^{(1)} = \mathbf{H}_3^{(3)} \cdot \mathbf{f}^{(0)} = (1, 1, 0, 1, 1, 1, 0, 1)^T$;

4. $j = 1$;

5. Compute the matrix product, $\mathbf{f}^{(2)} = \mathbf{K}_3^{(2)} \cdot \mathbf{f}^{(1)} = (1, 1, 1, 2, 1, 1, 0, 1)^T$;

6. $j = 2$;

7. Compute the matrix product, $\mathbf{f}^{(3)} = \mathbf{K}_3^{(1)} \cdot \mathbf{f}^{(2)} = (1, 2, 1, 2, 1, 1, 2, 3)^T$;

8. $j = 3$. STOP. The output column vector is

$$\mathbf{g} = \mathbf{f}^{(3)} = (1, 2, 1, 2, 1, 1, 2, 3)^T.$$

9.4 LOGICAL CORRELATIONS

9.4.1 INTRODUCTION

Generalization of the concepts of correlation and power spectrum for Walsh functions was done by Wiener and Paley [86]. Since 1960–1970 these concepts have found a large number of applications in signal processing (see [15–19, 21, 40, 51]).

The Walsh power spectrum has certain good properties. It can be sequence limited even when the corresponding time function is time limited [21]. This is in contrast to the behavior of the Fourier power spectrum where a time-limited function cannot have a band-limited power spectrum. Another property is that the Walsh power spectrum is more phase sensitive than the Fourier power spectrum.

The generalization of the Wiener–Khinchine theorem to Walsh spectra and the dyadic autocorrelation function was done by Gibbs and Gebbie [40]. Spectral analysis by an arbitrary base was proposed by Aizenberg [10].

We focus our attention on a study of general auto- and cross-correlation functions specified by any arithmetic or logical operation ρ. Then we describe a transform method for their computation in which we show the connection of logical correlations (ρ is any logical operation) to the binary polynomial transforms.

9.4.2 ARITHMETIC AUTO- AND CROSS-CORRELATION FUNCTIONS

Let $\{x(j)\} = \{x(0), \ldots, x(M + N - 1)\}$ and $\{y(j)\} = \{y(0), \ldots, y(M + N - 1)\}$ be sample functions of wide-sense stationary random processes, and let $\{x(j)\}$ and $\{y(j)\}$ be divided into M/N blocks (vectors) such that each block consists of $N \ll M$ points, where $M,\ N$ are natural numbers.

DEFINITION 9.4.1. The expected value

$$z(i) = A^{(x,y)}(i) = E\{A_s^{(x,y)}(i)\}, \quad i = 0, 1, \ldots, N-1, \tag{9–76}$$

of the *local (arithmetic) cross-correlation function*

$$A_s^{(x,y)}(i) = \frac{1}{N}\sum_{j=0}^{N-1} x(j+sN)y(j+sN+i), \tag{9–77}$$

$i = 0, 1, \ldots, N-1$, $s = 0, 1, \ldots, \frac{M}{N} - 1$, is called the *(arithmetic) cross-correlation function of* $\{x(j)\}$ *and* $\{y(j)\}$.

The cross-correlation function $A^{(x,y)}(i)$ can be estimated by its sample mean $\hat{A}^{(x,y)}(i)$ where

$$\hat{A}^{(x,y)}(i) = \frac{N}{M}\sum_{s=0}^{\frac{M}{N}-1} A_s^{(x,y)}(i) = \frac{1}{M}\sum_{j=0}^{M-1} x(j)y(j+i),$$
$$i = 0, 1, \ldots, N-1. \tag{9–78}$$

DEFINITION 9.4.2. Let $y(j) = x(j)$, $j = 0, 1, \ldots, M+N-1$. Then $z(i)$ defined by Eqs. 9–76 and 9–77 is called the *(arithmetic) autocorrelation function of* $\{x(j)\}$ and is denoted by $z(i) = A^{(x,x)}(i)$, $i = 0, 1, \ldots, N-1$.

9.4.3 LOGICAL AUTO- AND CROSS-CORRELATION FUNCTIONS

Let $\{x(j)\}$ and $\{y(j)\}$ be the random sequences defined above, and let $N = 2^n$.

Let $\oplus$ be the logical XOR-operation (modulo 2 sum), and $k \oplus j$, $k, j = 0, 1, \ldots, N-1$, be the bit-by-bit modulo 2 sum, i.e., $\mathbf{k} \oplus \mathbf{j} = (k_0 \oplus j_0, \ldots, k_{n-1} \oplus j_{n-1})$, where $\mathbf{k} = (k_0, \ldots, k_{n-1})$ and $\mathbf{j} = (j_0, \ldots, j_{n-1})$ are the corresponding binary representations of k and j, i.e.,

$$k = \sum_{r=0}^{n-1} k_r 2^{n-r-1}, \tag{9–79}$$

and

$$j = \sum_{r=0}^{n-1} j_r 2^{n-r-1}, \tag{9–80}$$

where $k_r, j_r \in \{0, 1\}$, $r = 0, \ldots, n-1$.

DEFINITION 9.4.3. The expected value

$$z(i) = B^{(\oplus,x,y)}(i) = E\{B_s^{(\oplus,x,y)}(i)\}, \quad i = 0, 1, \ldots, N-1, \tag{9–81}$$

of the *local dyadic (logical) cross-correlation function*

$$B_s^{(\oplus,x,y)}(i) = \frac{1}{N}\sum_{j=0}^{N-1} x(j+sN)y((j+sN)\oplus i), \tag{9–82}$$

$i = 0, 1, \ldots, N-1$, $s = 0, 1, \ldots, \frac{M}{N} - 1$, is called the *dyadic (logical) cross-correlation function* of $\{x(j)\}$ and $\{y(j)\}$.

The cross-correlation function $B^{(\oplus,x,y)}$ can be estimated by its sample mean $\hat{B}^{(\oplus,x,y)}(i)$, where

$$\hat{B}^{(\oplus,x,y)}(i) = \frac{N}{M}\sum_{s=0}^{\frac{M}{N}-1} B_s^{(\oplus,x,y)}(i) = \frac{1}{M}\sum_{t=0}^{M-1} x(t)y(t\oplus i),$$
$$i = 0, 1, \ldots, N-1. \tag{9–83}$$

DEFINITION 9.4.4. Let $y(j) = x(j)$, $j = 0, 1, \ldots, M+N-1$. Then $z(i)$ defined by Eq. 9–81 is called the *dyadic (logical) auto-correlation function of* $\{x(n)\}$ and is denoted by $z(i) = B^{(\oplus,x,x)}(i)$, $i = 0, 1, \ldots, N-1$.

9.4.4 GENERAL CORRELATION FUNCTION

Let ρ be an arbitrary, arithmetic or logical operation, and $\mathbf{x}$, $\mathbf{y}$ real random vectors of length $N = 2^n$ of the stationary random processes $\{x(s)\}$ and $\{y(s)\}$.

For a logical operation ρ let $\mathbf{k}\ \rho\ \mathbf{j}$, $k, j = 0, 1, \ldots, N-1$, be the bit-by-bit ρ operation, i.e., $\mathbf{k}\ \rho\ \mathbf{j} = (k_0\rho j_0, \ldots, k_{n-1}\rho j_{n-1})$, where $\mathbf{k} = (k_0, \ldots, k_{n-1})$ and $\mathbf{j} = (j_0, \ldots, j_{n-1})$ are corresponding binary representations of k and j, given by Eqs. 9–79 and 9–80, respectively. In Tables 9.4–9.6 the bit-by-bit ρ-operations ($\rho \in \{\oplus, \vee, \wedge\}$ for $\mathbf{k}\ \rho\ \mathbf{j}$, $k, j = 0, 1, \ldots, 7$, integers, are given.

Define the set of N matrices $\mathbf{\Gamma}^{(i;\rho)} = [\Gamma_{k,j}^{(i;\rho)}]$, $r = 1, \ldots, N$, where

$$\Gamma_{k,j}^{(i;\rho)} = \begin{cases} 1, & \text{if } \mathbf{i} = \mathbf{k}\ \rho\ \mathbf{j} \\ 0, & \text{otherwise,} \end{cases} \quad k, j = 0, 1, \ldots, N-1. \tag{9–84}$$

DEFINITION 9.4.5. The expected value

$$z(i) = B^{(\rho,x,y)}(i) = E\{B_s^{(\rho,x,y)}(i)\}, \quad i = 0, 1, \ldots, N-1, \tag{9–85}$$

Table 9.4. Bitwise modulo 2 addition, $j \oplus k$.

$k \oplus j$	0	1	2	3	4	5	6	7
0	0	1	2	3	4	5	6	7
1	1	0	3	2	5	4	7	6
2	2	3	0	1	6	7	4	5
3	3	2	1	0	7	6	5	4
4	4	5	6	7	0	1	2	3
5	5	4	7	6	1	0	3	2
6	6	7	4	5	2	3	0	1
7	7	6	5	4	3	2	1	0

Table 9.5. Bitwise disjunction, $j \vee k$.

$k \vee j$	0	1	2	3	4	5	6	7
0	0	1	2	3	4	5	6	7
1	1	1	3	3	5	5	7	7
2	2	3	2	3	6	7	6	7
3	3	3	3	3	7	7	7	7
4	4	5	6	7	4	5	6	7
5	5	5	7	7	5	5	7	7
6	6	7	6	7	6	7	6	7
7	7	7	7	7	7	7	7	7

Table 9.6. Bitwise conjunction, $j \wedge k$.

$k \wedge j$	0	1	2	3	4	5	6	7
0	0	0	0	0	0	0	0	0
1	0	1	0	1	0	1	0	1
2	0	0	2	2	0	0	2	2
3	0	1	2	3	0	1	2	3
4	0	0	0	0	4	4	4	4
5	0	1	0	1	4	5	4	5
6	0	0	2	2	4	4	6	6
7	0	1	2	3	4	5	6	7

of the *local ρ-cross-correlation function*

$$B_s^{(\rho,x,y)}(i) = \frac{1}{N} \sum_{k=0}^{N-1} \sum_{j=0}^{N-1} \Gamma_{k+sN,j+sN}^{(i;\rho)} x(k+sN)y(j+sN), \qquad (9\text{–}86)$$

where $i = 0, 1, \ldots, N-1$, $s = 0, 1, \ldots, \frac{M}{N} - 1$, is called the *$\rho$-cross-correlation function of $\{x(j)\}$ and $\{y(j)\}$.*

The cross-correlation function $B^{(\rho,x,y)}$ can be estimated by its sample mean $\widehat{B}^{(\rho,x,y)}(i)$, where

$$\widehat{B}^{(\rho,x,y)}(i) = \frac{N}{M} \sum_{s=0}^{\frac{M}{N}-1} B_s^{(\rho,x,y)}(i) = \frac{1}{M} \sum_{t=0}^{M-1} \sum_{r=0}^{M-1} \Gamma_{t,r}^{(i;\rho)} x(t)y(r), \qquad (9\text{–}87)$$

$i = 0, 1, \ldots, N-1$.

DEFINITION 9.4.6. Let $y(j) = x(j)$, $j = 0, 1, \ldots, M+N-1$. Then $z(i)$ defined by Eq. 9–85 is called the *ρ-autocorrelation function of* $\{x(n)\}$.

REMARK 9.4.1. The ρ-cross (auto)-correlation function contains as special cases:

– the arithmetic cross (auto)-correlation function, when ρ is the arithmetic operation of subtraction;

– the dyadic (logical) cross (auto)-correlation function, when ρ is the XOR ($\oplus$, addition modulo 2) operation.

Let $\mathbf{\Gamma}_k$, $k = 0, 1, \ldots, N-1$, be the following matrix

$$\mathbf{\Gamma}_k^{(\rho)} = \begin{pmatrix} \Gamma_{k,0}^{(0;\rho)} & \Gamma_{k,1}^{(0;\rho)} & \cdots & \Gamma_{k,N-1}^{(0;\rho)} \\ \Gamma_{k,0}^{(1;\rho)} & \Gamma_{k,1}^{(1;\rho)} & \cdots & \Gamma_{k,N-1}^{(1;\rho)} \\ \cdots & \cdots & \cdots & \cdots \\ \Gamma_{k,0}^{(N-1;\rho)} & \Gamma_{k,1}^{(N-1;\rho)} & \cdots & \Gamma_{k,N-1}^{(N-1;\rho)} \end{pmatrix}, \qquad (9\text{–}88)$$

where elements $\Gamma_{k,j}^{(i;\rho)}$, $i, j, k = 0, 1, \ldots, N-1$, are defined by Eq. 9–84.

EXAMPLE 9.4.1. Let $N = 2^2 = 4$, and ρ be the operation of conjunction $\wedge$. Then

$$\mathbf{\Gamma}_0^{(\wedge)} = \begin{pmatrix} 1 & 1 & 1 & 1 \\ 0 & 0 & 0 & 0 \\ 0 & 0 & 0 & 0 \\ 0 & 0 & 0 & 0 \end{pmatrix}; \quad \mathbf{\Gamma}_1^{(\wedge)} = \begin{pmatrix} 1 & 0 & 1 & 0 \\ 0 & 1 & 0 & 1 \\ 0 & 0 & 0 & 0 \\ 0 & 0 & 0 & 0 \end{pmatrix};$$

$$\mathbf{\Gamma}_2^{(\wedge)} = \begin{pmatrix} 1 & 1 & 0 & 0 \\ 0 & 0 & 0 & 0 \\ 0 & 0 & 1 & 1 \\ 0 & 0 & 0 & 0 \end{pmatrix};$$

and

$$\mathbf{\Gamma}_3^{(\wedge)} = \begin{pmatrix} 1 & 0 & 0 & 0 \\ 0 & 1 & 0 & 0 \\ 0 & 0 & 1 & 0 \\ 0 & 0 & 0 & 1 \end{pmatrix}.$$

Now let ρ be the operation of disjunction $\vee$. Then

$$\mathbf{\Gamma}_0^{(\vee)} = \begin{pmatrix} 1 & 0 & 0 & 0 \\ 0 & 1 & 0 & 0 \\ 0 & 0 & 1 & 0 \\ 0 & 0 & 0 & 1 \end{pmatrix}; \quad \mathbf{\Gamma}_1^{(\vee)} = \begin{pmatrix} 0 & 0 & 0 & 0 \\ 1 & 1 & 0 & 0 \\ 0 & 0 & 0 & 0 \\ 0 & 0 & 1 & 1 \end{pmatrix};$$

$$\mathbf{\Gamma}_2^{(\vee)} = \begin{pmatrix} 0 & 0 & 0 & 0 \\ 0 & 0 & 0 & 0 \\ 1 & 0 & 1 & 0 \\ 0 & 1 & 0 & 1 \end{pmatrix};$$

and

$$\mathbf{\Gamma}_3^{(\vee)} = \begin{pmatrix} 0 & 0 & 0 & 0 \\ 0 & 0 & 0 & 0 \\ 0 & 0 & 0 & 0 \\ 1 & 1 & 1 & 1 \end{pmatrix}.$$

DEFINITION 9.4.7. The column vector $\mathbf{z}$ of length $N = 2^n$ given by

$$\mathbf{z} = \mathbf{x} *_\rho \mathbf{y} = \sum_{k=0}^{N-1} x(k) \mathbf{\Gamma}_k \mathbf{y} \tag{9–89}$$

is called the *ρ-cross-correlation of the column vectors* $\mathbf{x} = (x(0), x(1), \ldots, x(N-1))^T$ *and* $\mathbf{y}$ of length $N = 2^n$.

EXAMPLE 9.4.2. $\wedge$-cross-correlation of the vector functions $\mathbf{x} = (0, 2, 1, 3)^T$ and $\mathbf{y} = (1, -5, -2, 3)$ is the vector $\mathbf{z} = \mathbf{x} *_\wedge \mathbf{y} = (-3, -19, -5, 9)^T$, because

$$\begin{aligned} z_0 &= x_0(y_0 + y_1 + y_2 + y_3) + y_0(x_1 + x_2 + x_3) + x_1 y_2 + x_2 y_1 = -3; \\ z_1 &= x_1 y_3 + x_3 y_1 + x_1 y_1 = -19; \\ z_2 &= x_2 y_3 + x_3 y_2 + x_2 y_2 = -5; \\ z_3 &= x_3 y_3 = 9. \end{aligned}$$

EXAMPLE 9.4.3. $\vee$-autocorrelation of the vector function $\mathbf{x} = (0, 2, 1, 3)^T$ is the vector $\mathbf{z} = \mathbf{x} *_{\vee} \mathbf{y} = (0, 4, 1, 31)^T$, because

$$\begin{aligned} z_0 &= x_0^2 = 0; \\ z_1 &= 2x_0x_1 + x_1^2 = 4; \\ z_2 &= 2x_0x_2 + x_2^2 = 1; \\ z_3 &= 2x_0x_3 + 2x_1x_2 + 2x_1x_3 + 2x_2x_3 + x_3^2 = 31. \end{aligned}$$

Let the column vector $\mathbf{e_i}$ of length $N = 2^n$ be defined by

$$\mathbf{e_i} = (\underbrace{0, \ldots, 0}_{i-1 \text{ times}}, 1, 0, \ldots, 0), \quad i = 0, 1, \ldots, N-1. \tag{9–90}$$

The following proposition holds.

PROPOSITION 9.4.1. *The ρ-cross-correlation of the vectors $\mathbf{e_k}$ and $\mathbf{e_j}$ is equal to*

$$\mathbf{e_k} *_{\rho} \mathbf{e_j} = \left(\Gamma_{k,j}^{(0)}, \ldots, \Gamma_{k,j}^{(N-1)}\right)^T, \tag{9–91}$$

where $\Gamma_{k,j}^{(i)}$, $i, j, k = 0, 1, \ldots, N-1$, are defined by Eq. 9–84.

PROOF. Let $\mathbf{z} = (z(0), \ldots, z(N-1))^T = \mathbf{e_k} *_{\rho} \mathbf{e_j}$. Then by the definition of the ρ-correlation we have

$$z(i) = \sum_{t=0}^{N-1} e_k(t) \sum_{r=0}^{N-1} \Gamma_{t,r}^{(i)} e_j(r) = \sum_{r=0}^{N-1} \Gamma_{k,r}^{(i)} e_j(r) = \Gamma_{k,j}^{(i)}. \quad \blacksquare$$

9.4.5 COMPUTATION OF GENERAL CROSS-CORRELATION

The results of this section are based on the theory of generalized shift operators developed by Levitan [63] and extended by Aizenberg et al. [10, 11, 13] (see also Matevosian [67]).

Let $N = 2^n$, where n is a natural number. Let $\mathbf{H} = [h_{k,j}]$, be a $(N \times N)$ matrix and denote the columns of $\mathbf{H}$ by $\mathbf{h}_j$, $j = 0, 1, \ldots, N-1$. For $j = 0, 1, \ldots, N-1$, write $\mathbf{\Lambda}^{(j)} = \mathrm{diag}(h_{0,j}, h_{1,j}, \ldots, h_{N-1,j})$, that is, the $(N \times N)$ diagonal matrix with diagonal elements $h_{0,j}, h_{1,j}, \ldots, h_{N-1,j}$. Let $\mathbf{\Gamma}_k$ be the matrix given by Eq. 9–88.

LEMMA 9.4.1. *For the matrix* $\mathbf{H}$ *the following relation holds*

$$\mathbf{H} \cdot (\mathbf{e_k} *_\rho \mathbf{e_j}) = (\mathbf{H} \cdot \mathbf{e_k}) \circ (\mathbf{H} \cdot \mathbf{e_j}) \tag{9–92}$$

if and only if

$$\mathbf{h_{k\rho j}} = \mathbf{h_k} \circ \mathbf{h_j}, \tag{9–93}$$

where $\circ$ *denotes the Hadamard (pointwise) product symbol and* $\mathbf{e_i}$ $(i = 0, 1, \dots, N-1)$ *is defined by Eq. 9–90.*

PROOF. Because $\mathbf{H} \cdot \mathbf{e_i} = \mathbf{h_i}$, then the right sides of Eqs. 9–92 and 9–93 are equal. Using the fact that by Proposition 9.4.1 $\mathbf{e_k} *_\rho \mathbf{e_j} = (\Gamma_{k,j}^{(0)}, \dots, \Gamma_{k,j}^{(N-1)})^T$, we show the equivalence of the left sides of Eqs. 9–92 and 9–93:

$$\mathbf{H} \cdot \left(\mathbf{e_k} *_\rho \mathbf{e_j}\right) = \mathbf{H} \cdot \left(\Gamma_{k,j}^{(0)}, \dots, \Gamma_{k,j}^{(N-1)}\right)^T = \sum_{i=0}^{N-1} \Gamma_{k,j}^{(i)} \mathbf{h_i} = \mathbf{h_{k\rho j}}. \qquad \blacksquare$$

The following theorem holds [11].

THEOREM 9.4.1. *The following statements are equivalent:*
1) $\mathbf{h_{k\rho j}} = \mathbf{h_k} \circ \mathbf{h_j}$,
2) *for any vectors* $\mathbf{x}$ *and* $\mathbf{y}$ *of length* N:

$$\mathbf{H} \cdot (\mathbf{x} *_\rho \mathbf{y}) = (\mathbf{H} \cdot \mathbf{x}) \circ (\mathbf{H} \cdot \mathbf{y}), \tag{9–94}$$

and

3)

$$\mathbf{\Gamma}_k = \mathbf{H}^{-1} \mathbf{\Lambda}^{(k)} \mathbf{H}. \tag{9–95}$$

PROOF. At first we show the equivalence of statements 2) and 3).

Let statement 2) hold. Then Eq. 9–92 also takes place. Hence using Eq. 9–91 we have, for arbitrary $j, k = 0, 1, \dots, N-1$:

$$\mathbf{H} \cdot \left(\Gamma_{k,j}^{(0)}, \dots, \Gamma_{k,j}^{(N-1)}\right)^T = \mathbf{h_k} \circ \mathbf{h_j} = \mathbf{\Lambda}^{(k)} \cdot \mathbf{h_j},$$

and therefore statement 3) holds.

Let statement 3) hold. Then using Eq. 9–89 we have

$$\mathbf{H}\cdot(\mathbf{x}*_\rho\mathbf{y})=\mathbf{H}\sum_{k=0}^{N-1}x(k)\mathbf{\Gamma}_k\mathbf{y}=\sum_{k=0}^{N-1}x(k)\mathbf{H}\mathbf{\Gamma}_k\mathbf{y}=\sum_{k=0}^{N-1}x(k)\mathbf{\Lambda}^{(k)}\mathbf{H}\mathbf{y}.$$

Because $\sum_{k=0}^{N-1}x(k)\mathbf{\Lambda}^{(k)}$ is a diagonal matrix with the entries of the diagonal forming the vector $\mathbf{Hx}$, we obtain statement 2). Thus statements 2) and 3) are equivalent.

Now we show the equivalence of statement 3) and Eq. 9–92, from which, with the formulation of Lemma 9.4.1, we establish the equivalence of statements 3) and 1).

Let statement 3) hold. Then statement 2) holds also, and as the particular case of statement 2) also Eq. 9–92. From Eq. 9–92, as we showed at the beginning of the proof, statement 1) follows. ■

As follows from Theorem 9.4.1, Eqs. 9–94 and 9–93 are equivalent. This provides the possibility of using the transform specified by the matrix **H** for the computation of the ρ-cross-correlation.

9.4.6 COMPUTATION OF LOGICAL CROSS-CORRELATION BASED ON ANY BOOLEAN OPERATION

We consider the problem of computation of ρ-auto- and cross-correlation functions for any logical operation ρ.

There is a close relation between the ρ-auto- and cross-correlation functions and binary polynomial transforms.

To each operation ρ in the ρ-cross-correlation function $\mathbf{z}=\mathbf{x}*_\rho\mathbf{y}$ we associate the triple of matrices $\mathbf{H}^{(\rho)}$, $\mathbf{H}'^{(\rho)}$ and $\mathbf{H}''^{(\rho)}$, in such a way that the following relation takes place:

$$\mathbf{H}^{(\rho)}\cdot\mathbf{z}=\left[\mathbf{H}'^{(\rho)}\cdot\mathbf{x}\right]\circ\left[\mathbf{H}''^{(\rho)}\cdot\mathbf{y}\right],\tag{9–96}$$

where $\circ$ denotes of element by element multiplication of vectors or, equivalently,

$$\mathbf{H}^{(\rho)}\cdot\mathbf{z}=\mathrm{diag}(\mathbf{H}'^{(\rho)}\cdot\mathbf{x})\cdot\mathbf{H}''^{(\rho)}\cdot\mathbf{y},\tag{9–97}$$

where $\mathrm{diag}(\mathbf{H}'^{(\rho)}\cdot\mathbf{x})$ is the diagonal matrix with the elements on the diagonal forming the vector $\mathbf{H}'^{(\rho)}\cdot\mathbf{x}$.

If the matrix $\mathbf{H}^{(\rho)}$ is nonsingular, formulae 9–96 and 9–97 allow us to compute the Boolean ρ-cross-correlation function $\mathbf{z} = \mathbf{x} *_\rho \mathbf{y}$ in the spectral domain by:

$$\mathbf{z} = \left[\mathbf{H}^{(\rho)}\right]^{-1} \cdot \mathrm{diag}\left(\mathbf{H}'^{(\rho)} \cdot \mathbf{x}\right) \cdot \mathbf{H}''^{(\rho)} \cdot \mathbf{y}. \tag{9–98}$$

PROPERTY 9.4.1. *Let ρ be an operation (given by the table $i\rho j$) and let $\mathbf{H} = \mathbf{H}^{(\rho)}$, $\mathbf{H}' = \mathbf{H}'^{(\rho)}$ and $\mathbf{H}'' = \mathbf{H}''^{(\rho)}$ be the triple of matrices corresponding to ρ and satisfying Eq.* 9–96. *Then*:

1) *The triple of matrices* $\mathbf{PH}$, $\mathbf{PH}'$ *and* $\mathbf{PH}''$ *also satisfies Eq.* 9–96, *where* $\mathbf{P}$ *is an arbitrary permutation matrix*;

2) *For the operation ρ' (respectively, ρ'') given by the table $k\rho' j = \bar{k}\rho j$ (respectively, $k\rho'' j = k\rho\bar{j}$) there corresponds the triple of matrices* $\mathbf{H}^{(\rho')} = \mathbf{H}$, $\mathbf{H}'^{(\rho')} = \mathbf{H}' \cdot \bar{\mathbf{I}}$ *and* $\mathbf{H}''^{(\rho')} = \mathbf{H}''$ ($\mathbf{H}, \mathbf{H}'$ *and* $\mathbf{H}'' \cdot \bar{\mathbf{I}}$, *respectively*), *satisfying Eq.* 9–96, *where* $\bar{\mathbf{I}}$ *is the opposite identity matrix.*

3) *To operation ρ', given by the table $k\rho' j = \mathbf{1} - k\rho j$ there corresponds the triple of matrices* $\mathbf{H}^{(\rho')} = \mathbf{H} \cdot \bar{\mathbf{I}}$, $\mathbf{H}'^{(\rho')} = \mathbf{H}'$ *and* $\mathbf{H}''^{(\rho')} = \mathbf{H}''$, *satisfying Eq.* 9–96.

PROOF. 1) is obvious.

2) Let $\Gamma^{(i)}_{k,j}$ defined by Eq. 9–84 correspond to the operation $\mathbf{i} = \mathbf{k}\rho\mathbf{j}$. Then $\Gamma^{(i)}_{N-k+1,j}$ and $\Gamma^{(i)}_{k,N-j+1}$ corresponds to the operations $\mathbf{i} = \mathbf{k}\rho'\mathbf{j} = \bar{\mathbf{k}}\rho\mathbf{j}$ and $\mathbf{i} = \mathbf{k}\rho''\mathbf{j} = \mathbf{k}\rho\bar{\mathbf{j}}$, respectively. Therefore, using the definition of ρ-cross-correlation, we have

$$\mathbf{x} *_{\rho'} \mathbf{y} = \left(\bar{\mathbf{I}}\mathbf{x}\right) *_\rho \mathbf{y}, \quad \text{and} \quad \mathbf{x} *_{\rho''} \mathbf{y} = \mathbf{x} *_\rho \left(\bar{\mathbf{I}}\mathbf{y}\right).$$

3) Let $\Gamma^{(i)}_{k,j}$ correspond to the operation $\mathbf{i} = \mathbf{k}\rho\mathbf{j}$. Then $\Gamma^{(N-i+1)}_{k,j}$ corresponds to the operation $\mathbf{i} = \mathbf{k}\rho'\mathbf{j} = \overline{\mathbf{k}\rho\mathbf{j}}$, which gives

$$\mathbf{x} *_{\rho'} \mathbf{y} = \bar{\mathbf{I}}(\mathbf{x} *_\rho \mathbf{y}). \qquad \blacksquare$$

Based on the properties given in Table 9.7, there are triples of matrices $\mathbf{H}^{(\rho)}$, $\mathbf{H}'^{(\rho)}$ and $\mathbf{H}''^{(\rho)}$, satisfying Eqs. 9–96 and 9–97 corresponding to all nontrivial logical operations ρ, where $\bar{\mathbf{I}}$ is the opposite identity matrix, $\mathbf{K}$ and $\mathbf{K}^T$ are conjunctive and conjunctive transposed matrices, respectively, $\mathbf{W}$ is the Walsh–Hadamard matrix and $\mathbf{J}$ is the matrix of ones.

Table 9.7. The triple of the matrices $\mathbf{H}^{(\rho)}$, $\mathbf{H}'^{(\rho)}$ and $\mathbf{H}''^{(\rho)}$, corresponding to all nontrivial logical operations ρ (here W is the Walsh matrix, K is the conjunctive matrix, $\overline{I}$ is the opposite identity matrix and J is the matrix of ones).

	$i\rho j$	Truth Table for $i\rho j$	$H^{(\rho)}$	$H'^{(\rho)}$	$H''^{(\rho)}$
1.	$i \wedge j$	$\begin{pmatrix} 0 & 0 \\ 0 & 1 \end{pmatrix}$	K^T	K^T	K^T
2.	$i \wedge \overline{j}$	$\begin{pmatrix} 0 & 0 \\ 1 & 0 \end{pmatrix}$	K^T	K^T	$K^T\overline{I}$
3.	i	$\begin{pmatrix} 0 & 0 \\ 1 & 1 \end{pmatrix}$	W	W	J
4.	$\overline{i} \wedge j$	$\begin{pmatrix} 0 & 1 \\ 0 & 0 \end{pmatrix}$	K^T	$K^T\overline{I}$	K^T
5.	j	$\begin{pmatrix} 0 & 1 \\ 0 & 1 \end{pmatrix}$	W	J	W
6.	$i \oplus j$	$\begin{pmatrix} 0 & 1 \\ 1 & 0 \end{pmatrix}$	W	W	W
7.	$i \vee j$	$\begin{pmatrix} 0 & 1 \\ 1 & 1 \end{pmatrix}$	K	K	K
8.	$i \downarrow j$	$\begin{pmatrix} 1 & 0 \\ 0 & 0 \end{pmatrix}$	$\overline{I}K\overline{I}$	$\overline{I}K$	$\overline{I}K$
9.	$i \equiv j$	$\begin{pmatrix} 1 & 0 \\ 0 & 1 \end{pmatrix}$	$\overline{I}W\overline{I}$	$\overline{I}W$	$\overline{I}W$
10.	j	$\begin{pmatrix} 1 & 0 \\ 1 & 0 \end{pmatrix}$	$\overline{I}W\overline{I}$	J	$\overline{I}W$
11.	$i \vee \overline{j}$	$\begin{pmatrix} 1 & 0 \\ 1 & 1 \end{pmatrix}$	K	K	$K\overline{I}$
12.	$\overline{i}$	$\begin{pmatrix} 1 & 1 \\ 0 & 0 \end{pmatrix}$	$\overline{I}W\overline{I}$	$\overline{I}W$	J
13.	$i \rightarrow j$	$\begin{pmatrix} 1 & 1 \\ 0 & 1 \end{pmatrix}$	K	$K\overline{I}$	K
14.	$i \mid j$	$\begin{pmatrix} 1 & 1 \\ 1 & 0 \end{pmatrix}$	K	$K\overline{I}$	$K\overline{I}$

REFERENCES

[1] Agaian, S. S., "Algorithm of orthogonal matrices fast transform", *Cyber. and Sys. Res., B.*, Vol. 8, pp. 317–321 (1981).

[2] Agaian, S. S., "A unified construction method of fast orthogonal transforms and signal processing problems", in *Proc. of 5th Intern. Congress of Cybernetics and Systems*, Mexico, pp. 17–22 (1981).

[3] Agaian, S. S., *Hadamard Matrices and Their Applications*, Lecture Notes in Math., 1168, Springer-Verlag, Berlin (1985).
[4] Agaian, S. S. and Gevorkian D. Z., "The complexity and parallel algorithms of discrete orthogonal transforms", *Cybernetics and Calculative Techniques*, Moscow, Issue 4 (in Russian), pp. 124–169 (1988).
[5] Agaian, S. S., Advances and Problems of Fast Orthogonal Transforms for Signal-Images Processing Applications (part 1), in the yearbook: *Pattern Recognition, Classification, Forecasting*, Nauka, Moscow, Issue 3 (in Russian), pp. 146–215 (1992).
[6] Agaian, S. S., Advances and Problems of Fast Orthogonal Transforms for Signal-Images Processing Applications (part 2), in the yearbook: *Pattern Recognition, Classification, Forecasting*, Nauka, Moscow, Issue 4 (in Russian), pp. 99–145 (1993).
[7] Astola, J., Agaian, S., Gevorkian D. and Egiazarian, K., "Parallel algorithms for binary polynomial transforms and their realization with a unified series of processors", Proc. SPIE 2421, pp. 240–251 (1995).
[8] Agaian, S. S., Astola, J. and Egiazarian, K., *Binary Polynomial Transforms and Nonlinear Digital Filters*, Marcel Dekker, New York (1995).
[9] Ahmed, N. and Rao, K. R., *Orthogonal Transforms for Digital Signal Processing*, Springer-Verlag, Berlin (1975).
[10] Aizenberg, N. N., "On spectrum of convolution of the discrete signals in arbitrary base", *Dokladi Akademii Nauk SSSR*, Vol. 241, No. 3 (in Russian), pp. 108–110 (1978).
[11] Aizenberg, N. N. and Trofimluk, O. T., "Shift, convolution and correlation function of the discrete signals in arbitrary base", *Dokladi Akademii Nauk SSSR*, Vol. 250, No. 1 (in Russian), pp. 75–90 (1980).
[12] Aizenberg, N. N. and Trofimluk, O. T., "The conjunctive transforms of discrete signals and their applications to finding tests and recognizing of monotonicity of the function of Boolean algebra", *Cybernetics*, Kiev, Vol. 5 (in Russian), pp. 138–139 (1981).
[13] Aizenberg, N. N. and Tsitkin, A. I., "On Fourier Transform for the convolution of signals in arbitrary base", *Application of Orthogonal Methods in Signal Processing and System Analysis*, Sverdlovsk, pp. 144–148 (1983).
[14] Aldroubi, A., "Oblique multiwavelet bases: examples", Proc. SPIE 2825, pp. 54–64 (1996).
[15] Proc. of Symp. *Applications of Walsh Functions*, Washington, D.C., p. 294 (1970).
[16] Proc. of Symp. *Applications of Walsh Functions*, Washington, D.C., p. 218 (1971).
[17] Proc. of Symp. *Applications of Walsh Functions*, Washington, D.C., p. 398 (1972).
[18] Proc. of Symp. *Applications of Walsh Functions*, Washington, D.C., p. 298 (1973).
[19] Proc. of Symp. *Applications of Walsh Functions*, Washington, D.C., p. 460 (1974).

[20] Artazjan, R. S. and Egiazarian, K. O., "The fast matrix algorithms of finding the abbreviated disjunctive normal forms, irredundant tests and their application to test pattern recognition", Dep. 62-Ar89 (in Russian) (1989).

[21] Beauchamp, K. G., *Walsh Functions and Their Applications*, Academic Press, San Diego, (1975).

[22] Bendat, J. S. and Piersol, G., *Engineering Applications of Correlation and Spectral Analysis*, Wiley, New York, p. 458 (1963).

[23] Besslich, Ph. W., "Spectral processing of switching functions using signal-flow transformations", in *Spectral Techniques and Fault Detection*, M. G. Karpovsky ed., Academic Press, pp. 91–142 (1985).

[24] Brown, R. D., "A recursive algorithm for sequence-ordered fast Walsh transforms", *IEEE Trans. Comput.*, Vol. 26, No. 8, pp. 819-822 (1977).

[25] Cameron, R. G. and Tabatobai, M., "Predicting the existence of the limit cycles using Walsh functions: some further results", *Int. J. Syst. Sci.*, Vol. 14, No. 9, pp. 1043–1064 (1983).

[26] Caspari, K., "Generalized spectrum analysis", In Proc. *Symp. on Walsh Functions*, Univ. of Maryland, USA, 1970, pp. 195–207 (1970).

[27] Cooley, J. W. and Tuckey, J. W., "An algorithm for the machine calculation of complex Fourier series", *Math. Comput.*, Vol. 19, April, pp. 297–301 (1965).

[28] Chen, W. Z. and Shih, Y. P., "Analysis and optimal control of time-varying linear systems via Walsh functions", *Int. J. Control*, Vol. 27, No. 6, pp. 917–932 (1978).

[29] Cheng, D. K. and Shankar, A. U., "Walsh-function representation and noise analysis of linear sequential circuits", *IEEE Trans. Circ. Syst.*, Vol. 32, No. 3, pp. 274–278 (1985).

[30] D'Alton, L. B., "Rademacher-Walsh diagnosis", *Elec. Eng.* (Gr. Brit.), Vol. 52, No. 643, pp. 83–87 (1980).

[31] Dwolatzky, B., "Intermediate domain system identification using Walsh transforms", *Identif. and Syst. Parameter Estim., 1982 Proc. 6th IFAC Symp.*, Washington, D.C., Vol. 2, Oxford eaa., pp. 1253–1258 (1983).

[32] Edwards, C. R., "The application of the Rademacher-Walsh transform to Boolean function classification and threshold logic synthesis", *IEEE Trans. Comput.*, pp. 48–62 (1975).

[33] Egiazarian, K. O. and Artazjan, R. S., *The Spectral Technique of Boolean Computation with Application in Testing Pattern Recognition*, Preprint, Armenian Academy of sciences, Institute of Problems of Informatics and Automation, Yerevan, p. 25 (in Russian) (1990).

[34] Egiazarian, K. O., "On new spectral approach to pattern recognition", in Proc. *I Int. Conf. Inform. Technologies for Image Analysis and Pattern Recognition (ITIAPR-90)*, Lviv, USSR (1990).

[35] Egiazarian, K. O., "On minimization of fuzzy logic functions by non-orthogonal transforms with application in logical pattern recognition", *Adv. Model. Anal.*, A, 11, No. 3, pp. 57–64 (1992).

[36] Elliott, D. F. and Rao, K. R., *Fast Transforms: Algorithms, Analyses, Applications*, Academic Press, San Diego (1982).

[37] Fine, N. J., "On the Walsh functions", *Trans. Am. Math. Soc.*, Vol. 3, pp. 372–414 (1949).

[38] Fino, B. J. and Algazi, V. R., "Unified matrix treatment of the fast Walsh–Hadamard transform", *IEEE Trans. Comput.*, Vol. C-25, pp. 1142–1146 (1976).

[39] Frick, P. A., "On Walsh noise: its properties and use in dynamic stochastic systems", *IEEE Trans. Sys. Man. Cybern.*, Vol. 9, No. 8, pp. 411–419 (1979).

[40] Gibbs, J. E. and Gebbie, H. A., "The application of Walsh functions to transform-spectroscopy", *Nature*, London, Vol. 224, pp. 1012–1013 (1969).

[41] Good, I. J., "The interaction algorithm and practical Fourier analysis", *J. Royal Stat. Soc.*, London, Vol. B-20, pp. 361–372 (1958).

[42] Gulamhusein, M. N., "Simple matrix theory proof of the dyadic convolution theorem", *Electron. Lett.*, Vol. 9, No. 11, pp. 238–239 (1973).

[43] Gulamhusein, M. N. and Fallside, F., "Short-time spectral and autocorrelation analysis in the Walsh domain", *IEEE Trans. Inform. Theory*, Vol. IT-19, pp. 615–623 (1973).

[44] Harmuth, H. F., "Applications of Walsh functions in communications", *IEEE Spectrum*, Vol. 6, pp. 82–91 (1969).

[45] Harmuth, H. F., *Transmission of Information by Orthogonal Functions*, 2nd ed., Springer-Verlag, Berlin (1972).

[46] Hsiao T. C. and Seth Sharad C., "An analysis of the use of Rademacher–Walsh spectrum in compact testing", *IEEE Trans. Comput.*, Vol. 33, No. 10, pp. 934–937 (1984).

[47] Hurst, S. L., *The Logical Processing of Digital Signals*, Crane, Russak, New York (1978).

[48] Hurst, S. L., Miller, D. M. and Muzio, J. C., *Spectral Techniques in Digital Logic*, Academic Press, San Diego (1985).

[49] Hurst, S. L. and Langheld, E., "Die spektrale darstellung binarer logikfunktionen", *Elektronik*, Vol. 30, No. 13, pp. 61–66 (1981).

[50] Hurst, S. L. and Langheld, E., "Die spektrale darstellung binarer logikfunktionen. 2 teil", *Elektronik*, Vol. 30, No. 14, pp. 69–74 (1981).

[51] Jia, X. and Nixon M. S., "Analysing front view face profiles for face recognition via the Walsh transform", *Pattern Recognition Letters*, Vol. 15, pp. 551–558 (1994).

[52] Johnston, C. P., "The lifting scheme and finite precision error free filter banks", Proc. SPIE 2825, pp. 307–316 (1996).

[53] Kaczmarz, S., *Uber ein Orthogonal System*, Comptes Rendus du I. Congres des Mathematiciens des Pays Slaves, Warsaw, pp. 189–192 (1992).

[54] Karanam, V. R., Frick., P. A. and Mohler, R. R., "Bilinear system identification by Walsh functions", *IEEE Trans. Autom. Contr.*, Vol. AC-23, No. 4, pp. 709–713 (1978).

[55] Karpovsky, M. G., *Finite Orthogonal Series in the Design of Digital Devices*, Wiley, New York (1976).

[56] Kawaji, S. and Toda, R. J., "Walsh series analysis in optimal control systems incorporating observers", *Int. J. Contr.*, Vol. 37, No. 3, pp. 455–462 (1983).

[57] Kendall, W., "A new algorithm for computing correlations", *IEEE Trans. Comput.*, Vol. 23, No. 1, pp. 88–90 (1974).
[58] Kitai, R., "Synthesis of periodic sinusoids from Walsh waves", *IEEE Trans. Instrum. Meas.*, Vol. 24, No. 4, pp. 313–317 (1975).
[59] Kouvaritakis, B. and Cameron, R. G., "The use of Walsh functions in multi-variable limit cycle prediction", *Automat.*, Vol. 19, No. 5, pp. 513–522 (1983).
[60] Kremer, H., "On the representation of Walsh functions and fast Walsh transform algorithms", *Angew. Inf.*, AI-1, pp. 7–20 (1973).
[61] Kumar, S. K. and Breuer, M. A., "Probabilistic aspects of Boolean switching functions via a new transform", *J. A.C.M.*, Vol. 28, pp. 502–520 (1981).
[62] Lechner, R., "A transform theory for functions of binary variables", in *Theory of Switching, Hardvard Computation Lab., Cambridge*, Mass., Progress Rept. BL-30, Sec-X, Nov., pp. 1–37 (1961).
[63] Levitan, B. M., *Theory of Generalized Shift Operators*, Moscow, p. 308 (in Russian) (1973).
[64] Lopresti, Ph. V., "A fast algorithm for the estimation of autocorrelation functions", *IEEE Trans. Acoust., Speech Sig. Process.*, Vol. ASSP-22, No. 6, pp. 449–453 (1974).
[65] Malugin, V. D., Kukharev, G. A., and Shmerko, V. P., *Transformation of the Polynomial Forms of Boolean Functions*, Problems of Control Institute, Preprint, Moscow (in Russian) (1986).
[66] Manley, H. J., Mattson, H. F., and Schatz, J. R., "Some applications of Good's theorem", *IEEE Trans. Inf. Theory*, Vol. 26, No. 4, pp. 475–476 (1980).
[67] Matevosyan, A. K., "The inverse problem of Karhunen–Loeve", *Cybern. Sys. Res.*, R. Trappl ed., No. 2, pp. 143–147 (1984).
[68] Moraga, C., "Ternary spectral logic", in Proc. *IEEE VII Intern. Multiple-Valued Logic Symp.*, pp. 7–12 (1977).
[69] Morgenstern, J., "Note on a lower bound of the fast Fourier transform", *J. Assoc. Comput. Math.*, Vol. 20, No. 2, pp. 305–306 (1973).
[70] Oh, S. Y., "A Walsh-Hadamard based distributed storage device for the associative search of information", *IEEE Trans. Patt. Anal. Mach. Intel.*, Vol. 6, No. 5, pp. 615–623 (1984).
[71] Palanisamy, K. R. and Arunachalam, V. P., "Analysis of bilinear systems via single-term Walsh series", *Int. J. Contr.*, Vol. 41, No. 2, pp. 541–547 (1986).
[72] Paley, R., "A remarkable system of orthogonal functions", *Proc. London Math. Soc.*, Vol. 34, pp. 241–279 (1932).
[73] Pichler, F. R., Technical Research Report R-70-11. Department of Electrical Engineering, University of Maryland, College Park, MD 20742, USA, August (1970).
[74] Rao, K. R., Devarajan, V., Vlasenko, V., and Narasimhan, M. A., "Cal-Sal Walsh-Hadamard Transform", *IEEE Trans. Acoust., Speech, Sig. Process.*, Vol. ASSP-26, No. 6, December, pp. 605–607 (1978).
[75] Rademacher, H., "Einige Satze von Allgemainen orthogonal funktionen", *Math. Ann.*, 87, pp. 122–138 (1922).

[76] Robinson, G. S., "Logical convolution and discrete Walsh and Fourier spectra", *IEEE Trans. Audio Electroacoust.*, Vol. AU-20, No. 4, October, pp. 271–280 (1972).

[77] Rubio Ayuso, A. J., "A note on a new Walsh transformation method for delta modulated functions", *Sig. Process.*, Vol. 3, No. 3, pp. 272–275 (1981).

[78] Sinha, M. S. P., Rajamani, V. S., and Sinha, A. K., "Identification of nonlinear distributed system using Walsh functions", *Int. J. Contr.*, Vol. 32, No. 4, pp. 669–676 (1980).

[79] *Spectral Techniques and Fault Detection*. Ed. M. Karpovsky, Academic Press, San Diego (1985).

[80] Stone, H. S., "Parallel processing with the perfect shuffle", *IEEE Trans. Comput.*, C-20, pp. 153–161 (1971).

[81] Swanson, M. D. and Tewfik, A. H., "A binary wavelet decomposition of binary images", *IEEE Trans. Imag. Process.*, Vol. 5, No. 12, pp. 1637–1650 (1996).

[82] Sweldens, W., "The lifting scheme: A custom-design construction of biorthogonal wavelets", *Technical report IMI* 1994:7, revised version (1994).

[83] *Theory and Applications of Spectral Techniques*. Third Intern. Workshop on Spectral Techniques, Dortmund, FRG, p. 168 (1988).

[84] Vetterli, M. and Kovačević, J., *Wavelets and Subband Coding*, Prentice Hall (1995).

[85] Walsh, J. L., "A closed set of orthogonal functions", *Am. J. Math.*, Vol. 55, pp. 5–24 (1923).

[86] Wiener, N. and Paley, E. A. C., "Analytic properties of the characters of infinite Abelian groups", *Verhandlungen des Internationalen Matematiker Kongress*, Zurich, II band, p. 95 (1932).

[87] Yablonski, S. V., *Introduction to Discrete Mathematics*, Moscow, Nauka (in Russian) (1986).

[88] Yuen, C., "Upper bounds of Walsh transforms", *IEEE Trans. Comput.*, Vol. C-21, pp. 1273–1280 (1972).

[89] Yuen, C. K., "An algorithm for computing the correlation functions of Walsh functions", *IEEE Trans.*, Vol. EMC-17, pp. 177–180 (1975).

CHAPTER 10

APPLICATIONS OF BINARY POLYNOMIAL TRANSFORMS

Karen O. Egiazarian
Jaakko T. Astola
Tampere University of Technology
Tampere, Finland

Sos T. Agaian
Univ. of Texas/Austin
San Antonio, Texas

Ruşen Öktem
Tampere University of Technology
Tampere, Finland

10.1 BINARY POLYNOMIAL TRANSFORMS IN NONLINEAR FILTERING

10.1.1 INTRODUCTION

Many digital signal processing applications cannot be satisfactorily handled using only linear methods. For example, if we are restoring an image corrupted by impulsive noise, a simple nonlinear filter (such as a median filter) may perform better than the optimal linear filter [67].

Stack filters are nonlinear digital filters which have been developed as an alternative to linear filters [99]. They are based on threshold decomposition and the stacking property. The output of a stack filter is obtained by decomposing the input signal into a set of binary signals, applying a positive Boolean function to each binary signal, and then summing up the results. The so-called stacking property [99] is satisfied when the applied Boolean function is positive.

Many nonlinear filters based on order statistics can be expressed as stack filters. The removal of the stacking property in the definition of a stack filter, i.e. allowing an arbitrary Boolean function instead of a positive Boolean function, enlarges the class of stack filters to Boolean filters [15] to so-called threshold Boolean filters [73]. Applications of threshold Boolean filters can be found in many areas, for instance, in the field of digital image processing and modern television systems, where median-type filters are used for noise suppression, image sequence coding, cross-effect elimination, and scan-rate conversion. These kinds of filters are important because of some excellent properties which cannot be achieved using traditional linear filters, especially in cases where the signal and the noise overlap in the frequency domain.

The development of fast methods for implementing threshold Boolean filters is very important because of the large amount of data involved in filtering and the necessity of real-time processing. Efficient designs for threshold Boolean filtering have been suggested in [73, 35].

Because median-type filters are nonlinear, it is very difficult to derive general results that would accurately describe the statistical behavior for a wide range of random signals as in the case for linear filters. Examples of characterizations that are useful in practice and possible to compute are white-noise attenuation and response to a noisy step or edge signal. Because the effects of a nonlinear filter on the noise and on the signal cannot be separated as with linear filters, a seemingly easy task, such as the analysis of filter response to a noisy edge, may be almost impossible. However, theoretical work and extensive simulations have provided a fairly good understanding of the statistical behavior of median-type filters.

The statistical properties of order statistics and generalizations were studied in the early 1930s by Pearson (1931) and more intensively in the 1960s to 1970s by Sarhan and Greenberg [90], David [35], and others. A significant part of these works concern distributions (including asymptotic distributions) of order statistics. The distribution function of an order statistic depends on the distribution function of input random variables, rank i of the statistic and sample size N.

For the description of a set of order statistics it is not sufficient to know only the marginal distribution function of each order statistic. Complete information about the statistical properties of order statistics is contained in the joint distribution.

Since stack filters are generalizations of order statistics filters, their statistical properties exhibit characteristics similar to those of order statistics. The statistical properties of stack filters have been studied in Justusson [67], Mallows [76], Yli-Harja [109].

Selection probabilities are recently introduced concepts for stack filters [86]. The output distribution functions for the case of independent and identically distributed (i.i.d.) inputs follow easily if rank selection probabilities are known. Sample selection probabilities in turn give information about the use of temporal-order information on stack filters (see [86, 71]). This temporal order information is valuable when the detail preservation properties of stack filters are evaluated. Sample selection probabilities also equal to the impulse response coefficients of the FIR filter, whose output spectrum is closest of all linear filters, to that of the stack filter for i.i.d. Gaussian inputs [76].

The selection probabilities can be found if so-called selection probability sets are known. They can be formed directly by listing all possible $N!$ permutations of N variables, where N denotes the size of the moving window. This is very time consuming and practically impossible for large windows. Thus there is a need for finding faster algorithms.

10.1.2 STACK FILTERS, THRESHOLD BOOLEAN FILTERS, AND EXTENDED THRESHOLD BOOLEAN FILTERS

Let $X = \{X(k),\ k = 1, 2, \ldots, L\}$ be an $(R+1)$-valued input signal ($X(k) \in \{0, 1, \ldots, R\}$) and let a window of size $M = U + V + 1$ slide across the signal $X(k)$ (with appended points $X(-t) = X(1),\ X(L+r) = X(L),\ t = 0, \ldots, U-1;\ r = 1, \ldots, V$) involving at time instant k the following vector

$$\mathbf{X}(k) = \left[X(k-U), \ldots, X(k), \ldots, X(k+V)\right] = \left[X_1^{(k)}, \ldots, X_M^{(k)}\right]. \tag{10–1}$$

By the operation of threshold decomposition, this vector can be represented as the sum of R binary signals:

$$\mathbf{X}(k) = \sum_{n=1}^{R} \boldsymbol{\sigma}_n\left(\mathbf{X}(k)\right), \quad k = 1, 2, \ldots, L, \tag{10–2}$$

where $\boldsymbol{\sigma}_n(\mathbf{X}(k)) = (\sigma_n(X(k-U)), \ldots, \sigma_n(X(k+V)))$

$$\sigma_n(X) = \begin{cases} 1 & \text{if } X \geqslant n, \\ 0 & \text{otherwise.} \end{cases}$$

DEFINITION 10.1.1 [15, 73]. The threshold filter (TF) $S_f(X)$ based on the M-variable discrete function $f(\cdot) : \{0, 1\}^M \to \Re$ maps the signal X into the output signal $Y = \{Y(k),\ k = 1, 2, \ldots, L\}$, where

$$Y(k) = S_f\left(\mathbf{X}(k)\right) = \sum_{n=1}^{R} f\left(\boldsymbol{\sigma}_n(\mathbf{X}(k))\right). \tag{10–3}$$

If $f(\cdot)$ is a Boolean function (BF) [i.e. $f(\cdot) : \{0, 1\}^M \to \{0, 1\}$], the threshold filter is called a threshold Boolean filter (TBF) [15, 73], and if $f(\cdot)$ is a positive Boolean function (PBF), the threshold filter is called a stack filter [99].

We can change the function f while the filter's window is sliding across the input signal. In this case instead of Eq. 10–3 we will have

$$Y(k) = S_{f^{(k)}}\big(\mathbf{X}(k)\big) = \sum_{n=1}^{R} f^{(k)}\big(\sigma_n(\mathbf{X}(k))\big). \tag{10–4}$$

Depending on the statistical properties of the input signal, changing the function f can be done with varying frequency.

Let $\mathbf{f}$ be the row vector of length 2^M containing the values $f(0,\ldots,0),\ldots,$ $f(1,\ldots,1)$ indexed by the binary representation of integers $0, 1, \ldots, 2^M - 1$.

Denote $\mathbf{x}_j = \sigma_j(\mathbf{X}),\ j = 1,\ldots,M$. Define an indicator

$$\chi(\mathbf{x}_j) = \big[\chi_0(\mathbf{x}_j),\ldots,\chi_{2^M-1}(\mathbf{x}_j)\big]; \quad j = 1,\ldots,M, \tag{10–5}$$

where

$$\chi_k(\mathbf{x}_j) = \begin{cases} 1, & \text{if } (k_{M-1},\ldots,k_0) = \mathbf{x}_j, \\ 0, & \text{otherwise,} \end{cases} \tag{10–6}$$

$j = 1,\ldots,M;\ (k_{M-1},\ldots,k_0)$ is the binary vector representation of the positive integer $k,\ k = 0,\ldots,2^M - 1$.

Denote by $\mathbf{h}$ the histogram vector of $\mathbf{x}_j,\ j = 1,\ldots,R$, i.e.

$$\mathbf{h} = \sum_{j=1}^{R} \chi(\mathbf{x}_j). \tag{10–7}$$

The matrix analogue of formula 10–3 is the following (index n is omitted)

$$y = \mathbf{f}\cdot\mathbf{h}^T, \tag{10–8}$$

where $\mathbf{h}^T$ is the transposed vector of the histogram vector $\mathbf{h}$ (10–7).

It is known [73] that any TBF can be expressed as a linear combination of stack filters:

$$S_f(X) = \sum_{i=1}^{Q} (-1)^{i-1} S_{f_i}(X), \tag{10–9}$$

where $1 \leqslant Q \leqslant M+1$, and each $f_i(\cdot)$ is a positive Boolean function. Functions f_i, $i=1,2,\ldots,Q$, are defined recursively by the following rule:

$$f_0'(\mathbf{x}) = f(\mathbf{x}), \quad V(f_i) = V\big(f_{i-1}'\big) \cup C\big(f_{i-1}'\big), \quad i=1,2,\ldots,Q, \tag{10–10}$$

where $V(f)=\{\mathbf{x}\in\{0,1\}^M \mid f(\mathbf{x})=1\}$ is the set of true vectors of the function f, $C(f)=\{\mathbf{y}\in\{0,1\}^M \mid \mathbf{y}\notin V(f)\ \&\ \exists\mathbf{x}<\mathbf{y},\ \mathbf{x}\in V(f)\}$, and f_i', $i=1,2,\ldots,Q$, such a Boolean function that $V(f_i')=C(f_{i-1}')$.

Expression 10–9 allows us to reduce the computation of the output of the threshold Boolean filter to the computation of the outputs of at most $M+1$ stack filters. This reduction can be efficient if simple procedures for evaluation of positive Boolean functions $f_i(\cdot)$ and for stack filtering are available.

Threshold Boolean Filters via Stack Filters

A procedure for evaluation of positive Boolean functions $f_i(\cdot)$, directly based on expression 10–10, is complicated in the sense of computational complexity. For finding each function f_i, $i=1,2,\ldots,Q$, this procedure requires forming sets $V(f_i)$ and constructing sets $C(f_{i-1}')$. In the worst case the cardinalities of these sets are of the order 2^M vectors of length M. For finding each set $C(f_{i-1}')$, bit-by-bit comparisons of every 2^M vector with all vectors from $V(f_{i-1}')$ are required. So in the worst case, the procedure for finding Boolean functions f_i, directly based on expression 10–10 require $T=O(M^2 2^{2M})$ operations and the area $A=M^2 2^M$. Below we present a spectral method for finding functions f_i, $i=1,2,\ldots,Q$, which requires $O(M^2 2^M)$ operations and the area $A=O(2^M)$. This method allows us to find both truth tables and minimal disjunctive normal form (MDNF) representations of functions f_i from Eq. 10–9 while the direct method allows us to find truth tables only. The method is highly parallelizable and is very suitable for VLSI implementation [12, 13].

For a column vector $\mathbf{h}=(h^{(0)},\ldots,h^{(2^M-1)})^T$ define the operator

$$\Delta(\mathbf{h}) = \big(\Delta(h^{(0)}),\ldots,\Delta(h^{(2^M-1)})\big)^T,$$

where

$$\Delta(h) = \begin{cases} 0 & \text{if } h=0, \\ 1 & \text{otherwise.} \end{cases}$$

THEOREM 10.1.1. *Let* $\mathbf{f}_i$, $i=1,2,\ldots,Q$, *be the truth tables of the PBFs defined by Eq.* 10–10. *Then*

$$\mathbf{f}_i = \Delta\big(\mathbf{K}_M^{(w)}\cdot\mathbf{f}_{i-1}'\big), \quad i=1,2,\ldots,Q, \tag{10–11}$$

where $\mathbf{f}'_0 = \mathbf{f}$, $\mathbf{f}'_{i-1} = \mathbf{f}_{i-1} - \mathbf{f}'_{i-2}$, $i = 2, \ldots, Q$. *The minimal disjunctive normal form of* f_i *is*

$$f_i(x_1, x_2, \ldots, x_M) = \bigvee_{t=1}^{T_i} x_1^{\delta_1^{(t,i)}} x_2^{\delta_2^{(t,i)}} \cdots x_M^{\delta_M^{(t,i)}}, \tag{10–12}$$

where $\delta_m^{(t,i)} \in \{0, 1\}$, *and* $\mathbf{d} = \mathbf{d}(t, i) = (\delta_1^{(t,i)}, \ldots, \delta_M^{(t,i)})$, $t = 1, 2, \ldots, T_i$, *are binary codes of all indices* d, *such that the* d*th component of the spectrum* $\mathbf{K}_M^{(w)} \cdot \mathbf{f}'_{i-1}$ *is equal to one.*

PROOF. First we prove that Eq. 10–11 defines the truth tables of the functions f_i, $i = 1, 2, \ldots, Q$. Let f'_{i-1} be the input Boolean function at the ith $i = 1, \ldots, Q$ iteration. Let

$$\mathbf{h} = \left(h_i^{(0)}, \ldots, h_i^{(2^M-1)}\right)^T = \mathbf{K}_M^{(w)} \cdot \mathbf{f}'_{i-1}.$$

If

$$h_i^{(\sigma)} = \sum_{\gamma=0}^{2^M-1} \left(K_M^{(w)}\right)_{\sigma,\gamma} \cdot f_{i-1}^{\prime(\gamma)} \geqslant 1$$

then, according to Proposition 2.1 in [46], there exists a number γ such that $\boldsymbol{\gamma} \leqslant \boldsymbol{\sigma}$ and $f_{i-1}^{\prime(\gamma)} = 1$, where $\boldsymbol{\sigma} = (\sigma_0, \sigma_1, \ldots, \sigma_{M-1})$ and $\boldsymbol{\gamma} = (\gamma_0, \gamma_1, \ldots, \gamma_{M-1})$ are binary codes of σ and γ, respectively. So, $\boldsymbol{\sigma} \in V(f'_{i-1}) \cup C(f'_{i-1})$ and it is a true vector of the function f_i defined by Eq. 10–10. If $h_i^{(\sigma)} = 0$, then $\sigma \notin V(f'_{i-1}) \cup C(f'_{i-1})$ and it is not a true vector of the function f_i. Thus $\Delta(h_i^{(\sigma)}) = 1$ if and only if σ is a true vector of the function f_i, i.e. $\mathbf{f}_i$, $i = 1, 2, \ldots, Q$, defined by Eq. 10–11 are the truth tables of the functions f_i defined by Eq. 10–10. $\mathbf{f}'_i$, $i = 1, \ldots, Q$, are the truth tables of the functions f'_i from Eq. 10–10 because $\mathbf{f}'_i = \mathbf{f}_i - \mathbf{f}'_{i-1}$ implies that $V(f'_i) = V(f_i) \setminus V(f'_{i-1}) = C(f'_{i-1})$, where the last equality follows from $V(f_i) = V(f'_{i-1}) \cup C(f'_{i-1})$ and $V(f'_{i-1}) \cap C(f'_{i-1}) = \emptyset$.

Now we prove the second part of the theorem. Let us consider the perfect DNF of the function f_i

$$f_i(x_1, x_2, \ldots, x_M) = \bigvee_{r=1}^{R_i} x_1^{\sigma_1^{(t,i)}} x_2^{\sigma_2^{(t,i)}} \cdots x_M^{\sigma_M^{(t,i)}},$$

which is formed by all true vectors $\mathbf{s}(r, i) = (\sigma_1^{(t,i)}, \sigma_2^{(t,i)}, \ldots, \sigma_M^{(t,i)})$ of the function f_i, i.e. by all numbers $s(r, i)$ such that $h_i^{(s(r,i))} \geqslant 1$. From Proposition 2.1 in [46] one can see that for each such a number there exists a number $d(t, i)$, $t = 1, 2, \ldots, T_i$, such that $\mathbf{d}(t, i) = (\delta_1^{(t,i)}, \ldots, \delta_M^{(t,i)}) < \mathbf{s}(r, i)$ and

$h_i^{(d(t,i))} = 1$. This means that for f_i, a DNF can be formed only by the vectors $\mathbf{d}(t,i)$, $t = 1, 2, \ldots, T_i$. According to Proposition 2.1 in [46] all vectors $\mathbf{d}(t,i)$, $t = 1, 2, \ldots, T_i$, are not comparable, i.e. $\forall t \in \{1, 2, \ldots, T_i\}$ $\nexists q \in \{1, 2, \ldots, T_i\}$ such that $\mathbf{d}(t,i) < \mathbf{d}(q,i)$. This is true because otherwise we will have $h_i^{(d(q,i))} \geqslant 1 + 2^{w(d(q,i),d(t,i))} > 1$ where w is the Hamming distance. Thus the DNF which is formed by the vectors $\mathbf{d}(t,i)$, $t = 1, 2, \ldots, T_i$. is the minimal DNF of f_i. The theorem is proved. ■

Theorem 10.1.1 gives the following:

ALGORITHM 10.1.1. This represents a Boolean function as a linear combination of PBFs.

Input. Truth table $\mathbf{f}$ of a Boolean function f.

Output. Truth tables $\mathbf{f}_i$ and MDNFs of PBFs f_i, $i = 1, \ldots, Q$, such that $f(X) = \sum_{i=1}^{Q} (-1)^{i-1} f_i(X)$.

Method.

Step 1. **Set** $i = 1$, $\vec{f}_0' = \vec{f}$.

Step 2. **Compute** $\mathbf{h}_i = \mathbf{K}_M^{(w)} \cdot \mathbf{f}_{i-1}' = (h_i^{(0)}, \ldots, h_i^{(2^M-1)})^T$.

Step 3. **Form** the MDNF 10–12 of the PBF f_i by forming the set of M-tuples $\mathbf{d} = \mathbf{d}(t,i) = (\delta_1^{(t,i)}, \ldots, \delta_M^{(t,i)})$, $t = 1, 2, \ldots, T_i$, for which $h_i(d) = 1$.

Step 4. **Set** $\mathbf{f}_i = \Delta(\mathbf{h}_i) = (\Delta(h_i^{(0)}), \ldots, \Delta(h_i^{(2^M-1)})^T$.

Step 5. **Set** $\mathbf{f}_i' = \mathbf{f}_i - \mathbf{f}_{i-1}'$, $i = i + 1$.

Step 6. **if** $\mathbf{f}_i' \equiv \mathbf{0}$ **then stop, else go to Step 2.**

EXAMPLE 10.1.1. Let $f(x_1, x_2, x_3) = x_1\overline{x}_2 + \overline{x}_2 + x_1 x_3$ (as in the example in [73]). Then the truth table $\mathbf{f} = (0, 1, 0, 0, 1, 1, 0, 1)^T$. Applying a Hamming-weighted conjunctive transform to $\vec{f}$ we find $\mathbf{h}_1 = (0, 1, 0, 2, 1, 5, 2, 11)^T$ where $h_1^{(d)} = 1$ only for $d = 1 = 001$ and $d = 4 = 100$. So we find the MDNF $f_1 = x_1 \vee x_3$. Applying operator Δ to $\vec{h}_1$ we find the truth table $\mathbf{f}_1 = (0, 1, 0, 1, 1, 1, 1, 1)^T$ and then the truth table $\mathbf{f}_1' = (0, 0, 0, 1, 0, 0, 1, 0)^T$. Applying the same procedure to $\vec{f}_1'$ we find $\mathbf{h}_2 = (0, 0, 0, 1, 0, 0, 1, 4)^T$, $f_2 = x_1 x_2 \vee x_2 x_3$, $\mathbf{f}_2 = (0, 0, 0, 1, 0, 0, 1, 1)^T$ and $\mathbf{f}_2' = (0, 0, 0, 0, 0, 0, 0, 1)^T$. Finally, applying the procedure to f_2' we find $\mathbf{h}_3 = (0, 0, 0, 0, 0, 0, 0, 1)^T$, $f_3 = x_1 x_2 x_3$, $\mathbf{f}_3 = (0, 0, 0, 0, 0, 0,$

$0, 1)^T$ and $\mathbf{f}'_2 = (0, 0, 0, 0, 0, 0, 0, 0)^T$. Thus we find $f(x) = x_1 \vee x_3 - x_1x_2 \vee x_2x_3 + x_1x_2x_3$.

Threshold Filters Using a Rademacher-Sorting Network

As we have seen before, any threshold filter can be realized as a linear combination of multilevel stack filters following from the realization of the threshold Boolean filter. The Hamming-weighted conjunctive transform can be efficiently used in this case.

Another realization of a threshold filter follows simply by modifying Eq. 10–8 to the form

$$y = \mathbf{h} \cdot \mathbf{H} \cdot \mathbf{g}^T, \tag{10–13}$$

with

$$\mathbf{g}^T = \mathbf{H}^{-1} \cdot \mathbf{f}, \tag{10–14}$$

where $\mathbf{H}$ is any nonsingular matrix. In fact, the realization of the threshold filter in [65] follows directly from Eqs. 10–13 and 10–14 when $\mathbf{H} = \mathbf{K}$ is the conjunctive (Reed–Muller) matrix.

Here we present a much simpler realization of this filter using a Rademacher-sorting network without computation of the threshold histogram $\mathbf{h}$. The flowchart structure ofa Rademacher-sorting network is similar to the flowchart of the fast Rademacher transform, with the change of addition operations to minimum operations. In a similar way, the FFT flowchart is changed to the FFT-sorting flowchart in [44]. In Fig. 10–1 an example of such a flowchart of the Rademacher-sorting network for the input vector $\mathbf{z} = [5, 1, 2]$ is presented. The histogram vector $\mathbf{h}$ for this example will be $\mathbf{h} = [0, 0, 0, 0, 3, 1, 0, 1]$, and $\mathbf{h} \cdot \mathbf{K} = [5, 2, 1, 1, 5, 2, 1, 1]$. As we see from Fig. 10–1 the Rademacher-sorting network gives the elements of $\mathbf{h} \cdot \mathbf{K}$ [the first element is not important for the filter output since $g(\mathbf{0}) = f(\mathbf{0}) = 0$]. Note that in this case we do not need to construct the histogram vector.

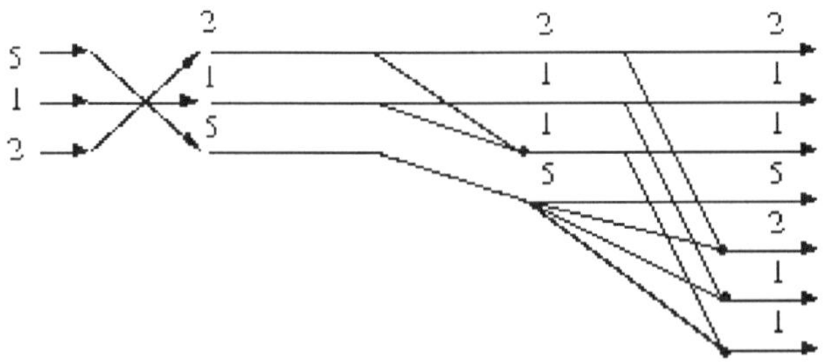

Figure 10–1. The Rademacher-sorting network of order 3.

An advantage of the computation of a threshold filter in the transform domain using Eqs. 10–13 and 10–14 compared with Eq. 10–8 will be in the case when **g** [formed from Eq. 10–14] has a sparse form for a given function **f**. In this case the problem of a spectral analysis of Boolean functions becomes important [43].

10.1.3 JOINT DISTRIBUTIONS OF STACK FILTERS

Notations and Definitions

Let $X(i),\ i \in \mathcal{Z}$, be a real-valued input signal, and let a window of width $N = 2M + 1$ slide across the signal. At time instant n we have the following vector

$$\begin{aligned}\mathbf{X}(n) &= [X(n-M), \ldots, X(n), \ldots, X(n+M)] \\ &= [X_1(n), X_2(n), \ldots, X_N(n)] \end{aligned} \qquad (10\text{–}15)$$

within the window, where $X_j(n) = X(n + j - M - 1),\ j = 1, 2, \ldots, N.$

The positive *logic function (LF)* is defined (by analogy with positive Boolean functions, the positive logic function is a logic function which contains no complements of input variables) by

$$F(X) = \max\big(\min(\Pi_1), \min(\Pi_2), \ldots, \min(\Pi_k)\big), \qquad (10\text{–}16)$$

where

$$\min(\Pi_p) = \min(X_{j(p,1)}, X_{j(p,2)}, \ldots, X_{j(p,r_p)}), \qquad (10\text{–}17)$$

$j(p, q)$ are the indices of the variables in Π_p in increasing order and r_p is the number of variables in Π_p.

The definition of a continuous stack filter is given in [106],

$$Y_n = S_F(\mathbf{X}(n)) = \max\big\{b \mid f\big(\sigma(\mathbf{X}(n), b)\big) = 1\big\}, \qquad (10\text{–}18)$$

where $f(x)$ is a positive Boolean function, obtained from a positive LF $F(X)$ by changing the operations of max, min to the binary $\vee,\ \wedge$, respectively, $\boldsymbol{\sigma}(\mathbf{X}(n), b) = (\sigma(X_1(n), b),\ \sigma(X_2(n), b), \ldots, \sigma(X_N(n), b))$, and $\sigma(X_j(n), b)$ is the following *threshold operator*:

$$\sigma(X_j(n), b) = \begin{cases} 1, & \text{if } X_j(n) \geqslant b, \\ 0, & \text{otherwise,} \end{cases} \qquad j = 1, 2, \ldots, N. \qquad (10\text{–}19)$$

There is also a "dual" definition of a stack filter:

$$Y(n) = S_F(\mathbf{X}(n)) = \min\big\{a \mid f\big(\eta(\mathbf{X}(n), a)\big) = 0\big\}, \qquad (10\text{–}20)$$

where $\boldsymbol{\eta}(\mathbf{X}(n), a) = (\eta(X_1(n), a), \eta(X_2(n), a), \ldots, \eta(X_N(n), a))$, and $\eta(X_j(n), a)$ is the following threshold operator:

$$\eta(X_j(n), a) = \begin{cases} 1, & \text{if } X_j(n) > a, \\ 0, & \text{otherwise}, \end{cases} \qquad j = 1, 2, \ldots, N. \tag{10–21}$$

Later when we consider SF on the fixed time instant n, we will omit n in $Y(n)$ and $X(n)$.

Let $F^1(X), F^2(X), \ldots, F^L(X)$ be logic functions with common independently distributed arguments $X_1, X_2, \ldots, X_N$, and $f^1(x), f^2(x), \ldots, f^L(x)$ the corresponding positive Boolean functions with corresponding Boolean arguments $x_1, \ldots, x_N$. Let Φ_i and Υ_i be distribution functions of inputs:

$$\Phi_i(t) = P(\{X_i \leqslant t\}), \quad i = 1, 2, \ldots, N, \tag{10–22}$$

and

$$\Upsilon_i(t) = P(\{X_i < t\}), \quad i = 1, 2, \ldots, N, \tag{10–23}$$

where t is a real number.

Let us write for $j = 1, 2, \ldots, L$ the following sets of Boolean vectors:

$$A^{(j)} = \left\{\boldsymbol{\alpha} = (\alpha_1, \ldots, \alpha_N) \in \{0, 1\}^N \mid f^j(\boldsymbol{\alpha}) = 0\right\} = \left(f^j\right)^{-1}(0) \tag{10–24}$$

and

$$B^{(j)} = \left\{\boldsymbol{\beta} = (\beta_1, \ldots, \beta_N) \in \{0, 1\}^N \mid f^j(\boldsymbol{\beta}) = 1\right\} = \left(f^j\right)^{-1}(1). \tag{10–25}$$

Elementary Symmetric Boolean Matrix and Transform

DEFINITION 10.1.2. The elementary symmetric Boolean matrix $\mathbf{S}_N$ is the rectangular $(N+1) \times 2^N$ matrix, defined by the following recursion:

$$\mathbf{S}_N = \begin{pmatrix} \mathbf{o}_{N-1} & \mathbf{S}_{N-1} \\ \mathbf{S}_{N-1} & \mathbf{o}_{N-1} \end{pmatrix}, \qquad \mathbf{S}_1 = \begin{pmatrix} 0 & 1 \\ 1 & 0 \end{pmatrix}, \tag{10–26}$$

where $\mathbf{o}_{N-1}$ is the row vector of zeroes of the length 2^{N-1}.

As an example

$$\mathbf{S}_2 = \begin{pmatrix} 0 & 0 & 0 & 1 \\ 0 & 1 & 1 & 0 \\ 1 & 0 & 0 & 0 \end{pmatrix}, \qquad \mathbf{S}_3 = \begin{pmatrix} 0 & 0 & 0 & 0 & 0 & 0 & 0 & 1 \\ 0 & 0 & 0 & 1 & 0 & 1 & 1 & 0 \\ 0 & 1 & 1 & 0 & 1 & 0 & 0 & 0 \\ 1 & 0 & 0 & 0 & 0 & 0 & 0 & 0 \end{pmatrix}. \tag{10–27}$$

The name *elementary symmetric Boolean matrix* $\mathbf{S}_N$ is connected with the fact that the $(i+1)$th row of this matrix is the truth table of the $(N-i)$th elementary symmetric Boolean function, in other words, the Boolean function $g^i(\mathbf{x})$, $i = 0, 1, \ldots, N$, such that

$$g^i(\mathbf{x}) = \begin{cases} 1, & \text{if } w_H(\mathbf{x}) = N - i, \\ 0, & \text{otherwise,} \end{cases} \qquad i = 0, 1, \ldots, N. \tag{10–28}$$

DEFINITION 10.1.3. Let $\mathbf{f}$ be a column vector of length 2^N. Then we call the column vector $\mathbf{S}_N \cdot \mathbf{f}$, the elementary symmetric Boolean transform of the vector $\mathbf{f}$.

DEFINITION 10.1.4. Let $\mathbf{\Omega}$ be a square matrix of order 2^N. Then we call the $(N+1) \times (N+1)$-matrix $\mathbf{S}_N \cdot \mathbf{\Omega} \cdot \mathbf{S}_N^T$ the two-dimensional elementary symmetric Boolean transform of $\mathbf{\Omega}$, where $\mathbf{S}_N^T$ is the transpose $\mathbf{S}_N$.

Distribution Function of a Stack Filter

We investigate the random variable defined by the output of a continuous stack filter where the inputs are real-valued random variables. We consider the positive Boolean function $f(\mathbf{x}) = f(x_1, \ldots, x_N)$. Let the inputs $X_1, \ldots, X_N$ be i.i.d. random variables with a common distribution function $\Phi(t)$.

The distribution function $\Psi(t)$ of Y can be written as

$$\Psi(t) = \sum_{i=0}^{N} A_i \big(1 - \Phi(t)\big)^i \Phi^{N-i}(t), \tag{10–29}$$

where the numbers A_i are defined by

$$A_i = \left|\{\mathbf{x} \in \{0,1\}^N \mid f(\mathbf{x}) = 0,\ w_H(\mathbf{x}) = i\}\right|, \tag{10–30}$$

where $|S|$ means the cardinality of the set S and $w_H(\mathbf{x})$ denotes the Hamming weight (i.e. the number of 1s in $\mathbf{x}$). To see this, note that

$$\Psi(t) = P\{Y \leqslant t\}$$

and the input space can be divided into 2^N mutually exclusive events of the form

$$G_1 \times \cdots \times G_N, \tag{10–31}$$

where G_j are $(-\infty, t]$ or (t, ∞), $j = 1, \ldots, N$. A typical event having i terms of type (t, ∞) and $N - i$ terms of type $(-\infty, t]$ has the probability

$$\left(1 - \Phi(t)\right)^i \Phi^{N-i}(t).$$

The event $\{Y \leqslant t\}$ is the union of exactly those events in Eq. 10–31 whose terms of type (t, ∞) match with 1's in some $\mathbf{x} \in \{0, 1\}^N$ with $f(\mathbf{x}) = 0$. As the events in Eq. 10–31 are mutually exclusive, we can write the probability of the event $\{Y \leqslant t\}$ as the sum in Eq. 10–29.

The spectral procedure for computing an output distribution of a stack filter is based on applying the transform by the elementary symmetric Boolean matrix $\mathbf{S}_N$ in order to get coefficients A_i from Eq. 10–30, and is given in the following:

ALGORITHM 10.1.2.

1. Form a column vector $\mathbf{f}$ of the length 2^N, the truth table of the given positive Boolean function.

2. Now the column vector $\mathbf{a} = (A_0, A_1, \ldots, A_N)^T$ of the coefficients defined in Eq. 10–30 is obtained as

$$\mathbf{a} = \mathbf{S}_N \cdot \mathbf{f}.$$

3. Calculate $\Psi(t)$ by Eq. 10–29.

EXAMPLE 10.1.2. To illustrate the spectral method, we compute the output distribution of the stack filter based on the positive Boolean function

$$f(x) = x_2x_3 \vee x_3x_4 \vee x_1x_2x_4 \vee x_1x_3x_5 \vee x_2x_4x_5.$$

1. Form the truth table of the Boolean function:

$$\mathbf{f} = (0, 0, 0, 0, 0, 0, 1, 1, 0, 0, 0, 1, 1, 1, 1, 1, 0, 0, 0, 0, 0, 1, 1, 1, 0, 0, 1, 1, 1, 1, 1, 1)^T.$$

2. The vector $\mathbf{a}$ has the form

$$\mathbf{a} = \mathbf{S}_N \cdot \mathbf{f} = (0, 0, 2, 8, 5, 1).$$

Joint Distribution of L Stack Filters

In this section we find a formula for the joint cumulative distribution function $\Psi(t_1, t_2, \ldots, t_L)$, of L stack filters $F^1, F^2, \ldots, F^L$ of independently distributed input values, namely

$$\Psi(t_1, t_2, \ldots, t_L) = P\big(\{F^1 \leqslant t_1, F^2 \leqslant t_2, \ldots, F^L \leqslant t_L\}\big). \tag{10–32}$$

Ordering the set $\{t_1, t_2, \ldots, t_L\}$ [of the bounds from Eq. 10–32] into nondecreasing order gives $t_{(1)} \leqslant t_{(2)} \leqslant \cdots \leqslant t_{(L)}$.

Let $t_{(j)} = t_s,\ j, s = 1, 2, \ldots, L$, and $f^{(j)}(\boldsymbol{\alpha}) = f^s(\boldsymbol{\alpha})$.

Let us define the following set

$$\Gamma = \{\boldsymbol{\gamma} = \boldsymbol{\alpha}^1 + \boldsymbol{\alpha}^2 + \cdots + \boldsymbol{\alpha}^L \mid \boldsymbol{\alpha}^{(1)} \geqslant \boldsymbol{\alpha}^{(2)} \geqslant \cdots \geqslant \boldsymbol{\alpha}^{(L)}\}, \tag{10–33}$$

of N-tuples $\boldsymbol{\gamma} = (\gamma_1, \gamma_2, \ldots, \gamma_N) \in \Gamma,\ 0 \leqslant \gamma_i \leqslant L,\ i = 1, 2, \ldots, N$, where $\boldsymbol{\alpha}^j \in (f^j)^{-1}(0)$ and $\boldsymbol{\alpha}^{(j)} \in (f^{(j)})^{-1}(0),\ j = 1, \ldots, L$.

The following proposition holds:

PROPOSITION 10.1.1. *With the above assumptions, the joint cumulative distribution function $\Psi(t_1, t_2, \ldots, t_L)$ of L stack filters of the sequence of independent input values can be calculated by the following formula*:

$$\Psi(t_1, t_2, \ldots, t_L) = \sum_{\gamma \in \Gamma} \prod_{i=1}^{N} \big[\Phi_i(t_{(\gamma_i+1)}) - \Phi_i(t_{(\gamma_i)})\big], \tag{10–34}$$

where Γ is defined by Eq. 10–33, *Φ_i is the distribution function of the ith input, as given in Eq.* 10–22, *and*

$$\boldsymbol{\alpha}^1 + \boldsymbol{\alpha}^2 + \cdots + \boldsymbol{\alpha}^L = (\alpha_1^1 + \alpha_1^2 + \cdots + \alpha_1^L, \ldots, \alpha_N^1 + \alpha_N^2 + \cdots + \alpha_N^L). \tag{10–35}$$

PROOF. The joint cumulative distribution function $\Psi(t_1, t_2, \ldots, t_L)$ defined by Eq. 10–32 is

$$\begin{aligned}\Psi(t_1, t_2, \ldots, t_L) &= P\{f^1(\eta(\mathbf{X}, t_1)) = 0,\ f^2(\eta(\mathbf{X}, t_2)) = 0, \ldots, \\ &\qquad f^L(\eta(\mathbf{X}, t_L)) = 0\}.\end{aligned} \tag{10–36}$$

Because $\{X_1, X_2, \ldots, X_N\}$ is a sequence of independent random variables, we have

$$\Psi(t_1, t_2, \ldots, t_L) = \sum_{\alpha^1,\ldots,\alpha^L} \prod_{i=1}^{N} P\{\eta(X_i, t_1) = \alpha_i^1, \eta(X_i, t_2) = \alpha_i^2, \eta(X_i, t_L) = \alpha_i^L\}$$

$$= \sum_{\alpha^1,\ldots,\alpha^L} \prod_{i=1}^{N} \mathcal{P}_i. \tag{10–37}$$

Let us consider the term $\mathcal{P}_i = P\{\eta(X_i, t_1) = \alpha_i^1, \eta(X_i, t_2) = \alpha_i^2, \eta(X_i, t_L) = \alpha_i^L\}$ in Eq. 10–37. Reorder the numbers $t_j,\ j = 1, 2, \ldots, L$, into nondecreasing order $t_{(1)}, t_{(2)}, \ldots, t_{(L)}$, and rewrite $\mathcal{P}_i$ as

$$\mathcal{P}_i = P\{\eta(X_i, t_{(1)}) = \alpha_i^{(1)}, \eta(X_i, t_{(2)}) = \alpha_i^{(2)}, \eta(X_i, t_{(L)}) = \alpha_i^{(L)}\},\quad i = 1, 2, \ldots, N. \tag{10–38}$$

The probability function $\mathcal{P}_i$ will have nonzero values on the following sequences of numbers:

$$(\alpha_i^{(1)}, \alpha_i^{(2)}, \ldots, \alpha_i^{(L)}) \in \{(0, 0, \ldots, 0, 0), (1, 0, \ldots, 0, 0), \ldots, (1, 1, \ldots, 1, 0), (1, 1, \ldots, 1, 1)\}.$$

In these $L + 1$ possible nonzero cases, $\mathcal{P}_i$ will have the following values, respectively:

$$\mathcal{P}_i = \begin{cases} \Phi_i(t_{(1)}), & \text{if } (\alpha_i^{(1)}, \alpha_i^{(2)}, \ldots, \alpha_i^{(L)}) = (0, 0, \ldots, 0, 0), \\ \Phi_i(t_{(2)}) - \Phi_i(t_{(1)}), & \text{if } (\alpha_i^{(1)}, \alpha_i^{(2)}, \ldots, \alpha_i^{(L)}) = (1, 0, \ldots, 0, 0), \\ \vdots & \vdots \\ \Phi_i(t_{(L)}) - \Phi_i(t_{(L-1)}), & \text{if } (\alpha_i^{(1)}, \alpha_i^{(2)}, \ldots, \alpha_i^{(L)}) = (1, 1, \ldots, 1, 0), \\ 1 - \Phi_i(t_{(L)}), & \text{if } (\alpha_i^{(1)}, \alpha_i^{(2)}, \ldots, \alpha_i^{(L)}) = (1, 1, \ldots, 1, 1). \end{cases} \tag{10–39}$$

Using notation Eq. 10–33 we can rewrite Eq. 10–39 as

$$\mathcal{P}_i = \Phi_i(t_{(\gamma_i+1)}) - \Phi_i(t_{(\gamma_i)}), \quad \text{if } \boldsymbol{\gamma} \in \Gamma, \tag{10–40}$$

from which and Eq. 10–37 we will have Eq. 10–34. The proposition is proved. ■

EXAMPLE 10.1.3. Let input variables X_1, X_2, X_3 be independent variables, i.e. $P(X_1 \leqslant t_1, X_2 \leqslant t_2, X_3 \leqslant t_3) = \prod_{i=1}^{3} P(X_i \leqslant t_i)$. Moreover, let X_1 and X_3 be uniformly distributed random variables with cumulative distributions $\Phi_1(t) =$

$P(X_1 \leqslant t) = t$, and $\Phi_3(t) = P(X_3 \leqslant t) = t$, respectively, and X_2 have the cumulative distribution $\Phi_2(t) = P(X_2 \leqslant t) = t^2$, $0 \leqslant t \leqslant 1$.

Calculate the joint cumulative distributed function $\Psi(0.5, 0.2, 0.3)$ of $F^{(1)}(X) = \min(X_1, X_2, X_3)$, $F^{(2)}(X) = \mathrm{med}(X_1, X_2, X_3)$ and $F^{(3)}(X) = \max(X_1, X_2, X_3)$. The corresponding positive Boolean functions are

$$f^{(1)}(x) = x_1 x_2 x_3, \tag{10–41}$$

$$f^{(2)}(x) = x_1 x_2 \vee x_1 x_3 \vee x_2 x_3 \tag{10–42}$$

and

$$f^{(3)}(x) = x_1 \vee x_2 \vee x_3. \tag{10–43}$$

We have

$$\begin{aligned}&\{\boldsymbol{\alpha}^{(1)} \mid f^{(1)}(\boldsymbol{\alpha}^{(1)}) = 0\}\\ &\quad = \{(0,0,0), (0,0,1), (0,1,0), (0,1,1), (1,0,0), (1,0,1), (1,1,0)\},\end{aligned} \tag{10–44}$$

$$\{\boldsymbol{\alpha}^{(2)} \mid f^{(2)}(\boldsymbol{\alpha}^{(2)}) = 0\} = \{(0,0,0), (0,0,1), (0,1,0), (1,0,0)\} \tag{10–45}$$

and

$$\{\boldsymbol{\alpha}^{(3)} \mid f^{(3)}(\boldsymbol{\alpha}^{(3)}) = 0\} = \{(0,0,0)\}. \tag{10–46}$$

After reordering the numbers from $\{t_1, t_2, t_3\} = \{0.5, 0.2, 0.3\}$ we have $(t_{(1)}, t_{(2)}, t_{(3)}) = (0.2, 0.3, 0.5) = (t_2, t_3, t_1)$.

We form the following set of matrices:

$$\begin{aligned}W &= \left\{ \begin{pmatrix} \alpha_1^{(1)} & \alpha_2^{(1)} & \alpha_3^{(1)} \\ \alpha_1^{(3)} & \alpha_2^{(3)} & \alpha_3^{(3)} \\ \alpha_1^{(2)} & \alpha_2^{(2)} & \alpha_3^{(2)} \end{pmatrix} \middle| \alpha_i^{(1)} \leqslant \alpha_i^{(3)} \leqslant \alpha_i^{(2)},\ i = 1,2,3 \right\}\\ &= \left\{ \begin{pmatrix} 0 & 0 & 0 \\ 0 & 0 & 0 \\ 0 & 0 & 0 \end{pmatrix}, \begin{pmatrix} 0 & 0 & 0 \\ 0 & 0 & 0 \\ 0 & 0 & 1 \end{pmatrix}, \begin{pmatrix} 0 & 0 & 0 \\ 0 & 0 & 0 \\ 0 & 1 & 0 \end{pmatrix}, \begin{pmatrix} 0 & 0 & 0 \\ 0 & 0 & 0 \\ 1 & 0 & 0 \end{pmatrix} \right\},\end{aligned} \tag{10–47}$$

from which it is easy to find the set Γ, defined in Eq. 10–33, by

$$\Gamma = \left\{ (1,1,1) \cdot \begin{pmatrix} 0 & 0 & 0 \\ 0 & 0 & 0 \\ 0 & 0 & 0 \end{pmatrix}; (1,1,1) \cdot \begin{pmatrix} 0 & 0 & 0 \\ 0 & 0 & 0 \\ 0 & 0 & 1 \end{pmatrix}; \right.$$

$$(1,1,1)\cdot\begin{pmatrix}0&0&0\\0&0&0\\0&1&0\end{pmatrix},(1,1,1)\cdot\begin{pmatrix}0&0&0\\0&0&0\\1&0&0\end{pmatrix}\Bigg\}.$$

Hence,

$$\Gamma=\{(0,0,0);\ (0,0,1);\ (0,1,0);\ (1,0,0)\}. \tag{10–48}$$

Now, by Proposition 10.1.1 we have:

$$\begin{aligned}\Psi(0.5,0.2,0.3) &= \Phi_1(0.2)\cdot\Phi_2(0.2)\cdot\Phi_3(0.2)\\ &\quad+\Phi_1(0.2)\cdot\Phi_2(0.2)\cdot[\Phi_3(0.3)-\Phi_3(0.2)]\\ &\quad+\Phi_1(0.2)\cdot[\Phi_2(0.3)-\Phi_2(0.2)]\cdot\Phi_3(0.2)\\ &\quad+[\Phi_1(0.3)-\Phi_1(0.2)]\cdot\Phi_2(0.2)\cdot\Phi_3(0.2)\\ &= 0.2\cdot0.04\cdot0.2+0.2\cdot0.04\cdot0.1+0.2\cdot0.05\cdot0.2\\ &\quad+0.1\cdot0.04\cdot0.2=0.052.\end{aligned}$$

From Proposition 10.1.1 we can find a formula for calculation of the joint cumulative distribution function $\Psi(t_1,t_2,\ldots,t_L)$, of L stack filters of independent *identically* distributed (i.i.d.) input values.

PROPOSITION 10.1.2. *The joint cumulative distribution function $\Psi(t_1,t_2,\ldots,t_L)$ of L stack filters of the sequence of independent identically distributed input values can be calculated by the following formula*:

$$\Psi(t_1,t_2,\ldots,t_L) = \sum_{\substack{\boldsymbol{\alpha}^{(1)},\ldots,\boldsymbol{\alpha}^{(L)}\\ \boldsymbol{\alpha}^{(1)}\geqslant\cdots\geqslant\boldsymbol{\alpha}^{(L)}}}\prod_{j=0}^{L}[\Phi(t_{(j+1)})-\Phi(t_{(j)})]^{(w_H(\alpha^{(L+1-j)})-w_H(\alpha^{(L-j)}))}, \tag{10–49}$$

where $w_H(\boldsymbol{\alpha})$ is the Hamming weight of the N-tuple $\boldsymbol{\alpha}$, $w_H(\boldsymbol{\alpha}^{(0)})=\Phi(t_{(0)})=0$, $w_H(\boldsymbol{\alpha}^{(L+1)})=N$, and $\Phi(t_{(L+1)})=1$.

PROOF. From Proposition 10.1.1 we know that the joint cumulative distribution function $\Psi(t_1,t_2,\ldots,t_L)$ given by formula 10–34, which in the case of identically distributed inputs [i.e. $\Phi_i(t)=\Phi(t)$ for all $i=1,2,\ldots,N$] can be written as

$$\Psi(t_1,t_2,\ldots,t_L)=\sum_{\boldsymbol{\gamma}\in\Gamma}\prod_{j=0}^{L}[\Phi(t_{(j+1)})-\Phi(t_{(j)})]^{\mu_j(\boldsymbol{\gamma})}, \tag{10–50}$$

where $\mu_j(\boldsymbol{\gamma})=|\{\gamma_k=j \text{ for all } k=1,2,\ldots,N\}|$, $\Phi(t_{(0)})=0$, and $\Phi(t_{(L+1)})=1$.

From Eq. 10–50 immediately follows Eq. 10–49. The proposition is proved. ■

Let us consider an example of the calculation of the joint cumulative distribution function in the case of i.i.d. inputs.

EXAMPLE 10.1.4. Let inputs X_1, X_2, X_3 be uniformly, independently and identically distributed random variables.

We calculate the joint cumulative distribution function $\Psi(0.5, 0.2, 0.3)$ of $F^{(1)}(X) = \min(X_1, X_2, X_3)$, $F^{(2)}(X) = \mathrm{med}(X_1, X_2, X_3)$ and $F^{(3)}(X) = \max(X_1, X_2, X_3)$.

Using Eq. 10–47 and calculating by 10–49 we find:

$$\begin{aligned}\Psi(0.5, 0.2, 0.3) &= [\Phi(0.2)]^3 + 3 \cdot [\Phi(0.3) - \Phi(0.2)][\Phi(0.2)]^2 \\ &= 0.008 + 3 \cdot 0.1 \cdot 0.04 = 0.02.\end{aligned}$$

COROLLARY 10.1.1. *The cumulative distribution function $\Psi(t)$ for one stack filter with i.i.d. inputs can be calculated by Proposition* 10.1.2, *and in the case of $L = 1$ by the following formula:*

$$\Psi(t) = \sum_{\alpha | f(\alpha)=0} [\Phi(t)]^{w_H(\bar{\alpha})}[1 - \Phi(t)]^{w_H(\alpha)}, \tag{10–51}$$

which coincides with the known formulae in [11].

Spectral Approach to the Calculation of the Joint Distribution of Stack Filters

Let $Y^1 = S_{F^1}(\mathbf{X})$, and $Y^2 = S_{F^2}(\mathbf{X})$ be two stack filters based on the positive logic functions F^1 and F^2, respectively [or on the corresponding positive Boolean functions $f^1(x)$ and $f^2(x)$, respectively]. The problem is to find their joint distribution function, i.e. a function

$$\Psi(t_1, t_2) = P(\{Y^1 \leqslant t_1, Y^2 \leqslant t_2\}), \tag{10–52}$$

where t_1, t_2 are real numbers.

In [109], the following formula for the joint distribution of two stack filters is found. Let $t_1 \leqslant t_2$, then

$$\Psi(t_1, t_2) = \sum_{\substack{\alpha^1, \alpha^2 \\ \alpha^1 \geqslant \alpha^2}} \prod_{i=1}^{N} \Phi_i(t_1)^{\bar{\alpha}_i^1 \bar{\alpha}_i^2} (\Phi_i(t_1) - \Phi_i(t_2))^{\alpha_i^1 \bar{\alpha}_i^2} (1 - \Phi_i(t_2))^{\alpha_i^1 \alpha_i^2}. \tag{10–53}$$

In the case of i.i.d. input, that is, $\Phi_i(t) \equiv \Phi(t)$ for all i, we can rewrite formula 10–53 as

$$\Psi(t_1, t_2) = \sum_{i=0}^{N} \sum_{j=0}^{N} A_{ij} \Phi(t_1)^i \big(\Phi(t_2) - \Phi(t_1)\big)^{N-i-j} \big(1 - \Phi(t_2)\big)^j, \tag{10–54}$$

where

$$\begin{aligned} A_{ij} = \Big| \big\{ (\mathbf{a}, \mathbf{b}) \in \{0, 1\}^N, \text{ such that } \mathbf{a} \geqslant \mathbf{b},\ f^1(\mathbf{a}) = f^2(\mathbf{b}) = 0, \\ w_H(\bar{\mathbf{a}} \cdot \bar{\mathbf{b}}) = i,\ w_H(\mathbf{a} \cdot \mathbf{b}) = j \big\} \Big| \end{aligned} \tag{10–55}$$

for $i, j = 0, 1, \ldots, N$, and $w_H(\mathbf{c})$ denotes the Hamming weight of the N-tuple $\mathbf{c}$, $\bar{\mathbf{a}}$ is the complement of $\mathbf{a}$, and $\mathbf{a} \cdot \mathbf{b}$ is the pointwise product of the N-tuples $\mathbf{a}$ and $\mathbf{b}$. As we see from formula 10–54, the main part of the calculation of the values of the function $\Psi(t_1, t_2)$ is finding coefficients A_{ij}, $i, j = 0, 1, \ldots, N-1$. Let's consider a matrix $\mathbf{A} = \|A_{ij}\|$, with elements given by Eq. 10–55. The following algorithm shows how to construct the matrix $\mathbf{A}$ by spectral methods using the initial positive Boolean functions $f^1(x)$ and $f^2(x)$ given in the minimal logical sum of products, i.e. written in form

$$f^k(x_1, x_2, \ldots, x_N) = \bigvee_{(i_1, i_2, \ldots, i_N)} \xi^{(k)}(i_1, i_2, \ldots, i_N) \cdot x_1^{i_1} x_2^{i_2} \cdots x_N^{i_N}, \tag{10–56}$$

for $k = 1, 2$, where $\xi(\cdot)$ takes values 0 or 1.

ALGORITHM 10.1.3.

1. From the positive Boolean functions $f^1(x)$ and $f^2(x)$, given in Eq. 10–56, we extract the corresponding characteristic functions $\boldsymbol{\xi}^1(\cdot)$ and $\boldsymbol{\xi}^2(\cdot)$, which are represented by binary vectors of the length 2^N.

By computing the conjunctive transforms of these vectors, we find the following spectra:

$$\mathbf{g}^k = \mathbf{K}_N \cdot \boldsymbol{\xi}^k, \quad k = 1, 2, \tag{10–57}$$

where $\mathbf{K}_N$ is the conjunctive matrix of the order 2^N.

2. Taking the following threshold operators

$$\mathbf{h}^k = \eta(\mathbf{g}^k, 1), \tag{10–58}$$

for $k = 1, 2$, where

$$\eta(g^k(i), 1) = \begin{cases} 0, & \text{if } g^k(i) \geqslant 1, \\ 1, & \text{otherwise}, \end{cases} \quad i = 1, 2, \ldots, N, \tag{10–59}$$

we have the vectors $\mathbf{h}^1$ and $\mathbf{h}^2$ of the length 2^N.

3. Form the diagonal $2^N \times 2^N$ matrices $\mathbf{D}^1$ and $\mathbf{D}^2$ by:

$$\mathbf{D}^k = \operatorname{diag}\big(h^k(1), \ldots, h^k(N)\big), \quad k = 1, 2. \tag{10–60}$$

Compute the following $2^N \times 2^N$ matrix Ω:

$$\Omega = \begin{cases} \mathbf{D}^1 \cdot \mathbf{K}_N \cdot \mathbf{D}^2, & \text{if } t_1 \leqslant t_2, \\ \mathbf{D}^2 \cdot \mathbf{K}_N \cdot \mathbf{D}^1, & \text{if } t_2 \leqslant t_1, \end{cases} \tag{10–61}$$

where t_1, t_2 are the parameters of joint distribution function 10–52 and $\mathbf{K}_N$ is the conjunctive matrix of order 2^N.

4. Computing the following two-dimensional transform of the matrix $\mathbf{\Omega}$, we obtain a $(N+1) \times (N+1)$ matrix

$$\mathbf{T} = \mathbf{S}_N \cdot \mathbf{\Omega} \cdot \mathbf{S}_N^T, \tag{10–62}$$

where $\mathbf{S}_N$ is the elementary symmetric Boolean matrix.

5. Inverting the columns of $\mathbf{T}$ in reverse order, we obtain the required matrix $\mathbf{A}$, i.e.

$$\mathbf{A} = \mathbf{T} \cdot \bar{\mathbf{I}}, \tag{10–63}$$

where $\bar{\mathbf{I}}$ is the opposite identity matrix of order $N+1$.

EXAMPLE 10.1.5. Let $f^1 = x_2 \vee x_1x_3$, and $f^2 = x_3 \vee x_1x_2$ be two positive Boolean functions, by which two stack filters are defined. Let us construct a matrix A by Algorithm 10.1.3.

1. The characteristic vectors of the minimal sum-of-product forms of the given PBF are the following:

$$\boldsymbol{\xi}^1 = [0,0,1,0,0,1,0,0]^T \quad \text{and} \quad \boldsymbol{\xi}^2 = [0,1,0,0,0,0,1,0]^T.$$

Then

$$(\mathbf{g}^1, \mathbf{g}^2) = \begin{pmatrix} 1&0&0&0&0&0&0&0 \\ 1&1&0&0&0&0&0&0 \\ 1&0&1&0&0&0&0&0 \\ 1&1&1&1&0&0&0&0 \\ 1&0&0&0&1&0&0&0 \\ 1&1&0&0&1&1&0&0 \\ 1&0&1&0&1&0&1&0 \\ 1&1&1&1&1&1&1&1 \end{pmatrix} \cdot \begin{pmatrix} 0&0 \\ 0&1 \\ 1&0 \\ 0&0 \\ 0&0 \\ 1&0 \\ 0&1 \\ 0&0 \end{pmatrix} = \begin{pmatrix} 0&0 \\ 0&1 \\ 1&0 \\ 1&1 \\ 0&0 \\ 1&1 \\ 1&1 \\ 2&2 \end{pmatrix}.$$

2. Take the threshold operators

$$\left(\mathbf{h}^1, \mathbf{h}^2\right) = \left(\eta(\mathbf{g}^1, 1), \eta(\mathbf{g}^2, 1)\right) = \begin{pmatrix} 1 & 1 & 0 & 0 & 1 & 0 & 0 & 0 \\ 1 & 0 & 1 & 0 & 1 & 0 & 0 & 0 \end{pmatrix}^T .$$

3. Form the matrix $\boldsymbol{\Omega}$. Let $t_1 \leqslant t_2$. Then

$$\boldsymbol{\Omega} = \mathrm{diag}\left(\mathbf{h}^1\right) \cdot \mathbf{K}_3 \cdot \mathrm{diag}\left(\mathbf{h}^2\right) = \begin{pmatrix} 1 & 0 & 0 & 0 & 0 & 0 & 0 & 0 \\ 1 & 0 & 0 & 0 & 0 & 0 & 0 & 0 \\ 0 & 0 & 0 & 0 & 0 & 0 & 0 & 0 \\ 0 & 0 & 0 & 0 & 0 & 0 & 0 & 0 \\ 1 & 0 & 0 & 0 & 1 & 0 & 0 & 0 \\ 0 & 0 & 0 & 0 & 0 & 0 & 0 & 0 \\ 0 & 0 & 0 & 0 & 0 & 0 & 0 & 0 \\ 0 & 0 & 0 & 0 & 0 & 0 & 0 & 0 \end{pmatrix} .$$

By analogy, if $t_2 \leqslant t_1$, then

$$\boldsymbol{\Omega} = \mathrm{diag}\left(\mathbf{h}^2\right) \cdot \mathbf{K}_3 \cdot \mathrm{diag}\left(\mathbf{h}^1\right) = \begin{pmatrix} 1 & 0 & 0 & 0 & 0 & 0 & 0 & 0 \\ 0 & 0 & 0 & 0 & 0 & 0 & 0 & 0 \\ 1 & 0 & 0 & 0 & 0 & 0 & 0 & 0 \\ 0 & 0 & 0 & 0 & 0 & 0 & 0 & 0 \\ 1 & 0 & 0 & 0 & 1 & 0 & 0 & 0 \\ 0 & 0 & 0 & 0 & 0 & 0 & 0 & 0 \\ 0 & 0 & 0 & 0 & 0 & 0 & 0 & 0 \\ 0 & 0 & 0 & 0 & 0 & 0 & 0 & 0 \end{pmatrix} .$$

4. Compute the two-dimensional $\mathbf{S}_3$ transform of the matrix $\boldsymbol{\Omega}$, where $\mathbf{S}_3$ is given by Eq. 10–27:

$$\mathbf{T} = \mathbf{S}_3 \cdot \boldsymbol{\Omega} \cdot \mathbf{S}_3^T = \begin{pmatrix} 0 & 0 & 0 & 0 \\ 0 & 0 & 0 & 0 \\ 0 & 0 & 1 & 2 \\ 0 & 0 & 0 & 1 \end{pmatrix} .$$

5. Finally,

$$\mathbf{A} = \mathbf{T} \cdot \bar{\mathbf{I}} = \begin{pmatrix} 0 & 0 & 0 & 0 \\ 0 & 0 & 0 & 0 \\ 2 & 1 & 0 & 0 \\ 1 & 0 & 0 & 0 \end{pmatrix} .$$

From the matrix **A** using formula 10–54 we have:

$$\begin{aligned}\Psi(t_1, t_2) &= \Phi^3(t_1) + 2\Phi^2(t_1)(\Phi(t_2) - \Phi(t_1)) + \Phi^2(t_1) \cdot \Phi(t_2) \\ &= \Phi^2(t_1)\big(3\Phi(t_2) - \Phi(t_1)\big).\end{aligned}$$

10.1.4 SELECTION PROBABILITIES OF STACK FILTERS

Any continuous logical function can be given by the table, where any ordering $\mathcal{O}_k$ (from $N!$ possible orderings, $k = 1, 2, \ldots, N!$) is associated with the corresponding output of the LF (or the corresponding continuous stack filter) $Y = X_j,\ j = 1, 2, \ldots, N$ (see example in Table 10.1). For a given ordering let the output Y be also the ith ranked (smallest) input $X_{(i)}$, i.e. $Y = X_j = X_{(i)},\ i, j = 1, 2, \ldots, N$.

For an ordering $\mathcal{O}_k$ and the corresponding output $Y = X_j = X_{(i)},\ i, j = 1, 2, \ldots, N$, the following *selection probability sets* G^k_{ij} and H^k_{ij} have been defined in [86]:

$$G^k_{ij} = \{X_{(n)} \mid n = 1, 2, \ldots, i-1,\ Y = X_{(i)} = X_j,\ \text{for the ordering } \mathcal{O}_k\} \quad (10\text{–}64)$$

and

$$\begin{aligned}H^k_{ij} = \{X_{(n)} \mid n = i+1, i+2, \ldots, N,\ Y = X_{(i)} = X_j,& \\ \text{for the ordering } \mathcal{O}_k\}.&\end{aligned} \quad (10\text{–}65)$$

In the following the superscript k will be omitted in most cases.

Naturally, if $X_{(i)}$ is the output of filter with the corresponding sets G_{ij} and H_{ij}, there exist $(i-1)!(N-i)!$ orderings for which $X_{(i)}$ is still the output of filter with the same sets G_{ij} and H_{ij}.

Table 10.1. CLF $F(X) = \max(\min(X_1, X_2), \min(X_1, X_3))$.

Region	Ordering	$F(X_1, X_2, X_3)$
1	$X_1 \leqslant X_2 \leqslant X_3$	X_1
2	$X_1 \leqslant X_3 \leqslant X_2$	X_1
3	$X_2 \leqslant X_1 \leqslant X_3$	X_1
4	$X_3 \leqslant X_1 \leqslant X_2$	X_1
5	$X_2 \leqslant X_3 \leqslant X_1$	X_3
6	$X_3 \leqslant X_2 \leqslant X_1$	X_2

Table 10.2. Sets G_{ij} and H_{ij} for a continuous stack filter. Defined by CLF $Y = f(\mathbf{x}) = \max(\min(X_1, X_2), \min(X_1, X_3))$.

Region k	Ordering	$Y = X_j = X_{(i)}$	G_{ij}^k	H_{ij}^k
1	$X_1 \leqslant X_2 \leqslant X_3$	$X_1 = X_{(1)}$	$\emptyset$	$\{X_2, X_3\}$
2	$X_1 \leqslant X_3 \leqslant X_2$	$X_1 = X_{(1)}$	$\emptyset$	$\{X_2, X_3\}$
3	$X_2 \leqslant X_1 \leqslant X_3$	$X_1 = X_{(2)}$	$\{X_2\}$	$\{X_3\}$
4	$X_3 \leqslant X_1 \leqslant X_2$	$X_1 = X_{(2)}$	$\{X_3\}$	$\{X_2\}$
5	$X_2 \leqslant X_3 \leqslant X_1$	$X_3 = X_{(2)}$	$\{X_2\}$	$\{X_1\}$
6	$X_3 \leqslant X_2 \leqslant X_1$	$X_2 = X_{(2)}$	$\{X_3\}$	$\{X_1\}$

In Table 10.2 the sets G_{ij}^k and H_{ij}^k for all possible orderings $\mathcal{O}_k$, $k = 1, 2, \ldots, 6$, for a continuous stack filter defined by the CLF $f(\mathbf{x})$ as in Table 10.1 are shown. Selection probabilities are defined by Prasad *et al.* [86].

DEFINITION 10.1.5. The ith *rank selection probability* is denoted by $P[Y = X_{(i)}]$, $1 \leqslant i \leqslant N$, and is the probability that the output $Y = X_{(i)}$. The *rank selection probability vector* is the row vector $\mathbf{r} = (r_1, r_2, \ldots, r_N)$, where $r_i = P[Y = X_{(i)}]$, $1 \leqslant i \leqslant N$.

DEFINITION 10.1.6. The jth *sample selection probability* is denoted by $P[Y = X_j]$, $1 \leqslant j \leqslant N$, and is the probability that the output $Y = X_j$. The *sample selection probability vector* is the row vector $\mathbf{s} = (s_1, s_2, \ldots, s_N)$, where $s_i = P[Y = X_j]$, $1 \leqslant j \leqslant N$.

DEFINITION 10.1.7. The i, jth *joint selection probability* is denoted by $P[Y = X_{(i)} = X_j]$, $1 \leqslant i, j \leqslant N$, and is the probability that the output $Y = X_{(i)} = X_j$. The *joint selection probability matrix* [71, 72, 86] is the $N \times N$ matrix $\mathbf{J} = \|J_{ij}\| = \|P[Y = X_{(i)} = X_j]\|$.

The number of distinct sets G_{ij}^k (or H_{ij}^k) for which $Y = X_{(i)} = X_j$ is denoted by C_{ij}. A matrix $\mathbf{C} = \|C_{ij}\|$, $i, j = 1, 2, \ldots, N$, called a *combination matrix* [86].

Selection probability vectors and the joint selection probability matrix are given by

$$r_i = P[Y = X_{(i)}] = \sum_{j=1}^{N} \frac{(i-1)!(N-i)!C_{ij}}{N!}, \quad i = 1, 2, \ldots, N, \qquad (10\text{–}66)$$

$$s_i = P[Y = X_j] = \sum_{i=1}^{N} \frac{(i-1)!(N-i)!C_{ij}}{N!}, \quad i = 1, 2, \ldots, N, \qquad (10\text{–}67)$$

Table 10.3. Selection Probabilities for a continuous stack filter. Based on the CLF $f(\mathbf{x}) = \max(\min(X_1, X_2), \min(X_1, X_3))$.

Rank selection probability vector **r**	Sample selection probability vector **s**	Combination matrix **C**	Joint selection probability matrix **J**
$1/3 \cdot [1\ 2\ 0]$	$1/6 \cdot [4\ 1\ 1]$	$\begin{bmatrix} 1 & 0 & 0 \\ 2 & 1 & 1 \\ 0 & 0 & 0 \end{bmatrix}$	$\begin{bmatrix} \frac{1}{3} & 0 & 0 \\ \frac{1}{3} & \frac{1}{6} & \frac{1}{6} \\ 0 & 0 & 0 \end{bmatrix}$

and

$$J_{ij} = P[Y = X_{(i)}, Y = X_j] = \frac{(i-1)!(N-i)!C_{ij}}{N!},$$
$$i, j = 1, 2, \ldots, N. \tag{10–68}$$

In Table 10.3 all selection probability matrices and vectors are based on the positive continuous function $f(\mathbf{x})$ from our "control" example in Table 10.1.

Construction of Selection Probability Sets for a Continuous Stack Filter

As has been shown in [45], the construction of sets G_{ij} and H_{ij} for a continuous stack filter based on PBF $f(x)$ can be obtained by means of partial derivatives of this PBF. We recall here the definition of a partial derivative of the Boolean function [83, 93].

DEFINITION 10.1.8. The *partial derivative* of a Boolean function $f(\mathbf{x}_N)$ of N variables with respect to the variable x_j $(1 \leqslant j \leqslant N)$ is the Boolean function $\partial f/\partial x_j$ defined by

$$\frac{\partial f(\mathbf{x}_N)}{\partial x_j} = f\left(\mathbf{x}_N^{(j,0)}\right) \oplus f\left(\mathbf{x}_N^{(j,1)}\right), \tag{10–69}$$

where $\oplus$ is the operation of addition by modulo 2,

$$\mathbf{x}_N = (x_1, x_2, \ldots, x_N), \tag{10–70}$$
$$\mathbf{x}_N^{(j,0)} = (x_1, \ldots, x_{j-1}, 0, x_{j+1}, \ldots, x_N) \tag{10–71}$$

and

$$\mathbf{x}_N^{(j,1)} = (x_1, \ldots, x_{j-1}, 1, x_{j+1}, \ldots, x_N). \tag{10–72}$$

Here we present a spectral algorithm for the construction of the selection probability sets. With this objective we give following definitions:

DEFINITION 10.1.9. The rectangular $(2^{N-1} \times 2^N)$ matrix Ξ^j, $j = 1, 2, \ldots, N$, is defined by

$$\Xi^j = \mathbf{I}_j \otimes (1, 1) \otimes \mathbf{I}_{N-j-1}, \tag{10–73}$$

where $\mathbf{I}_j$ is the identity matrix of order 2^j, and is called the jth *partial derivative matrix.*

DEFINITION 10.1.10. Let $\mathbf{f}$ be the truth table (vector column of the length 2^N) of an N-variable Boolean function $f(x)$. The following transform

$$\mathbf{f}'_\mathbf{j} = \Xi^j \cdot \mathbf{f}(mod\ 2) \tag{10–74}$$

is called the jth *partial derivative transform* (*spectrum*), where Ξ^j is the jth partial derivative matrix.

The name of the partial derivative transform is connected with the fact that $\mathbf{f}'_\mathbf{j}$ defined by Eq. 10–74 is the truth table (vector column of the length 2^{N-1}) of the partial derivative $\partial f(\mathbf{x}_N)/\partial x_j$ of the Boolean function $f(x)$ with respect to the variable x_j, $j = 1, 2, \ldots, N$.

Because each row in the partial derivative matrix Ξ^j (for arbitrary j, $j = 1, 2, \ldots, N$) has only two nonzero elements (equal to 1), for the computation of the jth partial derivative spectrum only 2^{N-1} additions are needed (more exactly, additions by modulo 2). The total number of additions for computation of all partial derivative transforms for $j = 1, 2, \ldots, N$ is $N \cdot 2^{N-1}$.

In Table 10.4 an example of the construction of truth tables of partial derivatives of the PBF $f(x) = x_1x_2 \vee x_1x_3$ with respect to variables x_j, $j = 1, 2, 3$, is shown.

Based on the result obtained in [45], we have the following formulae for construction of the sets G_{ij} and H_{ij}:

$$H_{ij} = \left\{D_X\left(\boldsymbol{\alpha}^{(j)}\right),\ w_H\left(\boldsymbol{\alpha}^{(j)}\right) = N - i,\ \mathbf{f}'_\mathbf{j}\left(\boldsymbol{\alpha}^{(j)}\right) = 1\right\} \tag{10–75}$$

and

$$G_{ij} = \left\{D_X\left(\boldsymbol{\beta}^{(j)}\right),\ w_H\left(\boldsymbol{\beta}^{(j)}\right) = i - 1,\ \mathbf{f}'_\mathbf{j}\left(\boldsymbol{\beta}^{(j)}\right) = 1\right\}, \tag{10–76}$$

where $i, j = 1, 2, \ldots, N$, $\boldsymbol{\alpha}^{(j)}$ and $\boldsymbol{\beta}^{(j)}$ are defined by:

$$\boldsymbol{\alpha}^{(j)} = (\alpha_1, \ldots, \alpha_{j-1}, \alpha_{j+1}, \ldots, \alpha_N)$$

Table 10.4. Partial derivatives $\mathbf{f}'_{\mathbf{j}}$ of the PBF $f(x) = x_1x_2 \vee x_1x_3$ with respect to variables x_j, $j = 1, 2, 3$.

x_1, x_2, x_3	$\mathbf{f}$	$\mathbf{f}'_1$	$\mathbf{f}'_2$	$\mathbf{f}'_3$
000	0	–	–	–
001	0	–	–	0
010	0	–	0	–
011	0	–	0	0
100	0	0	–	–
101	1	1	–	1
110	1	1	1	–
111	1	1	0	0

and

$$\boldsymbol{\beta}^{(j)} = (1 - \alpha_1, \ldots, 1 - \alpha_{j-1}, 1 - \alpha_{j+1}, \ldots, 1 - \alpha_N),$$

$w_H(\boldsymbol{\alpha}^{(j)})$ is the Hamming weight of the vector $\boldsymbol{\alpha}^{(j)}$, $D_X(\boldsymbol{\alpha}^{(j)})$ is the prethreshold operator, defined by

$$D_X(\boldsymbol{\alpha}^{(j)}) = \{X_m \mid \alpha_m = 0,\ m \in \{1, \ldots, j-1, j+1, \ldots, N\}\},$$

and $\mathbf{f}'_{\mathbf{j}}$ is the jth partial derivative transform vector, obtained by Eq. 10–74, $j = 1, 2, \ldots, N$.

Construction of Joint Selection Probability Matrix for a Continuous Stack Filter

Using the transforms by elementary symmetric Boolean matrix $\mathbf{S}_N$ and the partial derivative matrix Ξ^j, $j = 1, 2, \ldots, N$, we give a spectral algorithm for the construction of the joint selection probability matrix $\mathbf{J}$ for a continuous stack filter.

The idea of the algorithm is the following:

1. Find a table given representation $\mathbf{f}$ in the minimal disjunctive normal form PBF $f(x)$.

For this end we use conjunctive transform of the characteristic vector of minterms of $f(x)$. After thresholding this spectrum, we have a truth table $\mathbf{f}$.

2. Find all N partial derivative spectrums $\Xi^j \cdot \mathbf{f}$, $j = 1, 2, \ldots, N$, which give us all truth tables of the partial derivatives with respect to one variable.

3. Construct the combination matrix $\mathbf{C}$ of selection probabilities, for which we use transforms by the elementary symmetric Boolean matrix $\mathbf{S}_N$.

4. Construct the joint selection probability matrix $\mathbf{J}$ of selection probabilities, for which we take a product of a diagonal matrix with the matrix $\mathbf{C}$.

ALGORITHM 10.1.4. This algorithm consists of the following steps:

Step 1. Find the truth table $\mathbf{f}$ representation of the PBF $f(x)$:

1a) Form the characteristic vector of the minimal disjunctive normal form of $f(x)$. To form this we first code all M minterms $p_i = \bigwedge_{j \in P_i} x_j$ as binary vectors of length N, where the code w_i corresponds to p_i and w_i has ones on the components $j \in P_i$ and zeroes elsewhere. Denote $\Omega = \{w_1, w_2, \ldots, w_M\}$. Then we form the characteristic (indicator) vector χ_Ω of the set of N-tuples Ω by

$$\chi_\Omega(a) = \begin{cases} 1, & \text{if } a \in \Omega, \\ 0, & \text{otherwise.} \end{cases} \qquad (10\text{–}77)$$

Thus, we represent χ_Ω as a binary vector column of length 2^N, having ones on the places which correspond to the N-tuples from the set Ω, and zeroes on the other places.

1b) Compute the conjunctive transform of the characteristic vector χ_Ω, i.e. the following transform by the conjunctive matrix K_N:

$$\mathbf{X} = \mathbf{K}_N \cdot \chi_\Omega, \qquad (10\text{–}78)$$

1c) Threshold the vector $\mathbf{X}$:

$$\mathbf{f} = \sigma(\mathbf{X}, 1), \qquad (10\text{–}79)$$

where $\sigma(X_i, 1)$ is defined as in Eq. 10–19 with $t = 1$, $i = 0, 1, \ldots, 2^N - 1$, and $\mathbf{X} = (X_0, \ldots, X_{2^N-1})^T$ is the vector-column of the length 2^N, defined by Eq. 10–78.

Then $\mathbf{f}$ is the truth table of the given positive Boolean function $f(x)$.

Step 2. Find truth tables $\mathbf{f}'_\mathbf{j}$ of the partial derivatives $\partial f / \partial x_j$, $j = 1, 2, \ldots, N$, for which compute the following partial derivative transforms:

$$\mathbf{f}'_\mathbf{j} = \Xi^j \cdot \mathbf{f}(mod\ 2), \quad j = 1, 2, \ldots, N. \qquad (10\text{–}80)$$

Step 3. Construct the combination matrix $\mathbf{C}$ of selection probabilities, for which compute the following transform:

$$\mathbf{C} = \mathbf{S}_{N-1} \cdot \mathbf{F}', \tag{10–81}$$

where $\mathbf{F}'$ is the rectangular $(2^{N-1} \times N)$ matrix, the jth column of which are the vectors $\mathbf{f}'_{\mathbf{j}}$, $j = 1, 2, \ldots, N$, and $\mathbf{S}_{N-1}$ is an elementary symmetric Boolean matrix.

Step 4. Construct the joint selection probability matrix $\mathbf{J}$ of selection probabilities by:

$$\mathbf{J} = \mathbf{D}_N \cdot \mathbf{C}, \tag{10–82}$$

where $\mathbf{D}$ is the diagonal matrix of order N, with the elements $D_i = \left[i \cdot \binom{N}{i}\right]^{-1}$ on the diagonal, $i = 1, 2, \ldots, N$.

REMARK 10.1.1. *Algorithm 10.1.4 has the complexity $N(3 \cdot 2^{N-1} - N)$ operations of addition, 2^N operations of multiplication and 2^N comparisons for construction of the matrix J.*

PROOF. This estimate can be obtained as follows. For the computation of conjunctive transform and N partial derivative transforms are required by $N \cdot 2^{N-1}$ addition operations. Computation of the transformation by the elementary symmetric Boolean matrix $\mathbf{S}_{N-1}$ required $2^{N-1} - N$ additions, because the jth row of this matrix contains $\binom{N-1}{j-1}$ ones. ■

Consider an example illustrating this algorithm.

EXAMPLE 10.1.6. Let $f(x) = x_1x_2 \vee x_1x_3$.

Step 1: By this minimal form of the positive Boolean function $f(x)$ we find $w_1 = (1, 1, 0)$, $w_2 = (1, 0, 1)$, and $\Omega = \{(1, 1, 0), (1, 0, 1)\}$.

The characteristic vector equals $\chi_\Omega = (0, 0, 0, 0, 0, 1, 1, 0)^T$.

The conjunctive transform of χ_Ω equals $\mathbf{X} = \mathbf{K}_3 \cdot \chi_\Omega = (0, 0, 0, 0, 0, 1, 1, 2)^T$.

Thresholding of the $\mathbf{X}$ gives $\mathbf{f} = \sigma(\mathbf{X}, 1) = (0, 0, 0, 0, 0, 1, 1, 1)^T$.

By *Step 2* we obtain the following truth table vectors (see also Table 10.6):

$$\begin{aligned}
\mathbf{f}'_{\mathbf{1}} &= \Xi^1 \cdot \mathbf{f}(mod\ 2) = (0, 1, 1, 1), \\
\mathbf{f}'_{\mathbf{2}} &= \Xi^2 \cdot \mathbf{f}(mod\ 2) = (0, 0, 1, 0), \\
\mathbf{f}'_{\mathbf{3}} &= \Xi^3 \cdot \mathbf{f}(mod\ 2) = (0, 0, 1, 0).
\end{aligned}$$

Step 3: Now,

$$\mathbf{F}' = \begin{pmatrix} 0 & 0 & 0 \\ 1 & 0 & 0 \\ 1 & 1 & 1 \\ 1 & 0 & 0 \end{pmatrix},$$

and

$$\mathbf{C} = \mathbf{S}_2 \cdot \mathbf{F}' = \begin{pmatrix} 0 & 0 & 0 & 1 \\ 0 & 1 & 1 & 0 \\ 1 & 0 & 0 & 0 \end{pmatrix} \cdot \mathbf{F}' = \begin{pmatrix} 1 & 0 & 0 \\ 2 & 1 & 1 \\ 0 & 0 & 0 \end{pmatrix}.$$

Step 4: Finally, we obtain

$$\mathbf{J} = \mathbf{D} \cdot \mathbf{C} = \frac{1}{6} \cdot \begin{pmatrix} 2 & 0 & 0 \\ 0 & 1 & 0 \\ 0 & 0 & 2 \end{pmatrix} \cdot \mathbf{C} = \begin{pmatrix} 1/3 & 0 & 0 \\ 1/3 & 1/6 & 1/6 \\ 0 & 0 & 0 \end{pmatrix}.$$

Construction of the Sample Selection Probability Vector for a Continuous Stack Filter

The sample selection probability vector **s** can be found using weighted Chow parameters as we established in [45]. Here we consider a spectral approach to construction of vector **s**.

Let $\mathbf{R}_N$ be the rectangular $(2^N \times N)$ Rademacher (0, 1)-matrix of I-type. Owing to the property of this matrix, each ith column of R_N is the vector of the binary representation of the number i, $i = 0, 1, \ldots, 2^N - 1$. For example,

$$\mathbf{R}_2^T = \begin{pmatrix} 0 & 0 & 1 & 1 \\ 0 & 1 & 0 & 1 \end{pmatrix},$$

where $\mathbf{R}_N^T$ denotes the transposed matrix of $\mathbf{R}_N$.

Let $\mathbf{\Lambda}_N$ be a diagonal matrix of order 2^N with the elements on the diagonal

$$\lambda_i = (w_H(\mathbf{i}))!\big(N - 1 - w_H(\mathbf{i})\big)!, \quad i = 0, 1, \ldots, 2^N - 1, \qquad (10\text{–}83)$$

where $\mathbf{i} = (i_1, \ldots, i_N)$, is the binary representation of the number i, and $w_H(\mathbf{i})$ is the Hamming weight of the N-tuple $\mathbf{i}$.

The following proposition holds.

PROPOSITION 10.1.3. *Let* $\mathbf{f} = (f_0, f_1, \ldots, f_{2^N-1})^T$ *be a truth table of the N variables positive Boolean function* $f(x)$, *and* $\hat{\mathbf{f}} = (f_{2^N-1}, f_{2^N-2}, \ldots, f_1, f_0)^T$.

The sample selection probability vector $\mathbf{s}$ *of the stack filter based on this PBF can be computed by the following transform:*

$$\mathbf{s} = (N!)^{-1} \cdot \mathbf{R}_N^T \cdot \mathbf{\Lambda}_N \cdot (\mathbf{f} - \hat{\mathbf{f}}), \tag{10–84}$$

where $\mathbf{R}_N^T$ *is the transposed Rademacher* (0, 1)*-matrix, and* Λ_N *is a diagonal matrix with the elements on the diagonal defined in Eq.* 10–83.

PROOF. It follows from the algorithm for computation of the sample selection probability vector derived in [45]. ■

REMARK 10.1.2. *The sample selection probability vector* $\mathbf{s}$ *can be obtained by* $3 \cdot 2^N - N - 1$ *additions and* $2^N + N$ *multiplications.*

PROOF. This estimate is based on the fast Rademacher (0, 1) transform of I-type. ■

The computation of a sample selection probability vector is shown in the following example.

EXAMPLE 10.1.7. Let $f(x) = x_1x_2 \vee x_1x_3$, as in the earlier examples. Then $\mathbf{f} = (0, 0, 0, 0,\ 0, 1, 1, 1)^T$, $\hat{\mathbf{f}} = (1, 1, 1, 0, 0, 0, 0, 0)^T$, and $\mathbf{f} - \hat{\mathbf{f}} = (-1, -1, -1, 0, 0, 1, 1, 1)$.

By Eq. 10–84 we have:

$$\mathbf{s} = \frac{1}{6} \cdot \begin{pmatrix} 0 & 0 & 0 & 0 & 1 & 1 & 1 & 1 \\ 0 & 0 & 1 & 1 & 0 & 0 & 1 & 1 \\ 0 & 1 & 0 & 1 & 0 & 1 & 0 & 1 \end{pmatrix} \cdot \operatorname{diag}(2, 2, 2, 1, 2, 1, 1, 2)$$
$$\times (-1, -1, -1, 0, 0, 1, 1, 1)^T = \frac{1}{6} \cdot (4, 1, 1) = \left(\frac{2}{3}, \frac{1}{6}, \frac{1}{6}\right)^T.$$

Construction of a Rank Selection Probability Vector for a Continuous Stack Filter

Let $\mathbf{f}$ be a truth table of the positive Boolean function $f(x)$, $\mathbf{\Omega}$ be a diagonal matrix of order 2^N with the elements $\omega_i = 1/\binom{N}{i}$, $i = 0, 1, \ldots, 2^N - 1$, and $\mathbf{Q}$ be the following $N \times (N+1)$ matrix:

$$\mathbf{Q} = \begin{pmatrix} 1 & -1 & 0 & 0 & \ldots & 0 & 0 \\ 0 & 1 & -1 & 0 & \ldots & 0 & 0 \\ \vdots & \vdots & \vdots & \vdots & \ddots & \vdots & \vdots \\ 0 & 0 & 0 & 0 & \ldots & 1 & -1 \end{pmatrix}.$$

The following proposition holds:

PROPOSITION 10.1.4. *Let* $\mathbf{f} = (f_0, f_1, \ldots, f_{2^N-1})^T$ *be a truth table of the N variable positive Boolean function* $f(x)$. *The rank selection probability vector* $\mathbf{r}$ *of the stack filter based on this PBF can be computed by the following transform*:

$$\mathbf{r} = \mathbf{Q} \cdot \boldsymbol{\Omega} \cdot \mathbf{S}_N \cdot \mathbf{f}, \qquad (10\text{–}85)$$

where $\mathbf{S}_N$ *is an elementary symmetric matrix.*

PROOF. It is based on the algorithm for computation of the rank selection probability vector derived in [45]. ■

REMARK 10.1.3. *The rank selection probability vector* $\mathbf{r}$ *can be obtained by* 2^{N-1} *additions and N multiplications.*

An example of the computation of a rank selection probability vector follows.

EXAMPLE 10.1.8. Let $f(x) = x_1x_2 \vee x_1x_3$.

Then $\mathbf{f} = (0, 0, 0, 0, 0, 1, 1, 1)^T$, and by Eq. 10–85 we find:

$$\mathbf{r} = \begin{pmatrix} 1 & -1 & 0 & 0 \\ 0 & 1 & -1 & 0 \\ 0 & 0 & 1 & -1 \end{pmatrix} \cdot \begin{pmatrix} 1 & 0 & 0 & 0 \\ 0 & \frac{1}{3} & 0 & 0 \\ 0 & 0 & \frac{1}{3} & 0 \\ 0 & 0 & 0 & 1 \end{pmatrix}$$

$$\times \begin{pmatrix} 0 & 0 & 0 & 0 & 0 & 0 & 0 & 1 \\ 0 & 0 & 0 & 1 & 0 & 1 & 1 & 0 \\ 0 & 1 & 1 & 0 & 1 & 0 & 0 & 0 \\ 1 & 0 & 0 & 0 & 0 & 0 & 0 & 0 \end{pmatrix} \cdot (0, 0, 0, 0, 0, 1, 1, 1)^T = \left(\frac{1}{3}, \frac{2}{3}, 0\right)^T.$$

10.2 BINARY POLYNOMIAL TRANSFORMS IN GENETIC ALGORITHMS

10.2.1 INTRODUCTION

Wavelet packet (WP) transforms (or, equivalently, tree-structured transforms), which are powerful extensions of wavelets and multiresolution analysis, have recently received wide interest in the signal processing community. Based on tree-structure filter banks, WPs offer a rich family of orthonormal bases from which one can choose the "best" (under a certain criterion, such as entropy-based) basis [101, 97].

Genetic algorithms (GA) are search algorithms based on the mechanics of natural selection and natural genetics. A simple GA is composed of three operators: reproduction, crossover, and mutation [56]. An important problem of GA is how to analyze the difficulty of an objective (hence also fitness) function for the GA.

As a search algorithm one can use GA for the selection of the "best" basis (under a certain criterion) among all WP bases. Such kinds of problems occur, e.g. in image coding applications [103].

In this matter the following question arises: what kind of role does the WP transform play in the analysis of GA? A first attempt to apply orthogonal transforms to the analysis of genetic algorithms was made by Bethke [22], who discovered an efficient way for calculating schema average fitness values using the Walsh transform. The building blocks of GAs, which are combined to form optima or near optima, are short and low-order schemata with above-average fitness values. But the Walsh transform is only one transform from the class of the Haar-wavelet packets. How about other transforms, especially those based on rectangular basis functions?

In [69], an attempt was made to use the Haar transform with the GAs. It was based on the fact that in the case of a Haar transform there is a reduction of the computation time compared to the Walsh transform. Nevertheless, no general formula for schema average fitness on a Haar basis, and no general analysis of the complexity (such as the number of nonzero terms in the summation for schema average fitness) were presented in [69].

The main goal of this section is to investigate the application of wavelet packet transforms to the analysis of genetic algorithms. We derive an analytical expression for a GA average H-fitness matrix and average H-fitness cost vector (which are quantitative measures of the complexity of calculating schema average fitness values) corresponding to any basis H from the library of WP transforms.

10.2.2 THE SCHEMA THEOREM AND THE WALSH-SCHEMA TRANSFORM

First we recall some concepts of GA, the fundamental theorem of GA – a schema theorem, and the Walsh-schema transform [22, 57, 63].

We assume that GA processes n-bit strings, $\mathbf{x} = (x_{n-1}x_{n-2}\ldots x_0)$, $x_i \in \{0, 1\}$, corresponding to the decimal number $x = \sum_{i=0}^{n-1} x_i 2^i$.

A *schema* is a similarity subset containing strings with the similarity defined at some number of positions. For example, the subset $\{(001), (011)\}$ is the schema

$(0 * 1)$, where $*$ is the "don't care" character. Thus, a schema is

$$\mathbf{s} = (s_{n-1}s_{n-2}\dots s_0), \quad s_i \in \{0, 1, *\}.$$

If we represent an n-bit string as a node of the binary n-cube, a schema is the covering of corresponding nodes by intervals. There are in total 3^n different schemata.

The order $o(\mathbf{s})$ of a schema $\mathbf{s}$ is the number of fixed positions of similarity in the subset (the number of strings in this subset is $2^{n-o(\mathbf{s})}$). For example, $o(0 * 1) = 2$.

The length $\delta(\mathbf{s})$ of a schema is the distance between the outermost defining positions of a schema. For example, $\delta(0 * 1) = 2$, $\delta(00*) = 1$.

For a given GA problem we have a real-valued fitness function $f(g(\mathbf{x}))$, where $g(\mathbf{x})$ are decision variables. According to the *schema theorem* [56], under reproduction, simple crossover, and mutation, the expected number of representatives m of a particular schema $\mathbf{s}$ satisfies

$$m(\mathbf{s}, t+1) \geqslant m(\mathbf{s}, t)\frac{\tilde{f}(\mathbf{s})}{\overline{f}}\left[1 - p_c\frac{\delta(\mathbf{s})}{n-1} - p_m o(\mathbf{s})\right], \tag{10–86}$$

where $\tilde{f}(\mathbf{s})$ is the *schema average fitness* of the representatives of $\mathbf{s}$ in the current population, defined by

$$\tilde{f}(\mathbf{s}) = \frac{1}{|\mathbf{s}|}\sum_{\mathbf{x}\in\mathbf{s}} f(\mathbf{x}), \tag{10–87}$$

where $|\mathbf{s}|$ is the number of strings of the subset $\mathbf{s}$, $\overline{f}$ is the average fitness in the population, p_c and p_m are the crossover and the mutation probabilities, respectively, $\delta(\mathbf{s})$ is the length, and $o(\mathbf{s})$ is the order of the schema. This theorem says that a schema grows when it is short, of low order, and has above-average fitness [57].

The schema average fitness (in the Walsh transform domain) can be written as [22, 57]:

$$\tilde{f}(\mathbf{s}) = \sum_{j\in J(\mathbf{s})} c_j w_j\big(\beta(\mathbf{s})\big), \quad J(\mathbf{s}) = \big\{j : (\exists i) : (\mathbf{s} \subseteq \mathbf{s}_i(j))\big\}, \tag{10–88}$$

where

$$\beta(s_i) = \begin{cases} 0, & \text{if } s_i = 0, * \\ 1, & \text{is } s_i = 1, \end{cases} \qquad c_j = \frac{1}{2^n}\sum_{x=0}^{2^n-1} f(x)w_j(x),$$

$w_j(x)$ is the jth Walsh function.

The computation of Eq. 10–88 is called the *Walsh-schema transform* [57]. Note that the low order schemata are specified with a short sum and the high order schemata are specified with a long sum. With the schema theorem this shows why the Walsh transform is effective for GA.

10.2.3 AVERAGE FITNESS TRANSFORM MATRIX AND COST VECTOR

Let now $\mathbf{f} = [f(0), \ldots, f(2^n - 1)]^T$ be the vector form of a fitness function, $\tilde{\mathbf{f}} = [\tilde{f}(0), \ldots, \tilde{f}(3^n - 1)]^T$ be the vector form of an average fitness function, $\mathbf{H}_n$ be any nonsingular $2^n \times 2^n$ matrix, and

$$\mathbf{g} = \mathbf{H}_n \mathbf{f}. \tag{10–89}$$

We define a $3^n \times 2^n$ matrix $\mathbf{A}_n = \mathbf{A}(H_n)$ (depending on the matrix H) such that

$$\tilde{\mathbf{f}} = \mathbf{A}_n \mathbf{g}. \tag{10–90}$$

We call such a matrix $\mathbf{A}_n = \mathbf{A}(H_n)$ the *average H-fitness transform matrix*, and the transform 10–90 the *average H-fitness transform*.

It is easy to see that in the case $\mathbf{H}_n = \mathbf{I}_n$ (an identity matrix of order 2^n)

$$\mathbf{A}_n = \mathbf{A}(I_n) = \begin{pmatrix} 1 & 0 \\ 0 & 1 \\ 1 & 1 \end{pmatrix}^{\otimes n}, \tag{10–91}$$

where $\mathbf{B}^{\otimes n}$ is the nth Kronecker power of $\mathbf{B}$.

Note that the matrix $\mathbf{A}(I_n)$ plays the same role as the interval splicing matrix $\mathbf{A}_n$ (used for extracting all intervals of the n-cube [1]) and, in fact, coincides with it.

Since $\tilde{\mathbf{f}}$ is independent of $\mathbf{H}_n$ we have from Eqs. 10–90, 10–89 and 10–91:

$$\tilde{\mathbf{f}} = \mathbf{A}(H_n)\mathbf{H}_n\mathbf{f} = \mathbf{A}(I_n)\mathbf{f} = \begin{pmatrix} 1 & 0 \\ 0 & 1 \\ 1 & 1 \end{pmatrix}^{\otimes n} \mathbf{f},$$

therefore $\mathbf{A}(H_n)\mathbf{H}_n = \mathbf{A}(I_n)$, and

$$\mathbf{A}(H_n) = \begin{pmatrix} 1 & 0 \\ 0 & 1 \\ 1 & 1 \end{pmatrix}^{\otimes n} \mathbf{H}_n^{-1}. \tag{10–92}$$

For simplicity of comparison of the representation 10–90 for different matrices $\mathbf{H}_n$, we introduce an *average H-fitness cost vector* $\mathbf{r} = r(\mathbf{g}) = [r(0), \ldots, r(2^n - 1)]^T$, where $r(\alpha(s))$ shows the average number of nonzero terms $g(s)$ in the representation 10–90, where

$$\alpha(s) = \begin{cases} 0, & \text{if } s = 0, 1, \\ 1, & \text{if } s = *. \end{cases} \tag{10–93}$$

The H-average fitness cost vector gives a quantitative measure of the complexity of calculating schema average fitness values.

10.2.4 RECTANGULAR WAVELET PACKETS AND FITNESS AVERAGE MATRICES

The Haar wavelet packet of the Haar-like unitary transform $\mathbf{H}_n^{(P)}$ corresponds to the tree-structured filter banks (P is a binary tree) with the synthesis filter pairs (low-pass and high-pass) defined by (see e.g. [97]):

$$G_0(z) = \frac{1}{\sqrt{2}}(1 + z^{-1}), \qquad G_1(z) = \frac{1}{\sqrt{2}}(1 - z^{-1}). \tag{10–94}$$

An arbitrary pruned tree structure P of the full binary tree of depth n will give a family of Haar-like unitary bases $\{\mathbf{H}_n^{(P)}\}$. The extreme cases are the following: If P is the trivial tree (only the root), then $\mathbf{H}_n^{(P)}$ is the identity transform; if P is an octave-band tree (iterating only on the low-pass sections) then $\mathbf{H}_n^{(P)}$ is the Haar transform; and if P is the full tree, then $\mathbf{H}_n^{(P)}$ is the Walsh transform. Below we give an explicit formula for the average fitness transform matrix corresponding to an arbitrary basis from the Haar wavelet packet.

Let us first derive an analytic expression for the matrix $[\mathbf{H}_n^{(P)}]^{-1}$ (which is the inverse of the matrix $\mathbf{H}_n^{(P)}$ corresponding to a pruned tree P of the full binary tree of depth n). For this objective we code each node at the kth $(0 < k \leqslant n)$ level of the tree P (the root is on the level 0 and not coded) by a (0, 1)-vector of length k. Let some nonterminal node be coded by the binary vector $\mathbf{c}$. The descendants of this node will have the following codes: $\mathbf{c}0$ (the left one) and $\mathbf{c}1$ (the right one). To the codes of all terminal nodes which are not on the last, nth, level we add from the right the "don't care" character $*$.

Let P have t terminal nodes with the corresponding codes

$$\{\mathbf{c}^{(i)}\}, \quad \mathbf{c}^{(i)} = \left(c_1^{(i)}, \ldots, c_n^{(i)}\right), \tag{10–95}$$

where $c_j^{(i)} \in \{0, 1, *\}$, $i = 1, \ldots, t$.

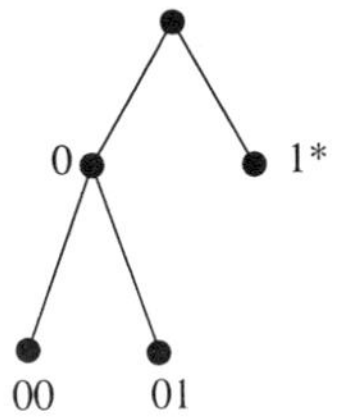

Figure 10–2. A pruned binary tree with corresponding codes of nodes.

PROPOSITION 10.2.1. *Let P be a pruned tree of the regular tree of depth n and have t terminal nodes with their codes given by Eq.* 10–95. *The inverse transform matrix $[\mathbf{H}_n^{(P)}]^{-1}$ for the Haar-like transform corresponding to the tree P will have the following form:*

$$[\mathbf{H}_n^{(P)}]^{-1} = 2^{-\frac{n}{2}}\left[2^{\frac{n_1}{2}}\mathbf{I}_{n_1} \otimes \left(\bigotimes_{j=1}^{k_1}\begin{pmatrix}1\\-1\end{pmatrix}^{\alpha(c_j^{(1)})}\right)\right.$$
$$\left.\cdots\ 2^{\frac{n_t}{2}}\mathbf{I}_{n_t} \otimes \left(\bigotimes_{j=1}^{k_t}\begin{pmatrix}1\\-1\end{pmatrix}^{\alpha(c_j^{(t)})}\right)\right], \qquad (10\text{–}96)$$

where I_k is the identity matrix of order 2^k, $n_i = n - o(\mathbf{c}^{(i)})$, $o(s)$ is the order of a schema s, $\alpha(c)$ is defined by Eq. 10–93,

$$\begin{pmatrix}1\\-1\end{pmatrix}^0 = \begin{pmatrix}1\\1\end{pmatrix} \quad \textit{and} \quad \begin{pmatrix}1\\-1\end{pmatrix}^1 = \begin{pmatrix}1\\-1\end{pmatrix}.$$

PROOF. It follows from the definition 10–94 of the Haar filter pair and the construction of tree codes. ■

As an example, the transform matrix corresponding to the tree P shown in Fig. 10–2 is represented analytically by

$$[\mathbf{H}_n^{(P)}]^{-1} = \frac{1}{2}\left[\mathbf{I}_{n-2} \otimes \begin{pmatrix}1\\1\end{pmatrix} \otimes \begin{pmatrix}1\\1\end{pmatrix} \quad \mathbf{I}_{n-2} \otimes \begin{pmatrix}1\\-1\end{pmatrix} \otimes \begin{pmatrix}1\\1\end{pmatrix}\right.$$
$$\left.\sqrt{2}\mathbf{I}_{n-1} \otimes \begin{pmatrix}1\\-1\end{pmatrix}\right].$$

PROPOSITION 10.2.2. *Let P be a binary tree as in Proposition* 10.2.1. *The average H-fitness transform matrix $\mathbf{A}(H_n)$ corresponding to the Haar-like transform*

$\mathbf{H}_n = \mathbf{H}_n^{(P)}$ *(defined by the tree P) will have the following form*:

$$\mathbf{A}(H_n) = 2^{-\frac{n}{2}}\left[2^{\frac{n_1}{2}}\mathbf{A}(I_{n_1}) \otimes \left(\bigotimes_{j=1}^{k_1} v(c_j^{(1)})\right) \cdots 2^{\frac{n_t}{2}}\mathbf{A}(I_{n_t}) \otimes \left(\bigotimes_{j=1}^{k_t} v(c_j^{(t)})\right)\right], \tag{10–97}$$

where $\mathbf{A}(I_n)$ *is defined by Eq.* 10–91,

$$v(0) = \begin{pmatrix} 1 \\ 1 \\ 2 \end{pmatrix}, \qquad v(1) = \begin{pmatrix} 1 \\ -1 \\ 0 \end{pmatrix},$$

and all other items are as in Proposition 10.2.1.

PROOF. Using Proposition 10.2.1, formula 10–92, and simple properties of the Kronecker product, we obtain an analytical expression for $\mathbf{A}(H_n)$. Vectors $v(0)$ and $v(1)$ are obtained by

$$v(0) = \mathbf{A}(I_1)\begin{pmatrix} 1 \\ 1 \end{pmatrix}, \qquad v(1) = \mathbf{A}(I_1)\begin{pmatrix} 1 \\ -1 \end{pmatrix}. \qquad \blacksquare$$

Using Eq. 10–92 it is easy to derive the Walsh-schema transform (the vector-matrix version of Eq. 10–88), or, equivalently, the Walsh-average fitness transform:

$$\tilde{\mathbf{f}} = \mathbf{A}(W_n)\mathbf{g},$$

where

$$\mathbf{A}(W_n) = 2^{-n/2}\left[\begin{pmatrix} 1 & 0 \\ 0 & 1 \\ 1 & 1 \end{pmatrix}\begin{pmatrix} 1 & 1 \\ 1 & -1 \end{pmatrix}\right]^{\otimes n} = 2^{-n/2}\begin{pmatrix} 1 & 1 \\ 1 & -1 \\ 2 & 0 \end{pmatrix}^{\otimes n} \tag{10–98}$$

and

$$\mathbf{g} = \mathbf{W}_n\mathbf{f}, \quad \mathbf{W}_n = 2^{-\frac{n}{2}}\begin{pmatrix} 1 & 1 \\ 1 & -1 \end{pmatrix}^{\otimes n} \text{ is the Walsh transform matrix.}$$

Similarly, using Eq. 10–97 one can derive other schema transforms, i.e. the Haar-like average fitness transforms.

Let us change the basis functions of Haar wavelets to another rectangular basis, which forms the class of nonorthogonal Reed–Muller wavelet packets. The name comes from the fact that in the extreme case (i.e. when we have a full binary tree decomposition) this construction leads to the Reed–Muller (or conjunctive) transform [1]. In the case of the octave-band tree, it leads to oblique wavelets (see [6]). Analysis and synthesis pairs of filters for the corresponding filter bank are:

$$H_0(z) = 1; \quad H_1(z) = z - 1; \quad G_0(z) = 1 + z^{-1}; \quad G_1(z) = z^{-1}.$$

First we give an analytic expression for the inverse matrix $[\mathbf{K}_n^{(P)}]^{-1}$ of the Reed–Muller packet matrix $\mathbf{K}_n^{(P)}$ corresponding to a pruned tree P of the binary tree of depth n.

PROPOSITION 10.2.3. *Let P be a pruned tree as defined in Proposition 10.2.2. The matrix $[\mathbf{K}_n^{(P)}]^{-1}$ corresponding to the decomposition tree P will have the following form:*

$$[\mathbf{K}_n^{(P)}]^{-1} = \left[I_{n_1} \otimes \left(\bigotimes_{j=1}^{k_1} \mathbf{h}_{\alpha(c_j^{(1)})} \right) \cdots I_{n_t} \otimes \left(\bigotimes_{j=1}^{k_t} \mathbf{h}_{\alpha(c_j^{(t)})} \right) \right], \tag{10–99}$$

where $\mathbf{h}_i$ is the ith column of the inverse Reed–Muller matrix

$$\mathbf{K}_1^{-1} = \begin{pmatrix} 1 & 0 \\ -1 & 1 \end{pmatrix},$$

and all other notations are similar to those in Proposition 10.2.1.

As an example, the transform matrix for the tree in Fig. 10–2 is:

$$[\mathbf{K}_n^{(P)}]^{-1} = \left[\mathbf{I}_{n-2} \otimes \begin{pmatrix} 1 \\ -1 \end{pmatrix} \otimes \begin{pmatrix} 1 \\ -1 \end{pmatrix} \quad \mathbf{I}_{n-2} \otimes \begin{pmatrix} 0 \\ 1 \end{pmatrix} \otimes \begin{pmatrix} 1 \\ -1 \end{pmatrix} \right.$$
$$\left. \mathbf{I}_{n-1} \otimes \begin{pmatrix} 0 \\ 1 \end{pmatrix} \right].$$

PROPOSITION 10.2.4. *Let P be the binary tree specified in Proposition 10.2.1. The average H-fitness transform matrix $\mathbf{A}(K_n)$ corresponding to the Reed–Muller-like transform $\mathbf{K}_n^{(P)}$ (defined by the tree P) will have the following form:*

$$\mathbf{A}(K_n) = \left[\mathbf{A}(I_{n_1}) \otimes \left(\bigotimes_{j=1}^{k_1} w(c_j^{(1)}) \right) \cdots 2^{\frac{n_t}{2}} \mathbf{A}(I_{n_t}) \otimes \left(\bigotimes_{j=1}^{k_t} w(c_j^{(t)}) \right) \right], \tag{10–100}$$

where $\mathbf{A}(I_n)$ *is defined by Eq.* 10–91,

$$w(0) = \begin{pmatrix} 1 \\ -1 \\ 0 \end{pmatrix}, \qquad w(1) = \begin{pmatrix} 0 \\ 1 \\ 1 \end{pmatrix},$$

and all other items are as in Proposition 10.2.1.

PROOF. It follows from Proposition 10.2.3 and formula 10–92. Vectors $w(0)$ and $w(1)$ are obtained by

$$w(0) = \mathbf{A}(I_1)\begin{pmatrix} 1 \\ -1 \end{pmatrix}, \qquad w(1) = \mathbf{A}(I_1)\begin{pmatrix} 0 \\ 1 \end{pmatrix}.$$

We have the following average-fitness matrix for the Reed–Muller transform:

$$\mathbf{A}(K_n) = \left[\begin{pmatrix} 1 & 0 \\ 0 & 1 \\ 1 & 1 \end{pmatrix}\begin{pmatrix} 1 & 0 \\ -1 & 1 \end{pmatrix}\right]^{\otimes n} = \begin{pmatrix} 1 & 0 \\ -1 & 1 \\ 0 & 1 \end{pmatrix}^{\otimes n}. \qquad (10\text{–}101)$$

Comparing this with the Walsh average-fitness matrix 10–98, one can see that there is no scaling factor 2 for low-order schemata, and it has a much shorter sum for specification of high-order schemata. ■

In order to compare the complexities of calculation of schema average fitness values, we analyze average fitness cost vectors corresponding to different rectangular wavelet packet transforms.

10.2.5 RECTANGULAR WAVELET PACKETS AND FITNESS AVERAGE COST VECTORS

Recall that each orthogonal Haar-like transform corresponds to a pruned tree structure. Fixing a pruned tree G, we will construct a pruned tree $G' = [G, j]$ just by adding branches to the jth node [i.e. the node with the code $j = (j_1, j_2, \ldots, j_p)$, $p < n$] of G.

PROPOSITION 10.2.5. *An average H-fitness cost vector* $\mathbf{r}(G')$ *corresponding to the tree* $G' = [G, j]$ *can be obtained from an average H-fitness cost vector* $\mathbf{r}(G)$ *corresponding to G according to the following procedure*:

$$\mathbf{r}(G') = \mathbf{r}(G) + \mathbf{r}'(j),$$

where, as an initialization step, the average I_n-fitness cost vector $\mathbf{r}(.)$ *(i.e. for the trivial tree) has the form* $\mathbf{r}(.) = (1\ \ 2)^{\otimes n}$, *and*

$$\mathbf{r}'(j) = \bigotimes_{i=0}^{p-1} (1\ \ \overline{j}_{p-i}) \otimes (1\ \ 2)^{\otimes(n-p-1)} \otimes (1\ \ -1),$$

$j = (j_1, j_2, \ldots, j_p)$, $(j_k \in \{0, 1\})$, $\overline{j_k}$ *is the negation of* j_k, $k = 1, \ldots, p$, *and* $(1\ \ 2)^{\otimes a}$ *is the a-th Kronecker power of* $(1\ \ 2)$.

PROOF. It follows from the fact that the transform matrix $\mathbf{H}_{G'}$ corresponding to the tree $G' = [G, j]$ can be constructed from the matrix $\mathbf{H}_G$ by $\mathbf{H}_{G'} = \mathbf{H}_G\mathbf{D}_j$, where $\mathbf{D}_j$ is the block-diagonal matrix $\mathbf{D}_j = \mathrm{diag}(\mathbf{B}_0, \ldots, \mathbf{B}_{2^p-1})$ and $\mathbf{B}_j = [\mathbf{I}_{n-p-1} \otimes (1\ \ 1)^T\ \mathbf{I}_{n-p-1} \otimes (1\ \ -1)^T]$ and other $\mathbf{B}_k, k \neq j$ are the identity matrices $\mathbf{I}_{n-p}$. ■

In Fig. 10–3 a recurrent construction of the average H_3-fitness cost vector $\mathbf{r}$ for the Haar wavelet packet transforms is shown. A similar construction for $\mathbf{r}(H)$ can be done also for the Reed–Muller wavelet packets.

Analyzing vectors $\mathbf{r}(H)$, one can see that the minimum number of terms for specifying the low-order schemata [corresponding to the kth elements of the vector $\mathbf{r}(H)$, $\mathbf{k} = (k_1, \ldots, k_n)$, such that $w_H(\mathbf{k})$, the Hamming weights of $\mathbf{k}$, are high] among the Haar wavelet packet transforms and Reed–Muller wavelet

Identity	1	2	2	4	2	4	4	8	+
-	1	-1	2	-2	2	-2	4	-4	+
1	1	0	-1	0	2	0	-2	0	
0	1	1	-1	-1	2	2	-2	-2	+
11	1	0	0	0	-1	0	0	0	
01	1	1	0	0	-1	-1	0	0	
10	1	0	1	0	-1	0	-1	0	
00	1	1	1	1	-1	-1	-1	-1	+
Walsh	8	4	4	2	4	2	2	1	Sum of all the rows from the table
Haar	4	3	4	2	5	3	5	1	Sum of the rows marked by "+"

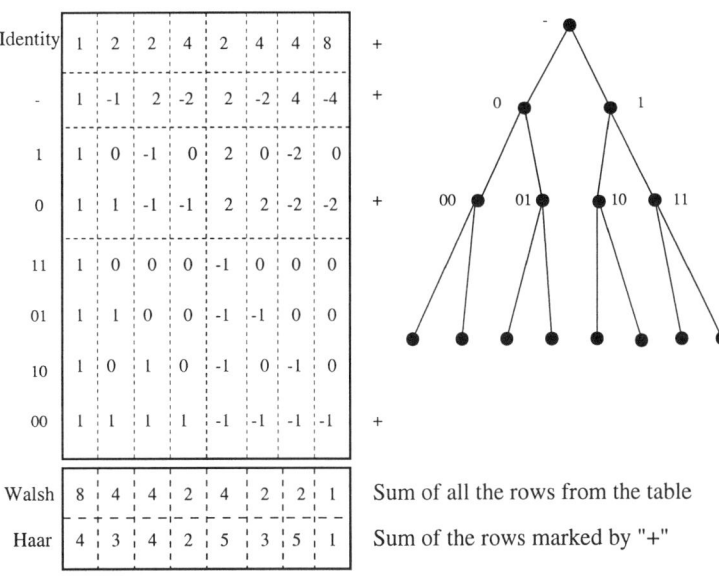

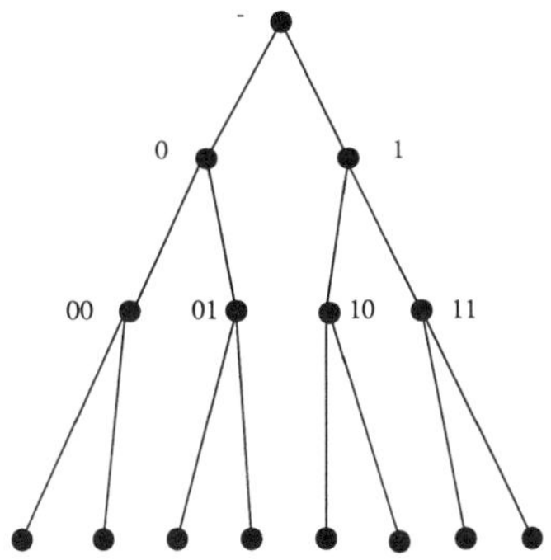

Figure 10–3. Recurrent construction of an average H-fitness cost vector **r** for the Haar wavelet packet transforms.

packet transforms is achieved for the Reed–Muller transform (since in this case $r_k = 1.5^{n-w_H(\mathbf{k})}$).

10.3 BINARY POLYNOMIAL TRANSFORMS AND CLASSIFICATION PROBLEM

The algorithms for distribution-free classification based on descriptors, tests of a table and their generalizations, testors, were introduced in [5, 20, 111] and they are at present used in many pattern recognition systems [88]. They are characterized by a high rate of correct classification in general and by absolutely correct classification for the training set. The main computational burden of these algorithms is in finding the unconditional table tests, testors, and descriptors. This means that for efficient implementation of these algorithms we need to find fast ways to construct minimal tests of a table.

The spectral algorithms for distribution-free classification based on tests and generalized tests of a table were presented in [37, 40]. In this section we consider spectral approaches to classification using generalized tests and descriptors.

10.3.1 CLASSIFICATION USING GENERALIZED TESTS

Description of the Classification Model

We briefly review the distribution-free classification model [88]. Let a set Ω of objects be given and assume that it is partitioned into s disjoint subsets (pattern classes) $C_1, \ldots, C_s$, i.e. $\Omega = \bigcup C_i$ and $C_i \cap C_j = \emptyset$ for $i \neq j$. Each object is characterized by n features $x_1, \ldots, x_n$, where $x_i \in \mathbf{Z}_{p_i}$. A priori information is given as a set of m objects $\omega_1, \ldots, \omega_m$ together with their feature vectors $\mathbf{z}_i = \Theta(\omega_i)$ and the class $C(\omega_i)$ to which ω_i belongs. The task is to classify a new object ω using its feature vector $\mathbf{z}(\omega)$ as well as possible.

The a priori information or training result can be presented as a table $\mathcal{T}$ showing the features and class division of the training objects. There are t_i objects in the class C_i, $i = 1, \ldots, s$, and the total number of training objects is $m = t^{(s)} = \sum_{i=1}^{s} t_i$ (see Table 10.5).

The recognition algorithm compares the features of the new object with those in the table and assigns the new object to one of the classes C_i. The classification is based on computing some measure of similarity (estimation) between $\mathbf{z}(\omega)$ and $\mathbf{z}(\omega_1), \ldots, \mathbf{z}(\omega_m)$.

Table 10.5. Table $\mathcal{T}$ of the training set.

Classes	**Objects**	x_1	x_2	...	x_n
C_1	ω_1	$z_1(\omega_1)$	$z_2(\omega_1)$	...	$z_n(\omega_1)$
	ω_2	$z_1(\omega_2)$	$z_2(\omega_2)$	...	$z_n(\omega_2)$
	...	...	...	...	...
	ω_{t_1}	$z_1(\omega_{t_1})$	$z_2(\omega_{t_1})$	...	$z_n(\omega_{t_1})$
⋮	⋮	⋮	⋮	⋮	⋮
C_s	$\omega_{t(s-1)+1}$	$z_1(\omega_{t(s-1)+1})$	$z_2(\omega_{t(s-1)+1})$	...	$z_n(\omega_{t(s-1)+1})$
	$\omega_{t(s-1)+2}$	$z_1(\omega_{t(s-1)+2})$	$z_2(\omega_{t(s-1)+2})$	...	$z_n(\omega_{t(s-1)+2})$
	...	...	...	...	...
	ω_m	$z_1(\omega_m)$	$z_2(\omega_m)$	...	$z_n(\omega_m)$
?	ω	$z_1(\omega)$	$z_2(\omega)$	...	$z_n(\omega)$

Step-by-step Description of the Classification Model

An algorithm $\mathcal{A}$ of pattern recognition (classification) based on the computation of estimations (ACE) consists of the following [88]:

1. The system $Q_{\mathcal{A}}$ of subsets of the set $\mathcal{N} = \{1, 2, \ldots, n\}$ is called the system of support sets of the algorithm.

2. For each support set Q and a training object ω_j, the following function is defined for objects

$$B_Q(\omega_j, \omega) = \begin{cases} 1, & \text{if } |z_i(\omega_j) - z_i(\omega)| < \varepsilon_i \text{ for } i \in Q, \\ 0, & \text{otherwise.} \end{cases} \tag{10–102}$$

The parameters ε_i are determined a priori by an expert.

3. The following similarity measure between training objects and ω with respect to the support set Q is formed

$$\Gamma_{\mathbf{Q}}(\omega_j, \omega) = \gamma_j B_Q(\omega_j, \omega)\left(\sum_{i \in Q} p_i\right), \tag{10–103}$$

where γ_i is a nonnegative parameter describing the importance of object ω_j, and p_i are nonnegative parameters describing the importance of the features x_i, $i \in \mathcal{N}$.

4. The following indices of the object ω with respect to the classes C_r, $r = 1, \dots, s$, are calculated

$$\Gamma_Q^{(r)}(\omega) = \frac{1}{t_r} \sum_{j=t^{(r-1)}+1}^{t^{(r)}} \Gamma_Q(\omega_j, \omega), \tag{10–104}$$

where $t^{(r)} = \sum_{i=1}^{r} t_i$, $t^{(0)} = 0$.

5. The index of the object for class C_r, $r = 1, \dots, s$, is now computed as

$$\Gamma(\omega, r) = \sum_{Q \in Q_{\mathcal{A}}} \Gamma_Q^{(r)}(\omega). \tag{10–105}$$

6. The algorithm $\mathcal{A}$ now has the following decision rule $F_{\mathcal{A}}$:

$$F_{\mathcal{A}}(\omega) = \begin{cases} v, & \text{if } \Gamma(\omega, v) > \Gamma(\omega, r) + \delta_2,\ r \neq v,\ 1 \leqslant r \leqslant s, \\ & \text{and } \Gamma(\omega, v) > \delta_1 \sum_{r=1}^{s} \Gamma(\omega, r), \\ 0, & \text{otherwise,} \end{cases} \tag{10–106}$$

where δ_1 and δ_2 are parameters given beforehand.

The crucial part of the classification algorithm $\mathcal{A}$ is the choice of a set $Q_{\mathcal{A}}$. If the set $Q_{\mathcal{A}}$ coincides with a set of all deadlock (generalized) tests [88] then the algorithm $\mathcal{A}$ is called the algorithm of pattern recognition (or classification) by (generalized) tests. In this case the index $\Gamma(\omega, r)$ of the object ω for class C_r $(r = 1, \dots, s)$ can be calculated by the following formula [88]:

$$\Gamma(\omega, r) = \frac{1}{t_r} \sum_{j=t^{(r-1)}+1}^{t^{(r)}} \gamma_j \sum_{Q \in Q_{\mathcal{A}}} \left(\sum_{i \in Q} p_i \right) B_Q(\omega_j, \omega).$$

In the following section we give a brief introduction to the theory of tests of tables and system of representatives, consider a generalization of tests, and present an efficient spectral algorithm for the construction of deadlock generalized tests of a table.

Unconditional Deadlock Tests of a Table

The notation of an unconditional test of a table was introduced by Yablonski [104] in 1955. Let some table (matrix) $\mathbf{T}_{mns}$ with the elements from $\{0, 1\}$ be given, which consists of m different rows and n columns such that the set of the rows is divided into s disjoint classes $\mathbf{C}_r$, $r = 1, 2, \dots, s$.

DEFINITION 10.3.1. The subset of columns of $\mathbf{T}_{mns}$, where any two rows belonging to different classes $\mathbf{C}_r$, $r = 1, 2, \ldots, s$, are different, is called an *unconditional test of the table* $\mathbf{T}_{mns}$.

The rows of the matrix $\mathbf{T}_{mns}$ can be interpreted as the objects and the columns as the features of these objects. An unconditional test is a collection of features by which the objects from different classes can be distinguished. The cardinality of this set is called the *cardinality of the unconditional test.*

DEFINITION 10.3.2. An unconditional test is called *deadlock unconditional test* if it does not contain a proper subset which is an unconditional test.

DEFINITION 10.3.3. A deadlock unconditional test of minimal cardinality is called a *minimal test.*

As an illustration consider the following example.

EXAMPLE 10.3.1. Let $\mathbf{T} = \mathbf{T}_{mns}$ be the table where $m = 3$, $n = 3$, $s = 2$:

$$\mathbf{T} = \begin{pmatrix} 0 & 1 & 0 \\ --- & --- & -- \\ 1 & 0 & 0 \\ 0 & 0 & 1 \end{pmatrix}, \tag{10–107}$$

the top of which is the class $\mathbf{C}_1$, and the bottom of which is the class $\mathbf{C}_2$. From the table $\mathbf{T}$ one can see that in the first and second columns every possible pair of rows from the different classes $\mathbf{C}_1$, $\mathbf{C}_2$ is different, i.e. $(0, 1) \neq (1, 0)$ and $(0, 1) \neq (0, 0)$. So the first and second columns form an unconditional test for the table $\mathbf{T}$.

However, the pair of first and second columns of the table $\mathbf{T}$ do not form an unconditional deadlock test, because the elements from the classes $\mathbf{C}_1$ and $\mathbf{C}_2$ in the intersection with the second column of $\mathbf{T}$ are pairwise different, i.e. the second column of $\mathbf{T}$ forms an unconditional deadlock test, which is also minimal. The pair of the first and third columns forms another deadlock test of the table $\mathbf{T}$.

Let us consider algorithms for finding unconditional deadlock tests of a table. Let $\mathbf{T} = \mathbf{T}_{mns}$ be an $(m \times n)$ matrix (table) with elements from $\{0, 1\}$, whose set of rows is partitioned into s subsets (submatrices) $\{\mathbf{C}_1, \ldots, \mathbf{C}_s\}$, and there are no similar rows in the different submatrices $\mathbf{C}_i$, $i = 1, 2, \ldots, s$.

Transform the matrix $\mathbf{T}$ into a new matrix $\mathbf{M}$, by

$$\mathbf{M} = \bigcup_{1 \leqslant i < j \leqslant r} (\mathbf{C}_i \oplus \mathbf{C}_j), \tag{10–108}$$

where $\mathbf{C}_i \oplus \mathbf{C}_j$ denotes the set of all componentwise modulo 2 sums of rows $\mathbf{a} \oplus \mathbf{a}'$, where $\mathbf{a} \in \mathbf{C}_i$, $\mathbf{a}' \in \mathbf{C}_j$.

DEFINITION 10.3.4. A subset of the set of columns of the matrix $\mathbf{M}$ is called the *system of representatives* of $\mathbf{M}$, if the submatrix $\mathbf{M}'$ specified by these columns does not contain the zero row (the row all elements of which are equal to zero).

DEFINITION 10.3.5. A system of representatives is called a *deadlock system of representatives* if it does not contain a proper (nonempty) subset which is a system of representatives.

EXAMPLE 10.3.2. Consider the table $\mathbf{T}$ given in Eq. 10–107. Form the matrix $\mathbf{M}$ by Eq. 10–108:

$$\mathbf{M} = \begin{pmatrix} 1 & 1 & 0 \\ 0 & 1 & 1 \end{pmatrix}. \qquad (10\text{–}109)$$

From the matrix $\mathbf{M}$ one can see that the second column of $\mathbf{M}$ contains no zeros and therefore forms the deadlock system of representatives, as well as the pair of first and third columns of $\mathbf{M}$. It is easy to see that the sets of (deadlock) unconditional tests for matrix $\mathbf{T}$ always coincide with the (deadlock) system of representatives of the matrix $\mathbf{M}$.

Denote by $\mathbf{f}_{C_j} = (f_{C_j}(0), \ldots, f_{C_j}(2^n - 1))$ and $\mathbf{f}_M = (f_M(0), \ldots, f_M(2^n - 1))$, respectively, the characteristic functions (indicators) of the matrices $\mathbf{C}_j$ and $\mathbf{M}$, $j = 1, 2, \ldots, s$, i.e.

$$f_{C_j}(k) = \begin{cases} 1, & \text{if } (k_1, \ldots, k_n) \in \mathbf{C}_j, \\ 0, & \text{otherwise} \end{cases} \qquad (10\text{–}110)$$

and

$$f_M(k) = \begin{cases} 1, & \text{if } (k_1, \ldots, k_n) \in \mathbf{M}, \\ 0, & \text{otherwise,} \end{cases} \qquad (10\text{–}111)$$

where $k = 0, 1, \ldots, 2^n - 1$; and $(k_1, \ldots, k_n)$ is the binary vector representation of k.

The following proposition gives a spectral algorithm for going from the table $\mathbf{T}$ specifying by classes $\mathbf{C}_j$, $j = 1, 2, \ldots, s$, to the table $\mathbf{M}$, or, in other words, a spectral algorithm for the computation of $\mathbf{M}$ by Eq. 10–108.

Let $\mathbf{W}_n$ be the Walsh–Hadamard matrix of order 2^n. Let $\delta_0(g_1, \ldots, g_{2^n}) = (\delta_0(g_1), \ldots, \delta_0(g_{2^n}))$ be the vectoral Kronecker delta function, defined by

$$\delta_0(g_j) = \begin{cases} 1, & \text{if } g_j = 0, \\ 0, & \text{otherwise.} \end{cases} \qquad (10\text{–}112)$$

PROPOSITION 10.3.1. *The characteristic function* $\mathbf{f}_M$ *can be expressed as*

$$\mathbf{f}_M = \delta_0^2\left[\sum_{i\neq j=1}^{s} \mathbf{f}_{C_i} *_{\oplus} \mathbf{f}_{C_j}\right] \tag{10–113}$$

or as

$$\mathbf{f}_M = \frac{1}{2^n}\delta_0^2 \mathbf{W}_n\left[\sum_{i\neq j=1}^{s} (\mathbf{W}_n\mathbf{f}_{C_i} \circ \mathbf{W}_n\mathbf{f}_{C_j})\right], \tag{10–114}$$

where $\circ$ *is the sign of componentwise (Hadamard) multiplication and* $*_{\oplus}$ *is the operation of dyadic (Walsh) convolution* [1].

The Generalized Unconditional Tests or ε-Tests

The definition of a generalized unconditional test of a table or $\boldsymbol{\varepsilon}$-test [88] is a natural extension of the definition of an unconditional test. Let (as before) a table (matrix) $\mathbf{T}_{mns}$ be given, which consists of m different rows and n columns such that the set of the rows is divided into s nonintersecting classes $\mathbf{C}_r,\ r = 1, 2, \ldots, s$.

DEFINITION 10.3.6. The subset $\{i_1, \ldots, i_k\}$ of columns of $\mathbf{T}_{mns}$ is called an $\boldsymbol{\varepsilon}$-test of the table $\mathbf{T}_{mns}$, if for any two rows $\mathbf{c}^{(r)} = (\alpha_{\cdot,1}^{(r)}, \ldots, \alpha_{\cdot,n}^{(r)})$ and $\mathbf{c}^{(v)} = (\alpha_{\cdot,1}^{(v)}, \ldots, \alpha_{\cdot,n}^{(v)})$ from the different classes $\mathbf{C}_r$, and $\mathbf{C}_v$, $r \neq v = 1, \ldots, s$, the system of inequalities

$$\left|\alpha_{\cdot,t}^{(r)} - \alpha_{\cdot,t}^{(v)}\right| \leqslant \varepsilon_t, \quad t = i_1, \ldots, i_k,$$

is incompatible.

If $\varepsilon_i = 0,\ i = 1, \ldots, n,$ an $\boldsymbol{\varepsilon}$-test becomes an ordinary unconditional test for the table.

DEFINITION 10.3.7. An $\boldsymbol{\varepsilon}$-test is called *deadlock* $\boldsymbol{\varepsilon}$*-test* if it does not contain a proper subset which is an $\boldsymbol{\varepsilon}$-test.

As an illustration consider the following example.

EXAMPLE 10.3.3. Let $\mathbf{T} = \mathbf{T}_{mns}$ be the table where $m = 3,\ n = 4,\ s = 2$:

$$\mathbf{T} = \begin{pmatrix} 3 & 2 & 3 & 1 \\ --- & --- & --- & --- \\ 1 & 1 & 0 & 2 \\ 1 & 0 & 0 & 3 \end{pmatrix}, \tag{10–115}$$

the top of which is the class $\mathbf{C}_1$, and the bottom of which is the class $\mathbf{C}_2$. Let us find the ε-test of the table $\mathbf{T}$, where $\boldsymbol{\varepsilon} = (\varepsilon_1, \varepsilon_2, \varepsilon_3, \varepsilon_4) = (2, 2, 1, 1)$. From the table $\mathbf{T}$ one can see that the third and fourth columns form a (2, 2, 1, 1)-test. But this test is not a deadlock test, because it contains another test. Thus, the third column of $\mathbf{T}$ forms the deadlock (2, 2, 1, 1)-test, since $|3 - 0| > 1$.

Spectral Algorithm for Finding Deadlock ε-Tests

We reduce the general case of finding an $\boldsymbol{\varepsilon}$-test $[\boldsymbol{\varepsilon} = (\varepsilon_1, \ldots, \varepsilon_n)]$ of a table to the finding of an ordinary test of a binary table by the following construction.

1. Take an element-by-element difference by absolute value of each pair of row vectors belonging to the different classes C_r, $r = 1, \ldots, s$, to form a matrix $\mathbf{P} = [P_{ij}]$, $i = 1, \ldots, t$; $j = 1, \ldots, n$;

$$t = \sum_{i \neq j = 1}^{s} t_i t_j, \tag{10–116}$$

where t_i is the number of objects in the class C_i.

2. Transfer a constructed matrix $\mathbf{P}$ to a matrix $\mathbf{M} = [M_{ij}]$ of the same size, by the following thresholding operator:

$$M_{ij} = \begin{cases} 1, & \text{if } P_{ij} > \varepsilon_j, \\ 0, & \text{otherwise.} \end{cases} \tag{10–117}$$

Let $\mathbf{T}$ be the matrix divided into groups of submatrices $\{\mathbf{C}_1, \ldots, \mathbf{C}_s\}$, $\mathbf{M}$ be the matrix determined by Eq. 10–117, and $\mathbf{f}_M$ be the characteristic function of the matrix $\mathbf{M}$ defined by Eq. 10–111.

We will use the following notations. Let $\mathbf{x} = (x_1, \ldots, x_n)$ be a binary vector of length n; we will denote $x = \sum_{i=1}^{n} x_i 2^{n-i}$. The relation $\mathbf{x} < \mathbf{y}$, where both are binary vectors, means componentwise operation $x_i < y_i$ for all i.

Let $\mathbf{K}_n = K(i, j)$ and $i, j = 0, \ldots, 2^n - 1$, be the Reed–Muller (conjunctive) matrix [1, 4] of order 2^n. The spectral algorithm for computation of deadlock $(\varepsilon_1, \ldots, \varepsilon_n)$-tests of the table $\mathbf{T}$ is the generalization of the result for ordinary tests [37].

THEOREM 10.3.1. *The set of the columns $\{i_1, \ldots, i_d\}$ forms a deadlock $(\varepsilon_1, \ldots, \varepsilon_n)$-test for the table* $\mathbf{T}$ *if and only if the column vector*

$$\mathbf{g} = \mathbf{K}_n^T \delta_0(\mathbf{K}_n \mathbf{f}_M) \tag{10–118}$$

takes the following values:

$$g(v) = 1, \quad g(b) > 1 \quad \textit{and} \quad g(w) = 0, \tag{10–119}$$

where $\mathbf{v} = (v_1, \ldots, v_n)$ *is a binary vector with zeros in the positions* $i_1, \ldots, i_d$, *and ones otherwise,* $\mathbf{b} < \mathbf{v} < \mathbf{w}$; $\mathbf{K}_n^T$ *is the transposed conjunctive matrix and* δ_0 *is the Kronecker delta function defined in Eq.* 10–112.

PROOF. The following notations will be used in the proof: $\mathbf{v}$ is the binary vector of length n with zeros in the positions $i_1, \ldots, i_d$; $\mathbf{a}$ and $\mathbf{w}$ are also binary vectors of length n, related by $\mathbf{a} \leqslant \mathbf{v} < \mathbf{w}$.

Necessity. Let zeros of the vector $\mathbf{v}$ form a deadlock ε-test of the table $\mathbf{T}$. Then by the construction of the matrix (table) $\mathbf{M}$ by Eq. 10–117, zeros of $\mathbf{v}$ will also form a deadlock test of the table $\mathbf{M}$. By the definition of test we have $f_M(a) = 0$.

We have to prove first that

$$h(x) = \begin{cases} 1, & \text{if } \mathbf{x} \leqslant \mathbf{v}, \\ 0, & \text{if } \mathbf{x} > \mathbf{v}, \end{cases} \tag{10–120}$$

where $h(x)$ is the xth element of the column vector $\mathbf{h} = \delta_0(\mathbf{K}_n \mathbf{f}_M)$.

The condition $h(a) = 1$ holds because $K(a, w) = 0$, $f_M(w) = 1$ and $f_M(a) = 0$ [and, therefore, $[K_n f_M](a) = 0$].

The condition $h(w) = 0$ follows from $K(w, w) = 1$, $f_M(w) = 1$. Thus Eq. 10–120 is proved.

Now, $g(w) = 0$ because $K^T(w, a) = 0$, $h(a) = 1$ and $h(c) = 0$ for $\mathbf{c} \geqslant \mathbf{w}$.

The condition $g(b) > 1$ holds since $K^T(b, a) = 1$, $h(a) = 1$ for $\mathbf{b} < \mathbf{a} \leqslant \mathbf{v}$.

Finally, $g(v) = 1$, since $K^T(v, v) = 1$, $K^T(v, b) = 0$, $h(a) = 1$ and $h(w) = 0$.

Sufficiency. Let the column vector $\mathbf{g}$ take the values from Eq. 10–119. We must prove that $\{i_1, \ldots, i_d\}$ forms a deadlock test of the table $\mathbf{M}$.

First, $h(a) = 1$ and $h(w) = 0$. This means that $[K_n f_M](a) = 0$ and $[K_n f_M](w) \geqslant 1$. Now $f_M(a) = 0$ and, therefore, any set of indices containing $i_1, \ldots, i_d$ forms a test of $\mathbf{M}$.

Let us suppose that $\{i_1, \ldots, i_d\}$ forms a test, but not a deadlock test of $\mathbf{M}$. This means that there exists a subset S of the set $\{i_1, \ldots, i_d\}$ which is the test of $\mathbf{M}$. Let $\mathbf{w}$ be the binary vector of length n having zeros on the indices from S, and ones otherwise. Obviously, $\mathbf{w} \geqslant \mathbf{v}$. But then the following relation holds: $f_M(x) = 0$ for all $\mathbf{x} \leqslant \mathbf{w}$, which contradicts $[K_n f_M](w) \geqslant 1$ [since $K(w, x) = 1,\ f_M(x) = 0$ and $K(w, y) = 0,\ \mathbf{y} > \mathbf{w}$].

Thus, $\{i_1, \dots, i_d\}$ forms a deadlock test of $\mathbf{M}$. By the construction of $\mathbf{M}$ [see Eq. 10–117] it follows that $\{i_1, \dots, i_d\}$ is also a deadlock $\boldsymbol{\varepsilon}$-test of $\mathbf{T}$. ■

ALGORITHM 10.3.1. To find deadlock $(\varepsilon_1, \dots, \varepsilon_n)$-tests for tables:

1. Construct a matrix $\mathbf{P} = [P_{ij}]$, where $P_{ij} = |\alpha^{(r)}_{i_1,j} - \alpha^{(k)}_{i_2,j}|,\ j = 1, \dots, n;\ 1 \leqslant i_1 \leqslant t_r,\ 1 \leqslant i_1 \leqslant t_k;\ r \neq k,\ r, k = 1, \dots, s.$

2. Construct a binary matrix $\mathbf{M} = [M_{ij}]$ from the matrix $\mathbf{P}$ by Eq. 10–117.

3. Form a characteristic vector $\mathbf{f}_M$ of $\mathbf{M}$ by Eq. 10–111.

4. Calculate the conjunctive spectrum of $\mathbf{f}_M$:

$$\mathbf{y} = \mathbf{K}_n \mathbf{f}_M.$$

5. Find the Kronecker delta function of $\mathbf{y}$:

$$\mathbf{w} = \delta_0(\mathbf{y}).$$

6. Calculate the conjunctive spectrum of $\mathbf{w}$:

$$\mathbf{g} = (g_0, \dots, g_{2^n-1})^T = \mathbf{K}_n \mathbf{w} = \mathbf{K}_n \delta_0(\mathbf{K}_n \mathbf{f}_M).$$

7. Find all deadlock $\boldsymbol{\varepsilon}$-tests of $\mathbf{T}$ by the following rule. Set $\{i_1, \dots, i_d\}$ forms a deadlock $\boldsymbol{\varepsilon}$-test of $\mathbf{T}$ if and only if $g_v = 1$, where the kth bit of the binary representation $\mathbf{v} = (v_n, \dots, v_1)$ of the number $v = \sum_{j=1}^{n} v_j 2^{j-1}$ has the form

$$v_k = \begin{cases} 1, & \text{if } k \in \{i_1, \dots, i_d\}, \\ 0, & \text{otherwise,} \end{cases} \quad k = 1, 2, \dots, n, \quad 1 \leqslant d \leqslant n. \tag{10–121}$$

REMARK 10.3.1. *The complexity of Algorithm* 10.3.1 *is equal to* $n2^n + nt$ *additions, and* $2^n + nt$ *comparisons, where* t *is defined by Eq.* 10–116.

For illustration of Algorithm 10.3.1 consider the following:

EXAMPLE 10.3.4. Let the table **T** have the following form:

$$\mathbf{T} = \begin{pmatrix} 2 & 1 & 2 \\ 3 & 1 & 0 \\ -- & -- & -- \\ 1 & 3 & 5 \\ 2 & 4 & 1 \\ 3 & 5 & 3 \end{pmatrix} \tag{10–122}$$

and $\varepsilon = (0, 2, 0)$. The problem is to find the deadlock (ε-tests of the matrix **T**, divided into two submatrices $\mathbf{T} = \mathbf{T}_{5,3,2} = \{\mathbf{C}_1, \mathbf{C}_2\}$.

First we form a matrix **P**:

$$\mathbf{P} = \begin{pmatrix} 1 & 2 & 3 \\ 0 & 3 & 1 \\ 1 & 4 & 1 \\ 2 & 2 & 5 \\ 1 & 3 & 1 \\ 0 & 4 & 2 \end{pmatrix},$$

after which, using thresholding with the vector $\boldsymbol{\varepsilon} = (0, 2, 0)$, we find a matrix **M**:

$$\mathbf{M} = \begin{pmatrix} 1 & 0 & 1 \\ 0 & 1 & 1 \\ 1 & 1 & 1 \\ 1 & 0 & 1 \\ 1 & 1 & 1 \\ 0 & 1 & 1 \end{pmatrix}.$$

Table 10.6. Illustration of Algorithm 10.3.1. The characteristic vectors $\mathbf{f}_{C_j}$ and their Walsh spectra.

x_1, x_2, x_3	$\mathbf{f}_M$	$\mathbf{K} \cdot \mathbf{f}_M$	$\delta_0(\mathbf{K} \cdot \mathbf{f}_M)$	$\mathbf{g}$	Deadlock (0, 2, 0)-tests
000	0	0	1	5	
001	0	0	1	1	$\{x_3\}$
010	0	0	1	2	
011	1	1	0	0	
100	0	0	1	2	
101	1	1	0	0	
110	0	0	1	1	$\{x_1, x_2\}$
111	1	3	0	0	

The next steps of Algorithm 10.3.1 are shown in Table 10.6.

Since the column vector **g** in Table 10.6 has a value of one on two ranges $(1, 1, 0)$ and on $(0, 0, 1)$, then the first and second columns of matrix **T** as well as the third column of the matrix **T** form a deadlock (0, 2, 0)-test.

10.3.2 EVOLUTIONARY HEURISTIC APPROACH TO GENERALIZED TESTS

We will propose an evolutionary heuristic approach to a minimal generalized unconditional test problem (MGUTP). It is related to the always "optimal" transform approach.

From Transform Approach to Evolutionary Heuristic Approach

According to the transform approach to MGUTP we did the following:

1. First found all $\boldsymbol{\varepsilon}$-tests (after Step 4 of Algorithm 10.3.1) by using conjunctive transform.

2. Found all deadlock $\boldsymbol{\varepsilon}$-tests (after Step 7 of Algorithm 10.3.1) by using conjunctive transposed transform.

3. The minimal generalized test was chosen as the one with the smallest cardinality among all such deadlock tests.

This strategy is used to obtain the formula for the fitness function that will be optimized by the genetic algorithm. For such an application we even can search for a minimal generalized test from all such generalized tests, i.e. the procedure can be shortened immediately after Step 4 of Algorithm 10.3.1.

Let us represent the minimal $\boldsymbol{\varepsilon}$-test V' by a binary vector $\mathbf{y} = (y_1, \ldots, y_N)$ (which is the binary code of decimal number y with the most significant bit y_1), where $y_i = 0$ if the ith column of T is in the V', and $y_i = 1$ otherwise.

Now in Eq. 10–89 using the fact that columns of the conjunctive matrix $\mathbf{K}_N$ are the truth tables of elementary conjunctions, we obtain the following expression for the elements of the column vector **g** presented by Eq. 10–89:

$$g(y) = \sum_{\alpha \in M} \prod_{i=1}^{N} y_i^{\alpha_i}, \qquad (10\text{–}123)$$

where summing is done by all rows $\boldsymbol{\alpha}$ of the matrix M (see Step 2 of Algorithm 10.3.1). Now if $g(y) = 0$, then $(y_1, \ldots, y_N)$ represents the $\boldsymbol{\varepsilon}$-test of T. Thus, for a minimal such test of T we need to *maximize* the following function (objective function for the genetic algorithm):

$$\psi(\mathbf{y}) = \sum_{i=1}^{N} y_i - N \cdot g(y), \tag{10–124}$$

where $g(y)$ is the element of the conjunctive transform vector given in Eq. 10–123 and plays the role of a penalty in a genetic algorithm.

Dual Fitness Function for Minimal Generalized Unconditioned Test Problem

In the previous section we developed the objective function $\psi(\mathbf{y})$ to be maximized by the genetic algorithm for MGUTP. Here we consider a dual fitness function to be minimized.

To this end let us again represent the minimal $\boldsymbol{\varepsilon}$-test V' by a binary vector $\mathbf{x} = (x_1, \ldots, x_N)$, where $x_i = 1$ if the ith node is in V', and $x_i = 0$ otherwise. Using this presentation, the fitness function presented in Eq. 10–124 will be transformed to the following dual fitness function $\eta(\mathbf{x})$ to be *minimized* by the genetic algorithm:

$$\eta(\mathbf{x}) = \sum_{i=1}^{N} x_i + N \cdot \text{penalty}(\mathbf{x}), \tag{10–125}$$

where

$$\text{penalty}(\mathbf{x}) = \sum_{\alpha \in M} \prod_{i=1}^{N} (1 - x_i)^{\alpha_i}, \tag{10–126}$$

and summing is carried out by all rows $\boldsymbol{\alpha}$ of the matrix M.

10.3.3 CLASSIFICATION BY DESCRIPTORS

The Model of Classification

We briefly review the descriptor-based distribution-free classification model [20]. Suppose that there are r different pattern classes $S_1, \ldots, S_r$. Each class S_i is characterized by an $(M_i \times N)$ binary (0, 1) matrix $\mathbf{T}_i = [t_{jk}^{(i)}],\ i = 1, \ldots, r$, where all the $M_1 + M_2 + \cdots + M_r$ rows of the matrices $T_1, \ldots, T_r$ are different. The columns of the matrices correspond to features and the rows correspond to (prototype) objects. For instance, $t_{jk}^{(i)} = 1$ means that the jth object in the ith class has feature number k. The objects are associated with the weights $\mu_j^{(i)},\ j = 1, \ldots, M_i,\ i = 1, \ldots, r$, and the features are associated with the weights $\nu_k,\ k = 1, 2, \ldots, N$.

DEFINITION 10.3.8. The elementary conjunction $C^{(i)}(\mathbf{x}) = x_{i_1}^{\gamma_1} \cdots x_{i_l}^{\gamma_l}$ is called *the descriptor* for the class S_i and if

$$\sum_{j=1}^{M_i} C^{(i)}\big(\mathbf{t}_j^{(i)}\big) = \sum_{j=1}^{M_i} C^{(i)}\big(t_{j1}^{(i)}, \ldots, t_{jN}^{(i)}\big) \geqslant \varepsilon,$$

and for $1 \leqslant m \leqslant r,\ m \neq i$,

$$C^{(i)}\big(\mathbf{t}_j^{(m)}\big) = 0, \quad j = 1, \ldots, M_m,$$

where $\varepsilon,\ 1 \leqslant \varepsilon \leqslant N$, is a fixed natural number.

The descriptor $C^{(i)}(\mathbf{x})$ is minimal if there is no descriptor $D^{(i)}(\mathbf{x})$ for the class i such that $C^{(i)}(\mathbf{x}) \leqslant D^{(i)}(\mathbf{x})$.

Let $\mathbf{z}$ denote a $(1 \times N)$ feature vector obtained from the observed signal. The discriminant function $g_i(\mathbf{z})$ for each class S_i is computed by the following formula:

$$g_i(\mathbf{z}) = |R_i|^{-1} \cdot \sum_{C \in R_i} g_i(C, \mathbf{z}), \tag{10–127}$$

where

$$g_i(C, \mathbf{z}) = C(\mathbf{z}) \cdot \nu(C) \cdot \sum_{j=1}^{M_i} \mu_j^{(i)} C\big(\mathbf{t}_j^{(i)}\big), \tag{10–128}$$

R_i is the set of minimal descriptors for the class S_i, $\nu(C) = \sum_{r=1}^{s} \nu_{k_r}$ for $C(\mathbf{x}) = x_{k_1} \cdots x_{k_s},\ i = 1, \ldots, r$.

The classification is made by the following decision rule:

$$g_i(\mathbf{z}) > g_j(\mathbf{z}) + \xi, \quad i \neq j, \quad \text{iff} \quad \mathbf{z} \in S_i,$$

where ξ is a predefined positive real number.

Outline of the Algorithm

The proposed algorithm for logical classification consists of the following steps:

1) Construct for each matrix $\mathbf{T}_i,\ i = 1, \ldots, r$, the corresponding characteristic (indicator) function $\mathbf{f}^{(i)}$, i.e. the column vector of length 2^N, such that $f^{(i)}(\mathbf{v}) = 1$ if and only if $\mathbf{v}$ is the row vector of $\mathbf{T}_i$.

2) Construct approximately minimal normal forms (disjunctive or conjunctive) of the Boolean functions specified by the truth tables $\mathbf{f}^{(i)}$, $i = 1, \ldots, r$.

3) Find the minimal descriptors (for fixed $\varepsilon \geqslant 1$) for each class S_i, $i = 1, \ldots, r$.

4) Calculate the local discriminant function $g_i(C, \mathbf{z})$ defined by Eq. 10–128.

5) Calculate the discriminant function $g_i = g_i(\mathbf{z})$ by Eq. 10–127 for all $i = 1, \ldots, r$.

6) Find the rth and $(r-1)$th order statistics of the set of elements $G = \{g_1, \ldots, g_r\}$. In other words, find the maximal element $g_{\max}$ of the set G and the maximal element g_{remax} of the set $G \setminus \{g_{\max}\}$. Let $g_{\max} = g_j$, and $g_{\text{remax}} = g_k$, $1 \leqslant j, k \leqslant r$.

The object $\mathbf{z}$ is assigned to class S_j if $g_j > g_k + \xi$; otherwise $\mathbf{z}$ is assigned to class S_k.

10.4 BINARY POLYNOMIAL TRANSFORMS IN COMPRESSION OF BINARY IMAGES

Binary image compression may find applications in compressing bilevel high-resolution images such as fax pages, scanned images or segmentation data and bit-plane by bit-plane compression of multilevel images in some specific cases. Recently, multiresolution analysis via wavelet transform has found efficient applications in multilevel image compression [97]. The success of multilevel image compression with wavelets has been followed by applications of wavelet theory over the finite field **GF**(2) [66, 91]. Swanson and Tewfik developed a theory of binary wavelet transform in terms of two-band perfect reconstruction filter banks in [91]. Their test results with binary image compression achieved promising results in terms of first-order entropy. Soon after, Johnston [66] used another approach (a lifting scheme [92]) to construct a binary wavelet transform which is a cascade of binary matrices composed of upper and lower unimodular binary blocks and a simple permutation matrix collecting even samples to the upper half and odd samples to the lower half. Here we show how to obtain similar binary transforms via binary polynomial matrices.

Consider the Reed–Muller matrix $\mathbf{K}_M$ of order M

$$\mathbf{K}_M = K_1^{\otimes M} = \begin{pmatrix} 1 & 0 \\ 1 & 1 \end{pmatrix}^{\otimes M} \tag{10–129}$$

and a binary vector $\mathbf{f}_M$ of length 2^M. The Reed–Muller transform of $\mathbf{f}_M$ can be defined as

$$\mathbf{f}_{RM} = \mathbf{K}_M \mathbf{f}_M,$$

Table 10.7. Compression results for thresholded 512×512 Lena and Ball, 128×128 Text and Bird Images.

	org. Lenna	S&P	S	BT	org.Bird	S	S&P	BT
no.nonzero	126742	14304	13163	15262	13208	951	830	817
bitrate	1.0	0.30	0.28	0.25	1.0	0.31	0.29	0.23
	org. Ball	S&P	S	BT	org. Text	S	S&P	BT
no.nonzero	29024	1858	1764	1747	3174	448	443	602
bitrate	1.0	0.05	0.05	0.04	1.0	0.14	0.14	0.14

where the output $\mathbf{f}_{RM}$ is computed in modulo 2 arithmetic. $\mathbf{f}_{RM}$ corresponds to a wavelet packet representation of $\mathbf{f}_M$ which can be obtained by a full wavelet tree decomposition. In this decomposition first and second rows of $\mathbf{K}_1$ act as low-pass and high-pass filters, respectively. The above transform can be generalized into a two-dimensional case for images as

$$\mathbf{F}_{RM} = \mathbf{K}_M \mathbf{F}^T \mathbf{K}_M^T, \tag{10–130}$$

where $\mathbf{F}$ is a binary matrix representing pixel values of a binary image. Note that Eq. 10–130 is an invertible transform where the inverse of $\mathbf{K}_M$ in $\mathbf{GF}(2)$ is equal to itself. We can obtain the best multiresolution decomposition of $\mathbf{F}$ by selecting a subdecomposition structure covered by $\mathbf{K}_M$. The subdecomposition structure can be obtained by forming a matrix such that

$$\mathbf{K}'_M = \begin{bmatrix} \mathbf{A} & \mathbf{0}_{M-1} \\ \mathbf{B} & \mathbf{C} \end{bmatrix}, \tag{10–131}$$

where $\mathbf{0}_{M-1}$ is the zero matrix with size $2^{M-1} \times 2^{M-1}$, and $\mathbf{A}$, $\mathbf{B}$, $\mathbf{C}$ can be either $\mathbf{0}_{M-1}$ or $\mathbf{K}'_{M-1}$ with the restriction that

$$\mathbf{K}'_1 = \mathbf{K}_1.$$

Some quantitative and visual results of the above binary transform are presented in Table 10.7 and Figs. from 10–4 to 10–9, respectively. The multiresolution decomposition structures can be seen from Figs. 10–5 and 10–8. The top-left decomposition block corresponds to slowly changing regions, and the rest corresponds to rapidly changing regions. The transform matrix $\mathbf{K}'_M$ is obtained by using the number of zeros as the best basis selection criterion. For all four test images, the logarithmically growing decomposition tree is found to give the best performance in terms of maximum number of zeros. Hence, the only overhead information needed is the depth of the tree. For comparison, the results of an S and S&P

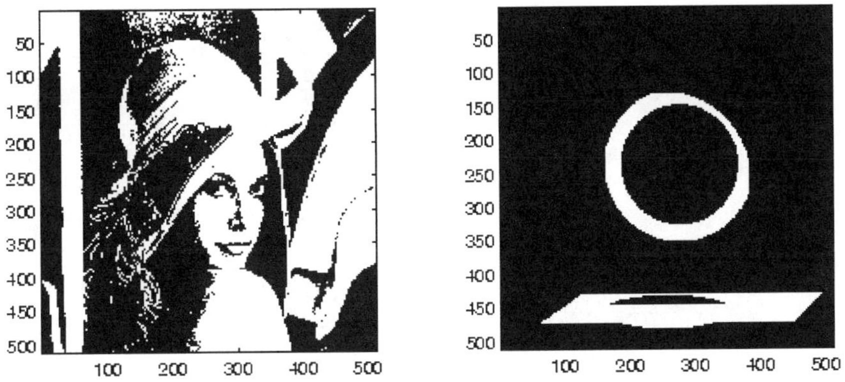

Figure 10–4. *Left:* Thresholded Lena image. *Right:* Thresholded Ball image.

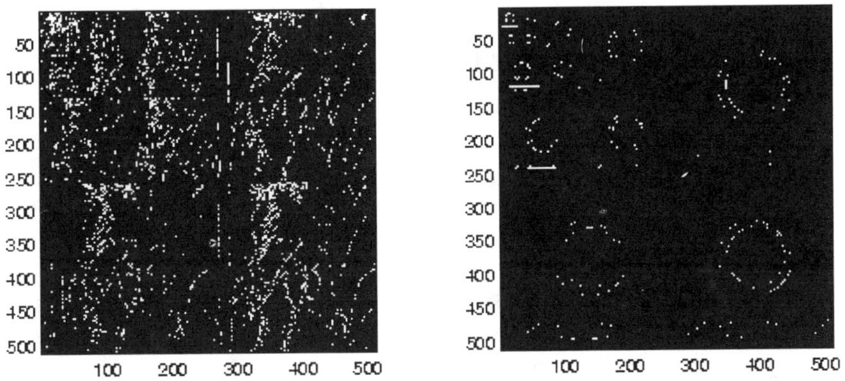

Figure 10–5. *Left:* Binary transformed Lena image. *Right:* Binary transformed Ball image.

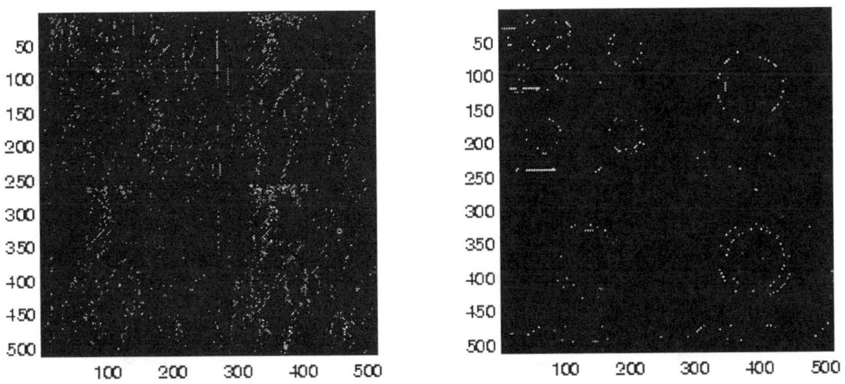

Figure 10–6. *Left:* SP transformed Lena image. *Right:* SP transformed Ball image.

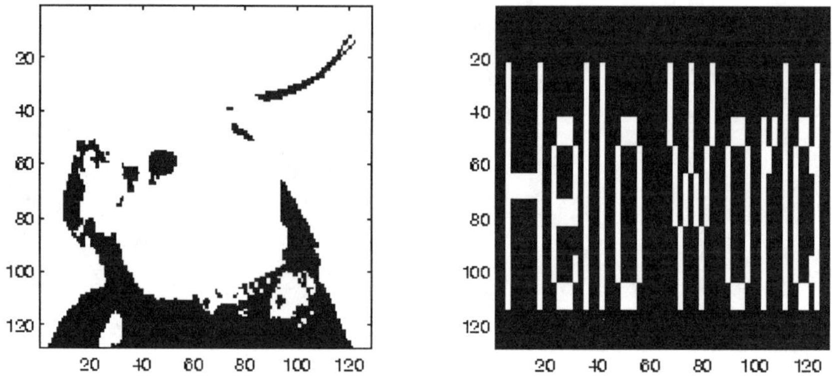

Figure 10–7. *Left:* Thresholded Bird image. *Right:* Thresholded Text image.

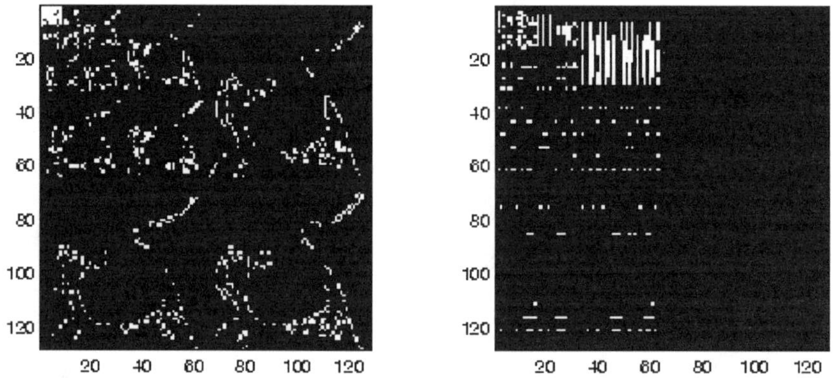

Figure 10–8. *Left:* Binary transformed Bird image. *Right:* Binary transformed Text image.

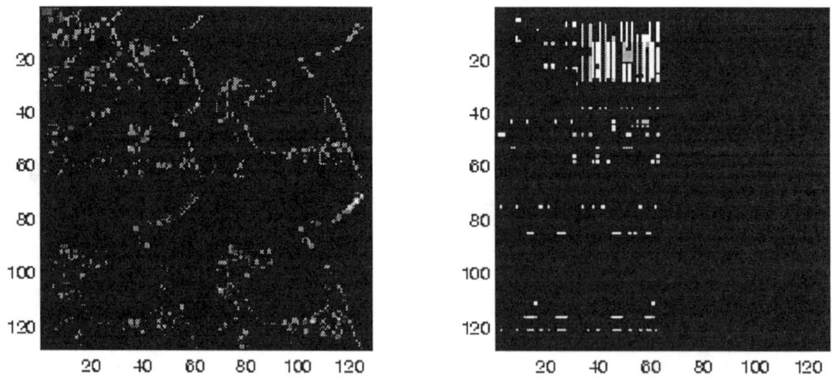

Figure 10–9. *Left:* SP transformed Ball image. *Right:* SP transformed Text image.

Transform [89] are also included. Although S and S&P are real wavelet transforms for lossy and lossless compression, they are also suitable for binary image compression since they produce integer outputs in a very limited range for binary data. The proposed binary transform is found promising for lossless binary image compression in terms of bit rate.

REFERENCES

[1] Agaian, S., Astola, J. and Egiazarian, K., *Binary Polynomial Transforms and Nonlinear Digital Filters*, Marcel Dekker, New York (1995).

[2] Agaian, S., Egiazarian, K. and Astola, J., "Spectral conversion algorithm from weighted median to stack filter", *Proc. SPIE* 2030, pp. 140–149 (1993).

[3] Ahlswede, R. and Wegener, I., *Suchprobleme*, B. G. Teubner, Stuttgart (1979).

[4] Aizenberg, N. N. and Trofimluk, O. T., "The conjunctive transforms of discrete signals and their applications to finding tests and recognizing of monotonity of the function of Boolean algebra", *Cybernetics*, Vol. 5, pp. 138–139 (in Russian) (1981).

[5] Aizenberg, N. N. and Tsitkin, A. I., "The questions of application of prime testors", *Kibernetika (Cybernetics)*, No. 1, pp. 135–141 (1974).

[6] Aldroubi, A., "Oblique multiwavelet bases: examples", *Proc. SPIE*, 2825, pp. 54–64 (1996).

[7] Arce, G. R., "Statistical threshold decomposition for recursive and nonrecursive median filters", *IEEE Trans. Inf. Theory*, Vol. IT-32, pp. 243–253 (1986).

[8] Arce, G. R., Gallagher, N. C., Jr. and Nodes, T. A., "Median filters: theory for one- and two-dimensional filters", in *Advances in Computer Vision and Image Processing*, Vol. 2, ed. T. S. Huang, JAI Press, Greenwich, CT, pp. 89–166 (1986).

[9] Arce, G. R. and Gallagher, N. C., Jr., "Stochastic analysis for the recursive median filter process", *IEEE Trans. Inf. Theory*, Vol. IT-34, pp. 669–679 (1988).

[10] Artazjan, R. and Egiazarian, K., "The fast matrix algorithms of finding the abbreviated disjunctive normal forms, irredundant tests and their application to test pattern recognition", *Dep. 62-Ar89*, March 10 (in Russian) (1989).

[11] Astola, J., Egiazarian, K., Akopian, D. and Gevorkian, D., "Threshold Boolean filtering based on pattern classification", in Proc. *IEEE Workshop on Nonlinear Signal and Image Processing*, Greece (1995).

[12] Astola, J., Akopian, D. and Egiazarian, K., "New processor architecture for running threshold Boolean filters", in Proc. SPIE 2501, pp. 366–377 (1995).

[13] Astola, J., Egiazarian, K., and Gevorkian, D., "Spectral methods for threshold Boolean filtering", in Proc. SPIE 2501, pp. 221–230 (1995).

[14] Astola, J., Egiazarian, K., and Huttunen, H., "Wavelet packets and genetic algorithms", *Proc. Int. Conf. ICASSP'97*, Vol. III, pp. 2117–2120 (1997).

[15] Astola, J., Koskinen, L., Yli-Harja, O., and Neuvo, Y., "Digital filters based on threshold decomposition and Boolean functions, Proc. SPIE 1199, pp. 461–470 (1989).

[16] Astola, J. and Neuvo, Y., "Optimal median type filters for exponential noise distributions", *Sig. Process.*, Vol. 17, pp. 95–104 (1989).

[17] Astola, J. and Neuvo, Y., "An efficient tool for analysing weighted median and stack filters", *IEEE Trans. Circ. Sys.*, Vol. 41, No. 7, pp. 487–489 (1994).

[18] Ataman, E., Aatre, V. K., and Wong, K. M., "Some statistical properties of median filtering", *IEEE Trans. Acoust. Speech, Sig. Process.*, Vol. ASSP-28, pp. 415–421 (1980).

[19] Atourian, S., Gevorkian, D., Egiazarian, K., and Astola, J., "Efficient nonlinear transform methods for image processing", *J. Electron. Imag.*, Vol. 5(3), pp. 323–334 (1996).

[20] Baskakova, L. V., and Zhuravlev, Yu. I., "Model of the recognizing algorithms with representative samples and systems of supported sets", *Jurnal Vichislitelnoy Matematiki i Matematicheskoy Fiziki* (Journal of Computational Math. Math. Phy.), Vol. 21, No. 5, pp. 1264–1276 (1981).

[21] Besslich, Ph. W., "Spectral processing of switching functions using signal-flow transformations" in *Spectral Techinues and Fault Detection*, ed. M. G. Karpovsky, Academic Press, New York, pp. 91–142 (1985).

[22] Bethke, A. D., *Genetic algorithms as function optimizers*, Doctoral dissertation, University of Michigan (1980).

[23] Boncelet, C. G., Jr., "Recursive algorithms and VLSI implementations for median filtering," in *Proc. IEEE Int. Symp. Circuits Syst.*, Espoo, Finland, pp. 1745–1747 (1988).

[24] Bovik, A. C., Huang, T. S., and Munson, D. C., Jr., "A generalization of median filtering using linear combinations of order statistics", *IEEE Trans. Acoust. Speech, Sig. Process.*, Vol. ASSP-31, pp. 1342–1350 (1983).

[25] Bovik, A. C., "Streaking in median filtered images", *IEEE Trans. Acoust. Speech, Sig. Process.*, Vol. ASSP-35, pp. 493–503 (1987).

[26] Bozoyan, Sh. E., "Certain properties of Boolean derivatives and activities of argument of Boolean functions", *Problemi Peredachi Informacii*, v. XIV, No. 1, pp. 77–89 (in Russian) (1978).

[27] Bozoyan, Sh. E. and Torossian, B. E., "Investigation of functions of the algebra of logic in terms of activity of joint variables", *Izvestia Akad. Nauk Arm. SSR*, Ser. Matematika, v. XIV, No. 2, pp. 124–141 (in Russian) (1979).

[28] Brownrigg, D. R. K., "The weighted median filter", *Commun. ACM*, Vol. 27, pp. 807–818 (1984).

[29] Chen, K., "Bit-serial realizations of a class of nonlinear filters based on positive Boolean functions", *IEEE Trans. Circ. Sys.*, Vol. CAS-36, No. 6, pp. 785–794 (1989).

[30] Chow, C. K., "On the characterization of threshold functions", *SCTLD, IEEE Special Publ.*, No. S134, pp. 34–38 (1961).

[31] Coyle, E. J. and Lin, J.-H., "Stack filters and the mean absolute error criterion", *IEEE Trans. Acoust. Speech, Sig. Process.*, Vol. ASSP-36, pp. 1244–1254 (1988).

[32] Coyle, E. J. and Gallagher, N. C., "Stack filters and neural networks", in *Proc. IEEE Int. Symp. Circ. Sys.*, Portland, OR, pp. 995–998 (1989).

[33] Coyle, E. J., Lin, J.-H., and Gabbouj, M., "Optimal stack filtering and the estimation and structural approaches to image processing", *IEEE Trans. Acoust. Speech, Sig. Process.*, Vol. ASSP-37, pp. 2037–2066 (1989).

[34] Danielsson, P. E., "Getting the median faster", *Comput. Graph. Imag. Process.*, Vol. 17, No. 1, pp.71–78 (1981).

[35] David, H. A., *Order Statistics*, Wiley, New York (1979).

[36] Egiazarian, K. and Artazjan, R., "The spectral technique of Boolean computation with application in testing pattern recognition", Rep. 90-8 (preprint), Armenian Academy of Sciences, Institute of Problems of Informatics and Automation, Yerevan, 25 p. (in Russian) (1990).

[37] Egiazarian, K. O., "On new spectral approach to pattern recognition", in *Proc. of I Int. Conf. on Information Technologies for Image Analysis and Pattern Recognition (ITIAPR'90)*, Lviv, Ukraine, pp. 175–179 (1990).

[38] Egiazarian, K. and Artazian, R., "Spectral methods of minimization of Boolean functions and related problems, in *Proc. 1st All-Union Conf. of Pattern Recogn. and Image Analyses*, Minsk, Vol. 3, pp. 52–54 (1991).

[39] Egiazarian, K. and Astola, J., "Spectral approach to the calculation of the joint distribution of stack filters", in *Proc. of TUT Symposium on Signal Processing'94*, Tampere University of Technology (1994).

[40] Egiazarian, K., Astola, J., and Agaian, S., "Spectral approach to classification based on generalized unconditional tests", *Proc. SPIE Meeting on Visual Communications and Image Processing*, Vol. 2501, Taiwan, pp. 119–128 (1995).

[41] Egiazarian, K., Astola, J., and Agaian, S., "Spectral approach to logical distribution-free classification problem", *IEEE International Symposium on Circuits and Systems*, ISCAS'95, Vol. III, pp. 2265–2268 (1995).

[42] Egiazarian, K., Huttunen, H., and Astola, J., "Evolutionary heuristic and spectral approach to minimal generalized unconditional test problem", *Proc. IPMU'96*, Granada, Spain (1996).

[43] Egiazarian, K. and Astola, J., "Transform domain analysis and implementation of extended threshold Boolean filters", *NORSIG'96*, Helsinki (1996).

[44] Egiazarian, K., Astola, J., Atourian, S., and Gevorkian, D., "Nonlinear filters based on ordering by FFT structure", *Proc. IS&T/SPIE Symp. on Electronic Imaging: Science & Technology, Nonlinear Image Processing VII*, San Jose, California, USA (1996).

[45] Egiazarian, K., Kuosmanen, P., and Astola, J., "Boolean derivatives, weighted chow parameters and selection probabilities of stack filters", *IEEE Trans. Sig. Process.*, Vol. 44, No. 7, pp. 1634–1641 (1996).

[46] Egiazarian, K., Astola, J., and Agaian, S., "Binary polynomial transforms and logical correlation", in *Nonlinear Filters for Image Processing*, eds. E. R. Dougherty and J. Astola, SPIE, Bellingham, WA (1997).

[47] Estola, K. P. and Sioranta, R., "A fast probabilistic median algorithm for integer arithmetics", in *Proc. 1st Gretsi Symp. Signal and Image Processing*, Juan-les-Pins, France, June, pp. 61–64 (1989).

[48] Fitch, J. P., Coyle, E. J., and Gallagher, N. C., Jr., "Median filtering by threshold decomposition", *IEEE Trans. Acoust. Speech, Sig. Process.*, Vol. ASSP-32, pp. 1183–1188 (1984).

[49] Fitch, J. P., Coyle, E. J., and Gallagher, N. C., Jr., "Root properties and convergence rates of median filters", *IEEE Trans. Acoust. Speech, Sig. Process.*, Vol. ASSP-33, pp. 230–240 (1985).

[50] Fitch, J. P., Coyle, E. J., and Gallagher, N. C., Jr., "Threshold decomposition of multidimensional ranked order operations", *IEEE Trans. Circ. Sys.*, Vol. CAS-32, pp. 445–450 (1985).

[51] Gabbouj, M., Haavisto, P., and Neuvo, Y., "Recent advances in median filtering", in *Communication, Control, and Signal Processing*, ed. E. Arikan, Ankara, Turkey: Elsevier, Vol. 11, pp. 1080–1094 (1990).

[52] Gabbouj, M., and Coyle, E. J., "Minimum mean absolute error stack filtering with structural constraints and goals", *IEEE Trans. Acoust. Speech, Sig. Process.*, Vol. ASSP-38, pp. 995–968 (1990).

[53] Gallagher, N. G., Jr. and Wise, G. L., "A theoretical analysis of the properties of median filters", *IEEE Trans. Acoust. Speech, Sig. Process.*, Vol. ASSP-29, pp. 1136–1141 (1981).

[54] Gilbert, E. N., "Lattice theoretic properties of frontal switching functions", *J. Math. Phys.*, Vol. 33, pp. 57–67 (1954).

[55] Gilbo, E. P. and Chelpanov, I. B., *Signal Processing on the Basis of Ordered Search*, Moscow, Sov. Radio, p. 344 (in Russian) (1975).

[56] Goldberg, D. E., *Genetic Algorithms in Search, Optimization, and Machine Learning*, Addison-Wesley, MA (1989).

[57] Goldberg, D. E., "Genetic algorithms and Walsh functions: Part I, a gentle introduction", *Complex Sys.*, Vol. 3, pp. 129–152 (1989).

[58] Goldberg, D. E., "Genetic algorithms and Walsh functions: Part II, deception and its analysis", *Complex Sys.*, Vol. 3, pp. 153–171 (1989).

[59] Gupta, H., *Selected Topics in Number Theory*, Wiley, New York (1975).

[60] Heinonen, P. and Neuvo, Y., "Smoothed median filters with FIR substructures", in *Proc. IEEE Int. Conf. Acoust. Speech, Signal Processing*, Tampa, FL, USA, pp. 49–52 (1985).

[61] Heinonen, P. and Neuvo, Y., "FIR-median hybrid filters", *IEEE Trans. Acoust. Speech, Sig. Process.*, Vol. ASSP-35, pp. 832–838 (1987).

[62] Heinonen, P. and Neuvo, Y., "FIR-median hybrid filters with predictive FSR substructures", *IEEE Trans. Acoust. Speech, Sig. Process.*, Vol. ASSP-36, pp. 892–899 (1988).

[63] Holland, J. H., *Adaptation in Natural and Artificial Systems*, University of Michigan Press, Ann Arbor (1975).

[64] Hurst, S. L., Miller, D. M., and Muzio, J. C., *Spectral Techniques in Digital Logic*, Academic Press, San Diego (1985).

[65] Jeong, H., Song, J., and Lee, Y. H., "Extended threshold Boolean filters: representation and design", *Proc. 1995 IEEE Workshop on Nonlinear Signal Processing*, ed. I. Pitas, Vol. II, pp. 911–914 (1995).

[66] Johnston, C. P., "The lifting scheme and finite precision error free filter banks", Proc. SPIE 2825, pp. 307–316 (1996).

[67] Justusson, B. I., "Median filtering: statistical properties", in *Topics in Applied Physics, Two-Dimensional Digital Signal Processing II*, ed. T. S. Huang, Springer-Verlag, Berlin, Vol. 43, pp. 161–196 (1981).

[68] Kassam, S. A. and Poor, H. V., "Robust techniques for signal processing: a survey", *Proc. IEEE*, Vol. 73 (1985).

[69] Khuri, S., "Walsh and Haar functions in genetic algorithms", *in Proc. 1994 ACM Symposium on Applied Computing (SAC'94)*, Phoenix AZ, ACM Press (1994).

[70] Khuri, S. and Bäck, T., "An evolutionary heuristic for the minimum vertex cover problem", *KI-94, 18th German Annual Conf. on Artificial Intelligence* (1994).

[71] Kuosmanen, P., Astola, J., and Agaian, S., "On rank selection probabilities", *IEEE Trans. Sig. Process.*, Vol. 42, No. 11, pp. 3255–3258 (1994).

[72] Kuosmanen, P., Egiazarian, K., and Astola, J., "Calculation of the sample selection probabilities of stack filters by using weighted chow parameters", in Proc. ICASSP-95, Detroit (1995).

[73] Lee, K. D. and Lee, Y. H., "Threshold Boolean filters", *IEEE Trans. Sig. Process.*, Vol. 42, No. 8, pp. 2022–2036 (1994).

[74] Lee, Y. H. and Kassam, S., "Generalized median filtering and related nonlinear filtering techniques", *IEEE Trans. Acoust. Speech, Sig. Process.*, Vol. ASSP-33, pp. 672–683 (1985).

[75] Liao, G. Y., Nodes, T. A., and Gallagher, N. C., Jr., "Output distributions of two-dimensional median filters", *IEEE Trans. Acoust. Speech, Sig. Process.*, Vol. ASSP-33, pp. 1280–1295 (1985).

[76] Mallows, C. L., "Some theory of nonlinear smoothers", *The Annal. Stat.*, Vol. 8, No. 4, pp. 695–715 (1980).

[77] Maragos, P., and Schafer, R., "Morphological filters-Part I, Their set theoretic analysis and relations to linear shift-invariant filters", *IEEE Trans. Acoust. Speech, Sig. Process.*, Vol. 35, Aug., pp. 1153–1169 (1987).

[78] Maragos, P. and Schafer, R., "Morphological filters-Part II, Their relations to median, order-statistics, and stack filters", *IEEE Trans. Acoust. Speech, Sig. Process.*, Vol. 35, Aug., pp. 1170–1184 (1987).

[79] Muroga, S., *Threshold Logic and Its Applications*, Wiley, New York (1971).

[80] Neejärvi, J., Koskinen, L., and Neuvo, Y., "Statistical analysis of median type and morphological filters", in Proc. SPIE 1818, pp. 366–375 (1992).

[81] Nodes, T. A. and Gallagher, N. C., Jr., "The output distribution of median type filters", *IEEE Trans. Commun.*, Vol. COM-32, pp. 532–541 (1984).

[82] Papadimitriou, C. H. and Steiglitz, K., *Combinatorial Optimization*, Prentice-Hall (1982).

[83] Petrosian, A. V., "Some differential characteristics of Boolean functions", *MTA Tanulmanyok*, Vol. 135, pp. 15–37 (in Russian) (1982).

[84] Pitas, I. and Venetsanopoulos, A. N., *Nonlinear Digital Filters*, Kluwer Academic Publishers, Boston, MA (1990).

[85] Pitas, I., "Marginal order statistics in color image filtering", *Opt. Eng.*, Vol. 29, No. 5, pp. 495–503 (1990).

[86] Prasad, M. K. and Lee, Y. H., "Stack filters and selection probabilities", in *Proc. IEEE 1990 Int. Symp. Circuits and Systems*, pp. 1747–1750, New Orleans, LA, USA, May 1–3 (1990).

[87] Riordan, J., *An Introduction to Combinatorial Analysis*, Wiley, New York (1958).

[88] Rjazanov, V. V., "On the construction of optimal recognition and classification algorithms for solving applicational problems", in Yearbook: *Pattern Recognition, Classification, Forecasting. Mathematical Techniques and Their Application*, Issue 1, ed. Yu. I. Zhuravlev, Moscow, Nauka (in Russian) (1988).

[89] Said, A. and Pearlman, W. A., "An image multiresolution representation for lossless and lossy compression", *IEEE Trans. Imag. Process.*, Vol. 5, No. 9, pp. 1303–1310 (1996).

[90] Sarhan, A. E., and Greenberg, B. G. eds., *Contributions to Order Statistics*, New York, Wiley (1962).

[91] Swanson, M. D. and Tewfik, A. H., "A binary wavelet decomposition of binary images", *IEEE Trans. Imag. Process.*, Vol. 5, No. 12, pp. 1637–1650 (1996).

[92] Sweldens, W., "The lifting scheme: a custom-design construction of biorthogonal wavelets", *Technical Report IMI* 7, revised version (1994).

[93] Thayse, A., *Boolean Calculus of Differences*, Springer-Verlag, Berlin (1981).

[94] Tukey, J. W., *Exploratory Data Analysis*, Addison-Wesley, Menlo Park, CA (1971).

[95] Tukey, J. W., "Nonlinear (nonsuperposable) methods for smoothing data", *Congr. Rec. EASCON*, pp. 673 (1974).

[96] Tyan, S. G., "Median filtering: deterministic properties", *Topics in Applied Physics, Two-Dimensional Digital Signal Processing II*, ed. T. S. Huang, Springer-Verlag, Berlin, Vol. 43, pp. 197–217 (1981).

[97] Vetterli, M. and Kovačević, J., *Wavelets and Subband Coding*, Prentice Hall (1995).

[98] Wendt, P. D., Coyle, E. J., and Gallagher, N. C., Jr., "Some convergence properties of median filters", *IEEE Trans. Circ. Sys.*, Vol. CAS-33, pp. 276–286 (1986).

[99] Wendt, P. D., Coyle, E. J., and Gallagher, N. C., Jr., "Stack filters", *IEEE Trans. Acoust. Speech, Sig. Process.*, Vol. ASSP-34, pp. 898–911 (1986).

[100] Wendt, P. D., "Nonrecursive and recursive stack filters and their filtering behavior", *IEEE Trans. Acoust. Speech, Sig. Process.*, Vol ASSP-38, pp. 2099–2107 (1990).

[101] Wickerhauser, M. V., *Adapted Wavelet Analysis from Theory to Software*, IEEE Press, A. K. Peters, Wellesley, MA (1994).

[102] Winder, R. O., "Threshold functions through $n = 7$", *Scientific Report No. 7, Air Force Cambridge Research Laboratory*, Contract AF19(604)-8423 (1969).

[103] Xiong, Z., Ramchandran, K., Orchard, M. T., and Asai, K., "Wavelet packets-based image coding using joint space-frequency quantization", *Proc. ICIP-94*, Vol. 3, Nov. 13–16, Austin, Texas (1994).

[104] Yablonski, S. V., *Introduction to Discrete Mathematics*, Moscow, Nauka (in Russian) (1986).

[105] Yli-Harja, O., Astola, J., and Neuvo, Y., "Generalization of the radix method of finding the median to order statistic, weighted median and weighted order statistic filters", Proc. SPIE 1001, pp. 69–75 (1988).

[106] Yli-Harja, O., Astola, J., and Neuvo, Y., "Analysis of the properties of median and weighted median filters using threshold logic and stack filter representation", *IEEE Trans. Sig. Process.*, Vol. SP-39, pp. 395–410 (1991).

[107] Yli-Harja, O., Astola, J., Heinonen, P., and Neuvo, Y., "Performance of edge preserving median filters in noisy conditions", *Proc IASTED Int. Symp. Appl. Control, Filt., Signal Proc.*, Geneva, Switzerland, pp. 68–72 (1987).

[108] Yli-Harja, O. and Astola, J., "Filters based on minimization of an error criterion", 6th *Scand. Conf., Image Analysis, 6, SCIA*. Oulu, Finland (1989).

[109] Yli-Harja, O., "Formula for the joint distribution of stack filters", *Sig. Process. Lett.*, Vol. 1, No. 9, pp. 129–130 (1994).

[110] Zeng, B., Gabbouj, M., and Neuvo, Y., "A unified design method for rank order, stack, and generalized stack filters based on classical Bayes decision", *IEEE Trans. Circ. Sys.*, Vol. CAS-38, pp. 1003–1020 (1991).

[111] Zhuravlev, Yu. I. and Nikiforov, V. V., "Algorithms of pattern recognition, based on the computation of estimations", *Kibernetika (Cybernetics)*, Vol. 3 (1971).

CHAPTER 11

RANDOM SETS IN VIEW OF IMAGE FILTERING APPLICATIONS

Ilya S. Molchanov
University of Glasgow
Glasgow, United Kingdom

11.1 SETS ARE BECOMING RANDOM

Everything can be random. In the everyday language, a random variable is a number that takes a value from some range depending on chance; a random person is a person encountered after an hour rambling on a crowded street; a random image is an image that appears if some pixels have random colors or gray-scale values; a random nonlinear filter is a filter picked at random from those described in other chapters of this volume.

Having this all in mind, there appears to be nothing mysterious in the concept of a random set. It is relatively easy to imagine a collection of all sets of interest and pick one at random. These sets of interest are usually closed in the topological sense, since in applications it is natural to consider sets which include their boundaries. In this survey we always consider sets in the d-dimensional Euclidean space $\mathbb{R}^d$, and denote by $\mathcal{F}$ the family of all closed subsets of $\mathbb{R}^d$.

To finish the definition, we must specify what it means to "pick at random" an element of $\mathcal{F}$. This procedure is equivalent to defining an $\mathcal{F}$-valued random element, or map $X : \Omega \to \mathcal{F}$, where Ω is an abstract space endowed with a measurable structure: a σ-algebra $\mathfrak{A}$ and a probability measure $\mathbf{P}$. Finally, we must say which information about X should be accessible, i.e. we need to define a family of measurable events that are related to X. This family of measurable (or observable) events is generated by the events

$$\{X \cap K \neq \emptyset\}, \qquad (11\text{–}1)$$

where K runs through the family $\mathcal{K}$ of compact sets in $\mathbb{R}^d$. Loosely speaking, a random set X is an $\mathcal{F}$-valued random element, such that we can observe if X hits or misses a compact set K for each $K \in \mathcal{K}$. Formally, X is a map from Ω to $\mathcal{F}$ such that $\{\omega \in \Omega\colon\ X(\omega) \cap K \neq \emptyset\} \in \mathfrak{A}$ for each $K \in \mathcal{K}$, see [32, 57].

The events of type given by Eq. 11–1 generate other (more complicated) events by taking finite or countable unions, intersections and complements of these elementary events. For example, the event

$$\{X \cap K = \emptyset, X \cap K_1 \neq \emptyset, \ldots, X \cap K_n \neq \emptyset\}$$

is measurable for all $K, K_1, \ldots, K_n \in \mathcal{K}$.

Sometimes *discrete random sets* are considered. Their definition follows the scheme above with $\mathbb{R}^d$ replaced by $\mathbb{Z}^d$ or any other finite or countable discrete carrier space. Note that discrete random sets are automatically closed.

The definition of a random closed set is broad enough to allow many examples of random sets that ensure its wide applicability. It is easy to see that all discrete random sets are measurable as long as the event "any given pixel is black" is measurable. In the continuous setup, a random singleton $X = \{\xi\}$, a random ball $X = B_\eta(\xi)$ (η is the radius, ξ is the centre), a random triangle $X = \Delta_{\xi_1,\xi_2,\xi_3}$ are random closed sets if the corresponding random vectors and random variables (used to define X) are measurable.

Clearly, random *binary* images correspond to random closed sets if the measurability condition holds (as we have seen, the measurability condition is straightforward in the discrete case). Random *gray-scale* images are related to random sets in one of the following ways (which are also applicable to random sets related to stochastic processes). A random *upper semicontinuous* function f defined on a window W yields several random closed sets:

$$\{f \geqslant t\} = \{x \in W\colon\ f(x) \geqslant t\} \tag{11–2}$$

which is the *excursion set* of f at the level t;

$$\operatorname{hypo} f = \{(x, t)\colon\ x \in W,\ f(x) \geqslant t\} \tag{11–3}$$

which is a subset of $W \times \mathbb{R}^d$ and is called the *hypograph* of f. Upper semicontinuity of f (it means $\limsup_{x \to a} f(x) \leqslant f(a)$ for all a) is necessary to ensure that the above defined sets are topologically closed. If, in addition, f is continuous, then also the *level* set

$$\{f = t\} = \{x \in W\colon\ f(x) = t\}$$

and the *graph*

$$\operatorname{graph} f = \{(x, t)\colon\ x \in W,\ f(x) = t\}$$

are random closed sets. The dual random sets to Eqs. 11–2 and 11–3 can be defined if f is lower semicontinuous. For instance,

$$\mathrm{epi} f = \{(x,t)\colon x \in W,\ f(x) \leqslant t\}$$

is called the *epigraph* of f. In other words, the epigraph is the set above the graph of f, and the hypograph is the set below this graph. Sometimes the hypograph of f is called the *umbra* [53].

It is possible to restrict attention to special random sets with realizations satisfying some conditions. For example, X is called a *random compact set* if X is a random closed set that takes values from the family $\mathcal{K}$ of compact sets. As $\mathcal{K}$ is a subfamily of $\mathcal{F}$, this means that the corresponding probability measure is concentrated on $\mathcal{K}$ rather than on the whole $\mathcal{F}$. Furthermore, X is a *random convex set* if its realizations are almost surely convex, so that $X \in \mathcal{C}$ almost surely, where $\mathcal{C}$ is the family of convex closed sets. A *random body* is a random closed set which is both compact and convex.

As everything can be random, everything can be fuzzy. The theory of fuzzy sets is, however, rather different from the theory of random sets, especially where its aims and techniques are concerned. A *fuzzy set* is completely defined by its membership function that associates each point with its "degree of membership" to a set. Clearly, a random set X gives rise to a fuzzy set with the membership function that equals the *coverage function* of X:

$$p(x) = \mathbf{P}\{x \in X\}. \qquad (11\text{–}4)$$

However, the converse is not always true. A function must satisfy some regularity condition to become a coverage function of a certain random closed set. While the membership function is the ultimate characteristic of a fuzzy set, the coverage function Eq. 11–4 does not determine the distribution of a random set, so that entirely different random sets can share the same coverage function.

Below we outline several directions in the theory of random closed sets chosen in view of possible applications to nonlinear image filters. Generally speaking, relationships between random sets and image filters remain unexplored so far. The reasons are numerous, e.g., scarcity of random set distributions, undeveloped analytical tools for random sets, considerable differences between the techniques applicable for discrete and continuous random sets, essential nonlinearity of the space $\mathcal{F}$ of closed sets, etc.

This survey deals with random subsets of the Euclidean space. From this setup it is possible to deviate in two opposite directions. First, one can study random sets in more general spaces, which are non-Hausdorff [44], or infinite-dimensional, or are general metric spaces [23, 45]. Another option is to "discretize" the setup and consider random subsets of a discrete space, e.g., the grid $\mathbb{Z}^d$, see [18, 54, 55]. However, the discrete approach makes it quite difficult to use techniques based on scaling and convex/integral geometry. The tools suitable to deal with discrete random sets usually come from mathematical morphology [15, 19].

The chapter is organized as follows. Section 11.2 introduces the formal definition of a random set and the main tool necessary to describe its distribution – the capacity functional of a random closed set. Section 11.3 outlines several general approaches to defining an expected value for a random set or average of a sample of sets. These general approaches are illustrated by a number of particular definitions of expectations. Section 11.4 deals with models of random compact sets, that, in principle, can be used as priors in Bayesian image analysis. The most important model of a stationary random set (the Boolean model) and basic ideas of statistical inference for the Boolean model are explained in Section 11.5. Finally, Section 11.6 introduces distances between distributions of random sets, which can be used to define distances between random images.

11.2 CAPACITIES AND DISTRIBUTIONS

A random set is a random element that takes values in the space $\mathcal{F}$ of all closed subsets of the given space (in our case $\mathbb{R}^d$). Its distribution corresponds to a probability measure on the space $\mathcal{F}$, so that for $\mathcal{A} \subset \mathcal{F}$,

$$\mathbf{P}\{X \text{ takes a value from a family } \mathcal{A}\} = \mathbf{P}(\mathcal{A}).$$

This probability measure is defined on families of sets $\mathcal{A}$ that are measurable. For instance, we should be able to define

$$\mathbf{P}(\{F : F \cap K \neq \emptyset\}) = \mathbf{P}\{X \cap K \neq \emptyset\}$$

for each compact set K. An important fact is that these probabilities determine the distribution of X. Heuristically, this can be explained by the fact that the corresponding events $\{X \cap K \neq \emptyset\}$ generate the underlying σ-algebra on the space $\mathcal{F}$ that has been used to introduce a random closed set. The functional

$$T_X(K) = \mathbf{P}\{X \cap K \neq \emptyset\} \tag{11–5}$$

is said to be the *capacity functional* (or the hitting functional) of X. We often omit the subscript X and write $T(K)$ instead of $T_X(K)$. The capacity functional T generalizes for random sets the well-known concept of the cumulative distribution function. For example, a random half-line $X = [\xi, \infty) \subset \mathbb{R}^1$ has the capacity functional

$$T(K) = \mathbf{P}\{\xi \leqslant \inf K\}.$$

The capacity functional T maps $\mathcal{K}$ into $[0, 1]$. It is clear that not every functional on the space $\mathcal{K}$ corresponds to a random closed set. The capacity functional satisfies the following properties.

(T1) $T(\emptyset) = 0$.

(T2) $T(K_1) \leqslant T(K_2)$ if $K_1 \subseteq K_2$, so that T is a *monotone* functional.

(T3) T is upper semicontinuous on $\mathcal{K}$, i.e. $T(K_n) \downarrow T(K)$ as $K_n \downarrow K$ in $\mathcal{K}$.

(T4) The functional T is *completely alternating*, i.e. the functionals given by

$$S_1(K_0; K) = T(K_0 \cup K) - T(K_0)$$
$$\cdots\cdots\cdots\cdots$$
$$S_n(K_0; K_1, \dots, K_n) = S_{n-1}(K_0; K_1, \dots, K_{n-1}) - S_{n-1}(K_0 \cup K_n; K_1, \dots, K_{n-1})$$

are nonnegative for all $n \geqslant 0$ and $K_0, K_1, \dots, K_n$ from $\mathcal{K}$.

The fundamental Choquet (or Choquet–Matheron–Kendall) theorem states that these properties single out those functionals that stem from distributions of random closed sets.

THEOREM 11–1 (see [32]). *Let* $T : \mathcal{K} \to [0, 1]$. *There exists a unique random closed set* X *in* $\mathbb{R}^d$ *such that* $\mathbf{P}\{X \cap K \neq \emptyset\} = T(K)$ *if and only if* T *satisfies conditions* **(T1)**–**(T4)**.

Note that **(T3)** follows from continuity of the probability measure $\mathbf{P}$. Property **(T4)** becomes trivial if we notice that the value of $S_n(K_0; K_1, \dots, K_n)$ is equal to the probability that X misses K_0 but hits $K_1, \dots, K_n$, i.e.

$$S_n(K; K_1, \dots, K_n) = \mathbf{P}\{X \cap K = \emptyset, X \cap K_1 \neq \emptyset, \dots, X \cap K_n \neq \emptyset\}.$$

In particular, T is increasing as soon as S_1 is nonnegative, so that **(T2)** follows from **(T4)**. The properties of T resemble those of the distribution function. Property **(T3)** is the same as the right continuity and **(T2)**, **(T4)** generalize the notion of monotonicity. However, in contrast to measures, functional T is not additive, but only *subadditive*, i.e.

$$T(K_1 \cup K_2) \leqslant T(K_1) + T(K_2)$$

for all compact sets K_1 and K_2. Note that T is not a measure on $\mathcal{K}$ for most examples of random sets.

EXAMPLE 11–1. Let $X = \{\xi\}$ be a single-point random set (random singleton) in $\mathbb{R}^d$. Then $T(K) = \mathbf{P}\{\xi \in K\}$ is the probability distribution of ξ. This is the only case when T itself is a measure.

EXAMPLE 11–2. If $X = B_1(\xi)$, then $T_X(K) = \mathbf{P}\{\xi \in K \oplus B_1(0)\}$, which is *not* a measure ($K_1 \oplus B_1(0)$ and $K_2 \oplus B_1(0)$ may be not disjoint for disjoint K_1 and K_2).

The random closed set X is called *stationary* if X has the same distribution as $X + a$ for each $a \in \mathbb{R}^d$. An *isotropic* random closed set is distribution-invariant with respect to all nonrandom rotations. Because the capacity functional uniquely determines the distribution of X, these invariance properties can be easily reformulated as invariance properties of the capacity functional.

The capacity functional $T(K)$ is originally defined for K from the class of compact sets. However, it can be extended first for sets G from the family $\mathcal{G}$ of open sets and then for all $M \subset \mathbb{R}^d$ by

$$\begin{aligned} T(G) &= \sup\{T(K)\colon\ K \subset G,\, K \in \mathcal{K}\}, \\ T(M) &= \inf\{T(G)\colon\ G \supset M,\, G \in \mathcal{G}\}. \end{aligned}$$

The resulting extension has the same probabilistic interpretation as the original functional, that is $T(M) = \mathbf{P}\{X \cap M \neq \emptyset\}$ for all $M \subset \mathbb{R}^d$.

As the word *capacity* suggests, the capacity functional should have something to do with electrostatics. Indeed, the classical Newton capacity corresponds to the distribution of a random set related to the trajectories of the Wiener process [32, 36].

Sometimes, it is more convenient to use other functionals to determine distributions of random closed sets. Such functionals are

- *avoiding functional*

$$Q(K) = 1 - T(K) = \mathbf{P}\{X \cap K = \emptyset\}, \quad K \in \mathcal{K};$$

- *containment functional*

$$C(F) = \mathbf{P}\{X \subseteq F\}, \quad F \in \mathcal{F};$$

- *inclusion functional*

$$I(K) = \mathbf{P}\{K \subseteq X\}, \quad K \in \mathcal{K}.$$

Clearly, $C(F) = Q(F^c)$, where $F^c = \mathbb{R}^d \setminus F$ is the complement to F. Furthermore, $I(K) = \mathbf{P}\{X^c \cap K = \emptyset\}$, so that the inclusion functional is related to the distribution of X^c. Containment functionals are useful for dealing with distributions of random bodies, since a distribution of a random body is completely determined by the containment functional $C(K)$ defined for all compact convex sets K, see [61]. The inclusion functional is well suited for coverage problems [17, 38] when the uncovered part of X (or its complement) is of interest.

It is easy to estimate the capacity functional T_X, if independent realizations $X_1, \ldots, X_n$ of X are available. The natural estimator

$$\hat{T}_{X,n}(K) = \frac{1}{n} \sum_{i=1}^{n} \mathbf{1}_{X_i \cap K \neq \emptyset} \tag{11–6}$$

is called the *empirical capacity functional*. This estimator is clearly consistent for each given K. Moreover, it is *uniformly consistent* for K from a certain class of sets $\mathcal{M} \subseteq \mathcal{K}$ [so that $\hat{T}_{X,n}(K)$ converges to $T_X(K)$ uniformly as $n \to \infty$ for $K \in \mathcal{M}$] if X and $\mathcal{M}$ satisfy some additional conditions, see [33, 40]. If X is stationary and ergodic [40], then the capacity functional can be estimated using a single realization of X inside the window W as

$$\hat{T}_{X,W}(K) = \frac{\mu_d((X \oplus \check{K}) \cap (W \ominus K))}{\mu_d(W \ominus K)}, \tag{11–7}$$

where $\check{K} = \{-x\colon\ x \in L\}$. This estimator is corrected for edge effects near the boundary of W and is called the *minus-sampling estimator*, since the effective observation area is restricted to $W \ominus K$.

11.3 AVERAGING

In the context of filtering, random sets appear within different frameworks. Since every binary image can be treated as a set and a gray-scale image corresponds to a set given by its hypograph, we consider below general random sets that (depending on context) can represent either binary images or hypographs of gray-scale images.

Several sets (or images) $X_1, \ldots, X_n$ can be interpreted as realizations of a random set X so that the aim could be to estimate the characteristics and the distribution (as the most exhaustive characteristics) of X from the sample of $X_1, \ldots, X_n$. The following particular cases are of special importance.

1. $X_1, \ldots, X_n$ represent "particles" (such as dust or sand grains) that are considered as realizations of a "typical" particle X with an unspecified location in the space (so that two congruent particles are treated as identical).

2. $X_1, \ldots, X_n$ are noisy or partially distorted observations of a single (binary or gray-scale) image that corresponds to a deterministic set K. The aim is to estimate K from the sample. Such a situation arises, for example, when one noisy variant of the image of interest is used to start several simulated annealing procedures. Sometimes the sample of sets appears if an image has been recorded many times with randomly fluctuating positions of the camera and optic parameters.

3. A random set X can model the noise that is superimposed with a "useful" (perhaps, random) image Y. The aim is to find some characteristics of Y from the observation of $X \cup Y$ (or $X \cap Y$, etc.).

The first question has its origin in the statistics of particles and shapes. A classical example concerns a collection of sand grains' profiles, so that $X_1, \ldots, X_n$ are particular grains. A reasonable problem would be to fit an appropriate probability model for this sample and to estimate its parameters. A standard approach refers to multivariate statistical methods [46, 59]. Clearly, it is difficult to apply multivariate methods *directly* to the sample of sets. The sets must be replaced by some numbers, so that the sample of sets becomes a sample of numbers, vectors or functions. For example, in the planar case we can measure the areas of the sets, so that $\mathsf{A}(X_1), \ldots, \mathsf{A}(X_n)$ becomes a univariate sample. Further possibilities are to measure their perimeters (boundary lengths) $\mathsf{U}(X_1), \ldots, \mathsf{U}(X_n)$ or calculate their *support functions*

$$h(X_i, u) = \sup\{\langle u, x\rangle : x \in X_i\}, \tag{11–8}$$

where $\langle u, x\rangle$ is the scalar product in $\mathbb{R}^d$. If the sets are star-shaped with respect to the origin (or another point), then it is very popular to work with their radius-vector functions

$$r(X_i, u) = \sup\{s\colon\ su \in X_i\}, \tag{11–9}$$

see [48, 59].

In addition, it is possible to combine the above quantities and further characteristics that come from mathematical morphology, e.g., the granulometry [1]. In general, the sample $X_1, \ldots, X_n$ is replaced by the sample $\varphi(X_1), \ldots, \varphi(X_n)$ for a function φ that maps $\mathcal{F}$ into $\mathbb{R}^m$ for some dimension m. This yields a sample of m-dimensional random vectors that can be analyzed by multivariate statistical methods. Clearly, a larger m improves the representation, but makes statistical inference more complicated and unstable. It should be noted that the distribution of $\varphi(X)$ is usually unknown or is fairly complicated, so that nonparametric (distribution-free) statistical methods are preferable. Several models discussed in Section 11.4 can be fitted by the method of moments. It is usually very difficult to work out the likelihood-based approach, so that Monte Carlo tests often remain the only tool for statistical inference.

If positions of sets are not important, then the functional φ should be motion-invariant. If in addition, the size influence should be eliminated, φ must be scale-invariant, so that $\varphi(cX) = \varphi(X)$ for each $c > 0$. Such functionals are called *shape ratios*; see [59] and [60]. For example,

$$\varphi(X) = 4\pi \mathsf{A}(X)/\mathsf{U}(X)^2$$

is a well-known shape ratio called the area-perimeter ratio of $X \subset \mathbb{R}^2$. It is equal to 1 for a circle and is less than 1 otherwise. Other shape factors determine how symmetrical the set in question is.

However, very often instead of analyzing some numerical values computed for the sample $X_1, \ldots, X_n$, it is more appealing to describe it using set-valued parameters. A natural set-valued parameter could be an *average* of a sample that serves as an empirical counterpart to an *expectation* of the underlying random set X. It should be noted that *the average is not a filter*. Indeed the averaging operation treats the whole set as important, so that the average looks more like a smoothing operation that provides a set-valued summary of a sample of sets or images. Many different definitions of an expectation of a random set can be found in the literature [59, 39]. We will use below a generic notation $\mathbf{E}X$ for an expectation of a random set X and different definitions of expectation will be denoted using different subscripts of $\mathbf{E}$. Each definition of the expectation has its counterpart as an average for a sample. From the point of view of typical image analysis applications it suffices to assume that X is a random *compact* set, although several definitions of expectation are applicable for unbounded random sets as well.

There are many different concepts of expectations for random compact sets, but no standard way so far. First of all, note that the general tools suitable for defining an expectation of a random element in a linear space are not applicable for $\mathcal{K}$-valued random elements (or random compact sets). The reason is that the space $\mathcal{K}$ is not linear. A common tool is to "linearize" $\mathcal{K}$ using a map (or maps) from $\mathcal{K}$ to a linear space, where it is easier to define an expectation. Usually, this linear space is chosen to be the real line $\mathbb{R}$.

Let $\Phi = \{\varphi_u\colon\ u \in U\}$ be a set of functionals that map $\mathcal{K}$ into $\mathbb{R}$, where U is a rather arbitrary parameter space. Assume that $\mathbf{E}\varphi_u(X)$ exists for each $\varphi_u \in \Phi$. If it does not exist for some φ_u, then we say that X has no expectation. If there exists a set $\mathbf{E}X$ such that

$$\varphi_u(\mathbf{E}X) = \mathbf{E}\varphi_u(X) \tag{11–10}$$

for all $\varphi_u \in \Phi$, then $\mathbf{E}X$ is the most natural candidate for the expectation of X. It certainly depends on the choice of the family Φ, and we call $\mathbf{E}X$ the Φ-*invariant expectation*. In principle, there might be several sets $\mathbf{E}X$ satisfying Eq. 11–10. However, it is possible to eliminate this ambiguity by assuming that $\mathbf{E}X$ belongs to some specified subfamily of $\mathcal{K}$, for example, assuming that $\mathbf{E}X$ is convex or star shaped or regular closed (coincides with the closure of its interior), etc.

Often the invariance equation 11–10 has no solution in the family $\mathcal{K}$. Then it is possible to weaken the invariance condition as follows

$$\varphi_u(\mathbf{E}X) \subseteq \mathbf{E}\varphi_u(X).$$

Then $\mathbf{E}X \subseteq \varphi_u^{-1}(\mathbf{E}\varphi_u(X))$, and the set

$$\mathbf{E}X = \bigcap_{u \in U} \varphi_u^{-1}\big(\mathbf{E}(\varphi_u(X))\big) \tag{11–11}$$

is called the Φ-*subinvariant expectation* (clearly, the expectation can be empty if the right-hand side of Eq. 11–11 is an empty set). This idea goes back to [36] and furthermore has been worked out in [20] for maps between two general lattices. In this general case a subinvariant expectation is given by

$$\bigwedge_{u \in U} \varepsilon_u\big(\mathbf{E}(\delta_u(X))\big),$$

where $(\varepsilon_u, \delta_u)$, $u \in U$, is a family of adjunctions between $\mathcal{K}$ and $\mathbb{R}$.

The "linearization" approaches based on the family of maps Φ can be illustrated by the following diagram.

$$\begin{array}{ccc}
X & \xrightarrow{\varphi_u,\ u \in U} & \varphi_u(X),\ \varphi_u \in \Phi \\
 & & \Big\downarrow \text{expectation} \\
\mathbf{E}X & \xleftarrow[\text{Eqs. 11–10 or 11–11}]{} & \mathbf{E}\varphi_u(X),\ \varphi_u \in \Phi.
\end{array}$$

If the invariance equation 11–10 has no solution, it is possible as well to define $\mathbf{E}X$ as the set K that minimizes

$$\big\|\mathbf{E}\varphi_u(X) - \varphi_u(K)\big\|_\infty = \sup_{u \in U} \big|\mathbf{E}\varphi_u(X) - \varphi_u(K)\big|,$$

which is the L^∞-distance between $\mathbf{E}\varphi_u(X)$ and $\varphi_u(K)$ as functions of $u \in U$, or

$$\left\|\mathbf{E}\varphi_u(X) - \varphi_u(K)\right\|_p = \left(\int_U \left(\mathbf{E}\varphi_u(X) - \varphi_u(K)\right)^p \mu(du)\right)^{1/p},$$

which is the L^p distance (for the latter, U must be equipped with a measure μ). In general, it is difficult to solve a minimization problem for a functional that depends on a compact set K, since the parameter space $\mathcal{K}$ is too rich. Then it is possible to restrict the whole range of sets K and assume that $K \in \mathcal{M}$ for some family of sets $\mathcal{M} \subset \mathcal{K}$. We call it the L^∞ (respectively, L^p) *minimization approach* over the class $\mathcal{M}$. Note that the invariant expectation (if it exists) appears as a particular case of the minimization approach.

Now consider several special definitions of expectations.

11.3.1 AUMANN EXPECTATION

Let Φ be the family of support functions, so that $U = \mathbb{R}^d$ and $\varphi_u(X) = h(X, u)$ is the support function of X (see Eq. 11–8). Note that $\mathbf{E}h(X, u)$ exists if the norm $\|X\| = \sup\{\|x\|\colon\ x \in X\}$ has a finite expectation. Then $\mathbf{E}h(X, u)$, $u \in \mathbb{R}^d$, is a support function of a compact convex set $\mathbf{E}_{\mathrm{A}}X$ called the *Aumann expectation* of X, so that the invariance equation 11–10 has a solution:

$$\mathbf{E}h(X, u) = h(\mathbf{E}_{\mathrm{A}}X, u) \quad \text{for all } u \in \mathbb{R}^d.$$

Deep mathematical reasons based on the theory of vector-valued measures and multivalued functions lead to an equivalent definition based on the concept of a selection [3, 62]. A random vector ξ is said to be a *selection* of X if $\xi \in X$ almost surely. Then $\mathbf{E}_{\mathrm{A}}X$ can be defined as the set of expectations of all integrable selections (it is assumed that the basic probability space Ω contains no atoms), i.e.

$$\mathbf{E}_{\mathrm{A}}X = \{\mathbf{E}\xi\colon\ \xi \in X \text{ almost surely, } \mathbf{E}\xi \text{ exists}\}.$$

The Aumann expectation appears as well in the strong law of large numbers for Minkowski sums of random compact sets [2]. If $X_1, X_2, \ldots$ are independent realizations of a random compact set X with $\mathbf{E}\|X\| < \infty$, then

$$\frac{X_1 \oplus X_2 \oplus \cdots \oplus X_n}{n} \to \mathbf{E}X \quad \text{as } n \to \infty$$

in the Hausdorff metric on $\mathcal{K}$.

It should be noted that the Aumann expectation is always convex. This property seriously restricts its applicability in image analysis, since most images are non-convex.

11.3.2 Doss Expectation

This expectation appears from the subinvariance definition, if $U = \mathbb{R}^d$ and $\varphi_u(X) = \rho_H(X, \{u\}) = \sup\{\|x - u\|: x \in X\}$ is the Hausdorff distance between $\{u\}$ and X. The Doss expectation is defined by

$$\mathbf{E}_D X = \bigcap_{u \in \mathbb{R}^d} \{x \in \mathbb{R}^d: \|x - u\| \leqslant \mathbf{E}\rho_H(X, \{u\})\}.$$

The Doss expectation is a convex set, because it appears as an intersection of balls. It is possible to show that both Doss and Aumann expectations are compatible with the standard definitions of expectation if X is a random singleton, while $\mathbf{E}_A X \subseteq \mathbf{E}_D X$ in general. In contrast to the Aumann expectation, the Doss expectation can be defined for random sets in a general metric space [22].

11.3.3 Radius-Vector Expectation

This expectation is rather popular in the engineering literature when analysing shapes of particles. The crucial assumption is that X must be star shaped with respect to the origin (or another fixed point). Then X is completely defined by its radius-vector function $\varphi_u(X) = r(X, u)$ (see Eq. 11–9), where u belongs to the unit sphere. Then $\mathbf{E}\varphi_u(X)$ is a radius-vector function of a star-shaped set $\mathbf{E}_{RV} X$ called the radius-vector expectation. This definition appears as a particular case of the L^2-minimization approach over the family of star shaped sets K.

11.3.4 Fixed Points and Quantiles

Now each $u \in U$ is identified with a compact set, so that $U = \mathcal{K}$. Let $\varphi_K(X)$ be 1 if X and K have a nonempty intersection and $\varphi_K(X) = 0$ otherwise. Then $\mathbf{E}\varphi_K(X) = T_X(K)$ is the capacity functional of X, and the subinvariant approach implies that $\mathbf{E}X$ is the set of all fixed points $x \in \mathbb{R}^d$ such that $x \in X$ almost surely. Along this line, it is possible to define a pth quantile of X as

$$M_p = \bigcup \{K \in \mathcal{M}: T_X(K) < p\},$$

where $0 \leqslant p \leqslant 1$, and $\mathcal{M}$ is a subclass of $\mathcal{K}$, see [34].

11.3.5 Vorob'ev Expectation

Let $U = \mathbb{R}^d$ and let $\varphi_u(X) = \mathbf{1}_{u \in X}$ be the *indicator function* of X, so that $\varphi_u(X) = 1$ if $u \in X$ and $\varphi_u(X) = 0$ otherwise. Then $\mathbf{E}\varphi_u(X) = p_X(u) = \mathbf{P}\{u \in X\}$ is the coverage function. Assume that $\mathbf{E}\mu(X) < \infty$, where μ is the Lebesgue measure in $\mathbb{R}^d$. In this case, the Φ-invariant expectation does not exist unless $p_X(\cdot)$ is an indicator function itself, which is possible only for a deterministic X. A set-theoretic mean $\mathbf{E}_V X$ is defined in [63, 59] by

$$\mathbf{E}_V X = X_t = \{u \in \mathbb{R}^d \colon\ p_X(u) \geqslant t\}$$

for t which is determined from the equation $\mathbf{E}\mu(X) = \mu(X_t)$ (if this equation has a solution), or, in general, from the inequality

$$\mu(X_s) \leqslant \mathbf{E}\mu(X) \leqslant \mu(X_t) \quad \text{for all} \quad s > t.$$

Note that the L^∞ minimization approach over the family X_t, $t \in [0, 1]$, yields the level set $X_{1/2}$, which is called the *median* of X, see [59]. The L^1 and L^2 minimization approaches yield apparently new definitions of expectations as X_t with t chosen to minimize

$$\|p_X(u) - \mathbf{1}_{u \in X_t}\|_1 = \mu(X_t) - \mathbf{E}\mu(X) + 2\int_{X_t^c} p_X(u)du$$

or

$$\|p_X(u) - \mathbf{1}_{u \in X_t}\|_2 = \int_{\mathbb{R}^d} p_X^2(u)du + \mu(X_t) - 2\int_{X_t} p_X(u)du.$$

However, in all cases singletons as well as sets of almost surely vanishing Lebesgue measures (such as curves in $\mathbb{R}^2$ or surfaces in $\mathbb{R}^3$) are considered as uninteresting, since their indicator functions vanish almost surely if the corresponding distributions are atomless.

11.3.6 DISTANCE AVERAGE

Let $U = \mathbb{R}^d$ and let $\varphi_u(X) = \rho(u, X)$ be the *Euclidean distance function* of X, i.e. $\rho(u, X)$ equals the Euclidean distance from u to the nearest point of X. A suitable level set of the expected distance function $\bar{d}(u) = \mathbf{E}\rho(u, X)$ serves as the mean of X. To find this suitable level, $\bar{d}(u)$ is thresholded to get a family of sets $X(t) = \{u \colon\ \bar{d}(u) \geqslant t\}$, $t > 0$. According to the L^∞ minimization approach, the *distance average* $\bar{X}$ is the set $X(t)$, where t is chosen to minimize

$$\left\|\bar{d}(\cdot) - \rho(\cdot, X(t))\right\|_\infty = \sup_{u \in \mathbb{R}^d} \left|\bar{d}(u) - \rho\big(u, X(t)\big)\right|.$$

See [6] for details, related concepts and further generalizations. In particular, rather good properties of the distance average can be achieved for the signed distance function $\varphi_u(X) = \rho(u, X) - \rho(u, X^c)$. The distance function approach allows us to deal with sets of zero Lebesgue measures, since, even in this case, the distance function is nontrivial (in contrast to the coverage function used to define the Vorob'ev expectation).

It is possible to list a number of properties that should be verified by a "reasonable" expectation of a random closed set. However, they are often noncompatible, and lead to restrictive definitions. For example, the property that $\mathbf{E}X \subseteq \mathbf{E}Y$ if $X \subseteq Y$ leads to the Aumann definition of the expectation, see [39]. If ψ is a function that maps compact sets into compact sets (e.g., an image filter), then the expectation $\mathbf{E}X$ is said to be *ψ-compatible*, if $\mathbf{E}\psi(X) = \psi(\mathbf{E}X)$. Clearly, the Φ-invariant expectation is compatible with each $\varphi \in \Phi$. All the above-mentioned expectations (apart from the radius-vector expectation) are compatible with Euclidean motions (translations and rotations). The Aumann expectation is compatible with dilations by convex sets. Other definitions of expectations are not compatible with the usual morphological operations. In general, the choice of particular expectation depends on the conditions assumed on the realizations (convex, star shaped, etc.) of the random set X and the required properties of the expectation.

Each of the above definitions of the expectation yields the corresponding average if a sample of sets is given. For example, if the distance average is chosen, then the sample $X_1, \dots, X_n$ can be used to calculate the empirical variant of $\bar{d}(u)$ as

$$\hat{d}(u) = \big(\rho(u, X_1) + \cdots + \rho(u, X_n)\big)/n.$$

The final step is to find its threshold $\hat{X}(t) = \{x\colon\ \hat{d}(x) \geqslant t\}$ in order to minimize $\|\hat{d}(\cdot) - \rho(\cdot, \hat{X}(t))\|_\infty$.

Note that when averaging we deal with a sample of independent identically distributed realizations of a random compact set. If positions of the sets are known, then we speak about statistics of *sets*, in contrast to statistics of *figures* when locations/orientations of sets are not specified. This means that the positions of the sets are irrelevant for the problem and the aim is to find the average *shape* of the sets in the sample. Such a situation appears in studies of particles (dust powder, sand grains, abrasives, etc.). Often also scale transformations are allowed, so that homothetic sets are considered to be identical. Similar problems appear when it is necessary to align several images [50] or to synchronize several functions using dynamic time warping [27].

If an observer deals with a sample of figures rather than sets, then the definitions of expectation of a random compact set are not really useful. For instance, the images of particles are isotropic sets, whence the corresponding set-expectations are balls or disks, so that the set-expectations provide only very limited information about the average shape of random sets.

The approach below can be found in [58]. It is inspired by the studies of landmark configurations and shapes for finite sets of points, see [7, 56]. Two compact sets are equivalent if they can be superimposed by a rigid motion (scale transformations are excluded). Then a sample of sets (which represent the corresponding figures) must be transformed in order to place them "close together", and then an appropriate set-theoretic mean should be determined for the transformed sets. Formally, this means identifying motions (or more general transformations) $g_1, \dots, g_n$ which minimize

$$\sum_{i,j=1}^{n} \mathbf{m}(g_i X_i, g_j X_j)^2, \qquad (11\text{–}12)$$

where $\mathbf{m}$ is a metric on the space $\mathcal{K}$, for example the Hausdorff metric or another metric, see [5]. If $g_1, \dots, g_n$ are chosen to minimize Eq. 11–12, then $X_1^* = g_1 X_1, \dots, X_n^* = g_n X_n$ are considered to be "closed together" or aligned [note that Eq. 11–12 is proportional to the inertia of the "physical" configuration of $g_1 X_1, \dots, g_n X_n$ if $\mathbf{m}(g_i X_i, g_j X_j)$ determines the distance between two unit masses attached to $g_i X_i$ and $g_j X_j$]. Therefore, it is possible to average the aligned sample of sets $X_1^*, \dots, X_n^*$ using one of the definitions of expectations for random sets.

Note that motions of sets correspond to transformations of some associated functions considered to be elements of a Hilbert space. For example, if convex compact sets are described by support functions, then the group of rigid motions (translations and rotations) of sets corresponds to appropriate actions on the space of support functions. It was proven in [58] that the "optimal" translations of convex sets $X_1, \dots, X_n$ superimpose their Steiner points

$$s(X_i) = \frac{1}{b_d} \int_{\|u\|=1} h(X_i, u) u \, du, \quad i = 1, \dots, n,$$

where b_d is the volume of the unit ball in $\mathbb{R}^d$. Thus, $X_1' = X_1 - s(X_1), \dots, X_n' = X_n - s(X_n)$ are to be superimposed using rotations only, cf. [14]. These rotations can be found using a stepwise algorithm described in [58]. It should be noted that, in general, minimization of Eq. 11–12 is a quite difficult numerical problem.

11.4 MODELS OR PRIORS

Statistical studies of compact random sets are difficult due to a lack of models (or distributions of random sets) that allow evaluations and provide sets of sufficiently variable shape. Clearly, simple random sets, like random singletons and balls, are ruled out if we would like to have models with really *random shapes*. This means that it is not possible to use the arsenal of distributions of random points usual in conventional statistical studies. They all are just points, which are of limited interest within the framework of random set theory. It should be noted that models of random sets can be used as prior distributions in image analysis. By now, most of priors stem from interacting point processes [30].

As we have seen, random points do not provide interesting models of random sets. A possible way to overcome this problem is to work with combinations of points. For example, three points define a triangle with a varying shape depending on the angles. More generally, consider a (measurable) map $M : \mathbb{R}^m \mapsto \mathcal{K}$, so that $M(u)$ is a compact subset of $\mathbb{R}^d$ for each $u \in \mathbb{R}^m$. To get nontrivial maps, the dimension m must be greater (and often much greater) than d. For example, a planar triangle can be represented as a map from $\mathbb{R}^6$ to $\mathcal{K}$ such that $M(u_1, \ldots, u_6)$ is a triangle with the vertices (u_1, u_2), (u_3, u_4) and (u_5, u_6). If ξ is a random point in $\mathbb{R}^m$, then $X = M(\xi)$ becomes a random compact set in $\mathbb{R}^d$. However, it is quite difficult to come up with analytical results in this case. Even for random triangles, the theory becomes rather involved [25, 29]. Related models appear as convex hulls of a finite number of points, so that $X = \text{conv}(\xi_1, \ldots, \xi_n)$, where $\xi_1, \ldots, \xi_n$ are random points in $\mathbb{R}^d$. However, the relevant distribution is quite complicated, and only asymptotic results for large n are known [51].

A "good" model of a random set must satisfy the following criteria:

- the model must provide realizations with variable shapes;
- the model should be flexible enough to allow changes in shape distribution in response to changes in its parameters;
- the distribution of the random set (the capacity functional or other related functionals) must be computable;
- the distributions (or at least the lower moments) of some important functionals (such as volume or surface area) must be computable;
- the model must be easy to simulate.

In general, it is very difficult to find a model which satisfies all these properties. A partial solution to this problem is provided by the theory of point processes. A *point process* is a locally finite random set X. In other words, $X = \{x_1, x_2, \ldots\}$ is a collection of points such that the number of points inside each bounded set is finite, although the total number of points may be infinite. It is often easier to define distributions of point processes than distributions of general random sets. The simplest point process is the Poisson point process, which is characterized by two basic properties:

- the numbers of points inside disjoint sets are independent;
- the number of points inside a set K has the Poisson distribution with the mean $\lambda(K)$, where λ is a measure on $\mathbb{R}^d$ called the *intensity measure*.

If λ is a finite measure, then the total number of points is almost surely finite. If λ has a density, then the density function is also denoted by λ and is called the *intensity function*. The corresponding point process is denoted by Π_λ. If the intensity measure λ is proportional to the Lebesgue measure, then the Poisson point process becomes stationary and its intensity function is constant. The capacity functional of the Poisson point process is easy to find as

$$T(K) = 1 - \exp\{-\lambda(K)\} = 1 - \exp\left\{-\int_K \lambda(x)dx\right\},$$

since $T(K)$ is exactly the probability that a Poisson random variable with a mean $\lambda(K)$ takes a positive value.

Two standard models of random polygons (Poisson and Poisson–Dirichlet polygons) are closely related to the Poisson point process. A Poisson point process in the space $[0, 2\pi) \times [0, \infty)$ generates an isotropic Poisson line process in $\mathbb{R}^2$, so that the points in $[0, 2\pi) \times [0, \infty)$ provide a natural parameterization of the lines through their normals and the distances to the origin. These lines divide the plane into convex polygons. These are shifted in such a way that their centers of gravity lay in the origin and are interpreted as realizations of the "typical" random polygon X called the *Poisson polygon* [32, 57, 59]. The inclusion functional of X is given by

$$I(K) = \mathbf{P}\{K \subseteq X\} = \exp\left\{-\int_{\|u\|=1} h(K, u)\nu(du)\right\},$$

where $0 \in K$, and ν is a symmetric measure on the unit sphere [for example, the $(d-1)$-dimensional Hausdorff measure in the isotropic case], see [32].

Consider the Dirichlet (or Voronoi) mosaic generated by a stationary Poisson point process of intensity λ. For each point x_i we construct the open set consisting of all points of the plane whose distance to x_i is less than the distances to other points. If shifted by x_i, the closures of these sets give realizations of the so-called *Poisson–Dirichlet polygon*, see [59].

Unfortunately, in the isotropic case the distributions of both the Poisson and Poisson-Dirichlet polygons depend on one parameter only (the intensity of the point process), which affects only their size distributions. If the size is eliminated by scaling, then the distributions of the corresponding scaled sets are fixed. That is why these two popular models have a restricted applicability for modelling compact sets.

Further models of random compact sets arise from limit theorems for scaled unions of random sets [36]. The idea is to take independent identically distributed random sets $X_1, \ldots, X_n$ with a simple distribution and perhaps not interesting shapes and take their unions normalized by an appropriate scale factor. Under some conditions the normalized set $a_n^{-1}(X_1 \cup \cdots \cup X_n)$ converges in distribution to a nontrivial random set called *union-stable*. It is possible as well to consider convex hulls $a_n^{-1}\mathrm{conv}(X_1 \cup \cdots \cup X_n)$ instead of the unions. Then in the limit one gets random closed sets that are called *convex-stable*. The statistical inference for convex-stable polygons (and for other models of compact random polygons) is discussed in [43].

Below we introduce a special class of union- and convex-stable random sets which are related to scale-invariant Poisson point processes. Consider a Poisson point process Π_λ in $\mathbb{R}^m$ with a *homogeneous* intensity function, so that

$$\lambda(cx) = c^{\alpha - m}\lambda(x), \tag{11–13}$$

for all $x \in \mathbb{R}^m$ and $c > 0$, where $\alpha > 0$ is a parameter of the distribution. Then Π_λ is called a scale-invariant Poisson point process. In the isotropic case, $\lambda(x) = \|x\|^{\alpha - m}a$, so that the distributions form a two-parametric family with the parameters α and a. Note that the point process with such an intensity function is non-stationary, and it has only a finite number of points outside any ball centered at the origin. Assume that $m = d$, and let $X = \mathrm{conv}(\Pi_\lambda)$ be the convex hull of Π_λ. The containment functional of X is easy to find as

$$C(F) = \mathbf{P}\{\Pi_\lambda \text{ has no points outside } F\} = \exp\left\{-\int_{F^c} \lambda(x)dx\right\},$$

where F is a convex closed set. The corresponding Aumann expectation is computable, as are the lower moments of the perimeter and the area of X, see [36, 43].

Another interesting and rather general random set model is defined as

$$X = \bigcup_{x_i \in \Pi_\lambda} M(x_i),$$

where $M : \mathbb{R}^m \mapsto \mathcal{K}$ is a function on $\mathbb{R}^m$ whose values are compact sets in $\mathbb{R}^d$, and Π_λ is a Poisson point process in $\mathbb{R}^m$ with the homogeneous intensity function satisfying Eq. 11–13. If M is homogeneous, i.e. $M(cx) = c^\eta M(x)$, then X is a union-stable random compact set and its convex hull $\mathrm{conv}(X)$ is convex-stable. Then

$$T_X(K) = 1 - \exp\left\{-\int_{x \in \mathbb{R}^m:\ M(x)\cap K = \emptyset} \lambda(x)dx\right\}$$

and, for all convex F,

$$\mathbf{P}\{X \subseteq F\} = \mathbf{P}\{\mathrm{conv}(X) \subseteq F\} = 1 - T_X(F^c).$$

In this case the parametric family of distributions is much wider. Interesting random sets appear if $m = 3$ and $M(u_1, u_2, u_3)$ is a disk in $\mathbb{R}^2$ with radius $u_1 \geqslant 0$ and the center (u_2, u_3), see [36]. Note that the easiest way to simulate such random sets is to use the limit theorem for unions as described in [36, 43].

One can use random polygons to obtain further models of random sets. For example, a *rounded polygon* is defined by $X \oplus \xi B$, where ξ is a positive random variable and B is the unit disk. Convex-stable random polygons have been applied in [31, 41] to model rather complicated star-shaped random sets that appear as planar sections of human lungs. Other possible models for compact random sets include radius-vector perturbed sets [31, 59], weak limits of intersections of half-planes [36, Sect. 8.5], and set-valued growth processes [59, pp. 138–139].

11.5 THE BOOLEAN MODEL

Poisson points are often used in image analysis as a simple noise model by either adding points to the background or deleting some pixels from the foreground. Such noise, however, can be easily eliminated by morphological filters if the image is sufficiently regular. For example, the additional points of the set $X \cup \Pi_\lambda$ are eliminated by an opening transform. It is possible as well to add marks (or gray-scale values) to the points of the Poisson process and use the marked process to model noisy gray-scale images.

A situation of a different type arises when the deterministic set (or binary image) X is accessible through Poisson points only. In this case one observes the intersection, $X \cap \Pi_\lambda$, of X and a Poisson point process. Then the aim is to restore the boundary of X or estimate X using only the set $X \cap \Pi_\lambda$. To solve this problem, one should make some assumptions on X, for example, that X is convex or has a sufficiently smooth boundary. The minimum assumption is that X is simply connected [9]. Such a problem for convex X has been considered in [49]. Then the natural estimator is the convex hull of $X \cap \Pi_\lambda$. This estimator is biased and always underestimates X. It is, however, possible to re-scale the convex hull to obtain an unbiased estimator. Many results of this type are discussed and treated systematically in [28].

The Poisson point process gives rise to a useful model of stationary random sets called the *Boolean model*. Consider a sequence of independent identically distributed random compact sets $X_1, X_2, \ldots$ called the *grains*. These sets are independent of Π_λ and usually chosen to have a simple distribution; for example, they can be random balls, random polygons, or even deterministic sets. Each set is associated with a point of the Poisson point process Π_λ in $\mathbb{R}^d$. In other words, each point $x_i \in \Pi_\lambda$ is marked with the corresponding set X_i. The Boolean model is a random set Ξ given by

$$\Xi = \bigcup_{x_i \in \Pi_\lambda} (x_i + X_i). \tag{11–14}$$

The points of Π_λ are called *germs*, and Ξ is sometimes called the *germ-grain model*. It also appears under the name of a coverage process [17] or a clump. The set X_0 which has the same distribution as each of the grains is called the *typical grain*. It is possible to relax the model assumptions [57], although then the model soon becomes intractable.

A specific feature of the Boolean model is that it can be used to model either the noise or the image itself. Over the course of time the Boolean model has been applied to various practical problems. Basic references are given in [17, 53, 57]. It is intuitively clear that the Boolean model can describe spatial patterns containing independent randomly located overlapping components. The list of applications includes bombing (germs are points of impact and grains are damaged regions caused by each separate bomb), the microstructure of paper (grains are elementary fibers which can be described as random rectangles), tumor growth, spatial patterns of heather in the countryside, geological deposits, crystallization in metals and polymers, the microstructure of dough, patterns in photographic emulsions, structure of materials, systems of water droplets and many others. An extensive list of references can be found in [40].

If the Boolean model describes the image of interest, then the principal aim is to estimate the parameters of the model (the intensity of the Poisson point process and the characteristics of the typical grain) from realizations of the set Ξ. A survey of recent statistical methods for the stationary Boolean model can be found in [40], see also [39]. Discrete random sets generated by Eq. 11–14 with $\{x_i\}$ being a binomial point process on the grid have been studied in [12, 54] under the name of discrete Boolean models. Such a discrete random set can be however more appropriately called the binomial germ-grain model, see [57].

Statistics of the Boolean model began with estimation of λ and mean values of the Minkowski functionals of the grain (mean area, volume, perimeter, etc.). In general, these values do not suffice to retrieve the distribution of the grain Ξ_0, although sometimes it is possible to find an appropriate distribution if a parametric family is given. Clearly, without the distribution of the grain, it is impossible to simulate the underlying Boolean model, and, therefore, to use tests based on simulations. For example, if the grain is a ball, then, in general, it is impossible to determine its radius distribution by the corresponding moments up to the dth order. However, if the distribution of the radius belongs to some parametric family, say log-normal, then it is determined by these moments. Even in the parametric setup, now most studies end with proof of strong consistency. Results concerning asymptotic normality are still rather exceptional, see [21, 42], and there are no theoretical studies of efficiency.

By the Choquet theorem, the distribution of X_0 is determined by the corresponding capacity functional $T_{X_0}(K)$. The capacity functional of the Boolean model Ξ can be evaluated as

$$T_\Xi(K) = \mathbf{P}\{\Xi \cap K \neq \emptyset\} = 1 - \exp\{-\lambda \mathbf{E}\mu_d(X_0 \oplus \check{K})\}, \tag{11–15}$$

see [17, 32, 57]. Here $\check{K} = \{-x \colon x \in K\}$. By the Fubini theorem, we get from Eq. 11–15

$$T_\Xi(K) = 1 - \exp\left\{-\lambda \int_{\mathbb{R}^d} T_{X_0}(K + x)dx\right\}. \tag{11–16}$$

The functional on the left-hand side of Eq. 11–16 is determined by the whole set Ξ and can be estimated from observations of Ξ, see [40, 39]. However, it is unlikely that the integral equation 11–16 can be solved directly.

The estimation methods are based on relationships between *observable* characteristics and parameters of the Boolean model, bearing in mind that replacing these observable characteristics by their empirical counterparts provides estimators for the corresponding parameters. In the simplest case, the accessible information about Ξ is the volume fraction p covered by Ξ. Because of stationarity and Eq. 11–15,

$$p = T_\Xi(\{0\}) = 1 - \exp\{-\lambda \mathbf{E}\mu_d(X_0)\}, \tag{11–17}$$

so that an estimator of $\lambda \mathbf{E}\mu_d(X_0)$ can be obtained if p is estimated by counting pixels or measuring covered areas.

Two-point covering probabilities determine the *covariance* of Ξ

$$C(v) = \mathbf{P}\{\{0, v\} \subset \Xi\} = 2p - T_\Xi(\{0, v\}).$$

Note that $C(v) = \mathbf{E}\zeta(0)\zeta(v)$, where $\zeta(v) = 1_{v \in \Xi}$ is the indicator function of Ξ. Then the function

$$q(v) = 1 + \frac{C(v) - p^2}{(1-p)^2} = \exp\{\lambda \mathbf{E}\mu_d(X_0 \cap (X_0 - v))\}$$

yields the set-covariance function of the grain $\gamma_{X_0}(v) = \mathbf{E}\mu_d(X_0 \cap (X_0 - v))$. It should be noted that the covariance is quite flexible in applications, since it only depends on the capacity functional on two-point compacts. For example, almost all covariance-based estimators are easy to reformulate for censored observations, see [35, 40]. Applications of the covariance function to statistical estimation of the Boolean model parameters were discussed also in [17, 53, 57].

Historically, the first statistical method for the Boolean model was the *minimum contrast method* for contact distribution functions or the covariance, see [53, 11]. Its essence is to determine the values of T_Ξ for some subfamilies of compact sets (balls, segments or two-point sets). Then the right-hand side of Eq. 11–15 can be expressed by means of known integral geometric formulas. In particular, if X_0 is a random body and $K = B_r(0)$ is a ball of radius r, then the Steiner formula gives the expansion of $\mathbf{E}\mu_d(X_0 \oplus B_r(0))$ as a polynomial of dth order whose coefficients are expressed through the Minkowski functionals of X_0 (which are the area and perimeter of X_0 in the planar case). The next step is to replace $-\log(1 - T_\Xi(K))$ (the transformed left-hand side of Eq. 11–15) by its empirical counterpart and approximate it by a polynomial, see [8, 17, 21]. Finally, the polynomial's coefficients yield estimators of the unknown parameters.

Now consider another estimation method which could be named the *method of intensities*. First, note that Eq. 11–17 is an equality relating spatial averages to parameters of the Boolean model. The volume fraction is the simplest spatial average. It is possible to consider other spatial averages, for example, mean surface area per unit volume, the specific Euler–Poincaré characteristics or, more generally, densities of extended Minkowski measures, see [65]. According to the method of intensities, estimators are chosen as solutions of the equations relating these spatial averages (their intensities) to estimated parameters.

A particular implementation of the method of intensities depends on the way the Minkowski functionals are extended onto the convex ring (the family of finite unions of convex compact sets). The *additive extension* [52] was used in [26, 64]. Another technique, the so-called *positive extension* [32, 52], has been applied to statistics of the Boolean model in [8, 53, 57]. It goes back to Haas, Matheron and Serra [16] and De Hoff [10].

In the planar case with convex grains, the latter approach is based on Eq. 11–17 and the following relationships

$$L_A = (1 - p)\mathbf{E}\mathsf{U}(X_0)\lambda, \tag{11–18}$$

$$N_A^+ = (1 - p)\lambda, \tag{11–19}$$

which are used to express the intensity λ, the mean perimeter of the grain $\mathbf{E}\mathsf{U}(\Xi_0)$ and the mean area $\mathbf{E}\mathsf{A}(\Xi_0)$ through the following observable values: the area fraction p, the specific boundary length L_A and the intensity N_A^+ of the point process of (say lower) positive tangent points. The specific boundary length is the expected length of the boundary of Ξ within a unit area. If we mark each lower-left tangent point of all grains, then some of these points will be covered by other grains, while others will be exposed. Then N_A^+ is the expected number of these exposed points in a unit area. This method yields biased but strong consistent and asymptotically normal estimators, see [42]. For instance, the asymptotic variance of the intensity estimator obtained from Eq. 11–19 is equal to $\lambda/(1-p)$. Higher-order characteristics of the point process of tangent points are ingredients for constructing estimators of the distribution of the grain, see [37].

Methods that deal with the situation when the Boolean model describes the noise are less developed. Initial studies can be found in [35]. Let X be a random set of interest. If Ξ is a Boolean model representing a union noise and independent of X, then we observe the random set $X \cup \Xi$ instead of X. This case is relatively easy to handle using the capacity functionals, since

$$T_{X\cup\Xi}(K) = T_X(K) + T_\Xi(K) - T_X(K)T_\Xi(K),$$

whence

$$T_X(K) = \frac{T_{X\cup\Xi}(K) - T_\Xi(K)}{1 - T_\Xi(K)}.$$

The capacity functional T_Ξ can be found from Eq. 11–15 if the distribution of Ξ is known, while $T_{X\cup\Xi}$ can be estimated using realizations of $X \cup \Xi$. For instance, if W is the window of observations, then $T_{X\cup\Xi}(K)$ can be estimated by

$$\hat{T}_{X\cup\Xi,W}(K) = \frac{\mu_d(((X \cup \Xi) \oplus \check{K}) \cap (W \ominus K))}{\mu_d(W \ominus K)}.$$

Compare this with Eqs. 11–7 and 11–6 if independent realizations of $X \cup \Xi$ are available. This yields a plug-in estimator for $T_X(K)$. For example, if Ξ is a Poisson point process of intensity λ, then $T_X(K)$ is estimated by

$$\hat{T}_{X,W}(K) = \hat{T}_{X\cup\Xi,W}(K)e^{\lambda\mu_d(K)} - e^{\lambda\mu_d(K)} + 1.$$

If X is random and is assumed to be a Boolean model, then knowing $T_X(K)$, one can estimate the model's parameters [35, 40]. On the other hand, if X is deterministic, then $T_X(K)$ is either 0 or 1, while the estimator $\hat{T}_{X,W}(K)$ may take values between 0 and 1. In this case, X can be interpreted as an expectation of another fictitious random set $\hat{X}$ with the capacity functional $T_{\hat{X}}(K) = \hat{T}_{X,W}(K)$. One can use, for instance, the Vorob'ev expectation or the quantile to find an expected value of $\hat{X}$ using its capacity functional.

The approach based on the capacity functionals works in more general situations than one using morphological filtering. Indeed, to get rid of Ξ in $X \cup \Xi$, we need to use opening and also to assume that the grains of Ξ are small enough and that the intensity λ is low. Otherwise, the grains of Ξ form large clumps which are difficult to filter out using morphological filters.

Another union noise setup has been considered in [61, 24]. Let $X = K \cup Y$, where K is a deterministic set and Y is a random set (which is not necessarily a Boolean model). If independent realizations $X_1, \ldots, X_n$ of X are available, then K can be estimated by $\hat{K}_n = X_1 \cap \cdots \cap X_n$, so intersections filter out the noise Y. This estimator $\hat{K}_n$ is consistent in the Hausdorff metric if $\mathbf{P}\{Y \subseteq K \oplus B_r(0)\} < 1$ for each $r > 0$. This condition is satisfied for all K if $\mathbf{P}\{Y \cap C = \emptyset\} > 0$ for every proper cone C.

A similar noise model appears if $X = K \oplus Y$. Then the intersection estimator $\hat{K}_n$ is consistent if $0 \in Y$ almost surely and $\mathbf{P}\{\|Y\| \leqslant r\} > 0$ for each $r > 0$.

Now consider $X \cap \Xi$, which is called the intersection noise model. Then it is almost impossible to write down the capacity functional $T_{X\cap\Xi}(K)$ in terms of $T_X(K)$ and $T_\Xi(K)$ for a general K, although it can be done in principle if K consists of several points. For example,

$$
\begin{aligned}
T_{X\cap\Xi}(\{x\}) &= T_X(\{x\})T_\Xi(\{x\}),\\
T_{X\cap\Xi}(\{x, y\}) &= \mathbf{P}\{x \in X \cap \Xi\} + \mathbf{P}\{y \in X \cap \Xi\} - \mathbf{P}\{\{x, y\} \subseteq X \cap \Xi\}\\
&= T_X(\{x\}) + T_\Xi(\{x\}) + T_X(\{y\})T_\Xi(\{y\})\\
&\quad - \big(T_X(\{x\}) + T_X(\{y\}) - T_X(\{x, y\})\big)\\
&\quad \times \big(T_\Xi(\{x\}) + T_\Xi(\{y\}) - T_\Xi(\{x, y\})\big).
\end{aligned}
$$

The most appropriate functional to deal with the intersection noise could be the inclusion functional, since

$$
I_{X\cap\Xi}(K) = \mathbf{P}\{K \subseteq X \cap \Xi\} = I_X(K)I_\Xi(K).
$$

Note, however, that the inclusion functional $I_\Xi(K)$ is equal to the probability that K is *covered* by Ξ, which is not known analytically for a general K if Ξ is a Boolean model. Some bounds can be found in [17]. Note that if the noise model $X \cap \Xi^c = X \setminus \Xi$ is considered, then

$$
I_{X\cap\Xi^c}(K) = I_X(K)I_{\Xi^c}(K) = T_X(K)\mathbf{P}\{K \cap \Xi = \emptyset\} = I_X(K)Q_\Xi(K),
$$

with the avoidance functional $Q_\Xi(K)$ being easily computable.

11.6 DISTANCE BETWEEN DISTRIBUTIONS OF RANDOM SETS

By now there is an extensive literature on metrics for binary images and some results for gray-scale images, see [5, 66]. In many cases the metrics for binary images admit straightforward generalizations for random closed sets. However, note that instead of measuring distances between realizations of sets, the distances between random closed sets determine how "far apart" their distributions are. This means that the distance is zero if two random sets are identically distributed, although they may have different realizations, depending on chance. Such distances between distributions of general random elements are usually called *probability metrics*, see [47].

Several approaches to defining probability metrics for random closed sets are described in [36]. Note that many probability metrics for random variables are defined through their distribution functions and densities. Although there is no concept of a density for random sets, the capacity functional in many cases can be used instead of the distribution function. Therefore, probability metrics for random sets correspond to distances between their capacity functionals. Several such distances are studied in [36] in view of applications to limit theorems for unions of random closed sets, see also [4]. However, in many cases probability metrics for random closed sets can be applied within an image analysis framework, for example, to assess differences between the input and output of image filters or to compare the performance of different image filters; see [13].

The *uniform distance* between the random sets X and Y is defined as

$$U(X, Y; \mathcal{M}) = \sup\{|T_X(K) - T_Y(K)|: \ K \in \mathcal{M}\},$$

where $\mathcal{M}$ is a subclass of $\mathcal{K}$. The *Lévy metric* is defined as

$$L(X, Y; \mathcal{M}) = \inf\{r > 0: \ T_X(K) \leqslant T_Y(K^r) + r, \ T_Y(K) \leqslant T_X(K^r) + r \\ \text{for all } \ K \in \mathcal{M}\},$$

where $K^r = K \oplus B_r(0)$ is the Minkowski sum of K and a ball of radius r centered at the origin. Both metrics depend on a class $\mathcal{M} \subseteq \mathcal{K}$, which should be chosen properly. A rich $\mathcal{M}$ leads to complicated calculations necessary to evaluate the metric, while a poor $\mathcal{M}$ makes the metric less sensitive, so it can vanish for a pair of rather different random sets. If X_n, $n \geqslant 1$, and X are random compact sets, then $L(X_n, X; \mathcal{K}) \to 0$ as $n \to \infty$ if and only if X_n converges to X in distribution, see [36]. In many cases, the family of all balls is a good example of the class $\mathcal{M}$. Further probability metrics are considered in [36].

Consider several simple examples of random sets and probability metrics.

EXAMPLE 11–3. Let $X = (-\infty, \xi]$ and $Y = (-\infty, \eta]$ be random subsets of $\mathbb{R}^1$, and let $\{\inf K: \ K \in \mathcal{M}\} = \mathbb{R}^1$. Then $U(X, Y; \mathcal{M})$ coincides with the uniform distance between distribution functions of ξ and η, and $L(X, Y; \mathcal{M})$ equals the Lévy distance between the distributions of ξ and η.

EXAMPLE 11–4. Let X and Y be the Poisson point processes in $\mathbb{R}^d$ with intensity measures Λ_X and Λ_Y, respectively. Then $U(X, Y; \mathcal{K})$ is not greater than the total variation distance between Λ_X and Λ_Y. If X and Y are stationary and have intensities λ_X and λ_Y, then $U(X, Y) = h(\lambda_X/\lambda_Y)$, where $h(x) = |x^c - x^{cx}|$ with $c = 1/(1-x)$.

EXAMPLE 11–5. If $X = \{\xi\}$ and $Y = B_r(\xi)$, then

$$U(X, Y; \mathcal{M}) = \sup\{\mathbf{P}\{\xi \in (K^r \setminus K)\}\colon K \in \mathcal{M}\}$$

which is sometimes called the concentration function of ξ.

EXAMPLE 11–6. Let Ξ and Ξ' be two planar Boolean models with intensities λ and λ' and convex typical grains X and X'. If $\mathcal{M}$ is the class of all balls, then

$$U(\Xi, \Xi'; \mathcal{M}) = \sup_{r \geqslant 0} \Big| \exp\{-\lambda(\mathsf{A} + 2\mathsf{U}r + \pi r^2)\} - \exp\{-\lambda'(\mathsf{A}' + 2\mathsf{U}'r + \pi r^2)\}\Big|,$$

where A and U (respectively, A' and U') are the mean area and perimeter of X (respectively, X'). This formula can be used to assess the performance of a filter that transforms the grains and the intensities of the Boolean model.

REFERENCES

[1] Archambault, S. and Moore, M., "Statistiques morphologiques pour l'ajustement d'images," *Int. Stat. Rev.* **61**, 283–297 (1993).

[2] Artstein, Z. and Vitale, R. A., "A strong law of large numbers for random compact sets," *Ann. Prob.* **3**, 879–882 (1975).

[3] Aumann, R. J., "Integrals of set-valued functions," *J. Math. Anal. Appl.* **12**, 1–12 (1965).

[4] Baddeley, A. J., "Hausdorff metric for capacities," Technical Report BS-R9127, Centrum voor Wiskunde en Informatica, Amsterdam (1991).

[5] Baddeley, A. J., "Errors in binary images and an L^p version of the Hausdorff metric," *Nieuw Archief voor Wiskunde* **10**, 157–183 (1992).

[6] Baddeley, A. J. and Molchanov, I. S., "Averaging of random sets based on their distance functions," *J. Math. Imag. Vis.* **8**, 79–92 (1998).

[7] Bookstein, F. L., "Size and shape spaces for landmark data in two dimensions (with discussion)," *Stat. Sci.* **1**, 181–242 (1986).

[8] Cressie, N. A. C., *Statistics for Spatial Data*, Wiley, New York (1991).

[9] Cuevas, A., "On pattern analysis in the nonconvex case," *Kybernetes* **19**, 26–33 (1990).

[10] DeHoff, R. T., "The quantitative estimation of mean surface curvature," *Trans. Am. Inst. Mining Metallurg. Petrol. Eng.* **239**, 617 (1967).

[11] Diggle, P. J., "Binary mosaics and the spatial pattern of heather," *Biometrics* **37**, 531–539 (1981).

[12] Dougherty, E. R. and Handley, J. C., "Recursive decomposition in discrete coverage processes," *Advances in Theory and Applications of Random Sets*, D. Jeulin, ed., pp. 215–230, World Scientific, Singapore (1997).

[13] Friel, N. and Molchanov, I. S., "Distances between grey-scale images," *Mathematical Morphology and its Applications to Image and Signal Processing*, H. J. A. M. Heijmans and J. B. T. M. Roerdink, eds., pp. 283–290, Kluwer, Dordrecht (1998).

[14] Galway, L. A., Statistical Analysis of Star-Shaped Sets, Ph.D. thesis, Carnegie-Mellon University, Pittsburgh, PA (1987).

[15] Goutsias, J., "Morphological analysis of discrete random shapes," *J. Math. Imag. Vis.* **2**, 193–215 (1992).

[16] Haas, A., Matheron, G. and Serra, J., "Morphologie mathematique et granulometries en place," *Ann. Mines* **11** and **12**, 736–753 and 767–782 (1967).

[17] Hall, P., *Introduction to the Theory of Coverage Processes*, Wiley, New York (1988).

[18] Handley, J. C. and Dougherty, E. R., "Maximum-likelihood estimation for discrete Boolean models using linear samples," *J. Microsc.* **182**, 67–78 (1996).

[19] Heijmans, H. J. A. M., "Discretization of morphological operators," *J. Vis. Commun. Imag. Rep.* **3**, 182–193 (1992).

[20] Heijmans, H. J. A. M. and Molchanov, I. S., "Morphology on convolution lattices with applications to the slope transform and random set theory," *J. Math. Imag. Vis.* **8**, 199–214 (1998).

[21] Heinrich, L., "Asymptotic properties of minimum contrast estimators for parameters of Boolean models," *Metrika* **31**, 349–360 (1993).

[22] Herer, W., "Esperance mathematique au sens de Doss d'une variable aleatoire a valeurs dans un espace metrique," *C. R. Acad. Sci., Paris, Ser. I* **302**, 131–134 (1986).

[23] Hess, Ch., "Loi de probabilité des ensembles aléatoires à valeurs fermées dans un espace métrique séparable," *C. R. Acad. Sci., Paris, Ser. I* **296**, 883–886 (1983).

[24] Jow, D., Some Contributions to the Theory of Random Sets, Ph.D. thesis, Claremont Graduate School, Claremont, CA (1983).

[25] Kendall, D. G., "Shape manifolds, procrustean metrics, and complex projective spaces," *Bull. London Math. Soc.* **16**, 81–121 (1984).

[26] Kendall, D. G., "Exact distributions for shapes of random triangles in convex sets," *Adv. Appl. Probab.* **17**, 308–329 (1985).

[27] Kneip, A. and Gasser, T., "Statistical tools to analyze data representing a sample of curves," *Ann. Stat.* **20**, 1266–1305 (1992).

[28] Korostelev, A. P. and Tsybakov, A. B., *Minimax Theory of Image Restoration*, Springer-Verlag, New York (1993).

[29] Le, H. and Kendall, D. G., "The Riemannian structure of Euclidean shape space: a novel environment for statistics," *Ann. Stat.* **21**, 1225–1271 (1993).

[30] Lieshout, M. N. M. van, *Stochastic Geometry Models in Image Analysis and Spatial Statistics*, Vol. 108, *CWI Tract*, Stichting Mathematisch Centrum, Centrum voor Wiskunde en Informatica, Amsterdam (1995).

[31] Mancham, A. and Molchanov, I. S., "Stochastic models of randomly perturbed images and related estimation problems," in K. Mardia and C. Gill, eds., *Image Fusion and Shape Variability Techniques*, K. Mardia and C. Gill, eds., pp. 44–49, Leeds University Press (1996).

[32] Matheron, G., *Random Sets and Integral Geometry*, Wiley, New York (1975).

[33] Molchanov, I. S., "Uniform laws of large numbers for empirical associated functionals of random closed sets," *Theory Probab. Appl.* **32**, 556–559 (1987).

[34] Molchanov, I. S., "Empirical estimation of distribution quantiles of random closed sets," *Theory Probab. Appl.* **35**, 594–600 (1990).

[35] Molchanov, I. S., "Handling with spatial censored observations in statistics of Boolean models of random sets," *Biomet. J.* **34**, 617–631 (1992).

[36] Molchanov, I. S., *Limit Theorems for Unions of Random Closed Sets*, vol. 1561, *Lect. Notes Math*, Springer-Verlag, Berlin (1993).

[37] Molchanov, I. S., "Statistics of the Boolean model: From the estimation of means to the estimation of distributions," *Adv. Appl. Probab.* **27**, 63–86 (1995).

[38] Molchanov, I. S., "A limit theorem for scaled vacancies of the Boolean model," *Stochast. Stochast. Rep.* **58**, 45–65 (1996).

[39] Molchanov, I. S., "Statistical problems for random sets," *Applications and Theory of Random Sets*, J. Goutsias, R. Mahler, and H. Nguyen, eds., pp. 27–45, Springer-Verlag, Berlin (1997).

[40] Molchanov, I. S., *Statistics of the Boolean Model for Practitioners and Mathematicians*, Wiley, Chichester (1997).

[41] Molchanov, I. S., "Unions of random sets and their applications," *Advances in Theory and Applications of Random Sets*, D. Jeulin, ed., pp. 35–48, World Scientific, Singapore (1997).

[42] Molchanov, I. S. and Stoyan, D., "Asymptotic properties of estimators for parameters of the Boolean model," *Adv. Appl. Probab.* **26**, 301–323 (1994).

[43] Molchanov, I. S. and Stoyan, D., "Statistical models of random polyhedra," *Comm. Stat. Stochast. Mod.* **12**, 199–214 (1996).

[44] Norberg, T., "Existence theorems for measures on continuous posets, with applications to random set theory," *Math. Scand.* **64**, 15–51 (1989).

[45] Papageorgiou, N. S., "On the theory of Banach space valued multifunctions I, II," *J. Multivar. Anal.* **17**, 185–206, 207–227 (1985).

[46] Pirard, E., "Roughness analysis on powders using mathematical morphology," *Acta Stereologica* **11** (Suppl. I), 533–538 (1992).

[47] Rachev, S. T., *Probability Metrics and the Stability of Stochastic Models*, Wiley, Chichester (1991).

[48] Réti, T. and Czinege, I., "Shape characterization of particles via generalised Fourier analysis," *J. Microsc.* **156**, 15–32 (1989).
[49] Ripley, B. and Rasson, J.-P., "Finding the edge of a Poisson forest," *J. Appl. Probab.* **14**, 483–491 (1977).
[50] Saxton, W. O., "Accurate alignment of sets of images," *J. Microsc.* **174**, 61–68 (1994).
[51] Schneider, R., "Random approximations of convex sets," *J. Microsc.* **151**, 211–227 (1988).
[52] Schneider, R., *Convex Bodies. The Brunn–Minkowski Theory*, Cambridge University Press, Cambridge (1993).
[53] Serra, J., *Image Analysis and Mathematical Morphology*, Academic Press, London (1982).
[54] Sidiropoulos, N. D., Baras, J. S. and Berenstein, C. A., "Algebraic analysis of the generating functional for discrete random sets and statistical inference for intensity in the discrete Boolean random-set model," *J. Math. Imag. Vis.* **4**, 273–290 (1994).
[55] Sivakumar, K. and Goutsias, J., "Morphologically constrained discrete random sets," *Advances in Theory and Applications of Random Sets*, D. Jeulin, ed., pp. 49–66, World Scientific, Singapore (1997).
[56] Small, C. G., *The Statistical Theory of Shape*, Springer-Verlag, New York (1996).
[57] Stoyan, D., Kendall, W. S. and Mecke, J., *Stochastic Geometry and Its Applications*, 2nd Ed., Wiley, Chichester (1995).
[58] Stoyan, D. and Molchanov, I. S., "Set-valued means of random particles," *J. Math. Imag. Vis.* **7**, 111–121 (1997).
[59] Stoyan, D. and Stoyan, H., *Fractals, Random Shapes and Point Fields*, Wiley, Chichester (1994).
[60] Tuzikov, A. V., Margolin, G. L. and Grenov, A. I., "Convex set symmetry measurement via Minkowski addition," *J. Math. Imag. Vis.* **7**, 53–68 (1996).
[61] Vitale, R. A., "Some developments in the theory of random sets," *Bull. Inst. Intern. Statist.* **50**, 863–871 (1983).
[62] Vitale, R. A., "An alternate formulation of mean value for random geometric figures," *J. Microsc.* **151**, 197–204 (1988).
[63] Vorob'ev, O. Yu., *Srednemernoje Modelirovanie (Mean-Measure Modelling)*, Nauka, Moscow (in Russian) (1984).
[64] Weil, W., "Expectation formulas and isoperimetric properties for non-isotropic Boolean models," *J. Microsc.* **151**, 235–245 (1988).
[65] Weil, W. and Wieacker, J. A., "Densities for stationary random sets and point processes," *Adv. Appl. Probab.* **16**, 324–346 (1984).
[66] Wilson, D. L., Baddeley, A. J. and Owens, R. A., "A new metric for grey-scale image comparison," *Int. J. Comput. Vis.* **24**, 5–17 (1997).

INDEX